The Emerald Handbook of Luxury Management for Hospitality and Tourism

The Emerald Handbook of Luxury Management for Hospitality and Tourism, from editors Anupama S. Kotur and Saurabh Kumar Dixit, investigates the challenges of creating luxury experiences in a post-pandemic world, as managers and guests alike increasingly embrace sustainability. From Oman to Delhi, Norway to the Caribbean, this compendium offers historical perspectives and of-the-moment insights on curating meaningful sensory experiences. Social sustainability and the role of technology are also discussed in detail. For those who create and manage luxury experiences, whether in Michelin-starred restaurants or resorts; or who engage in scholarly research in the field of luxury consumption, this comprehensive book is a must-read.

–**Professor Annamma Joy, PhD**, The University of British Columbia, Canada

Luxury Tourism and Hospitality are still a neglected issue, a privilege of few that is perceived with some prejudice. This *Handbook* is one more contribution to shape this trend and bring to the discussion a cutting-edge topic that will pave tourism development in forthcoming years. The *Handbook* looks at the concept and historical perspectives of luxury and introduces the reader to the challenges of managing and marketing luxury within a contemporary perspective of luxury where technology is the leading figure. This edited book shifts the negative perspective of luxury by relating it with sustainability and mirroring luxury trends after the lurking oddness the pandemic brought to the world. With contributions from international experts in the field, *The Emerald Handbook of Luxury Management for Hospitality and Tourism* provides a research-led perspective of luxury tourism and hospitality from different angles in major tourist newcomers; a must-have volume.

–**Professor Antonia Correia, PhD**, University of Algarve, Portugal

Waiting for resorts to open in space, luxury travelers are craving for unique sensory experiences. From wineries in Australia to fjords in Norway, this *Handbook* provides tourism and hospitality executives with 'Les Clefs d'Or' (the Golden Keys) to successfully target, attract and retain their luxury personas. A mandatory read!

–**Professor Diana Derval, PhD, EMBA**, Chair of Derval Research and Author of *Designing Luxury Brands: The Science of Pleasing Customers' Senses*

The Emerald Handbook of Luxury Management for Hospitality and Tourism

EDITED BY

ANUPAMA S. KOTUR

Symbiosis International (Deemed University), India

SAURABH KUMAR DIXIT

North-Eastern Hill University, India

emerald PUBLISHING

United Kingdom – North America – Japan – India – Malaysia – China

Emerald Publishing Limited
Howard House, Wagon Lane, Bingley BD16 1WA, UK

First edition 2022

British Library Cataloguing in Publication Data
A catalogue record for this book is available from the British Library

ISBN: 978-1-83982-901-7 (Print)
ISBN: 978-1-83982-900-0 (Online)
ISBN: 978-1-83982-902-4 (Epub)

Printed and bound by CPI Group (UK) Ltd, Croydon, CR0 4YY

In memory of late Dr Andy Hayes, whose contribution to the Handbook and to the domain of hospitality luxury is invaluable

Saurabh dedicates this book to Suchetan (his brother-in-law), who succumbed to COVID-19 on 7 May 2021

Anupama dedicates this book to her parents Sarojini and Somaling Kotur, and husband Ashwin Kaddi

Table of Contents

List of Figures and Tables

About the Editors

Anupama S. Kotur (Kaddi), PhD, is an Assistant Professor at the Symbiosis Centre for Management Studies, Symbiosis International (Deemed University), Pune. She is an expert in E and Mobile learning for travel and tourism professionals with an experience of working with several global leisure and tourism brands. She has also developed and delivered a number of training programs for luxury tourism and cruise line brands. Her research interests include Wine Tourism, Luxury Tourism and Destination Imagery studies. Anupama can be contacted at anukotur@gmail.com.

Saurabh Kumar Dixit, PhD, is an Associate Professor and founding Head of the Department of Tourism and Hotel Management, North-Eastern Hill University, Shillong (Meghalaya), India. His research interests include Consumer Behaviour, Gastronomic Tourism, and Experiential Management and Marketing in hospitality and tourism contexts. He has 15 books to his credit, including *The Routledge Handbook of Consumer Behavior in Hospitality and Tourism* (2017), *The Routledge Handbook of Gastronomic Tourism* (2019), *The Routledge Handbook of Tourism Experience Management and Marketing* (2020), *Tourism in Asian Cities* (2020), *Tourism in India: Marketing Perspectives* (2021) and *Consumer Behavior in Hospitality & Tourism* (2021). Presently he is editing *The Routledge Handbook of Wine Tourism and Teaching Cases in Tourism, Hospitality and, Events* (CABI) to be published in 2022. He is an editorial board member and guest editor of many international journals such as *International Journal of Contemporary Hospitality Management; British Food Journal; Journal of Research in Interactive Marketing; Tourism Recreational Research; Anatolia; Journal of Hospitality, Leisure, Sport & Tourism Education; Journal of Foodservice Business Research; Qualitative Market Research and Journal of Global Scholars of Marketing Science*, etc. He can be contacted at saurabh5sk@yahoo.com. His detailed profile can be seen at https://www.routledge.com/authors/i15903-saurabh-dixit.

About the Contributors

Cesare Amatulli is Associate Professor of Marketing at the University of Bari, Italy, and Adjunct Professor of Branding at LUISS University (Rome). He has taught at EMC Business School (Paris) and IUM (Monte Carlo). He has been Visiting Researcher at the Ross School of Business (USA). His research focuses on consumer behaviour, luxury consumption and sustainable consumption. He has published three books and several articles in major academic journals.

Arup Kumar Baksi is working as a Professor in the Department of Management and Business Administration, Aliah University, Kolkata, West Bengal, India. His area of research includes Customer Relationship Management, Service Quality Modeling and Tourism Branding. He has obtained his PhD from the University of Burdwan, West Bengal (India).

Adela Balderas-Cejudo, PhD, is a Lecturer at ESIC Business and Marketing School, and coordinator of the Management Area and coordinator of the Master's programmes at Basque Culinary Center, Spain. Adela Balderas-Cejudo is Research Fellow at Oxford University (UK). Dr Balderas completed executive training programmes in Professional Coaching and Leadership at New York University, Stanford University and Harvard University (USA). She is Visiting Lecturer at various international universities and the Ecole hôtelière de Lausanne, Switzerland.

Pierre Benckendorff is an Associate Professor and an award-winning educator and researcher specialising in tourist behaviour, experiences, design, technology, and tourism education and training. He is presently associated with the University of Queensland Business School, Australia.

Paridhi Brahma is a Research Scholar at the Department of Business Administration, Tezpur University, Assam, India. His research interests are in avitourism, community-based tourism and sustainable tourism.

Vinay Chauhan is a Professor at the Business School and Dean, Student Placement at the University of Jammu, India. He has also served as Dean (Planning and Development), Himachal Pradesh Technical University, besides holding several academic and administrative positions like UGC Nominee of SAP programme – Dibrugarh University; Visitor nominee – Assam Central University, etc. His interest areas include Tourism Management, Consumer Behaviour, Retail Marketing and Marketing Research. He has to his credit excellent research work in various refereed journals, edited books, funded research projects while also being on the editorial board of several journals.

P. Monica Chien holds a PhD in Marketing and is a Senior Lecturer within the University of Queensland Business School, Australia. Her scholarly interests lie in the field of Consumer Behavior in marketing, tourism and sport. She specialises in the examination of consumer information processing and behavioural change using experimental studies.

Barbara Czyzewska is Academic Dean and Head of Luxury Marketing and Brand Management Specialisation at Glion Institute of Higher Education in Switzerland. Prior to joining Glion, she served as Course Leader for MA in Luxury Hospitality Management at University of West London. Her research interests focus particularly on the history of luxury hotels as well as luxury brand management.

Anni Ding, MS, is a PhD candidate of the Conrad N. Hilton College of Hotel and Restaurant Management at the University of Houston. Her research focuses on branding, advertising and sustainability.

Mona Eskola is a doctoral student of tourism research at the University of Lapland, Faculty of Social Sciences, Multidimensional Tourism Institute (MTI), Finland. Her research interests are in the areas of experiential luxury, tourism and hospitality experiences, the body, and embodiment, sociomateriality, affect, and ethnographic methodologies.

Susana Fernández-Lores has a PhD in Marketing from Complutense University Madrid, UCM. She is a Senior Lecturer at ESIC Business and Marketing School, Head of the bachelor's degree in marketing at ESIC Business and Marketing School. She is Editor-in-Chief of *International Academic Journal of Communication Research*; author of several book chapters and academic papers; speaker at academic and professional conferences and co-author of the *Work Experiential Engagement* model. Her research is focused on branding, communication and new technologies.

José-Carlos García-Rosell, PhD, is a Senior Lecturer at the University of Lapland, Faculty of Social Sciences, Multidimensional Tourism Institute (MTI), Finland. He works in the fields of corporate social responsibility, business ethics and responsible management education. His current research agenda focuses on human–animal relations, sharing economy, tourism experiences and experiential learning.

Diana Gavilan has a PhD in Economics and a BA in Sciences Information (Extraordinary award) from Complutense University, Madrid, Spain. She is Assistant Professor of Marketing at Complutense University. She has co-authored *Work Experiential Engagement* model, two books and more than 40 published academic articles. She is a regular lecturer in professional forums. Her research interests are in the field of consumer experience and sensory marketing.

Chandan Goswami is a Professor of Marketing at Tezpur University, Assam, India. He has authored and co-authored many book chapters and research papers. He has more than 25 years of teaching and research experience. His research interest focuses on marketing, advertising and tourism.

Ulrike Gretzel is a Senior Fellow at the Centre of Public Relations, University of Southern California and Director of Research at Netnografica. Her research focuses on persuasion in human–technology interactions, smart technologies and intelligent systems, tourism experiences, online and social media marketing, adoption and use of technologies, as well as non-adoption and digital detox.

Gianluigi Guido is a Full Professor of Marketing at the University of Salento, Italy. He has been a professor at the University of Padua, Sapienza and LUISS University of Rome, Italy, and visiting researcher at the University of Florida at Gainesville, and Stanford University, USA. He has published 20 books and more than 250 articles in major scholarly journals in marketing and consumer behaviour.

Vikas Gupta is an alumnus of Amity University, Uttar Pradesh, India, and holds a PhD in Hospitality. He has rich and extensive teaching experience for more than 14 years in both India and abroad with some renowned names. His primary research areas include gastronomy, talent management and consumer behaviour. His work is widely published in national and international journals.

Deep Jyoti Gurung is Assistant Professor at the CHRIST Deemed to be University, Bengaluru, India. His research interest focuses on tourism marketing and consumer behaviour with special interest on the impact of social media on tourism. He has received his PhD from Tezpur University, Assam (India).

Minni Haanpää, PhD, is a Senior Lecturer at the University of Lapland, Faculty of Social Sciences, Multidimensional Tourism Institute (MTI). Her research interests are in the areas of tourism and event experiences, co-creation, affect and ethnographic methodologies, especially videography and autoethnography.

Tracy Harkison is the programme leader for the Bachelor of International Hospitality Management in the School of Hospitality and Tourism at Auckland University of Technology, Auckland, New Zealand. She has worked for 14 years in luxury hotels and lodges. Her research interests include co-creating luxury accommodation experiences, hospitality education and hospitality human resource management.

Andy Heyes holds numerous positions within organisations in the United Kingdom and mainland Europe. He is actively researching the luxury hospitality field and has published research in academic journals, authored and contributed numerous book chapters on hotel operations and strategic brand management, and presented at international conferences both in Europe and the Middle East.

Clare Hindley is a Professor in the Language and Communication Department at IU International University of Applied Sciences in Bad Honnef-Bonn, Germany.

She lectures primarily in research methods and communication modules and is a university research coordinator. Her international career in business and education has been diverse in location and focus, but language, culture and communication have remained the guiding lights with interest in cross-disciplinary studies. Her recent research publications focus on areas of hospitality, tourism, management education, culture and sociology.

Warinthorn Kachendecha is a Lecturer of MICE, Faculty of Hospitality and Tourism Industry, Rajamangala University of Technology Rattanakosin, Thailand. Her research interests lie in tourism marketing, event management and sustainability.

Sandeva Khajuria has an MBA in Hospitality and Tourism Management. She has qualified for the National Eligibility Test and pursued a PhD in Tourism Management, on the topic 'Risk Perception and Visiting Intentions of Tourist in North India'. She has also worked at School of Hospitality and Tourism Management, University of Jammu, India, as a contractual faculty. Her research papers have been published in various national and international journals. She has also presented papers at conferences and attended workshops and seminars.

Suvidha Khanna has been working as an Assistant Professor at SHTM, University of Jammu, India, since 2007. Before that, she worked in the Department as UGC-JRF for two years. She pursued a doctorate in the area of hospitality, relating to Gender issues and HRM. She also has degrees in PGDBA, PGDJMC and PGDHRM. Her articles and research papers have been publishing in various books, magazines and journals. She has also participated in national and international conferences and seminars. She has been secured the best paper award at the Indian Hospitality Congress, 2008.

Amnaj Khaokhrueamuang is an Associate Professor of Tourism at the School of Management and Information, University of Shizuoka, Japan. His research interest is a focus on rural tourism–related issues associated with community development, agricultural extension, culture and heritage, and sustainability.

Mohit Kukreti is former Program Director of the International Business Administration Program at the Ministry of Higher Education, Oman, DG of Applied Science Colleges. He is currently associated with the newly established Government University of Technology and Applied Sciences in Oman. His areas of interest are in Human Resources Development and Tourism Planning and Development.

Josef Kunc is an Associate Professor at the Department of Regional Economics and Administration, Faculty of Economics and Administration, Masaryk University, Czech Republic. His research activities primarily deal with consumer's preferences and shopping behaviour, special interest tourism and sustainable development, mainly from an economic and spatial perspective.

Gabriel Laeis is a Professor of Hospitality Management at the IU International University of Applied Sciences located in Bad Honnef, Bonn, Germany. He has a keen interest in gastronomy, food and sustainability. Gabriel holds a BA in Hotel Management and an MSc in Organic Agriculture and Food Systems and has worked for several restaurants, hotel chains and hospitality management consultancies. For his PhD in Development Studies at Massey University, he looked at the role of cuisine in tourism–agriculture linkages in Fiji. Gabriel has co-organised the first conference on tourism and the SDG in Auckland and published in leading hospitality journals and textbooks.

Kuan-Huei Lee is an Associate Professor at the Singapore Institute of Technology. She graduated from the University of Queensland, with a PhD in Tourism Management and holds an MBA from the University of Illinois at Urbana-Champaign. She accumulated many years of industry experience in hospitality businesses. Her research interest includes food tourism, food and beverage management and cross-cultural studies of tourists.

Tiffany S. Legendre, PhD, is an Assistant Professor in the Conrad N. Hilton College of Hotel and Restaurant Management at the University of Houston. Her research focuses on branding and sustainability, and her teaching concentrates on food and beverage management, sustainability and research methods.

Willy Legrand is a Professor of Hospitality Management at the IU International University of Applied Sciences located in Bad Honnef, Bonn, Germany. He is the lead author of *Sustainability in the Hospitality Industry: Principles of Sustainable Operations* as well as *Social Entrepreneurship in Hospitality: Principles and Strategies for Change* and the co-editor of the *Routledge Handbook of Sustainable Food and Gastronomy*. Prof. Legrand is Guest Chief Editor of the *Hotel Yearbook Sustainable Hospitality 2018* and *2020 Special Editions*. He chairs the *Hospitality Net World Panel on Sustainability in Hospitality*. This panel is supported by more than 80 experts across many fields, including food, energy, water, waste, facilities, communication, social and strategy.

Sandra Maria Correia Loureiro, PhD, is an Associate Professor at Instituto Universitário de Lisboa (ISCTE-IUL), Portugal, and she is the Director of the PhD in Management, specialising in Marketing and the PhD in Tourism Management. Her current research interests include relationships between marketing and tourism marketing issues connected to VR, AR and AI. Her papers have been published in a variety of peer-reviewed journals that include *Journal of Marketing*, *Journal of Retailing*, *Tourism Management*, *Tourism Management Perspectives* and *International Journal of Hospitality Management*. She has written several books and book chapters connected to tourism and hospitality.

Gema Martinez-Navarro has a PhD in Economics from Complutense University at Madrid and a BA in Business Administration from Carlos III University (Madrid). Currently, she is an Associate Professor of Marketing at Complutense University. She has authored several academic and professional articles. Recently she published *Fashion Marketing*. Her research interests are in the fields of fashion, tourism and new technologies.

Xiang Ying Mei is Associate Professor in Marketing at Inland Norway University of Applied Sciences, Norway. She holds a PhD in Tourism Management from the University of Queensland, Australia. Her research interests range from government policies, innovation, regional development, tourism management and development, experience economy, consumer behaviour and marketing, to digital learning tools and student engagement in higher education.

Dam Xuan Minh has a Master of International Tourism and Hospitality Management from James Cook University Singapore. He is working at the Consular Section of the Royal Embassy of Saudi Arabia in Hanoi. He has a research interest in overtourism and heritage tourism. With over five years of professional working experience in luxury hotels in various positions related to sales and marketing, he continues to be updated with local and international travel agents in Vietnam.

Katarzyna Minor is a Lecturer in Hospitality Management at Cardiff School of Management and a Researcher at Welsh Centre for Tourism Research, based at Cardiff Metropolitan University. Her research interests include hospitality and tourism management, focusing on digital media, user-generated content and social media in the hospitality industry. She is involved in a number of national and international collaborations researching, amongst others, skills gaps, digital applications and effects of discounts upon the hospitality and tourism industry.

Amitabh Mishra is the Head of the Business Administration Department at the newly established Government University of Technology and Applied Sciences in Oman. His interest areas are in Tourism Management and Marketing.

Johan Nordensvärd is a Senior Lecturer at the Department of Political Science, Uppsala University. Johan's research focuses on the intersection of environmental politics, international development and social policy, especially regarding low carbon development, innovation policy and energy policy. He is particularly interested in developing a broader understanding of environmental justice, identity and ecological citizenship concerning renewable energy technologies.

Markéta Novotná is an Assistant Professor at the Department of Regional Economics and Administration, Faculty of Economics and Administration, Masaryk University, Czech Republic, and a member of the Czech Association of Scientific Experts in Tourism. In her research, she focuses on regional tourism planning, consumer behaviour and the issue of sustainable tourism development.

Alessandro M. Peluso is Associate Professor of Business Management at the University of Salento, Lecce, Italy, and Adjunct Professor of Visual Marketing and Marketing Research at the LUISS University and the LUISS Business School, Rome, Italy. His research activity focuses on consumer behaviour, sustainable consumption and social communication. He has published three books and more than 50 scientific articles in major academic journals.

Senthilkumaran Piramanayagam is a Professor at the Department of Allied Hospitality Studies, Welcomgroup Graduate School of Hotel Administration,

Manipal Academy of Higher Education (University), Manipal, India. He has authored many book chapters, case studies, articles, reports and conference papers on consumer behaviour in tourism and hospitality. He has been selected as a Research Director for a research project funded by the Indian Council of Social Science Research (ICSSR), Government of India, on employing people with disability to manage high employee attrition in hospitality firms in India. His current main areas of interest include consumer behaviour in tourism and hospitality, branding of tourism services and hospitality educational administration.

Dr Anne Poelina is a Nyikina Warrwa Traditional Owner (Australia), Adjunct Professor Senior Research Fellow with Notre Dame University and a Research Fellow with Northern Australia Institute Charles Darwin University and Visiting Fellow Australian National University and Member of Water Justice Hub. Anne holds a Doctor of Philosophy (First Law), Doctor of Philosophy (Indigenous Wellbeing), Master Public Health and Tropical Medicine, Master Education, and Master of Arts (Indigenous Social Policy. Her current work explores First Law and the emergence of ancestral personhood, property rights, and equity through legal pluralism. Poelina advocates a Bio-regional Framework that focuses on new economies, regional governance, unity, and co-design in planning and decision-making.

Partho Pratim Seal has more than two decades of experience in academics and hospitality. An alumnus of IHM Chennai and postgraduate in Hotel Management and Management, he is pursuing his PhD in Hotel and Tourism Management. An avid wordsmith who believes that writing is the most influential and resourceful medium to disseminate information, he has authored three books, *Computers in Hotels: Concepts and Application*, *Food and Beverage Management* and *How to Succeed in Hotel Management Job Interviews*. His interest is in human resource management, entrepreneurship and food anthropology. He has published research papers in indexed international journals and is a reviewer of international journals.

Christian S. Ritter is lecturer in the Department of Media and Communication at Erasmus University Rotterdam. He was a research fellow in the Department of Social Anthropology at the Norwegian University of Science and Technology and at the Centre of Excellence in Media Innovation and Digital Culture at Tallinn University. His main research interests include cultures of influencers, locative media, digital labor, platformization, and tourism media. His latest research, which was funded by the Estonian Research Council between 2018–2020, assesses the emergence of travel influencers as a professional group in the global tourism industry. Christian's recent work appeared in the *Journal of Contemporary Religion, Qualitative Research and Anthropology of the Middle East.*

Namita Roy is a Researcher at the School of Management, Operations and Marketing, University of Wollongong. Her research focus is consumer culture in the field of tourism. She is interested in applying post-humanist perspectives to understand consumer experiences. Her doctoral thesis focused on tourist experiences of gastronomic trails in Australia.

Andrea Sestino is a PhD Candidate in Management and Marketing, at the University of Bari, Italy, and Assistant Researcher at the University of Salento, Italy. He holds an MSc in Business Management from Sapienza University, Rome (Italy). Additionally, he is also an R&D Consultant in the field of Applied Research to Digital Business, and his research focuses on technology as a phenomenon and related impact to companies and consumers in designing marketing strategies.

Johnnel Smith is a PhD Scholar at Griffith University, Australia, and the University of Otago, New Zealand. She has been a Lecturer in Tourism, Management and Marketing with Universities in the Caribbean and Pacific regions for over 13 years. Miss Smith holds a Master's and a Bachelor's degree (honors) in Tourism and Management from the University of the West Indies. Her expertise and scope of research areas include Big Data Mining, Netnography, Tourism and COVID-19, Gender and Luxury Management.

Jacqualine Spence is conducting research in tourism and business strategy within the University of Queensland Business School, Australia. A board member with services and retail companies, her research interests include strategic management, entrepreneurship and international business.

Pongtawat Srichamnong is a Lecturer of Tourism, Faculty of Hospitality and Tourism Industry, Rajamangala University of Technology Rattanakosin, Thailand. His particular interest concerns tourism and history.

K Thirumaran holds a PhD from the National University of Singapore. He is currently the Academic Head for the Business School at James Cook University, Singapore. He specialises in tourism and hospitality management. His varied research interests are primarily located in luxury and cultural tourism.

Acknowledgements

> No one who achieves success does so without acknowledging the help of others. The wise and confident acknowledge this help with gratitude.
>
> –Alfred North Whitehead

The Emerald Handbook of Luxury Management for Hospitality and Tourism, which began as a humble idea in late 2019, could not have been a reality without the help and assistance of many individuals. As editors of the Handbook, we owe them a huge debt of gratitude and take this opportunity to thank them immensely.

We would like to thank the contributors of the book who responded to our call to be a part of this endeavour even during these challenging times. At every stage, they were enthusiastic and extremely generous in accommodating our requests well within time. It has indeed been a privilege working with such a talented pool of experts, and we remain indebted for their support. We thank the team at Emerald Publishing, Fiona Allison, Publishing Editor, and David Jak Mulvaney, Content Development Editor, for their constant support and help. We also express our appreciation to the entire editorial team, production and marketing teams at Emerald Publishing for their valuable assistance. We thank the reviewers of the *Handbook,* whose advice has been helpful at every step of the way. Further, we owe our sincere gratitude to Prof. Antonia Correia, Prof. Annamma Joy and Prof. Diana Derval for sharing their generous endorsements for the *Handbook.* Finally, we thank our respective parents and families for keeping our spirits high and motivating us throughout this project.

Anupama S. Kotur (Kaddi), Pune and
Saurabh Kumar Dixit, Shillong
2021

Introduction

Anupama S. Kotur[1] *and Saurabh Kumar Dixit*[2]

> Offline is the new luxury.
>
> –Name of a popular documentary film

The title of a documentary, *Offline Is The New Luxury*, aired in 2016, interestingly, conveys volumes about the constant redefinition of what luxury means to individuals. People have surrounded themselves with the idea of luxury for hundreds of years, although the understanding of what it means has continued to evolve with the passage of time. The idea of luxury as well as luxury travel, throughout history, has been shaped and reshaped in response to changes in societal values, economic advancements and cultural shifts. The luxury tourism phenomenon has, over the years, undergone a continuous metamorphosis where one witness a mindset change in the perception of luxury 'from a transactional relationship to a holistic experience' (Atwal & Williams, 2007). The fluid, elastic and, in essence, subjective (Kapferer, 2014) concept of luxury has changed dramatically across time and culture (Yeoman & McMahon-Beattie, 2018). In the phenomena of 'luxurification' (Atwal & Williams, 2017) in societies, newer ideas of luxury continue to be embraced, making it even more interesting to explore 'where ordinary ends and luxury begins' (Tynan, McKechnie, & Chhuon, 2010; Yang & Mattila, 2016).

The Intriguing Mosaic of Luxury Definitions

Luxury has long been a subject of interest among academia, and, therefore, one finds a wealth of literature as a result of this constant examination. While some notable and comprehensive works (Conrady, Ruetz, & Aeberhard, 2020; Kanani, 2017; Swarbrooke, 2018; Thach & Yeung, 2019) have paved the way for discussions on luxury tourism, hospitality and culinary experiences, literature is still scarce (Correia, Kozak, & Del Chiappa, 2019) and evolving. With every successive study, newer dimensions of luxury tourism have emerged over time, lending a deeper insight into the meaning of luxury and its value among

[1]https://orcid.org/0000-0003-4202-4716
[2]https://orcid.org/0000-0002-9478-084X

The Emerald Handbook of Luxury Management for Hospitality and Tourism, 1–16

doi:10.1108/978-1-83982-900-020211028

consumers. Defining what luxury means in general has been a rather complex endeavour and more so within the tourism context. It has been widely acknowledged among scholars that there are multiple dimensions and interpretations to its definition. *Firstly*, much of the understanding about luxury is drawn from literature from wide-ranging disciplines such as history, economics, sociology, psychology, marketing, among others (Tynan, McKechnie, & Chhuon, 2010), where newer dimensions and themes of luxury have and continue to emerge (Correia et al., 2019). *Secondly*, the standpoint used to define luxury can be either product-centric, referred to as 'old luxury', where the inherent nature of the product makes it a luxury offering, or consumer-centric, referred to as 'new luxury', where luxury is more experiential and individualistic in nature (Florin, Callen, Mullen, & Kropp, 2007; Tynan, McKechnie, & Chhuon, 2010). *Thirdly*, the interpretation of luxury within the cultural and etymological precinct varies from one country to another. Kapferer (2014, p. 717) points out that the

> …world can be divided into two groups of countries: those for which the word luxury is part of their usual language and those where the word itself is not known.

This has been observed to influence the meaning, perception and value that individuals attach to luxury in different contexts and cultural settings (Kapferer, 2014). Lastly, luxury as a specific business strategy or business model (Kapferer, 2014) in different sectors continues to evolve, making it even more dynamic. Integrating these transient meanings of luxury into mainstream academic research and discussions, perhaps, is complex and time consuming. One can conclude that the phenomena of luxury is a grand mosaic that has endured through time and space, where every onlooker discovers a different interpretation, making it challenging to define what luxury means.

The Evolving Landscape of Luxury Tourism: Meaning and Interpretations

In its traditional most understanding, luxury, referred to as the 'old world luxury', was seen as a prerogative of the royalty and elite. Travel for pleasure was largely considered a preserve of the affluent, where travellers surrounded themselves with material comforts and extravagant indulgences while globe-trotting. The conventional definition of luxury, leaning towards conspicuous consumption (Veblen, 1899, 1992), was viewed as something with a symbolic value. In this seminal work on luxury consumption, based mostly on observation and experience, Veblen highlighted that wealthy individuals often consumed highly conspicuous goods and services in order to achieve greater status and prestige in their social circles (Bagwell & Bernheim, 1996). Leibenstein's work (1950) further contributed to 'Veblen effect' theory by adding two more dimensions, namely the snob effect and bandwagon effect in luxury consumption, where he demonstrated the relationship between luxury consumption and its value perception among the luxury class. Vigneron and Johnson (1999, 2004) proposed a framework delineating the value perception of luxury where they captured both personal (hedonism and quality) and non-personal (conspicuousness, uniqueness and social value) dimensions of luxury (Yang & Mattila, 2016). Kapferer (1998) associated attributes such as expensive and exclusivity of the product or experience, while Dubois, Laurent, and Czellar (2001) argued that besides conspicuousness, luxury could be perceived as anything that is scarce, aesthetics and superfluous. Tynan, McKechnie, and Chhuon (2010) associated attributes such as high quality, expensive and non-essential products and services with luxury. Correia et al. (2019), examining the antecedents and consequents of luxury tourism, proposed a framework depicting six factors (social value, conspicuousness, uniqueness, status, self-esteem and public display) that drive tourists to make a luxury-driven tourism trip (see Fig. 1). They argued that 'luxury has more a socially driven meaning than a self-constructed meaning' (p. 1). Dubois (2020, p. 76) highlights that through luxury consumption, 'consumers navigate social hierarchies and accompanies their rise and fall in the social hierarchy'. One may observe that this genre of luxury, categorised as 'absolute luxury' (D'Arpizio, 2007 in Brun & Castelli, 2013), influenced a growth of tourism trend where notions such as prestige, elite, status and special attention have come to be linked unavoidably and extensively with luxury tourism (Correia & Kozak, 2012; Correia, Kozak, & Reis, 2016; Correia & Moital, 2009; Jaworski & Thurlow, 2010; Riley, 1995; Thurlow & Jaworski, 2012).

While status and associated symbolism continue to drive luxury travel globally, yet another genre of luxury focussed on experiential value and personal fulfilment, sometimes also referred to as the 'feminisation of luxury' (Danziger, 2005; Israel, 2002; Gambler, 1997 in Yeoman, 2011), is reshaping the luxury consumer behaviour. The 'old world luxury consumption' leaning towards 'ultra-hedonism' (Yeoman, 2011), owing to constant shifts of socio-economic paradigms, has paved the way for 'new and individualistic' and a 'post-modernistic' (Atwal & Williams, 2017) form of luxury to emerge. Tourists, in an attempt to seek luxury beyond materialism, are investing in intangibles and 'experiences of time, space, authenticity, community individuality and well-being', allowing them to 'boost their social and cultural capital' (Yeoman & McMahon-Beattie, 2018, p. 205). Tourist experience has drawn much attention from academia (Pine & Gilmore, 1999; Ritchie & Crouch, 2003; Ritchie, Tung, & Ritchie, 2011; Swarbrooke, 2018) and its significance in creating a holistic tourism value universe. Creating a luxury tourism experience entails recognising the essence

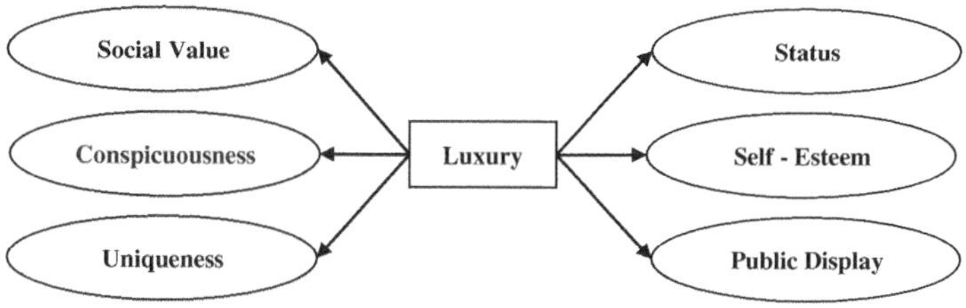

Fig. 1. Antecedents and Consequents of Luxury Tourism. *Source:* Correia et al. (2019).

of a tourist offering and 'amplifying it into a set of tangible, physical and interactive experiences' (Atwal & Khan, 2008, p. 341) that reinforce and augment the sense of luxury.

While experience remains the mainstay of 'new world' luxury tourism, some of the contemporary discussions highlight a growing trend among luxury tourists who seek to associate with luxury brands focussing on sustainable and ethical practices. The idea of sustainability is being built into luxury semiotics (Yeoman & McMahon-Beattie, 2018) and the brand narrative by a number of tourism and hospitality brands owing to a shift in tourist preference, motivated by rarity and authenticity of experience rather than conspicuous consumption (Moscardo & Benckendorff, 2010). Authors have argued that the sense of luxury for this segment of tourists emanates from curated tourist experiences, uniqueness of access to preserved sites, ethical care of the places people visit and equitable tourism development at destinations (Scheyvens, 2011; Poelina & Nordensvard, 2018). Griskevicius, Tybur, and Van den Bergh (2010) identified these phenomena as 'green conspicuousness' that is, perhaps, driving the trend of sustainable luxury today. This form of luxury tourism is identified by many terms such as green luxury, balanced luxury, deeper luxury, intelligent luxury (Ahn & Pearce, 2013; Moscardo & Benckendorff, 2010) and responsible luxury (Brenner & Aguilar, 2002). Many authors have also argued that the interrelationship between luxury tourism and sustainability is a rather complex one, as luxury based on conspicuous consumption is 'inherently rooted in social inequality' (Moscardo & Benckendorff, 2010) and, therefore, not sustainable. However, this interplay continues to be explored in order for the luxury tourism brands to remain leader versus mass, and 'luxury will have to be sustainable in social, economic and ecological terms' (Kapferer, 2010) for these brands to carve out a niche for themselves.

While a more conscious way of experiencing luxury tourism gains momentum worldwide, the idea of 'minimalistic luxury' (Hampton, 1998; Kapferer, 2010; Liu, Yildirim, & Zhang, 2018), on the same veins, is drawing attention from the scholars of luxury. Some of the notable hotel brands have been quick to embrace this trend and create hotel spaces with 'fresh and minimal' designs 'while maintaining an elevated level of luxury' (Luxury Travel Magazine, 2018). The 'minimalist aesthetic', or what is also known as 'purposeful limitation' (Fagan, 2017) approach to luxury, is popularised even in culinary circles using minimalist décor and menu to convey a sense of exclusivity and luxury. The idea of minimalistic luxury tourism is also being seen as a shift from 'eco-conspicuous' consumption to a more 'eco-conscious' consumption (Ramchandani & Coste-Maniere, 2018, p. 1).

Globally, yet another new segment of luxury tourism consumers is accelerating the luxury phenomena – the growing affluent middle classes in emerging economies. This segment, 'no longer at a financial distance from luxury' (Atwal & Khan, 2008), is aspiring to travel to exotic locales, experience premium hospitality and invest in exclusive experiences. One may observe that the chasm between unattainable luxury and 'accessible luxury' is gradually blurring, making luxury travel more 'democratised' (Lipovetsky & Roux, 2003 in Vigneron & Johnson,

2004; Brun & Castelli, 2013; Correia et al., 2019; Yeoman & McMahon-Beattie, 2018). This trend indicates a shift in the notion of traditional luxury where 'affordable luxuries and premiumised offerings' make luxury tourism more reachable. The concept of 'masstige' brands is also gaining recognition among academia (Kapferer, 2014; Paul, 2018; Truong, McColl, & Kitchen, 2009). Masstige brands are 'just a step below traditional luxury brands and a step above mid-priced brands' (Paul, 2018, p. 723). A number of traditional luxury hotel brands have experimented with their masstige brand extensions to cater to this growing tourist segment.

The idea of luxury is constantly journeying through time and evolving to acquire newer interpretations. Fig. 2 illustrates the various constructs that interpret the meaning of luxury based on examination of the literature. The luxury marketing space is witnessing a number of discussions and debates on how 'the next wave of luxury spenders', namely, the millennials and generation Z, will reshape the luxury market (de Kerviler & Rodriguez, 2019; Kapferer & Michaut-Denizeau, 2020). A number of researches have indicated that millennial luxury consumers avoid engaging with brands that lack ethics and integrity, while a Deloitte study (2017) argued that 'the sustainability consciousness of millennials was paradoxically quite low in luxury buying' (in Kapferer & Michaut-Denizeau, 2020). Within the luxury tourism space, one finds limited literature addressing the value that millennial tourists associate with luxury.

While luxury tourism and hospitality brands navigate their way through the 'attention economy, the aesthetic economy and the experience economy' (Kohrs, 2020), technology and innovation have undeniably changed the luxury landscape. Holmqvist, Wirtz, and Fritze (2020) observe that luxury industry has been slow and rather cautious to adopt technology-led change; however, many brands have

Fig. 2. Word Cloud of Meanings and Interpretation of Luxury.
Source: Authors.

also demonstrated that digitisation can be embraced by luxury brands successfully. Luxury hotel brands that rely heavily on heritage value have attempted to create a fine blend of technologically transformed spaces and bespoke experiences through personalisation while continuing to maintain human interactions at the core of their offering. Marketing communication has also transformed the brands' storytelling in order to engage better with the next generation of luxury travellers (Langer, 2020; Thirumaran, Jang, Pourabedin, & Wood, 2021) through social media networks. In a hyper-digitised, hyper-connected and hyper-viralised age that we live in, luxury brands are often faced with predicaments to find the right balance between exclusivity and ubiquity (Hennigs, Wiedmann, & Klarmann, 2012). For tourists whose idea of luxury leans towards status, prestige and public display, social media networks have created a window to showcase their luxury travel experiences. While technology and digitisation permeate through most travel activities today, there is a small but growing section of luxury tourists who seek to digitally isolate themselves. Digital or more specifically, social media privacy is increasingly being deemed as a luxury (Papacharissi, 2010; Papacharissi & Gibson, 2011) where tourists prefer to deliberately stay offline and refrain from posting about their luxury vacations in order to accentuate the feeling of exclusivity.

The global COVID-19 crisis brought tourism activities worldwide to a sudden and abrupt halt, affecting the tourism industry's demand and supply dynamics (Sigala, 2020). The COVID-19 crisis is being seen as a change trigger, bringing about a transition in tourism (Brouder, 2020; Cheer, 2020; Hall, Scott, & Gössling, 2020). Post-pandemic luxury tourism consumption is observed to be witnessing a realignment of consumer sentiments and behaviour with regard to how they view conspicuous spending. Luxury consumption patterns have altered, if only temporality, forcing luxury tourism brands to pivot and redesign tourist experiences in order to re-start tourism. This may further see newer meanings of luxury to emerge, redefining what luxury means to individuals.

Marketing Luxury Hospitality and Tourism

As complex as it is to define luxury, marketing luxury can be even more so. The marketing of luxury services has become increasingly composite because of its linkages with conveying an image of quality, performance and authenticity, and endeavouring to offer an experience by connecting it to the lifestyle constructs of consumers. Extant literature in luxury tourism and luxury, in general, is nuanced with a number of interpretations of luxury, perhaps, making it difficult to arrive at a precise definition. It would, therefore, be safe to assume that the business of luxury is unlike any other. Further, luxury consumer behaviour is constantly changing. The recent global COVID-19 crises have forced luxury travellers to 're-examine their priorities and, as a consequence, attitudes and behaviours towards luxury' (Yeoman, 2011). This would require unlearning the classical marketing principles, and approach and re-examine luxury tourism marketing from different angles (Kapferer, 2016).

The need for luxury is largely psychological (Derval, 2018; Dubois, Jung, & Ordabayeva, 2020; Sung, Choi, Ahn, & Song, 2015). While this is true, understanding luxury itself 'simply as something that is, or an object waiting to be analysed' is nearly impossible (Kapferer, 2016). Therefore, luxury tourism marketing will have to take into account a sociological, psychological, economic as well as other functional dimensions associated with luxury. Kapferer and Bastien (2012, in Kapferer, 2016) proposed a framework on functional perspective on luxury segmentation where the authors divide the luxury market segments into four quadrants made up of two variables forming a continuum between 'personal experience' to 'display for others' and 'dissociating from below' to 'integration above'. The resultant market luxury segments are as follows:

- Luxury as membership (luxury travellers who choose popular luxury destinations and brands in order to conform or mark their entry into the luxury segment);
- Luxury as an affirmation of power (luxury travellers choosing ultra-luxury destinations, uber-luxurious hotel brands in order to stand out and demonstrate power in their social circles);
- Luxury as a personal expression (avant-garde luxury travellers who seek unique, authentic and curated experiences);
- Luxury as an art of living (cultural connoisseurs who seek meaningful experiences that fulfil their personal aspirations and well-being).

These luxury traveller segments respond to their own unique stimuli, and luxury marketing will have to address them effectively by connecting the right tourism experiences with the right segments. Tourists, and more so the luxury tourists, have 'traditionally been experience collectors, iconic figures of the experience economy' (Dixit, 2020; Richards, 2021). Experience marketing is 'any form of customer-focused marketing activity that creates a connection to customers' by way of evoking experiences and focussing on enhanced customer engagements and interactions (Ketter, 2018; Schmitt, 2011). Experience marketing in luxury tourism and hospitality must also identify and adapt marketing strategies to the tourist 'experiential zones', namely, education, escapist and aesthetic values (Pine & Gilmore, 1999). Therefore, Atwal and Williams (2017, p. 345) propose that luxury experience marketing must focus on customer experiences and lifestyle that enhance 'sensory, emotional, cognitive and relational' values. Secondly, it must create effective 'synergies between meaning, perception, consumption and brand loyalty'. One may conclude that experience marketing in luxury tourism must highlight the richness of the experiences using the right marketing channels that can effectively deliver emotions and complex messages (Ketter, 2018; Schmitt, 1999).

The idea of luxury in tourism and hospitality is constantly being reimagined and reinterpreted. This has created an opportunity for a constant engagement between industry practitioners and academia to explore associated theoretical and practical issues. Through this book, it is our sincere attempt to offer a deeper and comprehensive insight into the ever-evolving phenomena of luxury tourism. The *Handbook* presents chapters authored by academicians and researchers in the field of luxury tourism and hospitality who discuss several examples and case studies drawn from current practice in order to help readers to understand clearly. This *Handbook* will definitely help researchers and practitioners to better appreciate luxury management concepts and practices through its rich literature comprising review of existing knowledge, upcoming paradigms and contemporary global case studies from luxury hospitality and tourism.

The Handbook Organisation

The *Handbook* consists of 27 chapters grouped under five thematic parts. Each part is assigned carefully a theme to offer a glimpse of both theoretical and practical dimensions of luxury management in tourism and hospitality. Part 1, 'Conceptualizing Luxury', attempts to examine the concept of luxury from various standpoints in order to create a foundation for the *Handbook* so as to establish the multi-disciplinarity of luxury. Part 2, 'Managing and Marketing Luxury Experiences', addresses the multidimensionality of marketing and managing luxury tourism, enabling the readers to appreciate the finer nuances of luxury experience creation and management. Part 3, 'Technology and other contemporary facets in Luxury', introduces that technological dimension to experience creation in luxury tourism and hospitality. Part 4, 'Sustaining Luxury', examines luxury tourism from a sustainable and ethical viewpoint, highlighting a shift among luxury consumers towards a more responsible luxury consumption. Part 5, 'Luxury in a post-pandemic world', attempts to give a glimpse of changes that the luxury tourism industry is witnessing since the onset of the COVID-19 pandemic.

Part 1 – Conceptualising Luxury

Part 1, comprising five chapters, builds a comprehensive theoretical base to the understanding of luxury tourism. Chapter 1, authored by Vinay Chauhan, Suvidha Khanna and Sandeva Khajuria, examines luxury tourism from a historical standpoint in an attempt to understand the historical progression. Authors highlight that the concept of luxury is constantly evolving and that the concept is not limited to one society or one culture. Authors conclude that any study of the progression of luxury tourism and its development must be viewed from a cross-cultural perspective. Chapter 2, authored by Mona Eskola, Minni Haanpää and José-Carlos García-Rosell, sets out to explore consumer-centred experiential luxury from the perspective of a human body. Using an autoethnographic approach, the authors investigate the tourist practices inducing luxury in the premises of a luxury hotel enriches the discussion of the co-creation between human bodies and the experience setting. The study proposes that luxury is sensed only through individuals' five external senses, and a deeper insight into the co-creation of luxury experiences could be gained through understanding the dynamic, situational and sensitive nature of luxury in the contemporary touristic experience. Authors Amnaj Khaokhrueamuang, Warinthorn Kachendecha and Pongtawat Srichamnong, in Chapter 3, examines the notion that contemporary tourists prefer luxury experiences rather than luxury products through their qualitative study conducted at a Japanese temple lodging in Koyasan, an experience-based accommodation. The authors present an original concept developed from demographic and psychographic factors where they conceptualise tourists' personalities regarding luxury tourism within three shades of grey, dark grey, grey and light grey, depending on the intensity of their interests in material concerns or learning experiences. In Chapter 4, authors, Dam Xuan Minh and K Thirumaran, venture into regime politics and luxury tourism and argue that luxury and communism are perceived to be of ideologically incompatible values. The study uses historical and periodised analysis to identify the relationship between state policies and tourism development activities, focussing on the intersection between socialist values and luxury gravitations in Vietnam. A critical assessment of contemporary luxury tourism in Vietnam reveals a favourable relationship with socialist principles. Authors of Chapter 5, Amitabh Mishra and Mohit Kukreti, explore the attractiveness and tourism infrastructure of destination Oman as a luxury destination. The Government of Oman, with its vision 2040, is actively tapping Oman's luxury tourism potential with improved tourism product development, increased projects funding, thoughtful marketing and brand promotion to reap benefits from tourism investments. The study also discusses the challenges faced by Oman in luxury tourism destination development.

Part 2 – Managing and Marketing Luxury Experiences

Part 2 consists of seven chapters that provide an extensive overview of the diverse marketing and management practices in luxury tourism, both from theoretical and practical dimensions. Markéta Novotná and Josef Kunc, in Chapter 6, examine the gradual shift of Czech travellers towards a new luxury travel model. In this survey-based study, the authors highlight that the new luxury model is associated with a high level of comfort and privacy, exclusive location and first-class services and new elements such as travelling, authenticity and sustainability. In Chapter 7, author Xiang Ying Mei explores the potential of Norway as a luxury tourism destination based on the elements of exclusivity in experiences and the importance of sustainability. The author also investigates luxury tourists and relevant topics associated with understanding 'luxury', as part of luxury tourism marketing efforts. The chapter stresses the need to market luxury by providing a balanced picture of the attractions and attributes that appeal to luxury tourists. Johnnel Smith, in Chapter 8, makes a unique contribution to the compendium of knowledge on luxury hospitality and tourism by providing insights from the supply side of luxury accommodation brands in the Caribbean and Pacific regions. The author examines the role of the Butler and the elements of Butler service in luxury accommodation through this primary data–based investigation. The study endeavours to fill the lacuna in existing literature on this topic by lending valuable insights into the role of training and retention within the supply side of luxury management and practical applications for the industry. Chapter 9 aims to understand marketing strategies that support luxury gastronomic tourism experiences. Authors Namita Roy and Ulrike Gretzel use in-depth analysis of the website of a destination marketing organisation in the Hunter Valley gastronomic region of Australia. This chapter highlights that the gastronomic tourism experience is marketed as bucolic luxury using marketing strategies of connection, congregation and repetition, all of which channel and maintain the effect of bucolic luxury. The human interaction component and the co-creation that occurs between staff and guests is an essential dimension of the industry. The influence of these interactions on guest experiences and the delivery of services is explored by Tracy Harkison, in Chapter 10. The author emphasises that, in the process of co-creating unforgettable and unique luxury accommodation experiences, hospitality professionals play a crucial role in anticipating and accommodating guests' requirements and desires. In Chapter 11, authors Senthilkumaran Piramanayagam and Parto Pratim Seal, using case study–based investigation, aim to analyse how a multi-brand hospitality firm differentiates its luxury brand from other luxury brands in the portfolio. Authors observe that most of the brand differentiation strategies rely on brand-specific features, traditions, services, location of the hotel and metadata on the Internet. Arup Kumar Baksi, in Chapter 12, focuses on developing a brand relationship scale, namely, TraveLux, in the context of luxury travel consumption and tested its robustness to explain the shared sentiments and emotions of travellers engaged in luxury travel across social media. The study highlights that TraveLux was found to capture the shared experience of travellers consuming luxury travel brands, thereby establishing a synch between the instrument constructs and manifested human cognition in a real-life situation.

Part 3 – Technology and Other Contemporary Facets in Luxury

This part consisting of four chapters endeavours to highlight the role of technology in creating and marketing luxury experiences in tourism and hospitality. Sandra Maria Correia Loureiro, in Chapter 13, intends to give insights into the use of technologies such as virtual reality (VR), augmented reality (AR) and artificial intelligence (AI) in tourism and hospitality. The author also provides examples about using such technologies and points out suggestions about further research in the interface of luxury and technologies in tourism and hospitality. Chapter 14, authored by Cesare Amatulli, Alessandro M. Peluso, Andrea Sestino and Gianluigi Guido, empirically studies the effects of openness to change and status consumption orientation in influencing luxury hotel guests' perceived usefulness of voice assistants (VOs) when integrated with their staying. Authors suggest that openness to changes and status consumption represents two potentially alternative aspects that managers could leverage to encourage hotel guests' perceived usefulness and vocal assistant devices. In Chapter 15, authors Vikas Gupta and Saurabh Kumar Dixit examine whether the branded luxury guest room amenities provided in five-star hotels of Delhi influence the guest's hotel purchase decisions. This study is a novel attempt by authors to find how the guests' hotel selection and purchase intentions are influenced by the placement of branded luxury amenities in guest rooms. Adela Balderas-Cejudo, Diana Gavilan and S Fernandez-Lores, in Chapter 16, aim to firstly understand Michelin star system and luxury gastronomy; secondly, analyse the potential link between top gastronomy and luxury gastronomy; and finally, listen to chefs' voices and perspectives on changes in consumer behaviour, perspectives and trends. The authors highlight that the findings of this study may help create innovative ways to address luxury tourism and hospitality.

Part 4 – Sustaining Luxury

Sustainability is generating much interest among scholars of luxury tourism and hospitality. This part, comprising 5 chapters, examines various themes associated with sustainable and ethical luxury tourism practices. Anni Din and Tiffany S. Legendre, in Chapter 17, use a case study approach to understand how a luxury brand communicates and maintains its sustainability through its various dimensions through the process of brand creation, communication and management. The chapter also provides practical implications for existing luxury hospitality brand organisations in the arena of customer service. Chapter 18, authored by Paridhi Brahma, Deep Jyoti Gurung and Chandan Goswami, attempts to present the existing knowledge about sustainable luxury tourism by implementing a systematic literature review. The authors highlight the opportunities and challenges associated with sustainable luxury tourism are being highlighted. This study also identifies the factors that need to be considered to promote sustainable luxury tourism. Johan Nordensvärd and Anne Poelina, in Chapter 19, explore reasons and means to integrate social sustainability into sustainable luxury tourism and its potential pitfalls. Authors also attempt to conceptualise nature-based and Indigenous tourism to empower local Indigenous communities and provide them with sustainable employment, economic development and community services. The chapter uses the case study of Martuwarra, Fitzroy River and its communities in Kimberley, Western Australia, to develop a socio-cultural sustainable luxury tourism framework that includes governance, legal and management, and social policy perspective. Authors of Chapter 20, Jacqualine Spence, Pierre Benckendorff and P. Monica Chien, firstly, examine the relationship between luxury tourism and yield concepts, and secondly, examine how luxury services and experiences have been covered in tourism and hospitality literature. Authors also synthesise a number of strands of inquiry across disparate bodies of literature to identify a research agenda. In Chapter 21, authors K. B. Minor and Andy Heyes critically examine the hotels' role in breaches of human rights and their possible involvement in illegal practices such as human trafficking and modern slavery. The authors deal with the often-omitted notion of the darker side of luxury hospitality and highlight the vulnerability of the luxury sector owing to such practices.

Part 5 – Luxury in a Post-pandemic World

Comprising 5 chapters, this part examines the early responses of the luxury tourism and hospitality industry to the global outbreak of the COVID-19 pandemic. The chapters also discuss notable shifts in the global luxury sector owing to the crisis. Christian S. Ritter, in Chapter 22, assesses how luxury travel imaginaries were modified in the aftermath of the outbreak of the coronavirus pandemic. The author, through this ethnographic investigation, identifies important responses of travel influencers to the current long-term crisis of tourism. The in-depth investigation suggests that travel influencers were in a position to create new powerful representations of luxury as safe travel since they acquired the skills to establish stable story worlds for their travel experiences, which attracted the attention of large platform audiences. In Chapter 23, Kuan-Huei LEE presents different types of luxury tourism, luxury lifestyle, luxury tourists' decision-making and luxury hospitality products. The luxury food and beverage business in Singapore is presented to illustrate the commercial environment during the pandemic. Clare Hindley, Willy Legrand and Gabriel C.M. Laeis, in Chapter 24, aim to establish the relation of luxury tourism to sustainability and investigate whether tourism in its current form is not itself a luxury. The authors, in this chapter, question whether tourism itself has become an unsustainable pandemic and an indefensible luxury. The author of Chapter 25, Barbara Czyzewska, discusses the role of embedding the culture of adaptability to change to ensure the business's longevity and resilience to crises. The chapter also deconstructs some of the industry's best practices to showcase the most creative and resourceful approaches to embracing the new reality and offering authentic experiences to discerning guests. In Chapter 26, authors, Diana Gavilan, Adela Balderas-Cejudo and Gema Martinez-Navarro, explore in-depth the particularities of the luxury travel industry during the coronavirus pandemic. The authors present the case of Fiji and the manner in which the destination tackled the pandemic to protect the health and safety of its inhabitants while preserving the local economy, which relies on tourism. In the concluding chapter entitled 'Conclusion: Way forward for the hospitality and tourism luxury', editors Saurabh Kumar Dixit and Anupama S. Kotur present a synthesis of the findings of the chapters in the *Handbook*. The editors also underscore some emerging avenues for future research in the domain of luxury hospitality and tourism.

The *Handbook* book positions itself as a comprehensive compendium and guide in luxury hospitality and tourism marketing not only to students of tourism management but also to industry practitioners. This volume is also relevant to those studying other disciplines such as resort management, cruise operations, etc. Practitioners are likely to find the *Handbook* of value, including those working in destination management and marketing organisations, visitor attractions and other private enterprises, which make up the tourism industry. We, as editors of the *Handbook*, are confident that our humble effort in bringing together diverse topics in luxury hospitality and tourism in the form of a comprehensive guide will add to the existing body of knowledge on luxury marketing and management and present a direction for future research in this domain.

What began as a humble idea two years ago could not have taken shape into this *Handbook*, if not for the contributing authors and experts. We are ever so grateful and indebted to all the chapter authors for agreeing to be a part of this journey and contributing their research works. It has been a great pleasure working with you all and a tremendous learning experience.

References

Ahn, Y. H., & Pearce, A. R. (2013). Green luxury: A case study of two green hotels. *Journal of Green Building*, *8*(1), 90–119.

Atwal, G., & Khan, S. (2008, February). Luxury marketing in India: 'Because I'm worth it'. *Admap*, February, pp. 36–38.

Atwal, G., & Williams, A. (2007). Experiencing luxury. *Admap*, March, pp. 30–32.

Atwal, G., & Williams, A. (2017). Luxury brand marketing – The experience is everything! In J. N. Kapferer, J. Kernstock, T. Brexendorf, & S. Powell (Eds.), *Advances in luxury brand management. Journal of brand management: Advanced collections*. Cham: Palgrave Macmillan. doi:10.1007/978-3-319-51127-6_3

Bagwell, L. S., & Bernheim, B. D. (1996). Veblen effects in a theory of conspicuous consumption. *The American Economic Review*, *86*(3), 349–373. Retrieved from https://msuweb.montclair.edu/~lebelp/BagwellVeblenEffAER1996.pdf

Brenner, L., & Aguilar, A. G. (2002). Luxury tourism and regional economic development in Mexico. *The Professional Geographer*, *54*(4), 500–520.

Brouder, P. (2020). Reset redux: Possible evolutionary pathways towards the transformation of tourism in a COVID-19 world. *Tourism Geographies*, *22*(3), 484–490.

Brun, A., & Castelli, C. (2013). The nature of luxury: A consumer perspective. *International Journal of Retail & Distribution Management*, *41*(11/12), 823–847. doi:10.1108/IJRDM-01-2013-0006

Cheer, J. M. (2020). Human flourishing, tourism transformation and COVID-19: A conceptual touchstone. *Tourism Geographies*, *22*(3), 514–524.

Conrady, R., Ruetz, D., & Aeberhard, M. (Eds.). (2020). *Luxury tourism: Market trends, changing paradigms, and best practices*. Cham: Springer Nature.

Correia, A., & Kozak, M. (2012). Exploring prestige and status on domestic destinations: The case of Algarve. *Annals of Tourism Research*, *39*(4), 1951–1967.

Correia, A., Kozak, M., & Del Chiappa, G. (2019). Examining the meaning of luxury in tourism: A mixed-method approach. *Current Issues in Tourism*, *23*(8), 952–970.

Correia, A., Kozak, M., & Reis, H. (2016). Conspicuous consumption of the elite: Social and self-congruity in tourism choices. *Journal of Travel Research*, *55*(6), 738–750.

Correia, A., & Moital, M. (2009). The antecedents and consequences of prestige motivation in tourism: An expectancy-value motivation. In M. Kozak & A. Decrop (Eds.), *Handbook of tourist behavior: Theory and practice* (pp. 16–34). Oxon: Routledge.

Danziger, P. (2005). *Let them eat cake: Marketing luxury to the masses-as well as the classes*. New York: Kaplan Publishing. Retrieved from https://www.google.co.in/books/edition/Let_Them_Eat_Cake/HHC0swEACAAJ?hl=en

Derval, D. (2018). Designing luxury brands. In *Designing luxury brands. Management for professionals*. Cham: Springer. doi:10.1007/978-3-319-71557-5_4

Dixit, S. K. (2020). *The Routledge Handbook of tourism experience management and marketing*. London: Routledge.

Dubois, D. (2020). Fulfilling social needs through luxury consumption. In F. Morhart, K. Wilcox, & S. Czellar (Eds.), *Research Handbook on luxury branding*. Cheltenham: Edward Elgar Publishing. doi:10.4337/9781786436351

Dubois, D., Jung, S., & Ordabayeva, N. (2020). The psychology of luxury consumption. *Current Opinion in Psychology*, *39*, 82–87. doi:10.1016/j.copsyc.2020.07.011

Dubois, B., Laurent, G., & Czellar, S. (2001). *Consumer rapport to luxury: Analysing complex and ambivalent attitudes* (No. 736). Jouy-en-Josas: HEC Paris.

Fagan, C. (2017). Minimalism: Another boring product wealthy people can buy. Retrieved from https://www.theguardian.com/lifeandstyle/2017/mar/04/minimalism-conspicuous-consumption-class

Florin, D., Callen, B., Mullen, S., & Kropp, J. (2007). Profiting from mega-trends. *The Journal of Product and Brand Management*, *16*(4), 220–225. doi:10.1108/10610420710763903

Griskevicius, V., Tybur, J. M., & Van den Bergh, B. (2010). Going green to be seen: Status, reputation, and conspicuous conservation. *Journal of Personality and Social Psychology*, *98*(3), 392.

Hall, C. M., Scott, D., & Gössling, S. (2020). Pandemics, transformations and tourism: Be careful what you wish for. *Tourism Geographies*, *22*(3), 577–598.

Hampton, M. P. (1998). Backpacker tourism and economic development. *Annals of Tourism Research*, *25*(3), 639–660.

Hennigs, N., Wiedmann, K. P., & Klarmann, C. (2012). Luxury brands in the digital age–Exclusivity versus ubiquity. *Marketing Review St. Gallen*, *29*(1), 30–35.

Holmqvist, J., Wirtz, J., & Fritze, M. P. (2020). Luxury in the digital age: A multi-actor service encounter perspective. *Journal of Business Research*, *121*, 747–756.

Israel, B. (2002). *Bachelor girl: The secret history of single women in the twentieth century*. London: William Morrow and Co. Retrieved from https://www.google.co.in/books/edition/Bachelor_Girl/0hldAAAACAAJ?hl=en

Jaworski, A., & Thurlow, C. (Eds.). (2010). *Semiotic landscapes: Language, image, space*. Edinburgh: A&C Black.

Kanani, R. B. (2017). *A wealth of insight: The world's best luxury hoteliers on leadership, management, and the future of 5-star hospitality*. Virginia Beach, VA: Koehler Books.

Kapferer, J. N. (1998). Why are we seduced by luxury brands? *Journal of Brand Management*, *6*(1), 44–49. doi:10.1057/bm.1998.43

Kapferer, J. N. (2010). All that glitters is not green: The challenge of sustainable luxury. *European Business Review*, *2*, 40–45.

Kapferer, J. N. (2014). The future of luxury: Challenges and opportunities. *Journal of Brand Management*, *21*(9), 716–726.

Kapferer, J. N. (2016). The challenges of luxury branding. In F. Dall'Olmo Riley, J. Singh, & C. Blankson (Eds.), *The Routledge companion to contemporary brand management* (pp. 505–523). London: Routledge. doi:10.4324/9781315796789

Kapferer, J. N., & Michaut-Denizeau, A. (2020). Are millennials really more sensitive to sustainable luxury? A cross-generational international comparison of sustainability consciousness when buying luxury. *Journal of Brand Management*, *27*(1), 35–47.

de Kerviler, G., & Rodriguez, C. M. (2019). Luxury brand experiences and relationship quality for Millennials: The role of self-expansion. *Journal of Business Research*, *102*, 250–262.

Ketter, E. (2018). It's all about you: Destination marketing campaigns in the experience economy era. *Tourism Review*, *73*(3), 331–343. doi:10.1108/TR-03-2017-0053

Kohrs, K. (2020). The language of luxury fashion advertising: Technology of the self and spectacle. *Journal of Fashion Marketing and Management: International Journal*, *25*(2), 257–273. doi:10.1108/JFMM-02-2020-0029

Langer, D. (2020). How technology will change the game for luxury brands. Retrieved from https://jingdaily.com/how-technology-will-change-the-game-for-luxury-brands/

Leibenstein, H. (1950). Bandwagon, snob, and Veblen effects in the theory of consumers' demand. *The quarterly journal of economics*, *64*(2), 183–207. doi:10.2307/1882692

Liu, Z. J., Yildirim, P., & Zhang, Z. J. (2018, September 16). A theory of minimalist luxury. *A Theory of Minimalist Luxury*. doi:10.2139/ssrn.3246729

Luxury Travel Magazine. (2018). Is minimalism travel's new luxury? Retrieved from https://www.luxurytravelmagazine.com/news-articles/is-minimalism-travels-new-luxury

Moscardo, G., & Benckendorff, P. (2010, November). Sustainable luxury: Oxymoron or comfortable bedfellows. In Proceedings of the 2010 international tourism conference on global sustainable tourism, Mbombela, South Africa (pp. 15–19).

Papacharissi, Z. (2010). Privacy as a luxury commodity. *First Monday*, *15*(8). Retrieved from https://journals.uic.edu/ojs/index.php/fm/article/download/3075/2581

Papacharissi, Z., & Gibson, P. L. (2011). Fifteen minutes of privacy: Privacy, sociality, and publicity on social network sites. In S. Trepte & L. Reinecke (Eds.), *Privacy online* (pp. 75–89). Berlin and Heidelberg: Springer. doi:10.1007/978-3-642-21521-6_7

Paul, J. (2018). Toward a 'masstige' theory and strategy for marketing. *European Journal of International Management*, *12*(5–6), 722–745.

Pine, B. J., & Gilmore, J. H. (1999). *The experience economy: Work is theatre & every business a stage*. Boston, MA: Harvard Business Press.

Poelina, A., & Nordensvard, J. (2018). Sustainable luxury tourism, indigenous communities and governance. In M. Gardetti & S. Muthu (Eds.), *Sustainable luxury, entrepreneurship, and innovation. Environmental footprints and eco-design of products and processes*. Singapore: Springer. doi:10.1007/978-981-10-6716-7_8

Ramchandani, M., & Coste-Maniere, I. (2018). Eco-conspicuous versus eco-conscious consumption: Co-creating a new definition of luxury and fashion. In S. Muthu (Ed.), *Models for sustainable framework in luxury fashion. Textile science and clothing technology*. Singapore: Springer. doi:10.1007/978-981-10-8285-6_1

Richards, G. (2021). Transforming tourism in a post-Covid world. *Vita Magazine*, *5*(May), 62–63. Retrieved from https://www.academia.edu/48963005/Transforming_tourism_in_a_post_Covid_world?email_work_card=title

Riley, R. W. (1995). Prestige-worthy tourism behavior. *Annals of Tourism Research*, *22*(3), 630–649.

Ritchie, J. B., & Crouch, G. I. (2003). *The competitive destination: A sustainable tourism perspective*. Wallingford: CABI.

Ritchie, J. B., Tung, V. W. S., & Ritchie, R. J. (2011). Tourism experience management research: Emergence, evolution and future directions. *International Journal of Contemporary Hospitality Management*, *23*(4), 419–438. doi:10.1108/09596111111129968

Scheyvens, R. (2011). The challenge of sustainable tourism development in the Maldives: Understanding the social and political dimensions of sustainability. *Asia Pacific Viewpoint*, *52*(2), 148–164.

Schmitt, B. (1999). Experiential marketing. *Journal of Marketing Management*, *15*(1–3), 53–67.

Schmitt, B. (2011). *Experience marketing: Concepts, frameworks and consumer insights*. Hanover, MA: Now Publishers Inc.

Sigala, M. (2020). Tourism and COVID-19: Impacts and implications for advancing and resetting industry and research. *Journal of Business Research*, *117*, 312–321. doi:10.1016/j.jbusres.2020.06.015

Sung, Y., Choi, S. M., Ahn, H., & Song, Y. A. (2015). Dimensions of luxury brand personality: Scale development and validation. *Psychology & Marketing*, *32*(1), 121–132.

Swarbrooke, J. (2018). *The meaning of luxury in tourism, hospitality and events*. Oxford: Goodfellow Publishers Limited.

Thach, L., & Yeung, P. (2019). *Luxury wine marketing: The art and science of luxury wine branding*. Oxford: Infinite Ideas Limited.

Thirumaran, K., Jang, H., Pourabedin, Z., & Wood, J. (2021). The role of social media in the luxury tourism business: A research review and trajectory assessment. *Sustainability*, *13*(3), 1216.

Thurlow, C., & Jaworski, A. (2012). Elite mobilities: The semiotic landscapes of luxury and privilege. *Social Semiotics*, *22*(4), 487–516.

Truong, Y., McColl, R., & Kitchen, P. J. (2009). New luxury brand positioning and the emergence of masstige brands. *Journal of Brand Management*, *16*(5), 375–382.

Tynan, C., McKechnie, S., & Chhuon, C. (2010). Co-creating value for luxury brands. *Journal of Business Research*, *63*(11), 1156–1163. doi:10.1016/j.jbusres.2009.10.012

Veblen, T. (1899). *The theory of the leisure class*. New York: Macmillan.

Veblen, T. (1992). *The theory of the leisure class* (1st ed.). London: Routledge. doi: 10.4324/9781315135373

Vigneron, F., & Johnson, L. W. (1999). A review and a conceptual framework of prestige-seeking consumer behavior. *Academy of Marketing Science Review*, *1*(1), 1–15.

Vigneron, F., & Johnson, L. W. (2004). Measuring perceptions of brand luxury. *Journal of Brand Management*, *11*(6), 484–506. doi:10.1007/978-3-319-51127-6

Yang, W., & Mattila, A. S. (2016). Why do we buy luxury experiences? *International Journal of Contemporary Hospitality Management*, *28*(9), 1848–1867. doi:10.1108/IJCHM-11-2014-0579

Yeoman, I. (2011). The changing behaviours of luxury consumption. *Journal of Revenue and Pricing Management*, *10*(1), 47–50. doi:10.1057/rpm.2010.43

Yeoman, I., & McMahon-Beattie, U. (2018). The future of luxury: Mega drivers, new faces and scenarios. *Journal of Revenue and Pricing Management*, *17*(4), 204–217.

Part 1
Conceptualising Luxury

Chapter 1

Historical Progression of Luxury

Vinay Chauhan,[1] Suvidha Khanna,[2] and Sandeva Khajuria[3]

Abstract

Throughout history, the concept of luxury has had different emphases. It was scaling its existence in tourism, from the lifestyles of ancient kingdoms to the Roman empire's witnessing sightseeing trips, tourist attractions in townships to business tours and luxury hotels in Paris to the British phenomenon of 'The Grand Tour' and the 'Golden Age of Travel'. The chapter also covers its spread across the small number of affluent societies and groups of wealthy consumers hailing from not only developed but developing economies in the early twentieth century.

Presently, the concept has become 'mass' as the Digital-First Millennial and Generation X, and also the affluent female market is all contributing to reshape the concept and the demand of luxury travel in the emerging scenario. Therefore, to understand it fully, it is imperative to review what has gone before and the influences of factors such as economic development, technological innovation and the consumption behaviour for luxury in different societies. Thus, the chapter is an attempt to understand the historical progression of luxury tourism for giving directions for future implications for the sustainable growth of this form of tourism across the globe and in India. The chapter concludes by noting the applied and social approaches in the development of luxury tourism, leaving many academic researchers and those in the tourism industry to resolve the real progression of luxury tourism and its development with the support of a cross-cultural perspective.

Keywords: Luxury; tourism; historical progression; cross-cultural; experiential-based approach; luxury experience

[1]https://orcid.org/0000-0002-52529550
[2]https://orcid.org/0000-0003-4345-2018
[3]https://orcid.org/0000-0002-8309-4684

The Emerald Handbook of Luxury Management for Hospitality and Tourism, 19–35

doi:10.1108/978-1-83982-900-020211001

Introduction

The increase in luxury tourism attracted the attention of academics and industry. Marketing and consumer behaviour literature suggest that luxury and non-luxury attitudes depend on consumers' judgement (Chang, Ko, & Leite, 2016; Dixit, 2017; Tynan, McKechnie, & Chhuon, 2010). Tourists are looking for the different/extraordinary element that leads to more services and inclusion of experiential factor to the physical factors. A leading example comprises an increased use of luxury products and services, particularly tourism, as tourists desire to visit unique places and take up memorable experience-filled activities (Veríssimo & Loureiro, 2013). According to Strong (2006), luxury travel can be explained as travels for the elites that involve luxurious first-class facilities and amenities and very important person (VIP) service, creating a lasting memory and transformative experience and enhancing the customers' lifestyle. According to Schänzel, Yeoman, & Backer (2012), luxury lies in limited edition, authentic, customised and kind products, and one of the main goals for all those marketers interested in positioning their offer in the luxury market is to create an illusion of acute personalisation. Purinton-Johnson (2013) defines luxury travel covers high-class tangible and intangible aspects of the product. Accordingly, luxury travel meant to take up a novel experience and visit a rare and exotic place, having exclusive services, staying at lavish and luxurious accommodation, having unmatched food, hiring well-informed and professional travel guides, etc. It is considered a stress-free way of travel that eliminates the pressure of daily lives, so that all expectations are not only met but exceeded.

The definition is highly relevant for cruise ship products or 5-star accommodation (Chen & Peng, 2014). Fitzsimmons (2017), supported by Schensul (2017), points out a perspective and explains that 'luxurious' is not about materials but about experience and service, something precious, rare, invested in emotional value, such as choosing one's room fragrance, blending your wine and enjoying an outstanding and personal collection of art, plants or other collectable artefacts.

However, at the same time, the level of luxury, one of the main motivations of tourists, also depends on the income and budget of each tourist (Riley, 1995; Yeoman & McMahon-Beattie, 2006). Later, Schänzel et al. (2012) introduced time, wellness and cultural enrichment as antecedents of luxury tourism. To summarise, 'luxury can be anything, accessible to mass markets, not traditional, but personal, authentic and experiential', but, after all, the old-world luxury of consumption and exclusiveness still keeps its original icon (Yeoman & McMahon-Beattie, 2006).

Historical Progression

Global Perspective

The concept has changed and evolved over a period of time with evident roots in history, dating back to about 200 or 300 years, although some would argue that its existence has a much longer history with its origin from empires and kingdoms; however, as a business practice, luxury hotels in Paris use to reflect the tastes of the seventeenth and eighteenth centuries. If the history of tourism is evaluated, it is found that travel was basically for pilgrimage, business, military, explorers and adventurers, and general travellers (sightseers) in earlier days. It is, however, true that for any form of tourism during those days, anything beyond just travel was considered a luxury. For example, for a poor person, travelling on a horse was considered to be a luxury. In the case of sightseeing travellers, it was a luxury affair as they would only visit a place if they have disposable income and free time, which comes in luxury's purview. So, it was mainly for the 'affluent class'. The Roman Empire witnessed people travelling to nearby places for sightseeing as well as to long distances. Not only were their liveable accommodations, but fine villas, scrumptious food and drinks, especially for the more affluent section, so that along with the comfort, they can enjoy themselves during their journeys. The towns had sightseeing places and attractions. But it was not seen as a luxury by many as they had similar luxuries at the homes. Yet both the Romans and the Greeks travelled in minimal numbers as tourists for a century or so ago.

There came the concept of a Grand Tour. The very British phenomenon called 'The Grand Tour' is perhaps one of the first embodiments of the idea of luxury travel in history. This era (1613–1785), which led to the luxury travel and tourism activities, started with the wealthy English and subsequently spread to become a trend among other individuals who had time and money. It was more like a vacation meant to enhance the knowledge and experience of the young aristocrats. As in today's time, as the students take a gap year, a Grand Tour too was a sabbatical time in which the young people, especially men, provide them with a broader perspective of education and art. But this tour was not for everyone. Being a luxury, it was taken only by those having disposable time and income. Also, being lasting for a period of two to three years, involving a lot of travelling, staying at fine accommodations, having good food and entertainment, it happened to an expensive event, and, thus, only elite could go for a Grand Tour. Grand Tour remained very popular in the seventeenth and eighteenth centuries. And by the end of the eighteenth century, as it was becoming more and more desirable, people with lesser disposable money also craved for a similar experience. Thus, a new type of Grand Tour was conducted for people who travelled for hedonism and sightseeing, rather than education and travelled for weeks rather than months and years. The Grand Tour was gradually considered a trend that took its peak during about 1660 until rail transport in the 1840s was the advent and was associated with a standard itinerary, especially for educational purposes. The concept of the Grand Tour was primarily associated with the British nobility and wealthy gentry, but during the second half of the eighteenth century, the young men from other countries, including North America, South America and nations from Northern Europe, started taking up these trips. The Grand Tour had become a regular feature of aristocratic education in Central Europe, although it was restricted to the higher nobility by the mid-eighteenth century. Along with the aristocracy but in ever more significant numbers, the wealthy people were at the heart of the 'Golden Age of Travel', a nostalgic term when travel was the preserve of a fortunate small minority with European and American societies. And gradually, the concept travelled to other parts of the world too in the nineteenth century, and, therefore, J. Gilford-Milne published a paper on this very subject in 1916, leading to the conclusion that the concept of luxury tourism has taken a gearing shape. The rise of the wealthy class that emerged from the grime and suffering of the industrial revolution in Europe and the United States (and the ones who earned wealth on their own) coincided with the railway and steamship age. Alongside the aristocracy, this concept grew in ever larger numbers during the early nineteenth century, and this period is known as the 'Golden Age of Travel'. Conspicuous consumption was necessary, whether through travelling first class on ocean liners, buying art or gambling at the casino.

Today, luxury tourism has more or less become 'mass' with a continuously increasing demand on the part of everyone who wants a touch of luxury in their lives, and this previously exclusive market is now being accessed by people from different levels of the societies and no longer confined to a few or very wealthy. According to Fedeli (2011), the concept of luxury is relative, based on a series of physical factors and the perceptions and personal evaluation that people attribute to them. Kapferer and Laurent (2015) support the idea to analyze the perception of consumers to detect the consumers' definition of luxury. The trend shows that Digital-First Millennial and Generation X and the affluent female market contribute to reshaping concept and the demand for luxury travel in the emerging scenario (Danziger, 2019). Not only that, the middle class and upper middle class are the upcoming markets in the luxury travel landscape, which is going to influence the supply dynamics (Jain, 2018). Therefore, to understand it fully, it is imperative to review what has gone before and the influences of factors such as economic development, technological innovation and the consumption behaviour for luxury in different societies. The conclusive global historical progression of luxury tourism is presented in the following Table 1.1.

Indian Perspective

Indian tourism and hospitality have been renowned for centuries. Indians have been treating their guests like a god for ages. With a history of about 5000 years, diverse culture, many religions and alternative medicine, the country fascinates all tourists' segments. Earlier, pilgrimage or pilgrim travel was the only important form of tourism. Around 250 BCE, Emperor Ashoka had set up rest houses and Dharamsala where travellers could rest. A large number of monasteries were also built for the pilgrims during this period. The trend was followed by emperor Harsha. It is said that even when Alexander the Great arrived in India in 326 BC, he found pretty well-maintained roads which were covered with shady trees. This showed that travel facilities were much efficient, and travelling was not a burdensome experience. Marco Polo was another great traveller who passed through India on his way back from China in 1271. All travellers were much interested in seeing India and her fabled riches for themselves, indicating that India was rich, prosperous and progressive in those days.

Traditionally, the royal court started travelling to the rivers and the hills seeking pleasure. These movements further attracted those who had business at the court to travel, escaping the cities to calmer retreats. During the Mughal

Table 1.1. Global Historical Progression of Luxury Tourism.

Age/year	Developmental Perspective
The era of kingdoms and empires	• Luxury tourism had existence but was not very much known.
200–300 years ago	• The concept was any form of tourism beyond just travel was a luxury.
Roman Empire	• Sightseeing trips were there over long distances too. • Although travel was existent with the luxury concept, the quality was similar to that of what they experienced at home.
Grand Tour in a phased manner (1613–1785)	• Phase 1: Luxury travel was for wealthy Englishmen who had time and money. • Phase 2: Extended vacation designed to be for 'young aristocrats' for 2–3 years. • Phase 3: The concept began to happen countless times. People with less spare time and income too started enjoying experiences of aristocracy but on a smaller scale. • Phase 4: It was for older people who travelled for hedonism and sightseeing, and it was the travel for weeks rather than months and years. • Phase 5: By the mid-eighteenth century, it had become a regular feature of aristocratic education in Central Europe.
Nineteenth century	• The concept travelled beyond Europe to other parts of the world, and J. Gilford-Milne published a paper on this subject in 1916.
Twentieth century	• 'Golden Age of Travel': The rise of the wealthy class that emerged from the grime and suffering of the industrial revolution in Europe and the United States (and the ones who earned wealth on their own) coincided with the railway and steamship age. Alongside the aristocracy, but in ever larger numbers. • Conspicuous consumption was important, whether it be through travelling first class on ocean liners, buying art or gambling at the casino.

Table 1.1. *(Continued)*

Age/year	Developmental Perspective
Twenty-first century	• Fedeli (2011): Leads to the conclusion that the concept of luxury is relative based on a series of physical factors as well as the perceptions and personal evaluation that people attribute to them. • Kapferer and Laurent (2015): The idea to analyze the perception of consumers to detect the consumers' definition of luxury. • Jain (2018): Middle class and upper middle class are the markets in the luxury landscape. • Danziger (2019): Digital-First Millennial and Generation X, and also affluent female market are all contributing to reshape the luxury market.

Source: Authors Adapted from Fedeli (2011), Kapferer and Laurent (2015), Danziger (2019), Jain (2018), Swarbrooke (2018), Sharma (2017), Skift and Peak Destination Management Company (2014), CBI (2018), and Fitzsimmons (2017).

Empire, the emperors used to travel extensively and developed resorts. Mughals, along with Afghans, are attributed to build 'caravanserais' along the highways and other travel routes. But with the fall of their empires, trade and commerce faced a setback. Thus, mobility of people was reduced, but pilgrims continued travelling. Eventually, the scenario changed, and a complex character of tourism surfaced. Furthermore, finally, the emergence of modern tourism and incorporation of luxury with tourism happened because of the growth of modern technology, rising incomes and improved facilities.

The rail network in India actually led to the need for recreation within reach of an increasingly large number of people who had leisure time and the means to enjoy it. Travel became more accessible and comfortable with the onset of railways. On 16 April 1853, the first passenger train ran between Bori Bunder (Bombay) and Thane, and after that, Indian railways never stopped connecting places, making it possible for the potential tourist to visit destinations they could not access earlier. India's first pan-country super-luxury train, 'Palace on Wheels', was started on 26 January 1982 and was refurbished and relaunched in August 2009 with more luxurious services. And with its high success, many more luxury trains came into existence.

Adding on to the luxury facilities for the high-spending tourists, Mr Jamsethji Tata, in 1903, opened the famous Taj Mahal Hotel in Mumbai. He was the first person to establish luxury hospitality for years until Mr Rai Bahadur Mohan Oberoi bought two Clarke's hotels in Delhi and Shimla in 1934. Later in the early

1960s, ITDC (India Tourism Development Corporation) was started to provide Western comforts to international tourists. It further contributed to the development of a modern superstructure for international visitors (e.g., The Ashoka Group of Hotels). By the end of the nineteenth century, various luxury hotels in South India opened up following Western styles to the core. West End, Palm Beach, Imperial, Elphinstone, Ritz, New Woodland are some of the few.

Indian Airline sector also made a progress and India's first scheduled airline, Tata Airlines, in 1932, was also launched by JRD Tata who piloted the first flight himself from Karachi to Mumbai via Ahmedabad on a single engine. By 1994, India repealed all laws regulating the formation of airlines and allowed the introduction of scheduled private airlines. This deregulation allowed for the formation of new airlines such as Jet Airways and ModiLuft, and laid the foundation for the future Indian aviation industry. Later in 1991, with the development of luxury accommodations in India and hosting of events like Commonwealth Indian MICE (meetings, incentives, conferences and exhibitions), tourism witnessed substantial growth. The focus of the MICE industry shifted from Europe to the Asia Pacific region, and India had a major breakthrough (Everything Experiential, 2020), and since then, MICE is constantly growing with an estimated annual rate of 15–20%. The Union Government aims to achieve 2% of the global MICE market share by 2025. The summarised historical progression of luxury tourism vis-a-vis the Indian perspective is concluded in Table 1.2.

Being a country abundant in tourism resources and a rich cultural blend, India has to date managed to continue its legacy in luxury tourism. Having unmatchable and unique destinations, a varied range of both heritage and modern luxury hotel chains, including well-qualified and trained tourism – hospitality professionals and welcoming people, India has earned the fame of being one of the most favourite luxury tourism destinations across the globe. Some of the most desired luxury travel experiences of India include the floating lake place of Udaipur, feeling like a maharaja in luxury trains, Ayurveda and wellness at the Himalayas, tea tourism at Darjeeling, staying like royals in castles and forts, spotting tigers on a luxury safari, sailing on a private houseboat and spa retreats in backwaters of Kerala, etc. While the classic destinations are still in demand, more contemporary travel concepts are being stipulated by a new tourist segment. These new and young luxury travellers are taking off-beat paths, seeking experiences that augment personal growth and self-actualisation. They need something more than just leisure. This segment of aspirational consumers willing to spend a substantial amount of money upon travelling settle on three major motivations including occasion, experience and fondness for luxury (Skift & Peak Destination Management Company, 2014). India is all capable of catering to this potential segment by offering a unique and different experience like there are diving trips to the Andaman Islands, photography tours to Leh and Ladakh, apple picking trails in Shimla, desert safaris and stars dinner at the sand dunes, lavish stay at a plush resort by taking a helicopter ride, a spectacular dinner below moon and stars at the sand dunes of Thar.

Consequently, the luxury travel market in India is expanding. Its annual growth rate from 2015 to 2025 has been predicted to be 12.8% (Shankaman, 2017). The global luxury travel market is expected to gear up to $1,154 billion by

Table 1.2. Historical Progression of Luxury Tourism in India.

Era/Year	Developmental Perspective
5000 years ago	Travel itself was a luxury.
250 BC	Emperor Ashoka had established rest houses, Dharamshala and monasteries for pilgrims.
Thirteenth century	Marco Polo passed through India on return from China (1271).
Era of Mughals	Travel for pleasure and luxury started. The emperors travelled extensively, and, therefore, the resorts were also developed.
Era of Afghans	Built 'caravanserais' along the highway and business routes.
Fall of Great Empires	Only pilgrimage tourism developed.
Development of rail network (16 April 1853)	The first passenger train was between Bori Bunder (Bombay) and Thane, and after that, the Indian travellers who had leisure time and means to visit destinations.
1903	The first luxury hotel, i.e. Hotel Taj Mahal, was built by Mr. Jamsethji Tata, and after that, many other luxury hotel brands also emerged subsequently.
First scheduled airlines in 1932	Tata Airlines was India's first scheduled airline, which piloted the first flight from Karachi to Mumbai via Ahmedabad, and it was on a single engine.
26 January 1982	India's first super-luxury train, 'Palace on Wheels', was started. And with its high success, many more luxury trains came into existence.
1991–2010	The development of the MICE concept in India led to the advent of luxurious accommodations, hosting large-scale events such as Commonwealth Games and also developing luxury travel in India.
2015 onwards till date	The luxury travel market in India is expanding with an annual growth rate of 12.8%.

Source: Authors adapted from Skift and Peak Destination Management Company (2014), Shankaman (2017), Parulekar (2019), MILT (2017), Hans (2020), Carmela (2016), and NTA (2014).

2022, and the Indian market will form a significant share of it, expecting a compounded annual growth rate (CAGR) of 13% (Parulekar, 2019).

Not only limited to offering luxury experience, but India is also producing luxury tourists for other destinations. Every year, approximately 3.6 million luxury Indian tourists take a trip abroad. The number could increase up to 8.6% a year. Their arrivals are quickly escalating in both nearby countries like Dubai as well as longer-haul destinations like Sweden. India Infoline had estimated India to annually generate 6.5 million outbound luxury and MICE tourists by the year 2020, strengthening the country's influence as a key source market for MICE and luxury travel (MILT, 2017).

Cross-cultural Perspective

Culture has remained a key determinant of the basic values of our society. The foundation of peoples' choices and behaviour somewhere lies in their cultural standards and ethnicity. It is evident that societal backdrop shapes consumers' purchasing motivations (Li, 2014). In travel and tourism also, tourists belonging to different cultural backgrounds exhibit different preferences. Luxury for one section of the tourist may not please the other. It is, therefore, that literature lacks a conclusive definition of luxury tourism. Yet to an extent, it can be said that idea of luxury tourism is centred on having an unfiltered experience. Luxury tourists prioritise value over price, which is measured in terms of experience (Ledezma, 2017). For tourists from Europe, luxury means 'superior', 'convenient" and 'engaging". There is a shift in value for Europeans now; rather than collecting luxurious possessions, they opt for luxurious experiences. Europe itself is the leading luxury tourism hub. European luxury travellers are said to often look for exclusive one-off experiences. They look for personalised service, reliable transport, positive and professional interaction with staff, and pay attention to details like quality of beds, quality of linen, the comfort of the seats when travelling, legroom, quality of food, wine, etc., or in short, they look for exclusivity (CBI, 2018). The traditional European luxury tourist looked for the most glamorous travel experiences and indulged in luxury because they wanted to enjoy to the fullest and indulge in luxury; for example, they would hire a luxury yacht for a group of friends or family. They used to advise luxury influencers/consultants to enhance their trips and put their spending on display. However, in today's context, European luxury tourists seek authentic and unique experiences rather than high-end luxury; these experiences may be relatively expensive due to their exclusivity. Those who wish to pamper themselves or want a novel experience go for a luxury yoga retreat or join a cookery weekend in Vietnam, and so on. The section which spends their saving on luxury tours reserved for special occasions like wedding proposal, honeymoon, etc., looks for 'wow factor" experiences. Another section is the one that demands high-end luxury products and services like staying at seven-star hotels, travelling in first-class or by private jets, having one-to-one service providers. There has been an inclination towards wellness tourism among European luxury tourists for the last few years. Surprisingly, the main markets for trips to developing countries, i.e. the United Kingdom, France and Germany, are also leading European luxury tourism markets.

The last few years have seen the travel industry experiencing ascend in Asian luxury tourists. It becomes mentionable that they are not all from China. As per the luxury signature (2020), Asian luxury travellers generally desire destinations that offer unique experiences. They want to do activities and tend to choose detentions where they have not been to before. Many of them are concerned about the environment too, and for them, their travel must not hurt the environment. As far as accommodation is concerned, most of them go for staying at four- to five-star hotels, and few choose luxury apartments. Most Asian luxury travellers want to do activities like discovering local cuisine, going to theme parks and shopping. Despite the rise of low-cost carriers in the Asian region, the luxury travellers here still prefer full-service carriers. Travellers from India and Japan are more likely to upgrade their air travel to either first or business class, while travellers from Taiwan, Hong Kong and China opt for the economy. A few of them opt for cruise tours (Carmela, 2016).

Talking about the United States, according to the NTA (2014), the luxury traveller takes up an average of three trips per year for an average of six days per trip. Their luxury travel's common motivations comprise exploring new destinations, seeking authentic experiences, relaxation, personal enrichment, etc. (NTA, 2014). These affluent travellers prioritise accommodations followed by the unique experience and air travel, respectively. They want customised and exclusive service, like cruising, yachting and small ship expeditions, adventure trips, luxury train trips, celebration travel, philanthropy trips, shopping trips, etc. In Latin America, the luxury segment is proliferating as the middle class is transforming into the ones who spend their savings on luxury trips (Gutierrez, 2020).

When it comes to luxury tourists from Australia, they are seen to be motivated by a desire to visit new places and experience new cultures. They do not prefer to repeat the destination. They intend to invest a handsome amount in their travel and also expect value in their travel. According to a recent survey by a luxury travel magazine (2019), luxury tourists from Australia are interested in history and heritage sites. They emphasise dining and food and like to visit arts/museums and culture. Their favourite destination is Europe, followed by North America. Classic European destinations, including London and Rome, have been the most liked ones, followed by Paris. Some of them are smitten with South East Asia as well. Their favoured method for booking outbound travel remains using face-to-face services of travel advisors. Very few of them opt for online booking.

The Applicative Perspective of Luxury Tourism

Luxury and Tourism

Luxury is considered to be a stipulation of immense ease and comfort, which adds to pleasure but is not needed. It is a subjective concept since its meaning is led by personal dimensions and perceptions of quality, comfort and elegance. In certain situations, limited and rare products are considered luxuries (Nueno & Quelch,

1998). However, the idea of luxury is centred on exclusivity and the notion that not everyone can afford it. It does not respond to the demand; it does not satisfy the desires of the masses nor seeks positioning. People buy luxuries for different reasons. Some people buy luxuries purely for their aesthetics. Some buy due to their essentialism. Some buy them for attaining deep pleasure. Another reason to buy them is to signal their richness to others. Sometimes a luxury purchase is made to raise ego instead of comfort or consumption (Veblen, 1899).

To satisfy the desire of the elite class customers, travel and tourism started providing luxury. Formally initiated and flourished in Europe, the concept has taken over the world. As luxury refers to all that can be got that is different and exclusive; likewise, in the field of tourism, luxury takes shape to the extent that, rather than price, puts the emphasis on value, and now in today's scenario, it is measured through the experience of the consumer. Riley (1995) pointed out that luxury tourism is very much there in the destination that is perceived as unique and different, but it depends on how the individuals choose the destination and way to travel. This means that tourism is very much related to luxury, whether it is by tourists' attitudes or by the destination or services they seek in for.

The luxury travel segment comes in all possible ways like some use luxury in their daily lives, others save up for unique or special occasions. The concept of luxury travel has been evolving. Gradually, luxury travel is witnessing a shift in values from materialism to experiential. Rather than accumulating luxurious possessions, people want to have memorable experiences (Dixit, 2020; Sharma, 2017). The idea of a luxury holiday is centred on the luxury of an authentic experience. It is the luxury of time and a true sense of place, which means experiencing destinations in inspirational, extraordinary ways, on one's terms (The Luxury Holiday, 2018). Though not new in themselves, the main dimensions, also called the 5 Cs of luxury travel, including Culture, Cuisine, Community, Content and Customisation, are much in focus today. Right from the classic luxury destinations like Italy, Paris, London, New York, Las Vegas, etc., to the contemporary cruise ride through the Nile river, sailing through the Komodo island, wildlife in South Africa, private sandbank tour in the Maldives, helicopter tour to the Grand Canyon, ship safari through the Ha Long Bay, ski experience with fine food, exploring islands or stay in the Mongolian wilderness, luxury in travel is diverse.

The practice of possessing luxury may be more significant to some people and not to others. Those who like to demonstrate their social status in terms of acquiring products and consumption of services chose luxury destinations when it comes to travel. They chose prestigious and conspicuous destinations to have a bet of the experience. However, with high-end luxury tourists, the phenomenon of experience-based travel is getting obsolete. They are heading towards emotional tourism (Satli, 2019). The luxury tourist segment seeks emotional, experimental and, of course, non-Google search trips and destinations, and not what everyone wants, and also, they do not take advantage of the new trends that provide them to have a little space to enjoy it. Simultaneously, the technology-loving tourist of this era, 'the millennial',' look for exclusivity with fun and want to communicate it to the world. They tend to display their travel activities on social media and connected online at all times. Another powerful concept in luxury tourism is that these tourists convey their experiences through social networks by creating and reinforcing the notoriety of destinations and brands.

Experiential-based Approach for Luxury Tourism

Customer experience management has become a very important aspect of studying because it has been found in studies that travellers' memories are a basic outcome variable of interest influenced by the travellers' affective experiences (Kim, Prideaux, & Chon, 2010; Manthiou et al., 2014; Sundbo & Dixit, 2020; Tung & Ritchie, 2011). Kemp (1998) pointed out that the perception of what is a necessity and what is luxury does vary from society to society despite the similar nature of basic human needs, and so, this idea could lead to the social perception of what is luxury or not for different people to agree or disagree. Furthermore, this holds true for the tourism business in the luxury context too. Strong (2006) argues that operating a luxury travel business means providing elite travels that create a memorable memory and transformative experience that would upgrade the life and the style of the customers. In the luxury market, some elements like price and status are very important and cannot be left behind, but other factors have also been identified (Husic & Cicic, 2009). According to Page (2009), by definition, luxury tourism is to consume high-quality and expensive travel services, especially those offering beautiful experiences, and following the idea that the characteristics of luxury experiences evolve during the course of time. The specifics of experiences in luxury tourism depend on the level of the demands of the travellers who usually wish to get authentic, exclusive and personalised experiences with the benefit of paying huge sums of money. Such experiences might include, e.g. individual excursions and trips, dinners at unique places (be it in renowned restaurants or a secluded place in the countryside) or a chance to experience traditional rituals with local inhabitants in a given destination, etc. (Mahika, 2011). For Lee and Hwang (2011), the concept of luxury is based on three dimensions: functionalism, experiential symbolism and symbolic interaction. Functionalism is the idea that luxury goods solve problems and necessities of customers; the experiential symbolism shows the idea of sensorial pleasure; and finally, the symbolic interaction represents the real satisfaction to customers' needs. According to Skift and Peak Destination Management Company (2014), people these days want to travel and have emotional and personal experiences while travelling. They are less interested in standardised holiday packages and rather desire to have adventurous, authentic and local activities oriented holidays. An affluent clientele starts to search for material standard and a specific added value of non-material nature (Koch, 2011). Purinton-Johnson (2013) presents a detailed discussion on luxury travel and points out that luxury travel is very much lead by the experiences that are real about a destination and the immersion into local culture; yet, the comfort factor cannot be ignored, and high standard of accommodation and dining will have to be the essential features. Hwang and Hyun (2016) found that luxury cruise passengers' perception of cruise lines' innovativeness (which can be seen as a proxy for novelty) and found that that innovativeness is an important factor influencing various aspects of cruise travellers' experience in the luxury market. Fitzsimmons (2017), supported by Schensul (2017), points out a perspective and explains that 'luxurious is not only about the physical aspect but about experience and service, something different and expensive, which is not always there, have an emotional value, like choosing one's own room fragrance, blending your own wine, enjoy an outstanding and personal collection of art, plants, or other collectable artefacts'. Her view resembles the 'classic elite' with less emphasis on materials, rather on the essence of travels, which is the 'experience'. This is highly relevant for cruise ship product or 5-star accommodations (Chen & Peng, 2014).

In today's context, luxury travel is all about the experience, not just the physical aspect. Luxury travel has to be glamorous, but it necessarily always meaningful. At the same time, local culture may not always be glamorous, but it always has to be meaningful (Chopra, 2016). The concept of luxury is varying concerning demographics and the nature of activities that the tourists take up, and the concept of 'luxury' is also evaluated differently by different consumer groups. For instance, in the case of millennials, while on the one hand, they may not like to go for very expensive services or accommodations, and on the other hand, they may spend good fortune for exclusive activities to a crated memorable experience (Koumelis, 2018).

Managerial Implications

The tourism industry has greatly been influenced by uniqueness, brand imaginary, high price levels and many more of such trends. That is why studies relating to trends in luxury tourism show a high degree of similarity with this industry (The Future of Luxury Travel). Therefore, the demand for luxury travel is expected to grow, which shows significant market research, especially in young customers, females, Generation X and the latest because even the middle class and the upper middle class would also go with this newly emerging form of tourism.

The current study also has a contributing effect on theoretical and managerial implications on luxury tourism. This study reflects that a lot of scopes still exists to develop this concept of tourism. From a marketing perspective, the luxury industry's marketing and consumption behaviour still need to be explored. The demand for luxury travel is expected to grow in the coming years, so it becomes imperative to understand the consumers in terms of the luxury perspective. The historical progression of this study clearly reflects the fact that the type of consumers in this niche area is continuously changing, and so this gives a platform to the marketers to evaluate the futuristic set of customers in this domain.

The study also has a major academic implication because it adds to the existing knowledge of luxury tourism. The study reflects that the concept also emerged with the cross-cultural perceptive, which further adds to the available literature.

Conclusive Evidence

To conclude, it can be said that luxury tourism has emerged as an upcoming tourism form concerning economic developments on a global scale. Not only that, it is very significantly impacted by the changes in the tastes and preferences of customers. Still, a view of what luxury looks like is not very clear, and also what is the course of the effective strategies for future sustainable luxury marketing remain unstudied. If the historical progression of both throughout the world and in India is evaluated, it can be pointed out that the meaning and concept of 'luxury tourism' is continuously changing. Not only is that, but the markets of this particular form are also constantly changing viz. demographics, and other socio-psychological factors, and so because of this, in this context, the market segmentation importance could be evaluated to attract people with specific characteristics, such as middle-aged, high-income and high-spending groups (Kim et al., 2010). The only thing that has not changed in the domain of luxury travel is 'experiences'. Today, the luxury travellers want the facilities, but he is looking for the 'luxurious experiences' related to the non-tangible aspect of tourism. The study also points out that this concept is not limited to one society or one culture, but it has extended its tentacles to almost all societies (as evident in the cross-cultural aspect of the study). Therefore, the concept of luxury tourism is to be developed in a true manner, and it is only possible with the involvement of tourists not only from the elite and upper middle class segments of the societies but should be developed keeping in view the developing market segments too as the case may be.

References

Carmela, J. (2016, July 29). Defining the Asian luxury traveler. *WTM Global Hub*. Retrieved from https://hub.wtm.com/defining-the-asian-luxury-traveller/. Accessed on August 11, 2020.

CBI, Ministry of Foreign Affairs. (2018, November). What are the opportunities for luxury tourism from Europe? Retrieved from https://www.cbi.eu/market-information/tourism/luxury-tourism. Accessed on August 12, 2020.

Chang, Y., Ko, Y. J., & Leite, W. L. (2016). The effect of perceived brand leadership on luxury service WOM. *Journal of Services Marketing*, *30*(6), 659–671. doi: 10.1108/JSM-01-2015-0005

Chen, A., & Peng, N. (2014). Examining Chinese consumers' luxury hotel staying behavior. *International Journal of Hospitality Management*, *39*, 53–56. doi:10.1016/j.ijhm.2014.01.002

Chopra, N. (2016). What is experiential luxury travel and why I choose it as a millennial. *The Cultureur*. Retrieved from https://thecultureur.com/what-is-luxury-travel/. Accessed on August 12, 2020.

Danziger, P. N. (2019). *Meet the Henrys: The Millennials that matter most for luxury brands* (1st ed.). Rochester, NY: Paramount Market Publishing, Inc.

Dixit, S. K. (Ed.). (2017). *The Routledge handbook of consumer behaviour in hospitality and tourism*. London: Routledge.

Dixit, S. K. (2020). *The Routledge handbook of tourism experience management and marketing*. London: Routledge.

Everything Experiential. (2020). Has India emerged to become a hot MICE destination? Retrieved from http://everythingexperiential.businessworld.in/article/Has-India-emerged-to-become-a-hot-MICE-destination-/04-01-2020-181480/

Fedeli, G. (2011). *The luxury side of tourism and China: A qualitative study of motivations and national values of the affluent mainland Chinese leisure traveller in Sydney, Australia*. Doctoral dissertation, NHTV Breda University of Applied Sciences. Researchgate. Retrieved from https://www.researchgate.net/. Accessed on August 7, 2020.

Fitzsimmons, A. (2017, January 11). 7 luxury travel trends for 2017. *National Geographic*. Retrieved from https://www.nationalgeographic.com/travel/top-10/seven-luxury-travel-trends-hotel/. Accessed on August 10, 2020.

Gutierrez, S. (2020, February 13). The potential of the ultra luxury segment in Latin American tourism. *Latin Trade*. Retrieved from https://latintrade.com/2020/02/13/the-potential-of-the-ultra-luxury-segment-in-latin-american-tourism/. Accessed on June 12, 2020.

Hans, A. (2020, January 4). Has India emerged to become a hot MICE destination? *Everything Experiential*. Retrieved from http://everythingexperiential.businessworld.in/article/Has-India-emerged-to-become-a-hot-MICE-destination-/04-01-2020-181480/. Accessed on May 25, 2020.

Husic, M., & Cicic, M. (2009). Luxury consumption factors. *Journal of Fashion Marketing and Management: An International Journal, 13*(2), 231–245. Retrieved from https://www.emerald.com/insight/content/doi/10.1108/13612020910957734/full/html?journalCode=jfmm. Accessed on May 15.

Hwang, J., & Hyun, S. S. (2016). Perceived firm innovativeness in cruise travelers' experience and perceived luxury value: The moderating effect of advertising effectiveness. *Asia Pacific Journal of Tourism Research, 21*(Suppl. 1), S101–S128. doi:10.1080/10941665.2015.1016051

Jain, V. (2018). Luxury: Not for consumption but developing extended digital self. *Journal of Human Values, 24*(1), 25–38. doi:10.1177/0971685817733570

Kapferer, J. N., & Laurent, G. (2015). Where do consumers think luxury begins: A cross cultural comparison over 21 luxury products in 7 countries. *Journal of Business Research, 69*(1), 332–430. doi:10.1016/j.jbusres.2015.08.005

Kemp, S. (1998). Perceiving luxury and necessity. *Journal of Economic Psychology, 19*(5), 591–606. doi:10.1016/S0167-4870(98)00026-9

Kim, S. S., Prideaux, B., & Chon, K. (2010). A comparison of results of three statistical methods to understand the determinants of festival participants' expenditures. *International Journal of Hospitality Management, 29*(2), 297–307. doi:10.1016/j.ijhm.2009.10.005

Koch, K. D. (2011). Luxury tourism – does this market segment still work? In R. Conrady & M. Buck (Eds.), *Trends and issues in global tourism 2011*. Berlin, Heidelberg: Springer.

Koumelis, T. (2018). Luxury travel becomes increasingly appealing and experiential. *Travel Daily News*. Retrieved from https://www.traveldailynews.com/post/luxury-travel-becomes-increasingly-appealing-and-experiential-says-globaldata. Accessed on July 20, 2020.

Ledezma, V. (2017). Globalization and fashion: Too fast, too furious. *Laurier Undergraduate Journal of the Arts, 4*(1), 9. Retrieved from https://scholars.wlu.ca/cgi/viewcontent.cgi?article=1052&context=luja. Accessed on June 22, 2020.

Ledezma, A. (2017, January 31). *What luxury tourism means*. Interface Tourism. Retrieved from https://interfacetourism.es/en/2017/01/31/what-luxury-tourism-means/. Accessed on May 6, 2020.

Lee, J. H., & Hwang, J. (2011). Luxury marketing: The influences of psychological and demographic characteristics on attitudes toward luxury restaurants. *International Journal of Hospitality Management*, *30*(3), 658–669. doi:10.1016/j.ijhm.2010.12.001

Li, M. (2014). Cross-cultural tourist research: A meta-analysis. *Journal of Hospitality & Tourism Research*, *38*(1), 40–77. doi:10.1177/1096348012442542

Mahika, E. C. (2011). Current trends in tourist motivation. *Cactus Tourism Journal*, *2*(2), 15–24. Retrieved from https://www.semanticscholar.org/paper/Current-trends-in-tourist-motivation-Mahika/b3301deb4b70bef7b6a0f9134e7a1b54631967eb. Accessed on May 5, 2020.

Manthiou, A., Lee, S. A., Tang, L. R., & Chiang, L. (2014). The experience economy approach to festival marketing: Vivid memory and attendee loyalty. *Journal of Services Marketing*, *28*(1), 22–35. doi:10.1108/JSM-06-2012-0105

MILT. (2017). Annual MICE India and luxury travel congress. Retrieved from http://www.miltcongress.com/. Accessed on August 11, 2020.

NTA. (2014). NTA and the luxury & affluent travel market. Retrieved from https://ntaonline.com/education/travel-market-guides/luxury/. Accessed on May 20, 2020.

Nueno, J. L., & Quelch, J. A. (1998, November). The mass marketing of luxury. *Business Horizons*, *41*(6), 61. Retrieved from https://go.gale.com/ps/anonymous?id=GALE%7CA55228757&sid=googleScholar&v=2.1&it=r&linkaccess=abs&issn=00076813&p=AONE&sw=w. Accessed on May 13, 2020.

Page, S. J. (2009). Current issue in tourism: The evolution of travel medicine research: A new research agenda for tourism? *Tourism Management*, *30*(2), 149–157. doi:10.1016/j.tourman.2008.04.011

Parulekar, D. (2019). Shifting notion of luxury in travel. Retrieved from https://www.businesstoday.in/lifestyle/travel/shifting-notion-of-luxury-in-travel/story/314611.html. Accessed on June 11, 2021.

Purinton-Johnson, E. F. (2013). Segmenting the mature market: Have marketers matured along with their market? *ASBBS Proceedings*, *20*(1), 579. Retrieved from http://asbbs.org/files/ASBBS2013V1/PDF/P/Purinton-JohnsonE(P579-588).pdf. Accessed on May 8, 2020.

Riley, R. W. (1995). Prestige-worthy tourism behavior. *Annals of Tourism Research*, *22*(3), 630–649. doi:10.1016/0160-7383(95)00009-U

Satli, J. (2019, August 22). Keep up with the latest trends or get left behind: How tourism professionals are reacting to digitalization. Retrieved from http://www.jamalsatli.com/en/nuevo-cambio-en-el-turismo-de-lujo-de-las-experiencias-a-las-sensaciones/. Accessed on June 5, 2020.

Schänzel, H., Yeoman, I., & Backer, E. (Eds.). (2012). *Family tourism: Multidisciplinary perspectives* (Vol. 56). Bristol: Channel View Publications.

Schensul, J. (2017, July 7). Vacations of the 1%: Trends from luxury travel agents. *USA Today*. Retrieved from https://www.usatoday.com/story/travel/destinations/2017/07/07/luxury-travel/455451001/. Accessed on June 14, 2020.

Shankaman, S. (2017, June 29). India's luxury sector poised for exponential growth. Skift. Retrieved from https://skift.com/2017/06/29/indias-luxury-sector-posed-for-exponential-growth/. Accessed on May 23, 2020.

Sharma, C. (2017, March 22). From material to experiential: here's how luxury travel has evolved over time. *The Economic Times*. Retrieved from https://economictimes.indiatimes.com/from-material-to-experiential-heres-how-luxury-travel-has-evolved-over time/articleshow/57749941.cms?utm_source=contentofinterest&utm_medium=text&utm_campaign=cppst. Accessed on July 4, 2020.

Skift, & Peak Destination Management Company. (2014). *The rise of experiential travel.* New York City, Melbourne: Skift, Peak DMC. [Online]. Retrieved from http://skift.com/wp-content/uploads/2014/06/skift-peak-experiential-traveler-report1.pdf. Accessed on June 28, 2020.

Strong, J. (2006). *Craving for travel: Luxury experiences for the sophisticated traveler.* Luxury Travel Books.

Sundbo, J., & Dixit, S. K. (2020). Conceptualisations in tourism experience. In S. K. Dixit (Ed.), *The Routledge handbook of tourism experience management and marketing* (pp. 15–26). London: Routledge.

Swarbrooke, J. (2018). *The meaning of luxury in tourism, hospitality and events.* Oxford: Goodfellow Publishers Limited.

The Luxury Holiday Company. (2018). What is luxury travel in 2020? Retrieved from https://www.theluxuryholidaycompany.com/stories/what-is-luxury-travel-in-2018. Accessed on July 11, 2020.

Tung, V. W. S., & Ritchie, J. B. (2011). Investigating the memorable experiences of the senior travel market: An examination of the reminiscence bump. *Journal of Travel & Tourism Marketing*, *28*(3), 331–343. doi:10.1080/10548408.2011.563168

Tynan, C., McKechnie, S., & Chhuon, C. (2010). Co-creating value for luxury brands. *Journal of business research*, *63*(11), 1156–1163. doi:10.1016/j.jbusres.2009.10.012

Veblen, T. (1899). *A teoria da classe ociosa. Um estudo económico das instituições* (3rd ed.). São Paulo: Pioneira.

Veríssimo, M., & Loureiro, S. M. C. (2013). Experience marketing and the luxury travel industry. *Tourism and Management Studies*, 296–302. Retrieved from http://www.tmstudies.net/index.php/ectms/article/viewFile/429/701. Accessed on June 8, 2020.

Yeoman, I., & McMahon-Beattie, U. (2006). Luxury markets and premium pricing. *Journal of Revenue and Pricing Management*, *4*(4), 319–328. doi:10.1057/palgrave.rpm.5170155

Chapter 2

Emerging Paradigms in Luxury: Understanding Luxury as an Embodied Experience in a Yoga Retreat Holiday

Mona Eskola,[1] Minni Haanpää,[2] and José-Carlos García-Rosell[3]

Abstract

This chapter sets out to explore consumer-centred experiential luxury from the perspective of a human body. We focus on the various practices related to a yoga retreat holiday experience in luxury hotel premises, such as encounters with hotel facilities, employees, nature and atmosphere besides yoga practice. Attention to bodily practices and affectivities on a yoga retreat holiday experience enables discussing intangible luxury beyond the traditional debate of luxury as related to product or brand features or experiential luxury focused only on the cognitive multisensory perceptions. The autoethnographic approach supports unwrapping the subtle affectual sensations building individual luxury in the experience setting. The data are gathered along with the first author's fieldwork during her three yoga retreat holidays in Thailand. The embodied investigation of tourist practices inducing luxury in the premises of a luxury hotel enriches the discussion of the co-creation between human bodies and the experience setting. It draws attention to the dynamic, situational and sensitive nature of luxury in the contemporary touristic experience of a yoga retreat holiday. It also advances the existing research on the body, practice and knowing by featuring the way luxury is emerging within the practice of yoga retreat holiday. By challenging the paradigm of luxury sensed only through our five external senses, our findings on the being, doing and moving body deepen the understanding of the co-creation and sensitiveness, affecting the subjective, transparent and embodied understanding of luxury experience.

[1]https://orcid.org/0000-0002-7967-1848
[2]https://orcid.org/0000-0003-3621-6957
[3]https://orcid.org/0000-0001-5649-8838

The Emerald Handbook of Luxury Management for Hospitality and Tourism, 37–56

doi:10.1108/978-1-83982-900-020211002

Keywords: Experiential luxury; embodiment; body; tourism experience; practice theory; autoethnography

Introduction

> Come along on a yoga holiday in Thailand. You are starting the day with energising and dynamic Flow yoga and ending the day with relaxing Yin yoga practice. During the day, there is a possibility of asana-clinic practices and relaxing massage. You can attend the yoga classes as often as you wish. Whether you wish to enjoy the sea, walking on the beach, good food, sunbathing, strengthening yoga practice, or relaxing massage, in this yoga holiday, you are free to do anything as much or as little as you wish.
>
> –Advertisement of a Finnish yoga studio, 2016

This chapter investigates the experiential nature of luxury emerging in various practices related to a yoga retreat holiday in luxury hotel premises. In foregrounding the embodied and subjective nature of luxury, we depart from the conventional understanding of luxury defined by the managerial and product- and brand-centric views (Kapferer & Bastien, 2012) towards a notion of luxury grounded in the tourist body and bodily practices. Our perspective complements the growing consumer-centred experiential luxury views. We aim to understand luxury as it appears in the embodied form in tourist practices, understanding of the contemporary world and luxury in the premises of a yoga retreat holiday. The first author's immersive embodied and affective experience of luxury urged her and the co-authors to wonder what it is that actually creates luxury in her yoga retreat holiday experience in luxury hotel premises and how luxury is constructed from the sensory, embodied perspective.

We reflect the bodily understanding of luxury within the setting of a yoga retreat holiday. In addition to that, we draw attention to the central role of being and doing as a body (Veijola & Jokinen, 1994), bodily moving (Chronis, 2015) and sensible knowing (Strati, 2007) in tourism experience. We demonstrate the co-creation of luxury experience through observational and autoethnographic data in the premises of a luxury boutique hotel in Thailand. Autoethnographic observations support the unwrapping of subtle affectual sensations that build individual luxury in the experience setting (Edensor, 2015; Gherardi, 2017). The yoga retreat holiday experience pertains to a set of daily bodily exercise and an assortment of entangling embodied practices where the human body engages within the experiential setting of the social, material and natural tourism environment. The given experience environment encloses encounters with other bodies (fellow yogis and guests, yoga teacher, hotel employees) and human-made (yoga shala 'place of yoga', hotel premises and service, other service providers) as well as natural materialities (weather, sea, wind, nature). To understand luxury as an embodied experience, we first need to examine experiential luxury and embodiment concepts.

Towards Embodied Luxury Experience

Luxury is a socio-cultural construct unfolding in diverse forms in various eras and societies (Berry, 1994). Traditionally, luxury is defined as a way of life associated with product/brand-based features of expensiveness, exclusiveness, uniqueness, quality and heritage (Chevalier & Mazzalovo, 2012; Ko, Costello, & Taylor, 2019). The rise of experiential luxury and the democratisation of luxury have contributed to shifting consumption from products and brands, from owning and experiencing (e.g., Cristini, Kauppinen-Räisänen, Barthod-Prothade, & Woodside, 2017; Silverstein & Fiske, 2005), towards unique and personalised experiences serving self-interest more than status (Bauer, von Wallpach, & Hemetsberger, 2011; Yeoman, 2011). The special moments affiliating a luxury feeling in consumers' minds are emerging in luxury research. Today's individual luxury consumption is characterised by having and using luxury brands in creative and personalised manners (Atwal & Williams, 2009; Seo & Buchanan-Oliver, 2019). Experiencing luxury implies not only featuring the experience as a static state (Cristini et al., 2017) but also emerging in performing luxury consumption (Seo & Buchanan-Oliver, 2019) and further, in various states of doing, being and becoming (Cristini et al., 2017; Llamas & Thomsen, 2016). Human beings do not solely have a body. They are their living body in experiencing the world and sensorials (Merleau-Ponty, 1992). These considerations set the scene for seeing a human body as inborn in luxury tourism experience, as a foundation of how we perceive and encounter the world. Thereby, the recognised intangible nature of luxury (Bauer et al., 2011) can exist in mundane objects in given contexts (Mortelmans, 2005). The nature of luxury consumption is transformed from something ostentatious towards being something very private yet savoured with distinction.

Over the years, scholars have elaborated the elements of luxury as well as customers' perceptions of luxury value, which determine, among other aspects, functional, financial, symbolic/sign, hedonic, individual, experiential and relational value (e.g., Hennigs, Wiedmann, Klarmann, & Behrens, 2015; Tynan, Mckechnie, & Chhuon, 2010; Sundbo & Dixit, 2020; Wiedmann & Hennigs, 2009). These typologies provide overlapping customer perceptions of luxury based on the features of a product/brand. Yang and Mattila (2016) observed that the earlier typologies of luxury value cannot be applied as such in hospitality based on the unique nature and intangible elements of hospitality experiences such as atmosphere, mindful service and the subtle influence of other customers. Additionally, in consumers' minds, luxury may also exist in a non-commercial context (Banister, Roper, & Potavanich, 2020; Kauppinen-Räisänen, Gummerus, von Koskull, & Cristini, 2019). Still, luxury is bundled with quality and scarcity (Kapferer & Bastien, 2012) – whether these would be related to the enhanced level of a product, service, uniqueness of experience or lived experience.

Experiential luxury defines the quality of life over status and prestige (Yeoman & McMahon-Beattie, 2018). In hectic modern societies, consumers see luxury as a bundle of intangible elements such as time, personal fulfilment, well-being and activities (see Yeoman & McMahon-Beattie, 2018) like rest, (re)connection with the self, self-development and seeking resilience in the modern complex world.

Senses and sensuality are recognised to be interwoven in experiences, and they build emotional favour such as pleasure or sensory gratification (Dubois & Laurent, 1994; Holbrook & Hirschman, 1982; Joy & Sherry, 2003). Luxury experiences are designed in a holistic multisensory manner to appeal to all bodily senses (see Hennigs, Wiedmann, & Klarmann, 2012; Wiedmann, Labenz, Haase, & Hennigs, 2018). In their studies, Wiedmann, Labenz, Haase, and Hennings (2016, 2018) and Yang and Mattila (2016) recognised the meaning of a multisensory, soothing environment, ambience and attentive service as invisible constituents of luxury in hospitality experiences. Luxury can be felt over our visual and cognitive perceptions. While our visuality, the tourist gaze (Urry & Larsen, 2011), is our predominant sense, the sensorium surpasses ocular perceptions, embracing the totality of human perceptions and the interpretation of it (Merchant, 2011). Indeed, rather than only 'seeing', tourists experience and perform with their whole bodies the multisensory environment, including various odours, tastes, haptics and movement (Chronis, 2015; Valtonen, Markuksela, & Moisander, 2010). Considering the subjective experiencing of the world in multiple manners, luxury can have various interpretations and embrace features out of the product/brand sphere (Sudbury-Riley, Hunter-Jones, Al-Abdin, Lewin, & Spence, 2020).

The recent, developing stream of academic discussion alludes to luxury as private and unnoticeable from others and suggests luxury as not being related to a brand/product (Kauppinen-Räisänen et al., 2019). Hemetsberger, von Wallpach, and Bauer (2012) unveiled intrinsic 'moments of luxury' and projects related to the self in studying the nature of everyday luxury through consumer diaries. Bauer et al. (2011) also related luxury consumption to the self and inward experience. Moreover, their study disconnected luxury from conspicuous consumption and showed the private side of consuming 'luxuries'. Scholars (Banister et al., 2020; Kauppinen-Räisänen et al., 2019) have hinted that luxury may unfold in a lived experience or relation to an activity. Luxurious states can be seized in consumers' mundane luxury practices even though the objects themselves do not correspond to traditional luxury (Banister et al., 2020; Bauer et al., 2011). The functional/sign values related to luxury product/brand give way to more subjective preferences and emotional and prudential value (Kauppinen-Räisänen et al., 2019). Recent studies have highlighted the cultural understanding and experiencing luxury in highly private ways in various contexts. Luxury is no longer understood only as features of a product/brand or being a static object. Instead, luxury consumption is comprehended as dynamic co-created construction in practices (Seo & Buchanan-Oliver, 2019; Tynan et al., 2010). Contemporary luxury has become a highly subjective matter that is relational, context-dependent, embodied in nature and always on the move (see Holmqvist, Diaz Ruiz, & Peñaloza, 2020; Kauppinen-Räisänen et al., 2019).

Case Study 2.1: Yoga Retreat Holiday

The tourism setting of this research is yoga retreat holidays. When discussing yoga, we refer to the modern postural yoga practice as distinct from the original Indian yoga tradition and more profound philosophy. Modern yoga is characterised as a mix of mental and physical exercise performed to accomplish enhancement in health, self-development, fitness and esotericism, among other aspects (Hauser, 2013). De Michelis (2008) featured modern yoga as more private, commodified and medicalised than pre-modern yoga. Aligning modern hybrid conceptualisation, the term 'yoga' here refers to a bundle of bodily practices, discourses and ideas that arose at the end of the nineteenth century in Europe and North America (Hauser, 2013). Smith and Kelly (2016) defined yoga retreats as yoga-based holidays with everyday yoga classes offering diverse yoga types (ashtanga, hatha, yin, kundalini, etc.). Speier (2020) made a distinction between yoga retreats and yoga holidays. According to Speier (2020, p. 71), yoga retreats denote 'a place for spiritual renewal, detoxing and pure yoga', while yoga holidays are associated with tourism elements and yoga. Yoga historian Rautaniemi (personal communication, 15 July 2020) stated that even though yoga is known as a strict and ascetic practice, today's yoga retreats are oases in which tourists are able to concentrate on yoga practice in the pursuit of feeling well. In contemporary societies, yoga has become a brand of a good life.

The concept of the embodiment provides new lenses to study the dynamic nature of luxury more deeply in focussing on the relationship between the human body and experiential luxury in the contemporary socio-cultural context. Experiencing in tourism settings through embodiment involves bodily being, moving and sensing the tourism place, space, activities, things and other bodies. Simultaneously, the body is a subject and an object, as the body is affecting and affected in engaging within the tourism world (Palmer & Andrews, 2020). In their pioneering work, 'The Body in Tourism', Veijola and Jokinen (1994) presented the experiencing and performing tourist body. According to them, the corporeal and embodied performances experienced through the body are foundational parts of the tourism experience. Later, in tourism research, the body has been increasingly recognised as substantial when understanding touristic experiences. There has been a growing body of studies in tourism dealing with the experiencing body for some time. Wang (1999) rethought the authenticity in tourism experiences and recognised the intra-personal layer of existential authenticity, highlighting the body as a primary organ and the inner fountain of feelings and sensual pleasure. Inauthenticity results in the abandonment of both the spiritual and the corporeal body. Since 2000, in tourism studies, the body has been related to gender or sex (Johnston, 2001), identity/self (Elsrud, 2001; Saxena, 2018) or tourism experience (Edensor, 2000; Li, 2000; Pritchard, Morgan, Ateljev, & Harris, 2007). The shift seems to show a gradual abandonment of the body's objectivity and take inspiration from the greater potential to discuss the body as a physical means. The human body is more and more seen in its corporeal entity as a co-creating actor that interacts with the world mentally, physically and internally, and through the visceral senses and movement. Academics have recognised the meaning of a performing human body, its relation with the surrounding place and space in the creation of experiences (Crouch & Desforges, 2003; Pons, 2003).

Tourism places surround tourists' bodies, and tourism places envelop bodies (Chronis, 2015). The knowing of luxury is essentially emerging in performing luxury consumption (Seo & Buchanan-Oliver, 2019), the tourism environment and its practices. Knowing arising in 'practice' foregrounds creative tourists who invent and constitute experience space not only in their rational minds but also through their corporeality, which allows them to form sensible knowledge (Crouch & Desforges, 2003; Strati, 2007, p. 66). The human body within its environment is recognised as a site of knowledge production also through sensual knowing (Gherardi, 2017).

To advance our understanding of the embodied luxury experience, we investigate bodily practices in a yoga retreat holiday in luxury hotel premises. We study how bodies, things, sayings and doings are understood in the practice of a yoga retreat holiday experience. Schatzki (2001, p. 2) defined practices as 'embodied, materially mediated arrays of human activity centrally organised around shared practical understanding'. Edensor (2000) observed that it is essential to understand tourism as a constant process of (re)designing praxis. Tourists apply their own imagination (Crouch & Desforges, 2003) in various ways of performing the tourism environment. In exploring the tourist's practices empirically, we concentrate on how practical understanding of luxury arises in yoga retreat holiday everyday practices and foregrounds the knowing body.

Today many luxury hotel chains – like Mandarin Oriental, Six Senses and Small Luxury Hotels – offer yoga or wellness retreat holidays (Fig. 2.1). Modern yoga holidays represent 'unplugging' from the demands of hectic Western life and are seen as luxurious compared to the modern lifestyle (Speier, 2020). Yoga has become a commercialised business. Yoga retreats and getaways are part of a bigger wellness travel trend and market covering 639 billion USD and growing at 8 percent a year, which is twice the growth of tourism globally (Global Wellness Institute, 2018). Koch (2013) found that yoga practitioners taking a yoga holiday were remarkably more affluent than their yogic coequals not attending yoga holidays. Yoga and yoga retreat holidays are especially favoured among the global educated middle class as a way to cope with their everyday life (Altglas, 2008). For modern women, contemporary yoga serves as an empowerment platform (Newcombe, 2013). Indeed, yoga holidays and getaways are seen as a luxurious form of well-being tourism (Speier, 2020) in providing holistic rest, recovery and inspiration.

A yoga retreat holiday is an embodied practice in which tourists use their own body as a means to co-create the yogic experience they are seeking. Still, modern yogis are not merely practising yoga every morning and evening or continuously occupied with related activities. As the opening vignette shows, yogic bodies are

Fig. 2.1. Flow Yoga at the Beach. *Source:* Aino Salonen.

invited to enjoy the holistic experience in yoga retreat holiday premises. Relaxing, socialising in the inspiring company, eating healthy, sleeping, having 'me-time' and reconnecting with oneself are common rejuvenating practices in immersing into the yogic Shangri-La. The embodied experience reaches outside the borders of yoga practice. Touristic bodies perform in the holistic tourism environment in entangling practices with social, material and natural surroundings. The holistic experience in luxury hotel premises sets various expectations for overall high quality, nuances and service encounters and the bodily and sensory perspective in the creation of a luxury experience.

Smith and Kelly (2016) stated that yoga retreat holidays take place in a suitable setting in their typology of retreats. Lea (2008) drew attention to the retreat environment, nature, healing properties and retraction. The tired bodies of travelling yogis might be quite sensitive to the qualities of the material, immaterial, social and natural environment. Yogis might be suffering from jet-lag symptoms, tired of yoga practice, or they might usually prefer calmness around them. These conditions require sensitivity regarding the design of the environment to fulfil yogis' dreams of the healing sphere in the retreat holiday. Healing places include four environmental elements: natural, built, symbolic and social (Gesler, 2003). Merging a natural environment and yoga provides an extra layer for tourists searching for popular mind–body experiences. Besides providing a purposeful natural environment, the setting of a yoga retreat holiday in luxury hotel premises sets expectations for the built, symbolic and social environment.

> Designed with relaxation in mind amidst the gentle waters of the Gulf of Thailand. The beautiful beach getaway offers just the right blend of relaxation and revitalisation, with stunning tropical surroundings and a choice of tantalising indulgences. While

> complete tranquillity comes first and foremost, there are also still plenty of invigorating activities to intrigue the mind, awaken the body, and satisfy the soul.
>
> –Extract from a luxury hotel website (anonymized research data)

This and similar excerpts illuminate that many yoga retreat holidays in Asian luxury hotels or resorts are situated in the middle of stunning natural beauty and represent a carefully designed human-made environment. The environment usually stirs the soul and emanates something almost spiritual. As opposed to the marble and gold of traditional luxury hotel decoration, the curative interiors embrace boutique hotel style with the beauty of imperfection in the spirit of wabi-sabi (traditional Japanese aesthetics): beauty, rough natural materials, organic features and cultural artefacts. The soothing multisensory premises (Wiedmann et al., 2016) welcome drained modern bodies to find inner peace not only through yoga, mindfulness, nature, healing treatments and healthy nutrition but moreover with their caring encounters. As Kreuzer, Cado, and Raïes (2020) and Sudbury-Riley et al. (2020) have found in their studies, authentic luxury is more about experiencing humanity in encounters than having fawning service. Sustainability is a constituent of quality in luxury (Kapferer & Michaut, 2015), and it materialises in various ways. Responsible service providers contribute to the sustainable development of local culture and espouse sustainable development goals (UN, 2015). This resonates with the multinational clientele, as they see themselves giving importance to health and well-being as well as to environmental and social responsibility to promote a responsible way of living and good for all (Strauss & Mandelbaum, 2013).

Methodology

In this chapter, we reflect on luxury through the lenses of embodiment in the context of a yoga retreat holiday experience. The vignettes result from the first author's autoethnographic fieldwork conducted from January 2016 to June 2019 during her three yoga retreat holidays in Thailand. The accommodation was provided in a hotel belonging to a luxury boutique chain. During these week-long yoga retreat holidays, she performed tourism practices and experienced luxury in the tourism setting through her body, using her body as an instrument (Longhurst, Ho, & Johnston, 2008) and applying the methodological rigour of affective autoethnography (Gherardi, 2018; Haanpää, 2017). The first-person narration illustrates the experience of tourism practices affecting the notions of luxury, experiencing luxury through our body and beyond our conscious mind. Therefore, the tourist's – and here the researcher's – body can be considered as a central instrument in participating in the entanglements and practices in the tourism environment (Longhurst et al., 2008). To analyse the experiences, we carefully read the first author's fieldwork diaries and studied the visual material gathered in the yoga retreat holidays. We then discussed and reflected together to elicit the memories of experiencing luxury by the first author.

It is challenging to illustrate luxury from the bodily perspective, as luxury unfolds as an intrinsic, multilayered experience. We all have our inner world inside our physical bodies, which undoubtedly affects our experiences. Words seem not to be enough to unveil the depth of the shadowed experiences. The vignettes in this chapter exemplify some sense of the first author's experience and knowing of luxury as her body remembers them. Wrestling with these methodological challenges opens the search for a suitable approach to scrutinise embodied knowing. Simply listing the practices seems not to open the essence of luxury felt through the body. How can one describe the intrinsic resonance arising from bodily practices? The regular autoethnographic depiction may result in short in providing affective attunement. Therefore, the first author grasps into the affective ethnography. Gherardi (2018, p. 2) depicted affective ethnography as a style of ethnography 'relying on [the] researcher's capacity to affect and be affected to produce an interpretation that may transform the things they interpret'. This style of research practice allows an immerse way to 'be with' the phenomena under study. It recognises that all actors, including materialities, texts and language, are interweaved in tangled ways and 'should be read in their intra-action…as data that move' (Gherardi, 2018, p. 2). This dynamic style of ethnography allows us to seize the intangible in experiencing and studying the nature of luxury in bodily practices.

According to the postmodern understanding, knowledge can also be gathered through sensory input (Strati, 2007; Valtonen et al., 2010). The first author gave particular focus to the role of her body, bodily movements and sensorials to explore luxury in the tourist experience in a holistic manner. This kind of approach allows in-depth access and understanding into the inner bodily feelings and sensations of luxury in the tourist experience, which might otherwise be challenging through the use of other methods. The use of different methods, like interviews or observations, in exploring bodily luxury sensations might result in a more superficial level of understanding of luxury in the tourist's experience.

Attending a yoga retreat holiday establishes the motivation for the experience depicted in her field notes. For her, the main reasons to attend yoga holidays were related to the restorative qualities of the experience to renew the body–mind–soul into the formulation of the wellness of being and recover from the demands of the hectic Western life. The situational, spatial, social and material practices have an important influence on how the meanings of luxury are co-created in a given environment. Luxury hotel premises provide trust and confidence in the form of a high-quality experience with calming surroundings, but they also bring peace of mind for a woman travelling alone and appreciating privacy.

The first author gathered four kinds of documentary material: ethnographic field notes, pictures, video recordings and hotel brochures. Various materialities are predominant in the production of the meanings attached to the experience (see Haanpää, 2017). When making the field notes, she paid attention to the body(-ies) in the social setting and materialities of places, spaces, senses, affects, movement and stillness (Lefebvre, 1991; Thrift, 2008). To analyse the nature of luxury and various ways of experiencing luxury, we have closely read the collected material, the first author's personal ethnographic story, through the concept of embodied knowing. In the description of practices, she took careful note of the sense perceptions (seeing, hearing, smelling, touching, tasting) (Roberts, 2013) and affective sensations of her own body (feelings, affects, movement, and stillness) (Jóhannesson & Lund, 2017). The embodied knowing of luxury is critically explored through her sensual body and bodily practices (see Valtonen & Haanpää, 2018). The aim is not to classify practices emitting bodily luxury but to attempt to capture the dynamic and situational nature of luxury (Barnes, 2001; Berry, 1994) to create a collaboratively constructed embodied understanding of contemporary luxury in the context of the yoga retreat experience. The conceptual exploring of tourist practices centralises in the discussions on embodied practices of a tourist towards embodied luxury experience.

Experiencing Luxury through Embodiment

> I begin to dress for the first evening yin yoga. Our bodies have been forced to sit in a specific stable position in the aircraft slowing down our metabolism. Therefore, the teacher has chosen a more dynamic yin with short asanas to stretch our bodies from the restraints of the long journey to activate our metabolism and to remove the metabolic waste. From my earlier stays, I remember that there is a complimentary set of a light outfit hanging in the room's wardrobe – ideal for this state. The loose outfit pampers my tired body. As always, I slowly walk along the seashore to the evening yoga. The warm, humid breezes of the wind almost traverse my whole body and mind while the loose outfit is flowing in the wind. My body still remembers these memories from my earlier years' stays. I feel safe even though the night is falling, and I am walking alone as a woman, my feet halfway in the sea. Only some local dogs making acquaintance from time to time. Feels almost like home.

The above vignette results from the first author's second-year stay in a yoga retreat holiday in luxury hotel premises (Fig. 2.2). She describes the embodied performances related to preparation for the first evening yoga. The vignettes presented in this chapter are written from her experience. Accordingly, the analysis below changes to first-person narration. The collection of the vignettes results from the years 2016–2019.

The vignette illustrates the experience from the bodily perspective in appreciable ways. First, it spotlights the effect of all bodily senses in tourist practice (Agapito, Mendes, & Valle, 2013; Veijola & Valtonen, 2007) and sees bodily being in the world as an elementary practice of human beings (Csordas, 1999). Recently, Holmqvist et al. (2020) mentioned the embodied aspect of experience in their study of salsa dancing, dressing up and the moment of luxury as hedonic escapism and aesthetics. Still, body and experiencing do not merely end at the

Fig. 2.2. Complimentary Loose Outfit in the Room. *Source:* Mona Eskola.

skin. The vignette demonstrates that bodily being is affected by the natural functioning of the intestines and viscera, which guide the dressing practices and, thus, affects the experience. My stomach was irritated and swollen because of the long flying experience, so the loose outfit was merciful. Second, the vignette demonstrates how the performance of various actors is attuned to the customer's bodily being in the world for a soothing experience. The model of the outfit the hotel provided was loose, not tight; it did not force the body to any shape. Rather, the model was approbative, letting the body be in its relaxing natural state. Moreover, the texture of the outfit was feather-light and of the softest cotton. The professional practice of the yoga teacher is to appraise the style of yin yoga suitable for turgid body–minds suffering from jet-lag. A dynamic practice awakens the bodily fluids, after which the physical–mental–spiritual body feels lively. From the traditional luxury research view, the hotel's complimentary outfit was valuable for many reasons. In offering a complimentary outfit, the hotel provides customers ease and time savings (Danziger, 2005). Packing my luggage back home was a bit easier since I remembered the hotel providing the purposeful loose outfit suitable for many occasions during this retreat.

Moreover, the texture of the fabric is so light and soft that it would not even hurt burned skin. Taking care of the details is one primary element in luxury experiences (Kapferer & Bastien, 2012). In addition, the pace of bodily movement, dogs, weather and climate as natural materialities influenced the bodily experience (see Rantala, Valtonen, & Markuksela, 2011). The repetitive rhythm of the slowly walking body and welcoming acquaintances in the familiar environment completed the feeling of being safe.

> I do not open my eyes, but I hear and feel my co-yogis silently roll out their yoga mats at the beach garden. With whispering voice and in our own pace, the yoga teacher advises us to take the child's posture. No need to open our eyes. The child's posture grounds ourselves to the coming practice. She routes us to direct our attention to the roaring sea. Healing theme music and its salt whiff... This evening's restorative yin the asanas last 3–5 minutes. I enjoy staying long in asanas [body posture in Yoga], which allows me the time to concentrate on my body and study my bodily feelings. I am allowed to let go of all other thoughts.... At the end of the practice, we are lying down on our back in shavasana, resting position. With a soft voice, the yoga teacher is guiding us to put our right hand on the top of our heart and left hand on the top of the lower abdomen. I feel the low beating of my heart. My abdomen goes up and down, following the rhythm of my breathing – calming yin energy circulating around my body. Soon I hear the magical sounds of a koshi chime [a small bell similar to wind chimes]. The yoga teacher walks around us, holding two koshi chimes by their cord. The chimes play faintly with the wind, and their fairylike melody carries us into another world. Koshi chimes melodies are the most grounding sounds. Halfway to good night's sleep. Namaste.

In experiences, immediate feelings and affectivities of the body happen through embodiment and senses beyond our visual perception. The body knows before the cognitive perception in the form of a pre-reflective knowing in experience setting (Merleau-Ponty, 1992). The practice of respectful silence is part of the ambience that allows yoga participants to settle down in their yogic space. Intangible elements such as ambience are focal in luxury experiences (Bauer et al., 2011; Yang & Mattila, 2016). Nature as well as social and material actors co-produce various affectivities that yield a luxurious atmosphere. The yogic space is co-performed with yoga mats and bodily movements.

As the vignette reveals, I kept my eyes closed most of the time during the yin practice. As restorative yin is slow in pace, the whole practice is easy to perform without opening my eyes. Keeping my eyes closed most of the time during the practice allows me to fully concentrate on my body and to delve into my bodily feelings and inner world. Aligning these notions, it can be stated that experiencing overlaps our visual (Urry & Larsen, 2011) and cognitive perceptions, and our skin, to the inner world. This understanding widens the understanding of experiencing as being something signalled or communicated to others (Holmqvist et al., 2020).

On the contrary, in setting the human body as a foundational actor in experiencing, the vignette departs from the traditional understanding of luxury. It reveals luxury as something that is utterly private and invisible, an emerging feeling inside of a body affected by yoga postures. Although von Wallpach, Hemetsberger, Thomsen, and Belk (2020) noted the sphere of self-transcendence, we point out the living body's dynamic, postural and moving role through yoga asanas in reaching a self-transcendent state. These hidden luxury moments (Hemetsberger et al., 2012) and feelings of luxury in private and special space (Bauer et al., 2011) are invoked along yoga practice in relation to the surroundings. The teacher's nuanced performance choreographs yoga practice. The purposeful guiding creates a suitable atmosphere for the particular style of yoga to enable the body–mind to relax and, thus, allow the participant to reach the sacred space of transcendental experience.

The type of weather and nature-based elements affect yoga practice through the senses. During the first visits, the strong wind typical of the district disturbed me. I felt the movement of the loud sea and swinging palm trees around me. With time, I got used to it. Actually, I began to imagine the wind blowing away all the burdens from my shoulders, providing a positive force to push me through everything. In the vignette, taking a particular body posture and the koshi chimes provide a grounding experience relevant to experiencing intra-personal existential authenticity. This understanding is congruent with Banister et al. (2020), who described the help of various props and how the associated performances transformed something quite ordinary into something experientially special. However, we here centralise the attention more tightly around the bodily performances and how various bodily postures conjoined with the props create luxury in the experience. Furthermore, the sounds of koshi chimes open an avenue to a good night's sleep, allowing the yin experience to resonate in the body after the actual yoga practise has ended. All this allowed me to relax and apply imagination to create my own intrinsic luxury space along with the practice.

> The night is falling, and housekeeping knocks carefully at the door. The lady takes off her shoes and makes turnaround service for the room: closes the curtains, sets the evening ambience with indirect soft lighting, opens the bed halfway, sets the slippers beside the bed, makes sure that there are enough water bottles, and inquires if I need anything. Quietly, with a shy smile, she exits the room. I feel I am taken good care of.... While going to bed, I fluff my own pillow, which I have brought from home. I know that the hotel is providing a pillow menu. Still, I prefer my own one. If I sleep with a different kind of pillow than I'm used to, I may get a

> neck ache, affecting my following day – my mood, doings, and concentration. I do not want to waste a day because of bad sleep.

The vignette sheds light on the sleeping practices' co-created bodily, social, material and natural elements (Valtonen & Moisander, 2012). Sleeping is part of the holistic tourism experience and inborn to our bodily being also during holidays (Valtonen & Veijola, 2011). In fact, in the modern 24/7 lifestyle, sleeping well – or at all – has become a luxury. As sleeping preconditions the tourism performance, it is hard to imagine a luxury experience without the luxury of sleeping. In foregrounding the quality of sleeping, the vignette understands experience as surpassing space, embracing also the subconscious state while asleep. As time is considered luxury (Yeoman & McMahon-Beattie, 2018), wasting time is not desirable during precious 'me-time'. We widen the attention to the quality of the time used in the subconscious state of sleeping from the bodily perspective. The subconscious state of sleeping is one layer in our bodily being in the world, which affects the quality of my experience, and how I am able to use my retreat holiday time. Time would be wasted if the body were to ache because of bad sleeping, and I would be compelled to skip a yoga practice because of a headache or neck pain, or I would perform with low energy.

Further, the vignette unfolds the practices related to sleeping and highlights the co-creational performances related to a pillow from the bodily perspective. Consumer-defined luxury experiences and consumption practices are conventionally understood as related to traditional and high-priced luxury items (Bauer et al., 2011; Seo & Buchanan-Oliver, 2019). Some studies (Banister et al., 2020; Hemetsberger et al., 2012) have shifted this conventional view and connect luxuries with consumers' mundane objects and practices related to them. Similarly, with earlier vignettes here, we bond the everyday objects with embodiment and embodied practices. The pillow, as a material, is of low value and does not carry any luxury status as such. Still, the practices related to the pillow unveil the special luxury meaning hidden in the item. Carrying one's own pillow from home is a practice that ensures the holistic tranquilising experience. One's own pillow has suitable softness, size and shape to ensure a good night's sleep as well as comfort on the following day.

While turnaround service is considered as a basic component in a luxury hotel's practices, aligning with earlier mentions, it is a caretaking practice with various layers affecting the tourist's bodily being. Moreover, turning to our biological body, dimming the amount of light affects the body by prompting it to produce more melatonin, which is a signal to the body to prepare for going to sleep. Simultaneously in turnaround service, the bodies of the housekeeper and the customer are interacting in a very private place (hotel room). This demands a great deal of discretion of both bodies over discursive expressions. The housekeeper's soft bodily moving, the pace of the steps, the gentle movements, the bodily posture and the lenient facial expressions affect the quality of the turnaround service, showing how, with several small bodily actions, the service can be considered as not disturbing one's privacy but as a nursing practice.

Conclusions

This chapter posed an alternative approach to analyse consumer-centred experiential luxury from the perspective of a body. Understanding luxury as an embodied experience moves the focus from product/brand features and complements the customer-centred experiential luxury discussions. We reflected on the embodied and sensory nature of luxury in a tourist yoga retreat holiday experience. As a co-created experience, embodied luxury emerges in multiple practices, reinforcing the totality of a multisensory experience and ambience (see Wiedmann et al., 2018; Yang & Mattila, 2016). The yogic bodies are enwrapped with all the sensualities and affectivities in the luxury hotel experience environment in addition to the yoga practice. In tourism, bodies are affected by various actors, social and material, and their quality, such as tone of voice, micro-expressions, haptics, odour and general caretaking practices, convey humanity. Sensing occurs beyond the five external senses and cognitive perceptions. Human bodies are sensible bodies – being, doing and moving bodies. Being in the world implies various states of bodily being. A body is not a static object. Rather, the body is a biological, socio-cultural entity, an imaginative co-creative actor in the experience setting. Indeed, the meanings of luxury are reinterpreted and co-created by contemporary customers rather than defined by the brand (Tynan et al., 2010). The embodied and subjective perspective together allowed a deeper analysis of the sensitivenesses, building luxury in the practices of a contemporary tourist in search of the ideal state of being in a yoga retreat holiday.

Luxury experiences have not previously been analysed through the notion of embodiment. Our findings can be applied in tourism experience design, management and marketing to understand better how luxury meanings, production and consumption are shaped. Imaginative tourists redefine luxury meanings arising in contemporary societies. Well-being holidays promoting mental and bodily recovery, such as yoga retreats, are seen as luxurious by contemporary tourists living in hectic modern life. The emergence of luxury in the experience comes into existence in the holistic experience setting and encounters in addition to the yoga practice. Encounters remain not only with static features of humans and materials and sensitivities of social and material and immaterial practices pertaining to a sensual human body. Sensitivenesses such as a body-gentle outfit, carefully guided yoga practice, a specific way of talking or moving and gestures in performing, for example, housekeeping practices, affect the understandings of luxury. Natural elements such as wind and seawater create affectivities in the experience setting. Further, experience-related embodied activities, and material items such as yoga, the magical sound of a koshi chime and mundane items like the tourist's own, physically suitable pillow, may facilitate the consumption of the experience to emerge beyond cognitive perceptions as in our imagination, in a meditative state or while asleep. Our five external senses are entangled with the sensory experience of a being – moving and living bodies encountering each other and producing various affectivities and rhythms in tourism space. Finally, as we are corporeal human beings, the existence of embodiment is inherent in various experiences. Therefore, our study is not only applicable in a yoga retreat holiday, the luxury hotel premises of this chapter, but it can be applied further to other tourism, hospitality and service fields.

References

Agapito, D., Mendes, J., & Valle, P. (2013). Exploring the conceptualisation of the sensory dimension of tourist experiences. *Journal of Destination Marketing and Management*, *2*(2), 62–73. doi:10.1016/j.jdmm.2013.03.001

Altglas, V. (2008). Indian gurus and the quest for self-perfection among the educated middle classes. In J. Stolz (Ed.), *Salvation goods and religious markets: Theory and applications* (pp. 211–234). Bern: Peter Lang.

Atwal, G., & Williams, A. (2009). Luxury brand marketing – The experience is everything! *Journal of Brand Management*, *16*(5/6), 338–346. doi:10.1057/bm.2008.48

Banister, E., Roper, S., & Potavanich, T. (2020). Consumers' practices of everyday luxury. *Journal of Business Research*, *116*, 458–466. doi:10.1016/j.jbusres.2019.12.003

Barnes, B. (2001). Practice as collective action. In T. Schatzki, K. Knorr Cetina, & E. von Savigny (Eds.), *The practice turn in contemporary theory* (pp. 25–36). London and New York, NY: Routledge.

Bauer, M., von Wallpach, S., & Hemetsberger, A. (2011). My little luxury: A consumer-centered experiential view. *Marketing ZFP – Journal of Research and Management*, *1*(11), 57–67. doi:10.15358/0344-1369-2011-1-57

Berry, C. J. (1994). *The idea of luxury: A conceptual and historical investigation.* Cambridge: Cambridge University Press.

Chevalier, M., & Mazzalovo, G. (2012). *Luxury brand management: A world of privilege* (2nd ed.). Singapore: Wiley.

Chronis, A. (2015). Moving bodies and the staging of the tourist experience. *Annals of Tourism Research*, *55*, 124–140. doi:10.1016/j.annals.2015.09.008

Cristini, H., Kauppinen-Räisänen, H., Barthod-Prothade, M., & Woodside, A. (2017). Toward a general theory of luxury: Advancing from workbench definitions and theoretical transformations. *Journal of Business Research*, *70*, 101–107.

Crouch, D., & Desforges, L. (2003). The sensuous in the tourist encounter. *Tourist Studies*, *3*(1), 5–22.

Csordas, T. (1999). Embodiment and cultural phenomenology. In G. Weiss & H. Haber (Eds.), *Perspectives on embodiment* (pp. 143–162). New York, NY: Routledge.

Danziger, P. (2005). *Let them eat cake. Marketing luxury to the masses – as well as the classes.* Chicago, IL: Dearborn Trade Publishing.

De Michelis, E. (2008). Modern yoga: History and forms. In M. Singleton & J. Byrne (Eds.), *Yoga in the modern world: Contemporary perspectives* (pp. 17–35). London: Routledge.

Dubois, B., & Laurent, G. (1994). Attitudes toward the concept of luxury: An exploratory analysis. *Asia-Pacific Advances in Consumer Research*, *1*, 273–278.

Edensor, T. (2000). Staging tourism: Tourists as performers. *Annals of Tourism Research*, *27*(2), 322–344. doi:10.1016/S0160-7383(99)00082-1

Edensor, T. (2015). Producing atmospheres at the match: Fan cultures, commercialisation and mood management in English football. *Emotion, Space and Society*, *15*, 82–89.

Elsrud, T. (2001). Risk creation in traveling: Backpacker adventure narration. *Annals of Tourism Research*, *28*(3), 597–617. doi:10.1016/S0160-7383(00)00061-X

Gesler, W. (2003). *Healing places*. Oxford: Rowman & Littlefield.

Gherardi, S. (2017). One turn … and now another one: Do the turn to practice and the turn to affect have something in common? *Management Learning*, *48*(3), 1–14.

Gherardi, S. (2018). Theorising affective ethnography for organisation studies. *Organisation*, *26*(6), 741–760. doi:10.1177/1350508418805285

Global Wellness Institute. (2018, November). *Global wellness tourism economy*. Global Wellness Institute. Retrieved from https://globalwellnessinstitute.org/wp-content/uploads/2018/11/GWI_GlobalWellnessTourismEconomyReport.pdf

Haanpää, M. (2017). *Event co-creation as choreography: Autoethnographic study on event volunteer knowing* (Acta Universitatis Lapponiensis 358). Doctoral dissertation. University of Lapland, Rovaniemi, Finland.

Hauser, B. (2013). Introduction: Transcultural yoga(s). Analysing a traveling subject. In B. Hauser (Ed.), *Yoga traveling: Bodily practice in transcultural perspective* (pp. 1–34). doi:10.1007/978-3-319-00315-3

Hemetsberger, A., von Wallpach, S., & Bauer, M. (2012). Because I'm worth it: Luxury and the construction of consumers' selves. *Advances in Consumer Research*, *40*(1982), 483–489.

Hennigs, N., Wiedmann, K.-P., & Klarmann, C. (2012). Luxury brands in the digital age – Exclusivity versus ubiquity. *Marketing Review St. Gallen*, *29*, 30–35. doi: 10.1007/s11621-012-0108-7

Hennigs, N., Wiedmann, K. P., Klarmann, C., & Behrens, S. (2015). The complexity of value in the luxury industry: From consumers' individual value perception to luxury consumption. *International Journal of Retail and Distribution Management*, *43*(10–11), 922–939. doi:10.1108/IJRDM-07-2014-0087

Holbrook, M. B., & Hirschman, E. C. (1982). The experiential aspects of consumption: Consumer fantasies, feelings, and fun. *Journal of Consumer Research*, *9*(2), 32–40.

Holmqvist, J., Diaz Ruiz, C., & Peñaloza, L. (2020). Moments of luxury: Hedonic escapism as a luxury experience. *Journal of Business Research*, *116*(October 2019), 503–513. doi:10.1016/j.jbusres.2019.10.015

Jóhannesson, G. T., & Lund, K. A. (2017). Aurora borealis: Choreographies of darkness and light. *Annals of Tourism Research*, *63*, 183–190.

Johnston, L. (2001). (Other) bodies and tourism studies. *Annals of Tourism Research*, *28*(1), 180–201. doi:10.1016/S0160-7383(00)00012-8

Joy, A., & Sherry, J. F., Jr (2003). Speaking of art as embodied imagination: A multisensory approach to understanding aesthetic experience. *Journal of Consumer Research*, *30*(2), 259–282. doi:10.1086/376802

Kapferer, J., & Bastien, V. (2012). *The luxury strategy: Break the rules of marketing to build luxury brands* (2nd ed.). London: Kogan Page.

Kapferer, J. N., & Michaut, A. (2015). Luxury and sustainability: A common future? The match depends on how consumers define luxury. *Luxury Research Journal*, *1*(1), 3. doi:10.1504/lrj.2015.069828

Kauppinen-Räisänen, H., Gummerus, J., von Koskull, C., & Cristini, H. (2019). The new wave of luxury: The meaning and value of luxury to the contemporary consumer. *Qualitative Market Research*, *22*(3), 229–249. doi:10.1108/QMR-03-2016-0025

Koch, A. (2013). Yoga as a production site of social and human capital: Transcultural flows from a cultural economic perspective. In B. Hauser (Ed.), *Yoga traveling: Bodily practice in transcultural perspective* (pp. 225–248). doi:10.1007/978-3-319-00315-3

Ko, E., Costello, J. P., & Taylor, C. R. (2019). What is a luxury brand? A new definition and review of the literature. *Journal of Business Research*, *99*, 405–413. doi:10.1016/j.jbusres.2017.08.023

Kreuzer, M., Cado, V., & Raïes, K. (2020). Moments of care: How interpersonal interactions contribute to luxury experiences of healthcare consumers. *Journal of Business Research*, *116*(November 2019), 482–490. doi:10.1016/j.jbusres.2019.11.033

Lea, J. (2008). Retreating to nature: Rethinking "therapeutic landscapes". *Area*, *40*(1), 90–98. doi:10.1111/j.1475-4762.2008.00789.x

Lefebvre, H. (1991). *The production of space*. Oxford: Blackwell.

Li, Y. (2000). Geographical consciousness and tourism experience. *Annals of Tourism Research*, *27*(4), 863–883. doi:10.1016/S0160-7383(99)00112-7

Llamas, R., & Thomsen, T. U. (2016). The luxury of igniting change by giving: Transforming yourself while transforming others' lives. *Journal of Business Research*, *69*, 166–176. doi:10.1016/j.jbusres.2015.07.030

Longhurst, R., Ho, E., & Johnston, L. (2008). Using 'the body' as an 'instrument of research': Kimch'i and pavlova. *Area*, *40*(2), 208–217.

Merchant, S. (2011). Negotiating underwater space: The sensorium, the body and the practice of scuba-diving. *Tourist Studies*, *11*(3), 215–234. doi:10.1177/1468797611432040

Merleau-Ponty, M. (1992). *Phenomenology of a perception*. New York, NY: Routledge.

Mortelmans, D. (2005). Sign values in processes of distinction: The concept of luxury. *Semiotica*, *157*(2005), 497–520. doi:10.1515/semi.2005.2005.157.1-4.497

Newcombe, S. (2013). Magic and Yoga: The role of subcultures in transcultural exchange. In B. Hauser (Ed.), *Yoga traveling: Bodily practice in transcultural perspective* (pp. 57–80). doi:10.1007/978-3-319-00315-3

Palmer, C., & Andrews, H. (2020). Tourism and embodiment. Animating the field. In C. Palmer & H. Andrews (Eds.), *Tourism and embodiment. Routledge advances in tourism and anthropology* (pp. 71–85). New York, NY: Routledge.

Pons, P. O. (2003). Being-on-holiday: Tourist dwelling, bodies and place. *Tourist Studies*, *3*(1), 47–66. doi:10.1177/1468797603040530

Pritchard, A., Morgan, N. J., Ateljev, I., & Harris, C. J. (Eds.). (2007). *Tourism and gender: Embodiment, sensuality and experience* (pp. 1–12). Wallingford: CABI. doi:10.1079/9781845932718.0158

Rantala, O., Valtonen, A., & Markuksela, V. (2011). Materialising tourist weather: Ethnography on weather-wise wilderness guiding practices. *Journal of Material Culture*, *16*(3), 285–300. doi:10.1177/1359183511413646

Rautaniemi, M. (2020, July 15). *Jooga ja Zen*. Helsinki: Era Nova Bookshop.

Roberts, R. A. (2013). How do we quote black and brown bodies? Critical reflections on theorising and analysing embodiments. *Qualitative Inquiry*, *19*, 280–287.

Saxena, G. (2018). Scarborough based study on bodies' affective capacities. *Annals of Tourism Research*, *68*(November 2017), 100–110. doi:10.1016/j.annals.2017.12.002

Schatzki, T. (2001). Introduction: Practice theory. In T. Schatzki, K. Cetina, & E. von Savigny (Eds.), *The practice turn in contemporary theory* (pp. 10–23). London: Routledge.

Seo, Y., & Buchanan-Oliver, M. (2019). Constructing a typology of luxury brand consumption practices. *Journal of Business Research*, *99*(November 2016), 414–421. doi:10.1016/j.jbusres.2017.09.019

Silverstein, M. J., & Fiske, N. (2005). *Trading up. Why consumers want new luxury goods – and how companies create them.* New York, NY: Penguin Group Portfolio.

Smith, M. K., & Kelly, C. (2016). Journeys of the self: The need to retreat introduction. In *The Routledge handbook of health tourism, 2006*, (pp. 138–151). doi: 10.4324/9781315693774

Speier, A. (2020). Yoga as an embodied journey towards flexibility, openness and balance. In C. Palmer & A. Hazel (Eds.), *Tourism and embodiment. Routledge advances in tourism and anthropology* (pp. 71–85). New York, NY: Routledge.

Strati, A. (2007). Sensible knowledge and practice-based learning. *Management Learning*, *38*(1), 61–77. doi:10.1177/1350507607073023

Strauss, S., & Mandelbaum, L. (2013). Consuming Yoga, conversing the environment: Transcultural discourses on sustainable living. In B. Hauser (Ed.), *Yoga traveling: Bodily practice in transcultural perspective* (pp. 175–200). doi:10.1007/978-3-319-00315-3

Sudbury-Riley, L., Hunter-Jones, P., Al-Abdin, A., Lewin, D., & Spence, R. (2020). Conceptualising experiential luxury in palliative care: Pathographies of liminal space, cathedral, and community. *Journal of Business Research*, *116*, 446–457. doi: 10.1016/j.jbusres.2019.12.004

Sundbo, J., & Dixit, S. K. (2020). Conceptualisations in tourism experience. In S. K. Dixit (Ed.), *The Routledge handbook of tourism experience management and marketing* (pp. 15–26). London: Routledge.

Thrift, N. (2008). *Non-representational theory. Space, politics, affect.* London: Routledge.

Tynan, C., Mckechnie, S., & Chhuon, C. (2010). Co-creating value for luxury brands. *Journal of Business Research*, *63*(11), 1156–1163. doi:10.1016/j.jbusres.2009.10.012

UN. (2015). *Sustainable development goals.* Retrieved from https://www.undp.org/content/undp/en/home/sustainable-development-goals.html

Urry, J., & Larsen, J. (2011). *The tourist gaze 3.0.* London: Sage.

Valtonen, A., & Haanpää, M. (2018). The body in autoethnography. In H. Syrjälä & A. Norgrann (Eds.), *Multifaceted autoethnography: Theoretical advancements, practical considerations and field illustrations* (pp. 125–146). New York, NY: Nova Science Publishers.

Valtonen, A., Markuksela, V., & Moisander, J. (2010). Doing sensory ethnography in consumer research. *International Journal of Consumer Studies*, *34*(4), 375–380. doi: 10.1111/j.1470-6431.2010.00876.x

Valtonen, A., & Moisander, J. (2012). Great sleep as a form of hedonic consumption. *Advances in Consumer Research*, *40*(1982), 436–441.

Valtonen, A., & Veijola, S. (2011). Sleep in tourism. *Annals of Tourism Research, 38*(1), 175–192. doi:10.1016/j.annals.2010.07.016

Veijola, S., & Jokinen, E. (1994). The body in tourism. *Theory, Culture and Society, 11*, 125–151.

Veijola, S., & Valtonen, A. (2007). The body in tourism industry. In A. Pritchard, N. Morgan, I. Ateljevic, & C. Harris (Eds.), *Tourism and gender: Embodiment, sensuality and experience* (pp. 13–31). Wallingford: CABI.

von Wallpach, S., Hemetsberger, A., Thomsen, T. U., & Belk, R. W. (2020). Moments of luxury – A qualitative account of the experiential essence of luxury. *Journal of Business Research, 116*, 491–502. doi:10.1016/j.jbusres.2019.10.060

Wang, N. (1999). Rethinking authenticity in tourism experience. *Annals of Tourism Research, 26*(2), 349–370. doi:10.1016/S0160-7383(98)00103-0

Wiedmann, K., & Hennigs, N. (2009). Value-based segmentation of luxury consumption behavior. *Psychology and Marketing, 26*(7), 625–651. doi:10.1002/mar.20292

Wiedmann, K. P., Labenz, F., Haase, J., & Hennigs, N. (2018). The power of experiential marketing: Exploring the causal relationships among multisensory marketing, brand experience, customer perceived value and brand strength. *Journal of Brand Management, 25*(2), 101–118. doi:10.1057/s41262-017-0061-5

Wiedmann, K.-P., Labenz, F., Haase, J., & Hennings, N. (2016). Soothe your senses: A multisensory approach to customer experience management and value creation in luxury tourism. *The European Business Review*. Retrieved from https://www.europeanbusinessreview.com/soothe-your-senses/?__cf_chl_jschl_tk__=4846b0f98ea7bafc22b939ec6f6621cc4d85bb44-1611382021-0-ATCMcF3EwpYx0JqXsjzAyARzEi1svK2e8FAA5cv2onnyEtBFiiiISGCkQg0fXr8Fk8zyFOaH2SI_hSUi-Poo5nGC6HSskmqUHYmCsFDLVSc_4zwAF0fBtsNX7SWsqc7QNVPVq8GV9mRGSgx0JIXMBX7FQi8GZti3fqKBuufKK7q43TKDEMGTjWv0E63w8yXun68UM6f18rnckrpEh_s5Y6BOzY874nl3xMRhJ6hiHfxBMT4RZrixqgwXYc7JE4kgT8VATc3afHdPFPza_uSdxhYqNqbZGvfOam2XhgDQCJCW3F8rG7YHYBUJIpL6GMIgwGJ1SMPXV3sFnToClVeNtxg

Yang, W., & Mattila, A. (2016). Why do we buy luxury experiences? Measuring value perceptions of luxury hospitality services. *International Journal of Contemporary Hospitality Management, 28*(9), 1848–1867. doi:10.1108/IJCHM-11-2014-0579

Yeoman, I. (2011). The changing behaviours of luxury consumption. *Journal of Revenue and Pricing Management, 10*(1), 47–50. doi:10.1057/rpm.2010.43

Yeoman, I., & McMahon-Beattie, U. (2018). The future of luxury: Mega drivers, new faces and scenarios. *Journal of Revenue and Pricing Management, 17*(4), 204–217. doi:10.1057/s41272-018-0140-6

Chapter 3

Koyasan's Temple Lodging: Shades of Grey in the Luxury Tourism Spectrum

Amnaj Khaokhrueamuang,[1] *Warinthorn Kachendecha,*[2] *and Pongtawat Srichamnong*[3]

Abstract

This chapter examines the notion that contemporary tourists prefer luxury experiences rather than luxury products by testing the opinions and attitudes of middle-class tourists as a critical target for global luxury tourism through the lens of experience-oriented accommodation. Notions from the study were used in conceptualising a spectrum to describe the shades of grey within luxury tourism. A Japanese temple lodging known as 'shukubo' in Koyasan, an experience-based accommodation, was used as a case to investigate the preferences of two groups of middle-class Thai tourists: non-pilgrims and pilgrims. The results, which revealed positive opinions and attitudes towards the programme, confirm that both groups were satisfied with the luxury tourism experience programme. However, non-pilgrims, described as psychocentric tourists (represented in black at the left end of the spectrum), tended to embed them with luxury materials, such as the accommodation's facilities and amenities. Conversely, pilgrims, identified as allocentric travellers (represented in white at the right end of the spectrum), were inclined to define 'simplicity' as a luxury. They were also interested in learning experiences at both 'off-the-beaten-track' and famous branded destinations. The differences found in such an example as described above conceptualised tourists' personalities regarding luxury tourism within three shades of grey: dark grey, grey and light grey, depending on the intensity of their interests in material concerns or learning experiences. Findings from this study are general; however, it presents an original concept developed

[1]https://orcid.org/0000-0001-5501-7633
[2]https://orcid.org/0000-0003-4005-9114
[3]https://orcid.org/0000-0001-6959-6285

The Emerald Handbook of Luxury Management for Hospitality and Tourism, 57–82

doi:10.1108/978-1-83982-900-020211003

from demographic and psychographic factors to broaden the understanding of luxury tourism, which is undergoing a paradigm shift.

Keywords: Temple lodging; luxury experience; luxury tourism spectrum; grey areas; tourist typology; Koyasan

Introduction

A notion related to the trends of the global luxury market posits that customers want a luxury experience rather than luxury products and services (Dykins, 2020; Batat, 2019; Demir & Saribas, 2014; Knutson, Beck, Kim, & Cha, 2006; Mtapuri, 2017; Swarbrooke, 2018; Yeoman, 2010). This phenomenon is seen in the number of people who can afford to pay extra for luxury having decreased, but the number of people who want to experience luxury consistently being on the rise, resulting in the growth of income and international travel (Park, Reisinger, & Noh, 2010). Thus, travel market sectors (e.g., accommodation, transportation, experiences and travel agents and tour operators) offer a wide range of luxury products and travel options (Bakker, 2005). Luxury tourism ranges from traditional mass market products and services, such as five-star resorts, to a large variety of personal, unique, authentic, and exclusive experiences (Popescu & Olteanu, 2014; Stankova & Kaleichev, 2013; Sundbo & Dixit, 2020).

As a critical element of the tourism industry, accommodation is also shifting from a material-based commodity to an experience-based commodity. Temple lodging is one of the accommodation types defined as 'theme-based tourist attractions' (Ross, 2019). In addition, with the rise of middle-class consumers in developing economies, who are also a key target for the tourism industry (Tutek, Gebbie, Chan, & Durand, n.d.), luxury becomes less exclusive, and materialism becomes less important (Yeoman, 2010). On the contrary, time and enrichment create more value (Dykins, 2020). The new era of luxury travel emphasises authentic experiences and requires brands (Dixit, 2020; Kim, 2017) to constantly monitor customers' expectations.

Towards this shift in consumer tastes and market trends, luxury tourism has become more conceptually diverse and blurred, having 'grey areas' of different shades. When focussing on the accommodation sector, the luxury accommodation experience is significantly different from standard accommodation experiences (Harkison, Hemmington, & Hyde, 2017). This notion raises the question of how to clarify the intensity of luxury consumption shifting from material concerns to an experience-based focus, particularly when targeting middle-class tourists. This study aimed to question the notion of 'luxury experience' in the context of an experience-based accommodation through the perspectives of middle-class tourists. When testing the opinions and attitudes of middle-class tourists, how does the luxury experience shade the grey areas in luxury tourism to assess its application to luxury marketing? Although this study is general in terms of tourism-related knowledge, its findings include an original concept that can contribute to a broader understanding of luxury tourism's changes as a grounded theory.

Research Objectives

The study aimed to examine the notion that contemporary tourists prefer luxury experiences rather than luxury products by testing the opinions and attitudes of middle-class tourists through the lens of experience-oriented accommodation. Another purpose was to conceptualise a spectrum of luxury tourism in describing a range of intensity of interest in and meanings ascribed to the luxury experience to understand that there are shades of grey within luxury tourism.

The study of middle-class tourists' opinions and attitudes used a case study of Japanese temple lodgings, or 'shukubo', an experience-based accommodation that promotes Japan as a global luxury tourism destination (Demir & Saribas, 2014). The survey of attitudes was conducted with middle-class Thai visitors (as one of Japan's priority inbound markets) staying overnight at Yochi-in Temple in Koyasan, a UNESCO World Heritage Buddhist monk town connected to one of the Buddhist pilgrimage routes of Kumano Kodo in the Kii Mountain Range.

Literature Review

Koyasan: Luxurious Japanese Temple Lodging in a Sacred Landscape

Koyasan, a Japanese Buddhist town in Wakayama Prefecture, is located in an alpine basin surrounded by eight mountain peaks at an altitude of about 1,000 m above sea level. It was founded by the great Buddhist monk Kobo Daishi Kukai in 816. There were about 2,000 ancient Buddhist temples built since the ninth century, of which 117 presently remain.

Currently, Koyasan is home to about 3,000 people and is a famous tourist destination receiving 1,200,000 visitors every year (Kametani, 2014). In 2004, six areas of the town were designated part of Japan's 12th UNESCO World Heritage Site, the Sacred Sites and Pilgrimage Routes in the Kii Mountain Range connecting in the south to the ancient shrines of Kumano Kodo. As the holy mountain, Koyasan is recommended on the list of '100 places to see before you die' (Carter, 2015).

Shukubo: A Luxurious Experience of Japanese Temple Lodging

Unlike ordinary inns and hotels, a shukubo is a sacred place of esoteric Buddhist training. As part of the pilgrimage for royal families and aristocrats since the Heian period (AD 794–1185), the temples in Japan that served as accommodations for pilgrims are now called shukubo, which is the root of the traditional Japanese hotel 'ryokan' (UNWTO, 2016). There are many charms of shukubo, which offer guests different experiences from other accommodations. Among the 117 temples in Koyasan, 52 temples are open as shukubo with meals and religious activities provided. These temple lodgings, which range from small to large and luxurious, are located between Daimon (the great gate) and Okunoin (the graveyard). The smallest shukubo accommodates 15 individual guests and has around 5 rooms. The large temple lodgings are available for over 100 individual guests with more than 25 rooms (Koyasan Shukubo Association, 2011). The luxurious shukubo offers large numbers of rooms and halls with garden views, 'onsen' (a Japanese hot spring) and superb Buddhist cuisine 'shojin ryori', such as Fukuchi-in (Harada, 2016). Another example of a luxurious shukubo was described by a foreigner as follows (Nicoloff, 2008, p. 19).

> Soji-in, a temple that burned a few years back, has the most luxurious facilities. Each of its guest apartments has air-conditioning, a private Western-style toilet, a closet with hangers, a colour T.V. Public baths are gorgeous, with thick towels, herbal soaps, and fresh orchids floating on the steaming surface.

Although the shukubo has an image of being a sacred Buddhist place, it also offers a brand of sake and beer. One of the unique factors of temple lodging is hospitality services from young priests or apprentices, such as welcoming guests as a receptionist, making the bed as a housekeeper and serving meals as a waiter. Senior monks direct guests to participate in Buddhist activities, such as morning chants, transcribing sutras and deep meditation.

Shojin Ryori: Simple but Luxurious Japanese Buddhist Cuisine

One of the unique services of shukubo is the Buddhist vegetarian cuisine called shojin ryori, a simple but beautiful meal. Shojin ryori means cooking ascetics in pursuit of enlightenment (Carter, 2015). Meals are skillfully cooked with traditional organising principles: preparing mild delicacies without any use of meat, fish, cheese, milk, eggs or gelatine, excluding ingredients with a pungent taste or smell, and the 'system of five', which comprises five basic tastes, five cooking methods and five colours (Nicoloff, 2008). Over the course of a 1000 years, the recipe of shojin ryori has been transmitted as a popular cuisine influencing the traditional Japanese diet 'washoku', which was added to UNESCO's Lists of Intangible Culture Heritage in 2013 (Cang, 2015). Shojin ryori represents Koyasan's luxurious services and is served on a red tray table by priests. This service makes tourists feel like highly honoured guests, as described in the following examples.

> We are eating like gods tonight.
>
> (Nicoloff, 2008, p. 22)

> Breakfast transforms me once again into a princess. A princess eating vegan food on miniature red trays in a temple after morning prayers.
>
> (Harada, 2016)

Yochi-in Temple Lodging: A Luxury Experience Programme

Yochi-in is one of 52 temple lodgings dating back through a 900-year history. It was originally built by Imperial Prince Kakuho-Shinno, the fourth son of Emperor Shirakawa, in 1127. This temple is not only prestigious but also luxurious in services, underpinning the concept of 'omotenashi', which reflects the sense of hospitality with a politeness that makes guests feel valued and respected.

The temple provides 24 Japanese-style rooms with one large hall which can accommodate 60 individual guests or 80 guests in a group. The room rate is 11,000 Japanese yen (approximately $100 USD) per person and includes two shojin ryori meals. Guests with food allergies can also arrange special meals. Dinner is served at 5:30 p.m. with tea, but other beverages are available at an additional cost, including beer and sake. Breakfast starts at 7:00 a.m. after the chanting ceremony ends. Lodging facilities include public Western-style toilets and mutual Japanese-style Jacuzzi hot tubs with shampoo and liquid soap that are open until 9:00 p.m. or 8:00 p.m. in winter. Rooms are provided with Japanese futon mattresses and other amenities, such as towels and Japanese bathing gowns, heaters and fans and televisions.

The temple requests that guests participate in the morning sutra chanting in the elegant ceremonial hall free of charge. Three Buddhist activities are offered at an additional cost: hand-copying of sutras, tracing Buddha's image and esoteric Buddhist meditation. There is an English tour guide service. A monk who is also the manager can communicate well in English while young monks and apprentices take care of the guests by serving the shojin ryori (Yochi-in, n.d.).

Conceptualising Shades of Grey in the Luxury Tourism Spectrum

The major characteristics of luxury are exclusivity, high quality with premium pricing, unique design, extra value and very limited supply (Harkison et al., 2017; Park et al., 2010). However, the concept of luxury changes dramatically across time and cultures. Luxury is not only a status, and it is more than simply monetary value (Yeoman, 2010). What is very luxurious for one person might be mainstream for another (Park et al., 2010). Thus, the word 'luxury' has different definitions for different people (McKone, Lewis, & Smith-George, 2017). This notion raises the question of how to define a clear concept of luxury tourism evolving from the past to the present and future through the lens of consumer segmentation.

According to the L.E.K. 2017 Luxury Travel Study, a survey of nearly 2,000 US travellers, the luxury travel market can be largely segmented by the average money spent for travel per person into three broad groups: *elite, aspirational and prudent* (McKone et al., 2017). These groups conform to the three 'As' of luxury, comprising *absolute luxury, aspiring luxury and accessible luxury* (Horwath HTL, 2011). These categories align with the generational divides (i.e., silver, baby boomers, Generation X and millennials) as follows:

(1) Elite (absolute luxury) – Travellers are high on the 'indulgence spectrum'; they tend to require luxury across their travel experience (McKone et al., 2017).

This category is assumed to be associated with the silver or silent generation and baby boomers. According to the Pew Research Center, the silent generation (born 1928–1945) and baby boomers (born 1946–1964) are between 55 and 91 years old at present (Dimock, 2019). The elite category also includes the young generation. This group prefers customised and private or tailor-made luxury.

(2) Aspirational (aspiring luxury) – Travellers are willing to indulge in high-end experiences in specific situations. This group comprises Generation X and millennials, which is a working population assumed to be the middle-income group or middle-class tourists. Based on the generations defined by the Pew Research Center, Generation X (born 1965–1980) and millennials (born 1981–1996) range between 23 and 54 years old at present (Dimock, 2019).

(3) Prudent (accessible luxury) – Travellers are more likely to be budget stretchers; they will partake in limited luxury travel from time to time. This group refers to Generation Z, which prefers adventure and modern luxury and embraces the concept of the 'travel splurge' (McKone et al., 2017). The Pew Research Center defined this generation as being those born between 1997 and 2012 (ages 7–22) (Dimock, 2019). However, this paper considered the post-millennials, who are between 23 and 30 years old, to be the potential tourists of this group.

Besides the three categories above, tourists in the luxury tourism market can be described based on their personalities as psychographic types. This paper considered the Psychographic Tourist Typology by Stanley Plog (1974), the most popularly cited tourism theory study, even there has been validity and applicability (Piuchan, 2018). Plog (1974) classified tourists with the concept of the allocentrism–psychocentrism spectrum into five types: psychocentrics, near-psychocentrics, mid-centrics, near-allocentrics and allocentrics. Allocentrics are individuals who feel that what happens to them is largely under their control. They thus prefer adventure or seek new experiences with risks. Conversely, psychocentrics believe that what happens to them is largely beyond their control. They thus try to make safe, consistent choices in terms of popular things and places. In the middle of the spectrum, mid-centrics are those travellers who accept both risk and safety. In addition, tourists might lean towards one direction or the other on the continuum, either as near-allocentrics or near-psychocentrics, without being fixed in the extremes (Cruz-Milán, 2017).

In summary, there are two important factors contributing to constructing the luxury tourism spectrum: demographic and psychographic factors. Demographic factors (e.g., age, gender, education and income) influence tourists' opinions regarding the motivation for purchasing luxury tourism products and experiences. Psychographic factors, such as lifestyles, personalities, beliefs and values, influence tourists' attitudes as determinants in consuming luxury travel; these factors can lead to significant differences between consumers groups and market segments (Batat, 2019). These two factors conform to the equation of the purchase decision process in tourism: 'motivators + determinants = purchase decision' (Swarbrooke, 2018).

Research Framework

In order to conceptualise the shades of grey in luxury tourism, the study formulated two hypotheses:

H1. There is a positive opinion regarding the offered luxury experience for middle-class Thai tourists.

H2. The attitude before and after offering the luxury experience is positive for middle-class Thai tourists.

H1 examined the positive opinions of middle-class Thai tourists. *H2* tested their attitude changes. Both hypotheses determined the notion derived from many scholars that middle-class tourists prefer luxury experiences more than luxury products (Dykins, 2020; Batat, 2019; Demir & Saribas, 2014; Knutson et al., 2006; Mtapuri, 2017; Swarbrooke, 2018; Yeoman, 2010).

After participating in the luxury tourism programme, opinions were counted in frequency to determine the positive and negative aspects. Attitudes before (pre-staying) and after (post-staying) participating in the luxury experience programme were conducted using a *t*-test. Factors contributing to attitude changes in both positive and negative directions were analysed with opinions to compare the differences influenced by these factors. Differences in opinions and attitudes of middle-income tourists were conceptualised to shade the grey areas across the luxury tourism spectrum. The research design is outlined in Fig. 3.1.

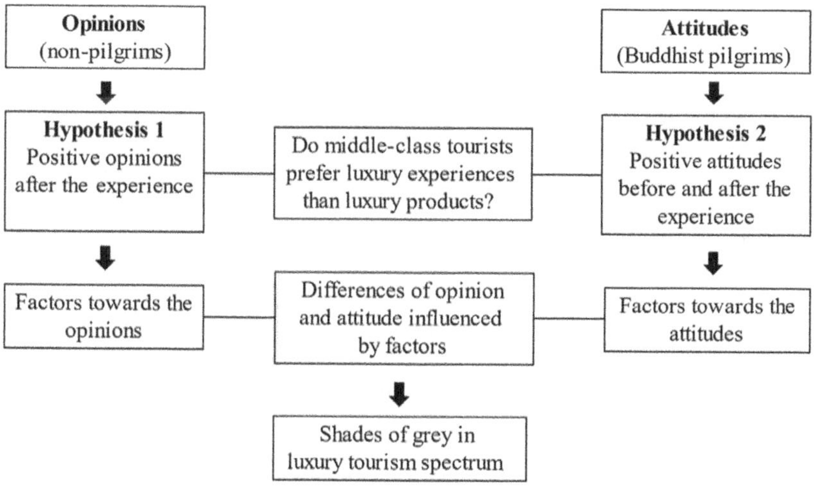

Fig. 3.1. Research Framework for Conceptualising Shades of Grey in Luxury Tourism. *Source:* Authors.

Research Methodology

Instrument

This paper used the opinions and attitudes of the middle-class tourists as qualitative data to determine the intensity of their preferences regarding the luxury experience programme. The opinion study was conducted after the end of the programme through a semi-structured interview with 20 questions concerning three aspects: (1) the temple's physical surroundings and services, (2) the temple's activities and (3) the potency and expectations of the guests (see questions in Table 3.5).

The attitude test, by means of a paper-based questionnaire, was conducted before the programme's start (pre-test) and after its end (post-test). The questionnaire (attitude test) comprised 20 items using a series of adjectives listed in the semantic differential technique of Osgood, Suci, and Tannenbaum (1957). Each attitude is measured on a bipolar adjectival 7-point Likert scale ranging from negative (1–3) to neutral (4) to positive (5–7). Pairs of bipolar adjectives reflect three major dimensions: an evaluative dimension (e.g., clean–dirty, luxurious–humble), an activity dimension (e.g., new–familiar, expected–unexpected) and a potency dimension (e.g., acceptable–upsetting, comfortable–shy). Most paired adjectives listed in the questionnaire were related to questions asked in the interviews (see Table 3.1).

Data Collection and Analysis

The survey of the tourists' opinions and attitudes towards the luxury tourism experience in Koyasan's temple lodging was conducted with 35 Thai tourists classified by their income levels as middle-class consumers. The sample participated in experiencing the Japanese temple lodging programme of Yochi-in Temple for two days and one night between September and November 2017. The programme offered the following luxury experiences.

(1) Having shojin ryori in the historical banquet hall the ancient emperor used.
(2) Bathing in a Japanese-style bathroom with a shared jacuzzi.
(3) Sleeping on a mattress in an authentic Japanese-style room.
(4) Participating in the morning chant (specially offered in the Pali language for Thai tourists, as similar to the Buddhist sect exists in Thailand).
(5) Communicating and sharing Buddhist-related culture with Japanese monks.
(6) Experiencing the atmosphere of the Japanese garden, architecture, and arts of the temple buildings, including other ancient sites in the World Heritage zone (e.g., the stupa, the Great Buddha and the graveyard).

The programme was organised twice. First, it was offered to (Buddhist) non-pilgrims from 12 to 13 September 2017. The participants were 10 scholars from the Rajamangala University of Technology Rattanakosin, Prachuap Khiri Khan Province, Thailand. After the end of the programme, semi-structured interviews

Table 3.1. List of a Series of Adjectives for Measuring Attitudes Related to Interview Questions.

No. (Adj. Group) Statements	Attitudes: Positive (7–5)	(4)	Negative (3–1)	Related Interview Questions
1. (ad) staying	Impressive	–	Unimpressive	1. Were you impressed by an experience while staying in Yochi-in, Koyasan's temple lodging?
2. (ad) staying	Desirable	–	Undesirable	2. How much did you want to try temple lodging?
3. (ed) surround	Cozy	–	Uncomfortable	3. Did the temple stay at Yochi-in offer you comfort, pleasure, peace of mind and a spiritual feeling?
4. (ed) surround	Pleasure	–	Displeasure	
5. (ed) surround	Calm, sacred	–	Desacralised	
6. (ed) surround	Moderate	–	Awful	
7. (ed) surround	Beautiful	–	Unaesthetic	4. Did you find Yochi-in, the temple lodging, beautiful and clean?
8. (ed) surround	Clean	–	Dirty	
9. (ed) guest room	Luxurious	–	Humble	5. Did you find the guest room at Yochi-in luxurious?
10. (ad) cuisine	Delicious	–	Unsavoury	6. Did you find the meal cooked by the temple delicious?
11. (ad) morning chant	Exciting	–	Boring	7. Were you excited/interested to join the morning chanting? 8. How did you feel when hearing the Japanese priest chanting in Pali?

Table 3.1. *(Continued)*

No. (Adj. Group) Statements	Attitudes: Positive (7–5)	(4)	Negative (3–1)	Related Interview Questions
12. (pd) shared bathrooms	Naked	–	Shy	9. Did you find the shared-bathroom/toilet experience unpleasant?
13. (pd) shared toilets	Inoffensive	–	Disgusting	
14. (pd) shared bedrooms	Acceptable	–	Upset	10. Can you sleep in a shared bedroom?
15. (pd) regulations	Adjustable	–	Unadaptable	11. Did you have a problem with Yochi-in's rules and regulations?
16. (ed) money	Valuable	–	Worthless	12. Was staying at Yochi-in worth your money?
17. (ad) learning	New	–	Familiar	13. Have you learnt new knowledge or did you benefit from an experience at Yochi-in?
18 (ad) staying	Beneficial	–	Useless	
19. (ad) practicing dharma	Expected	–	Unexpected	14. Were you inspired to start practicing the dharma after experiencing the temple stay?
20. (ad) staying	Repeated	–	Banned	15. Will you come back for a temple-stay experience at Koyasan again?

Note: ed, evaluative dimension; ad, activity dimension; pd, potency dimension.
Source: Authors.

were conducted to collect their opinions. Data on positive and negative opinions were counted and presented via descriptive statistics in terms of frequency and percentages.

The same programme was conducted again with the second group from 7 to 8 November 2017. The participants were 25 Thai Buddhist pilgrims. They included 3 Buddhist monks and 22 volunteers devoted to Rai Chuern Tawan (International Meditation Center and Organic Farm) in Chiang Rai Province, Northern Thailand. Participants were asked to complete a structured questionnaire that measured attitudes before (pre-test) and after (post-test) experiencing the programme. Data were analysed using the *t*-test to determine significant changes in attitudes before and after the programme. Pairs of bipolar adjectives that had *t*-test scores equal to or less than the 0.05 level ($p < 0.05$) were considered significant. This means that the attitudes changed before and after having the experience.

The results from analysing the opinions and measuring the attitudes of the participants were discussed with factors influencing the differences of opinions and attitudes to conceptualise a luxury tourism spectrum.

Findings

Thai Tourists' Opinions regarding Koyasan's Temple Lodging Programme

The participants were scholars from the hospitality industry with different specialisations. Four of them were from the tourism industry, another four were from the hotel industry and the last two were from the MICE (Meeting, Incentive, Convention, Exhibition) industry. Eight participants were female, and the other two were male. Their age range was 32–54, categorising them as Generation X and millennials. Generation X consisted of four participants (Participant no. 1, 6, 8 and 10). Millennials comprised six participants (Participant no. 2, 3, 4, 5, 7 and 9). As for their experience travelling to Japan, six scholars who participated in this study had travelled to Japan for the first time, three had been to Japan 2–3 times and one had travelled to Japan more than 10 times (see Table 3.2).

The 10 participants had different tourist behaviours, which have been classified into three categories (see Table 3.3) as follows:

(1) Interest in religion: Four participants showed an interest in learning about religions and were also strict Buddhists, whereas the other six stated that they were not religious or were not interested in learning about their own religion or other religions.
(2) Vegetarian food: Seven participants ate vegetables regularly and could accept vegetarian food, whereas the other three showed a strong dislike for vegetables and did not eat vegetarian food.
(3) Familiarity with Japanese culture: Only two participants in this study had familiarity with Japanese culture, whereas the other eight had limited knowledge about Japan and were not familiar with the Japanese culture at all.

Table 3.2. Demographic Information.

Participant	Specialisation	Gender	Age	Generation	Frequency in Visiting Japan
1	Tourism	Female	43	Gen X	2
2	MICE	Female	38	Millennials	1
3	Hotel	Female	35	Millennials	1
4	Hotel	Female	36	Millennials	1
5	Hotel	Female	32	Millennials	1
6	Tourism	Male	43	Gen X	3
7	Tourism	Female	32	Millennials	1
8	Hotel	Female	54	Gen X	1
9	MICE	Female	33	Millennials	>10
10	Tourism	Male	54	Gen X	2

Source: Authors.

Table 3.3. Tourist Behaviours of 10 Participants.

Participant	Interest in Religion	Vegetarian Food	Familiarity with Japanese Culture
1	✗	✓	✗
2	✓	✓	✗
3	✗	✓	✗
4	✓	✓	✗
5	✓	✗	✗
6	✓	✓	✓
7	✗	✗	✗
8	✗	✓	✗
9	✗	✓	✓
10	✗	✗	✗

Source: Authors.

According to the results of the interviews, the answers to 17 questions (85%) showed positive opinions of the temple lodging at Koyasan in terms of the overall perspectives of the temple's physical surroundings and services, temple activities and the potency and expectations of the guests. Regarding the physical surroundings and services of the temple, the participants were impressed and desired to stay, experienced feelings of comfort, pleasure, peace of mind and spirituality and enjoyed the beautiful and clean accommodations.

> At first, I thought it was just like staying in a temple, like in Thailand. I didn't think it's going to be a special experience, but it turned out staying in Yochi-in was beyond my expectation.
> (Participant no. 4)

> I think Yochi-in has both facilities and amenities as much as other hotels can offer. The atmosphere in the temple and its surroundings made me think of Japan in the old-time.
> (Participant no. 6)

For offered activities, half of the participants enjoyed the delicious taste of shojin ryori, but the others did not prefer it due to a desire for meat.

> I already like Japanese food, so I have no problem with the meal cooked by the temple. Plus, their ingredients were high quality; they were already delicious without much seasoning.
> (Participant no. 8)

They felt excited and surprised about the morning chant due to a feeling of religious performance and the surprise of chanting in Pali.

> It wasn't the same feeling when I heard it in Thailand, but it made me feel closer to the ceremony.
> (Participant no. 6)

Regarding the guests' potency and expectations, half of them accepted sleeping in a shared Japanese-style bedroom, but the others were worried about safety. They were able to adjust to the strict regulations, benefit from the new knowledge and thought the experience was worth the money. They wanted to stay in Koyasan's temple lodging again to seek different experiences and compare its services with others.

> It was worth the money we've paid because I consider an experience staying at Koyasan as a rare opportunity for travellers.
> (Participant no. 6)

> I want to go back to Koyasan and stay in a different temple so I can compare their services.
> (Participant no. 9)

However, answers to three questions (15%) expressed negative opinions (question no. 9, 13 and 19). The negative opinions regarded the guest rooms, shared bathroom and the inspiration to practice the Buddhist doctrine (dharma).

All the participants thought the guest rooms were rather plain instead of being of a luxurious style (question no. 9).

> The room was not luxurious like a 5-star hotel, it was very plain.
> (Participant no. 4)

Regarding the shared bathroom, being naked in a Japanese-style bathroom was an unpleasant experience to accept at first (question no. 9). Seven participants had a strongly negative attitude, finding it embarrassing to expose their naked bodies to strangers.

> I really don't like the shared bathroom at Yochi-in. I think it was disgusting.
> (Participant no. 3)

Regarding the inspiration to practice the dharma, not all participants were inspired to practice it after experiencing the programme (question no. 14).

> I did not feel like chanting or practising dharma; the Japanese chanting ceremony for me was more like a show for tourists.
> (Participant no. 5)

Thai Tourists' Attitudes towards Koyasan's Temple Lodging Programme

The attitude tests of group 2 (Buddhist pilgrims) showed statistically significant differences in total ($P = 0.022$) between the pre-programming ($\bar{x} = 5.71$) and the post-programming ($\bar{x} = 6.47$) enquiries, although the attitude question no. 5, 17 and 18 were insignificantly different ($P > 0.05$) due to slight changes between the pre-programming and the post-programming enquiries. Overall, of 20 attitudinal statements, 19 (95.0%) showed a change from low to high positivity levels. Only the attitude question no. 12 showed a change from a negative level to a positive level. The attitude during the pre-programming was negative ($\bar{x} = 3.88$), whereas the post-programming attitude was positive ($\bar{x} = 5.24$) (see Table 3.4). All statements of attitudes of group 2 shown in Table 3.4 were summarised and compared with the frequency of opinions of group 1, as shown in Table 3.5.

Hypothesis Test and Comparison of Opinions and Attitudes

According to Table 3.5, in total, 17 questions (85%) revealed positive opinions and 3 questions (15%) revealed negative opinions about the programme. This finding indicates that *H1* (*H1*. There is a positive opinion regarding the offered luxury experience for middle-class Thai tourists) is acceptable. The attitudes before and after the programme show that there has been a positive change.

Attitudes	Pre-programme $\overline{x}$	Pre-programme s	Post-programme $\overline{x}$	Post-programme s	t-Test	df	P
1. Impressive – unimpressive (staying)	6.16	0.99	6.88	0.33	−3.392*	24	0.002
2. Desirable – undesirable (staying)	6.12	1.01	6.64	0.64	−2.243*	24	0.034
3. Cozy – uncomfortable (surround)	5.36	1.29	6.44	0.77	−3.311*	24	0.003
4. Pleasure – displeasure (surround)	6.08	0.86	6.72	0.61	−2.874*	24	0.008
5. Clam, sacred – desacralised (surround)	6.44	0.82	6.80	0.50	−1.890	24	0.071
6. Moderate – awful (surround)	6.04	1.21	6.76	0.52	−2.688*	24	0.013
7. Beautiful – unaesthetic (surround)	6.12	1.20	6.84	0.37	−2.688*	24	0.013
8. Clean – dirty (surround)	6.12	1.13	6.84	0.37	−2.753*	24	0.011
9. Luxurious – humble (guest room)	5.16	1.43	6.12	0.93	−2.832*	24	0.009
10. Delicious – unsavoury (cuisine)	5.32	1.14	6.08	0.86	−2.728*	24	0.012
11. Exciting – boring (morning chant)	5.08	1.25	6.08	0.81	−4.201*	24	0.000
12. Naked – shy (shared bathrooms)	3.88	2.08	5.24	1.98	−3.209*	24	0.004
13. Inoffensive – disgusting (shared toilets)	4.76	1.45	5.92	1.35	−3.412*	24	0.002
14. Acceptable – upset (shared bedrooms)	5.60	1.00	6.28	1.10	−2.527*	24	0.018
15. Adjustable – unadaptable (regulations)	6.24	0.97	6.72	0.54	−2.493*	24	0.020
16. Valuable – worthless (money)	5.56	1.26	6.44	0.71	−2.724*	24	0.012
17. New – familiar (learning)	6.40	0.76	6.72	0.61	−1.693	24	0.103
18. Beneficial – useless (staying)	6.32	0.75	6.68	0.69	−1.984	24	0.059
19. Expected – unexpected (practicing dharma)	5.92	1.19	6.44	0.87	−2.116*	24	0.045
20. Repeated – banned (staying)	5.48	1.41	6.52	0.65	−3.375*	24	0.003
Total	*5.71*	*1.16*	*6.47*	*0.76*	*−2.757*	*24*	*0.022*

Note: *The t-test is significant at the 0.05 level ($P < 0.05$).
Source: Authors.

Table 3.5. The Comparison of Opinions and Attitudes of Thai Tourists.

No.	Questions	Opinions (Frequency)	Meaning	Attitude Statements	Attitudes	Meaning
1	Were you impressed by an experience staying in Yochi-in, Koyasan's temple lodging?	Yes (10)	Positive	1. Staying	Impressive	Positive
2	How much did you want to try a temple lodging?	Desire to stay (6)	Positive	2. Staying	Desirable	Positive
3	Did the temple stay offer you comfort?	Yes (10)	Positive	3. Surrounding	Cozy	Positive
4	Did the temple stay offer you pleasure?	Yes (10)	Positive	4. Surrounding	Pleasure	Positive
5	Did the temple stay offer you peaceful mind?	Yes (10)	Positive	5. Surrounding	Clam, sacred	Positive
6	Did the temple stay offer you spiritual feeling?	Yes (10)	Positive	6. Surrounding	Moderate	Positive
7	Did you find Yochi-in, the temple lodging, beautiful?	Yes (10)	Positive	7. Surrounding	Beautiful	Positive
8	Did you find Yochi-in, the temple lodging, clean?	Yes (10)	Positive	8. Surrounding	Clean	Positive
9	Did you find the guest room at Yochi-in luxurious?	No but plain (10)	*Negative*	9. Guest room	Luxurious	Positive
10	Did you find the meal cooked by the temple delicious?	Yes (5)	Positive	10. Cuisine	Delicious	Positive
11	Were you excited/interested to join the morning chanting?	Yes (10)	Positive	11. Morning chant	Exciting	Positive
12	How did you feel when hearing the Japanese priest chanting in Pali?	Surprised (10)	Positive			

13	Did you find a shared-bathroom experience unpleasant?	Yes (7)	*Negative*	12. Shared bathroom	Naked	Positive
14	Did you find a shared-toilet experience unpleasant?	No (10)	Positive	13. Shared toilet	Inoffensive	Positive
15	Can you sleep in a shared bedroom?	Yes (5)	Positive	14. Shared bedroom	Acceptable	Positive
16	Did you have a problem with Yochi-in's rules and regulations?	No (10)	Positive	15. Regulations	Adjustable	Positive
17	Was staying at Yochi-in worth your money?	Yes (7)	Positive	16. Money	Valuable	Positive
18	Have you learnt new knowledge or were you benefited from an experience at Yochi-in?	Yes (10)	Positive	17. Learning 18. Staying	New Beneficial	Positive
19	Were you inspired to start practicing the dharma after experiencing the temple stay?	No (10)	*Negative*	19. Practicing the dharma	Expected	Positive
20	Will you come back for a temple-stay experience at Koyasan again?	Yes (6)	Positive	20. Staying	Repeated	Positive
Summary result of opinions			**Positive** (85%)	Summary result of attitudes		**Positive** (100%)

Source: Authors.

This result indicates that *H2* (*H2*. The attitude before and after offering the luxury experience is positive for the middle-class Thai tourists) is acceptable.

However, there are three aspects showing different results between opinions (question no. 9, 13 and 19 regarding opinions) and attitudes (question no. 9, 12 and 19 regarding attitudes). Namely, three aspects of opinions were negative, but three aspects of attitudes were positive.

Discussion

Conceptualising Shades of Grey in the Luxury Tourism Spectrum

With the notion that luxury tourism is increasingly seen as an experience rather than the consumption of products and services (Dykins, 2020; Batat, 2019; Demir & Saribas, 2014; Knutson et al., 2006; Mtapuri, 2017; Swarbrooke, 2018; Yeoman, 2010), the results from examining the opinions and attitudes of middle-class Thai tourists show agreement that the luxury tourism market is inclined to raise the interest in an experience-based tourism product. This phenomenon results in changing travel behaviours of luxury consumption across time, culture (Yeoman, 2010) and technology.

In addition to the middle-class tourist group, the change in luxury consumption is also seen in the elite tourist group. Described by tourist typology identified by Plog (1974), the traditional rich people tend to be psychocentrics who prefer to keep to the things and places they know (Swarbrooke, 2018). However, the modern elite tourists tend to seek new experiences at an 'off-the-beaten-track destination' (Lux Life, 2019) as allocentrics (Plog, 1974). The material-based product consumption of psychocentric elite tourists has shifted to the experience-based tourism products consumed by near-allocentric and allocentric elite travellers. However, both rich psychocentric and rich allocentric consumers like to go on tailor-made, expensive tours organised by famous professional operators or branded travel agencies.

Furthermore, the current luxury consumption has also expanded from middle-aged individuals (40–50 years old) to younger segments (who are mostly budget travellers) and is showing rapid growth in the global market (Kang & Park, 2016). From this evolution, contemporary luxury travel has a gap between the preferences for material concerns and experiences, and the changes mainstream consumers are making in terms of the characteristics of luxury tourism consumption are various and different in intensity. In describing the concept, material-based luxury consumption can be identified as 'black', whereas the luxury experience preference is represented by 'white'. Thus, 'grey' is used in reference to the intensity of luxury tourism presented in different shades due to the proximity to either side between material-based and experience-based preferences.

Based on this idea, the luxury experience programme of temple lodging in Koyasan examining the different opinions (non-pilgrims) and attitudes (pilgrims) of middle-class Thai tourists helps to shade the grey areas in the luxury tourism concept. The programme offered a luxury experience in three ways: (1) the opportunity to enjoy the sacred landscape of a Buddhist monk town; (2) the opportunity to engage in Buddhism-related activities and (3) the pleasure and benefits of learning new things. They indicate that the interests of the middle-class tourist participants ranged from the built landscape and environment (the material) to the learning activities (the experience); in other words, the research identifies a grey area of the luxury tourism spectrum, which ranges from black to white, as illustrated in Fig. 3.2.

According to Fig. 3.2, black represents traditional luxury travel, characterised by a material orientation and providing a passive experience with a focus on extravagance and a renowned destination for psychocentric luxury tourists. On the opposite side, white symbolises the future trend of luxury travel, which shifts from material-based consumption to a focus on values and experiences. Customers tend to be allocentric travellers who seek both a rare experience and a unique destination (Plog, 1974) and raise the awareness of sustainability. For example, they prefer to purchase environmentally friendly, intelligent and healthy products (Yeoman, 2010) and to spend money in an ethically sustainable way (Ross, 2019). In addition, the grey shades between the black and white areas describe the mid-centric tourists, who are the largest target of the contemporary luxury market.

Focussing on the middle-class tourists in this study, they had different opinions and attitudes in three areas: (1) the decoration of the guest rooms, (2) the shared bathroom experience and (3) the inspiration to start practicing the dharma, indicating a range of intensity of tourist behaviours and interests in material concerns versus experiences. With regard to the tourists' attitudes towards the

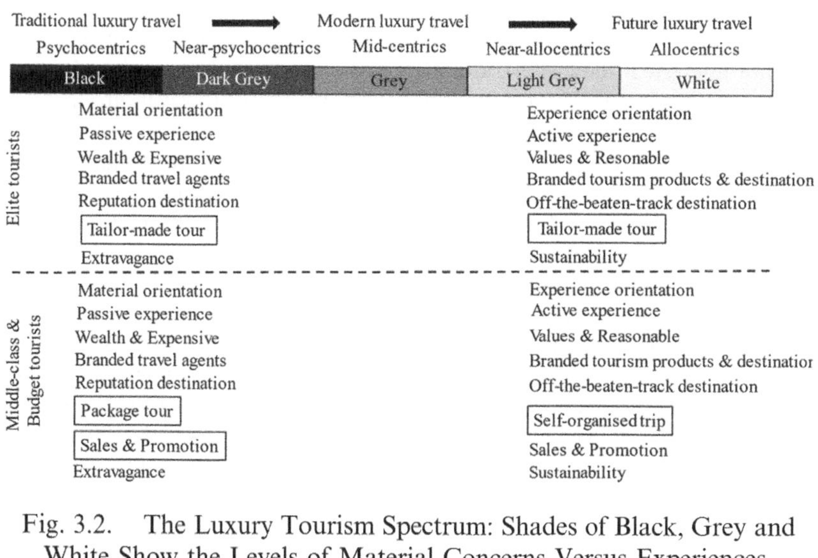

Fig. 3.2. The Luxury Tourism Spectrum: Shades of Black, Grey and White Show the Levels of Material Concerns Versus Experiences. *Source:* Authors.

guest rooms, the dark grey area describes the negative opinions of group 1 (non-pilgrims), who thought that the rooms were simple rather than luxurious. This finding indicates that this group demonstrated a near-material orientation close to the traditional image of luxury. On the other hand, the light grey area describes the positive attitudes of group 2 (pilgrims), who identified the simplicity of the rooms as luxurious.

The luxury concept was viewed differently between groups 1 and 2, although they are defined as being within the same segment by income (middle income) and generation (Generation X and millennials) but different in terms of psychographic factors, such as lifestyle and personality (Swarbrooke, 2018). Group 1 worked in an academic context influencing their lifestyle as being rather embedded with luxury material-based leisure as a social status symbol or a form of prestige (Batat, 2019; Harkison et al., 2017; Park et al., 2010; Yeoman, 2010). Thus, they are described as near-psychocentrics (dark grey) to mid-centrics (grey). Group 2, in contrast, devoted themselves to religious works, with Buddhism influencing their views of luxury being a simple life, and their interests in experiencing the Japanese Buddhist culture tended to result in them being defined as near-allocentrics (light grey) to allocentrics (white).

The shared bathroom was an unpleasant experience for group 1 (dark grey), who visited Japan for the first time and were unfamiliar with Japanese bathing. Regarding this finding, alongside the psychocentrics, this group belongs to what Sukmawati, Ernawati, and Nadra (2018) explained as being the conditional view of luxury tourism heightened by top-class facilities and amenities. This definition is a contradiction to the attitude of group 2 (light grey) that the shared bathroom experience was acceptable because of familiarity with mutual facilities and their adaptability skill.

With regard to starting to practice the dharma, group 1 (grey to dark grey) were not interested (dark grey), although participating in the morning chant was exciting because it was similar to watching a show (grey). However, group 2 (light grey), who were mostly temple volunteers, were inspired to start practicing the dharma. This finding aligns with Harkison et al.'s (2017) argument that 'luxury' is defined by guests' expectations and strong emotional experiences.

Unlike a group of wealthy consumers, whose luxury vacationing is a way of life (Swarbrooke, 2018), middle-class and budget tourists cannot afford to pay for tailor-made tours (customised individual travel). As Yeoman (2010) stated, consumers extensively search for bargains and consider using a package tour, including promotions and sales, to experience luxury at the destination. These groups, whether they are cash-poor but time-rich (e.g., students) or cash-poor and time-poor (e.g., workers on lower salaries), tend to make luxury travel experiences a special occasion or even a once-in-a-lifetime activity (Swarbrooke, 2018). Thus, alongside allocentric, the self-organised trip is growing due to information technology and social media.

As Batat (2019) stated, customers select the 'experiential journey' beyond the 'shopping journey'; luxury travel, therefore, concerns the consumption process in three stages: before, during and after. Thus, tourists define luxury in their own words and still have a sense of luxury brand consumption. The millennials and young generations particularly embrace the concept of the 'travel splurge' (McKone et al., 2017) to show off their status via social media, such as Facebook.

Implications and Limitations of the Concept in Marketing

Unlike traditional luxury marketing, experiential luxury marketing is based on the 'experiential segmentation' of markets as consumers define them (Batat, 2019). Thus, the concept of grey areas in offering luxury experiences can apply on both the supply and demand sides by considering the differences between tourist segments and their behaviours towards the characteristics of luxury consumption. On the demand side, for example, offering a luxury experience of temple lodging in Koyasan to a group of non-pilgrim middle-class Thai tourists should focus more on luxury facilities and amenities, such as the 'wow' moment of luxurious art decorations or shojin ryori with the created story of Japanese Buddhist cuisine rather than the religious-related activities. This idea contradicts the notion that materialism becomes less important (Yeoman, 2010). However, the communication and exchange programme between Japanese and Thai Buddhist pilgrims can offer a luxury experience by their definition.

For instance, on the supply side, the luxury experience should include considering the destination as one of the luxury tourism products, such as the landscape, gastronomy or sacred places. Thus, branding the destination is a crucial element in terms of luxury tourism. However, this notion was argued against by other scholars, such as Swarbrooke (2018), who excluded the destination from the concept of the tourism product. The concept of grey areas in luxury tourism proposes the difference because the branded destination can be a pull factor in luxury travel consumption, such as UNESCO World Heritage Sites.

However, this conceptual model has a limit in terms of explaining geographic factors, such as nationality and region; for example, middle-class Chinese or Asian tourists' behaviours differ from those of tourists from Europe or the Americas. Furthermore, there have been some curious concerns about Plog's tourist typology application that cannot be used to predict the selection of the destination. Destinations can target different tourist segments via other factors, such as physical attractiveness or the political and social environments of the destinations themselves (Piuchan, 2018). However, this concept can broadly explain the luxury tourism trend shifting from a material to an experience-based focus.

Conclusion

This study represents an effort to conceptualise the shades of grey in luxury tourism through the lens of demographic and psychographic factors of middle-class tourists by providing a broad understanding of changes in luxury travel consumption by examining the luxury experience programme offered in Japanese temple lodging in Koyasan. As one of the tourism industry's segments, such experience-based accommodation can create a luxury experience for tourists within three shades of grey – dark grey, grey and light grey – depending on the intensity of their interests in material concerns or learning experiences and their personal backgrounds. Indeed, marketing luxury tourism to different tourist types and segments should be based on an understanding of the grey areas within the luxury tourism spectrum.

Case Study 3.1: The Monastery Stays in Spain: A Dual Pilgrimage Route with Koyasan

In Northern Spain, the sacred landscape of monasteries is characterised by wineries, as monasteries play a crucial role in preserving the culture of gastronomy and wine. Various monasteries have become tourist attractions and offer tourists the opportunity to stay, taste food and wine and visit the vineyards in the luxurious landscape of religious and culinary attractions. Numerous monasteries have been incorporated into wine routes (Aulet, Mundet, & Vidal, 2017), and there are some monasteries branded UNESCO World Heritage Sites.

Monasteries along the original pilgrimage routes to the Camino Santiago de Compostela are example sites that offer a worthwhile stopover for pilgrims and tourists who desire to experience the luxury arts of a historical religious building, gastronomy and wine. The way of St. James (Camino de Santiago) has been promoted as a 'Dual UNESCO World Heritage Pilgrimage Site' with the Japanese pilgrimage route in Kumano Kodo and Koyasan.

Hostería San Millán (n.d.), one of the luxurious monastery destinations, gives travellers the chance to sleep in the remarkable, thousand-year-old Monastery de Yuso, a UNESCO World Heritage Site within a mountain valley in La Rioja, a famous wine region of Spain. The ancient fortifications hide a surprisingly modern luxury hotel that takes up a wing of the monastery with 25 rooms. The monastery offers three experience programmes during the stay: gastronomy, wine and cultural tourism. Guests can experience the food and drinks provided with a two-night stay in a double room with a special dinner of Rioja cuisine in both a traditional and a modern-style fusion dish. Wine tourism arranges the experience of professional wine tasting in the old kitchen of the Monastery de Yuso and a visit to a winery. Cultural tourism focusses on a visit to the Monastery de Yuso and the Abbey of San Salvador de Canas as an authentic experience of the 'Heritage of Humanity'. Targeted customers are mainly middle class to elite allocentric tourists due to the unique location and the atmosphere of wineries and rural Spain. This target supports the idea presented in the spectrum of luxury tourism that allocentrics seek to experience 'off-the-beaten-track destinations'.

Walking along the Camino pilgrimage trail from La Rioja to Galicia in northwestern Spain, pilgrims will reach Santiago de Compostela, the destination city for pilgrimages. This location, close to the cathedral of Santiago de Compostela, where St. James is buried, is also the location of the Parador Santiago de Compostela, probably the oldest hotel in the world. The hotel was originally a Royal Hospital and dates back to 1499. It provides an unforgettable lodging experience for pilgrims who want to reward themselves after or before a long walk, as the following comments suggest.

> We arrived at The Parador after walking the Camino for 9 days… it was our reward. The super comfortable beds, wonderful linens and huge bathtub were the best that we could wish for.
>
> We stayed at this hotel before doing the Camino (on the premise that if we died on the hike, at least we would have experienced the hotel.) It was outstanding. The service was impeccable, and our room was gorgeous. We had dinner at the restaurant, and it was magnificent…Truly a once in a lifetime experience.
>
> Congratulations to the Parador Group on managing to upgrade the rooms and restaurants to high modern standards while keeping the authentic essence of this remarkable historic building. Well worth the extra cost of staying here.
>
> (Parador, n.d.)

The Parador Group is a hotel chain conceived in 1910 by the Spanish government to improve Spain's image abroad and preserve the country's heritage. The chain mostly consists of redeveloped monasteries and castles. Apart from accommodations, the hotels offer other tourism products, such as visits to monasteries and wine routes (Aulet et al., 2017). They offer special promotions for young guests between 18 and 30 years old; these guests can enjoy a 15% discount off the official accommodation rate and up to a 15% discount off the breakfast menu rate. This promotion belongs to the campaign 'Young Persons' Gateway Promotion 2020' (Parador, n.d.), aiming to draw the attention of not only pilgrims but also young psychocentric tourists. This promotion indicates that young travellers (mainly referring to budget tourists) need sales and bargains, showing both sides of psychocentrics and allocentrics in the proposed luxury tourism spectrum.

The luxury experience of staying at the Parador attracts not only psychocentrics, who prefer visiting famous places, such as Santiago de Compostela, but also targets allocentrics (pilgrims), who usually travel to unknown locations. In this case, however, because UNESCO brands the destination, it also motivates allocentric to experience what Kim (2017) referred to as the new era of luxury tourism, which requires branding. This notion has been marked in the spectrum of luxury tourism proposed by this study that allocentric tend to visit both 'off-the-beaten-track' and branded destinations.

Acknowledgements

This research is deeply indebted to all priests and staff of Yochi-in Temple. Special thanks go to all participants of the tour programme mentioned in this paper.

References

Aulet, S., Mundet, L., & Vidal, D. (2017). Monasteries and tourism: Interpreting sacred landscape through gastronomy. *Brazilian Journal of Tourism Research, 11*(1), 175–196. doi:10.7784/rbtru.v11i1.1221

Baker, M. (2005). *Luxury and Tailor-made holidays, travel and tourism analyst*. Chicago, IL: Mintel International Group Ltd.

Batat, W. (2019). *The new luxury experience: Creating the ultimate customer experience*. Cham: Springer.

Cang, V. (2015). Unmaking Japanese food: Washoku and intangible heritage designation. *Food Studies: An Interdisciplinary Journal*, *5*(3), 49–58.

Carter, J. (2015, December 3). Five reasons to visit Koya-san, Japan. *South China Morning Post*. Retrieved from http://www.scmp.com/lifestyle/travel-leisure/article/1886349/five-reasons-visit-koya-san-japan. Accessed on April 12, 2020.

Cruz-Milán, O. (2017). Plog's model of typologies of tourists. In L. L. Lowry (Ed.), *The SAGE international encyclopedia of travel and tourism* (pp. 954–956). Thousand Oaks, CA: SAGE.

Demir, C., & Saribas, O. (2014). Luxury tourism in Turkey. *International Journal of Contemporary Economics and Administrative Sciences*, *4*(1–2), 1–20.

Dimock, M. (2019, January 17). Defining generations: Where Millennials end and Generation Z begins. Fact Tank. Retrieved from https://www.pewresearch.org/fact-tank/2019/01/17/where-millennials-end-and-generation-z-begins/. Accessed on July 15, 2020.

Dixit, S. K. (2020). *The Routledge handbook of tourism experience management and marketing*. London: Routledge.

Dykins, R. (2020). Shaping the future of luxury travel: Future traveler tribes 2030. Amadeus. Retrieved from https://amadeus.com/documents/en/travel-industry/report/shaping-the-future-of-luxury-travel-future-traveller-tribes-2030.pdf. Accessed on July 20, 2020.

Harada, M. (2016, July 5). Where monks make your futon. Trip out. Retrieved from http://tripout.jp/stay/the-shukubo-experience. Accessed on July 15, 2020.

Harkison, T., Hemmington, N., & Hyde, K. F. (2017). Luxury accommodation-significantly different or just more expensive? *Journal of Revenue and Pricing Management*, *17*(4), 231–243.

Horwath H. T. L. (2011, June). The future of luxury travel: A global trends report first findings for ILTM Asia. *International Luxury Travel Market*. Retrieved from https://invattur.softvt.com/ficheros/noticias/102093840Luxury_Travel_Trends_Report.pdf. Accessed on July 20, 2020.

Hostería San Millán. (n.d.). Hotel with charm in La Rioja. Hostería San Millán. Retrieved from https://hosteriasanmillan.com/en/. Accessed on July 20, 2020.

Kametani, Y. (2014). Conservation and possible relocation for the traditional town of Koyasan, based on the surveys of residents and visitors. *Journal of Asian Architecture and Building Engineering*, *13*(3), 571–578.

Kang, Y. J., & Park, S. Y. (2016). The perfection of the narcissistic self: A qualitative study on luxury consumption and customer equity. *Journal of Business Research*, *69*(9), 3813–3819.

Kim, S. S. (2017). Authenticity, brand culture, and templestay in the digital era: The ambivalence and in-betweenness of Korean Buddhism. *Journal of Korean Religious*, *8*(2), 117–146.

Knutson, B. J., Beck, J. A., Kim, S., & Cha, J. (2006). Identifying the dimensions of the experience construct. *Journal of Hospitality & Leisure Marketing*, *15*(3), 31–47. doi:10.1300/J150v15n03_03

Koyasan Shukubo Association. (2011). Shukubo temple lodging. Koyasan Shukubo Association. Retrieved from http://eng.shukubo.net/temple-lodging.html. Accessed on October 16, 2018.

Lux Life. (2019). Luxury travel: How rich people travel the world. *Lux Life*. Retrieved from https://www.lux-review.com/luxury-travel-how-rich-people-travel-the-world/. Accessed on August 3, 2020.

McKone, D., Lewis, A., & Smith-George, N. (2017). Selective indulgence: The changing market for premium travel. *Executive Insights*, *19*(59), 1–4.

Mtapuri, O. (2017). A conceptual coalescence: Towards luxury community-based tourism. *African Journal of Hospitality, Tourism and Leisure*, *6*(3), 1–14.

Nicoloff, P. (2008). *Sacred Koyasan: A pilgrimage to the mountain temple of saint Kobo Daishi and the great sun Buddha*. New York, NY: State University of New York Press.

Osgood, C. E., Suci, G. J., & Tannenbaum, P. H. (1957). *The measurement of meaning*. Urbana, IL: University of Illinois Press.

Parador. (n.d.). Parador Santiago de Compostela, probably the oldest hotel in the world, offers 5 stars deluxe accommodation. Parador Santiago de Compostela. Retrieved from https://www.paradores-spain.com/spain/pscompostela.html. Accessed on August 3, 2020.

Park, K., Reisinger, Y., & Noh, E. (2010). Luxury shopping in tourism. *International Journal of Tourism Research*, *12*(2), 164–178. doi:10.1002/jtr.744

Piuchan, M. (2018). Plog's and Butler's models: A critical review of psychographic tourist typology and the tourist area life cycle. *Turizam*, *22*(3), 95–106.

Plog, S. (1974). Why destination areas rise and fall in popularity. *Cornell Hotel and Restaurant Administration Quarterly*, *14*, 54–58.

Popescu, I. V., & Olteanu, V. (2014). Luxury tourism characteristics and trends of the behaviour of purchase. *SEA-Practical Application of Science*, *2*(4), 319–324.

Ross, S. L. (2019). Temple stay as transformative travel: An experience of the Buddhist Temple Stay program in Korea. *Journal of Tourism Insights*, *9*(1). Article 3. doi:10.9707/2328-0824.1090

Stankova, M., & Kaleichev, S. (2013). The cutting edge between the mass and luxury tourism. *Romanian Economic and Business Review*, *8*(4.1), 50–55.

Sukmawati, N., Ernawati, N., & Nadra, N. (2018). Luxury tourism: A perspective of facilities and amenities. *International Journal of Applied Sciences in Tourism and Events*, *2*(1), 32–37.

Sundbo, J., & Dixit, S. K. (2020). Conceptualisations in tourism experience. In S. K. Dixit (Ed.), *The Routledge handbook of tourism experience management and marketing* (pp. 15–26). London: Routledge.

Swarbrooke, J. (2018). *The meaning of luxury in tourism, hospitality and events*. Oxford: Goodfellow.

Tutek, E., Gebbie, M., Chan, K. Y., & Durand, S. (n.d.). Tourism megatrends: 10 things you need to know about the future of tourism. Horwath HTL. Retrieved from http://corporate.cms-horwathhtl.com/wp-content/uploads/sites/2/2015/12/Tourism-Mega-Trends4.pdf. Accessed on August 12, 2020.

UNWTO. (2016). *Case studies of traditional cultural accommodations in the Republic of Korea, Japan and China*. Madrid: UNWTO.

Yeoman, I. (2010). The changing behaviors of luxury consumption. *Journal of Revenue and Pricing Management*, *10*(1), 47–50. doi:10.1057/rpm.2010.43

Yochi-in. (n.d.). Shukubo. Yochi-in. Retrieved from http://yochiin.com. Accessed on July 20, 2018.

Chapter 4

Luxury Tourism in Vietnam: A Political Economy Analysis

K Thirumaran[1] *and DamXuanMinh*[2]

Abstract

Luxury and communism are perceived to be ideologically incompatible values. Prior to the end of the Cold War, luxury tourism and socialist economies had an accommodating relationship and were rationalised at the national level for foreign exchange revenues, showcasing national development in some instances and finally promoting leisure and political education. However, contemporary scholarship on political ideology and luxury tourism is limited given changes in the political and tourism domain over the last two decades. The start of luxury travel and increasing private promotion in Vietnam comes at a time where the communist regime is hurriedly developing the economy at a breakneck speed. Key cities like Hanoi and Ho Chi Minh City have become the forefront of luxury travel for Vietnam. Luxury malls, fine dining and five-star hotel resorts have sprung over the years in tandem with Doi Moi (economic restructuring). Tourism literature has seldom ventured into regime politics and luxury tourism. Using a historical approach, the chapter traces the preamble of luxury tourism in the age of Vietnam's modernity. A critical assessment of contemporary luxury tourism in Vietnam reveals a favourable relationship with socialist principles.

Keywords: Luxury tourism; communism; Vietnam; tourism development; economic restructuring; hospitality

[1]https://orcid.org/0000-0002-8399-1545
[2]https://orcid.org/0000-0002-3754-9123

The Emerald Handbook of Luxury Management for Hospitality and Tourism, 83–100

doi:10.1108/978-1-83982-900-020211004

Introduction

Soviet socialism perceived tourism as a right of every worker and was state organised (Tsipursky, 2014). The Western democracies promoted tourism as a private activity and left those with the wherewithal to expend on travel. Apart from improving the physical attributes of young children and adolescents, the Soviet government structures aimed to promote the Stalinist brand of communism and patriotism through leisure camps and activities (Szczerski & Wolańska, 2011; Tsipursky, 2014). For over 70 years of Soviet communism, people perceived luxury as something associated with the elite class (De Barnier & Vallete-Florence, 2013).

Since the era of Deng Xiaoping's market economic reforms in China and the fall of the Soviet Union in the 1990s, Vietnam has been sagacious to adopt economic restructuring to expand the state and private enterprises to nurture efficiency in resource implements and production to attract foreign direct investments (Asia News Monitor, 2014; Norton & Solomon, 2000; Selvaraj, Xuan Dam, DuBos, Jang, & Thirumaran, 2021). Vietnam is also one of the fastest-growing countries in economic development, with the advantage of a hardworking population, rich in cultural heritage and natural resources, and a long coastal edge of 3,260 km length. The country has made tourism a priority industry by issuing the Decree 08 NQ/TW dated 16 January 2017, which focuses on the development of attractions, infrastructure improvements and the promotion of Vietnam's rich tradition to compete with neighbouring countries in Southeast Asia.

The fact that luxury resorts and hotels are booming in Vietnam suggests the existence of wealthy travellers, both domestic and international. The local ultra-wealthy class of people holding USD 30 million and more in Vietnam is expected to rise by 170% to 540 people by 2025, the highest growth in percentage terms by worldwide comparison (Knight Frank, 2017, p. 67). Knight Frank (2017) also reports that the millionaire numbers are expected to increase to 38,600 during the same period. This implies that the commensurate development of Vietnam's capitalist style economy will produce a newly rich class of high net worth individuals patronising or desiring exclusive resorts or high-end shopping (Hutton, 2019). This chapter will expound on some of these emerging landscapes in the analysis section.

Geographically located in the arch of Indochina's eastern coast, Vietnam is the roaring dragon of Southeast Asia with often double-digit economic growth in the last decade and a half. A visit to Hanoi, the capital city located in the North and the original stronghold of the communist regime during the Vietnam War era, and observations of the massively busy Ho Chi Minh City, formerly the capitalist city and revived to similar heights, suggest that Vietnam is building two strong economic pillars that then extend outwards to the regions. The phenomenal wealth accumulation in the cities and the spillover effects in resort areas reflect a robust private–public economic relationship that augurs well for the tourism sector.

On 23 February 2016, Vietnam launched the report titled 'Vietnam 2035: Towards Prosperity, Innovation, Equity and Democracy' (World Bank, 2016). The pronouncements contained in the report imply that Vietnam wants to achieve greater people participation in the economy by 2035 as 'a state of the people, by the people and for the people', to transform Vietnam into a high-income country or upper middle-income country by 2035. The strategies are to develop the lower middle-income class. This economic project is very ambitious. It mostly intends to strengthen the local private sector. The report also said Vietnam would ensure social equity and inclusion while at the same time strengthening the state capacity and accountability. From the example of China's successful experiment with private enterprise, it is evident that in a communist country with capitalist virtues, luxury living by a certain segment of the Chinese citizenry, allocation of resources and access to the private sector creating luxury travel experiences, is a prospect for Vietnam as well.

This chapter on Vietnam's luxury tourism focuses on the intersection between socialist values and luxury gravitations in a developing economy. After a series of historical evaluation of luxury in the period of royal rules prior to 1955, pre- and post-war years (World War II and Vietnam War of the 1970s) and subsequent years into market reforms provide us with a rich framework to understand the nature of the relationship between the contemporary socialist state and luxury tourism. Occupying between these two opposing poles of regime governance and economic outcomes, the chapter attempts to record the luxury phenomenon in the tourism space.

The scholarship is scant in the area of government and economic systems vis-à-vis luxury tourism. Past studies indicate a positive relationship prior to the end of the Cold War, but contemporary research suggests that more understanding is needed in the space of inverse relationships between socialist ideals and luxury tourism. The literature section examines the relationship between communism and luxury tourism. The methodology section explains the usefulness of a periodised analysis to identify the relationship between state policies and tourism development activities in the last decade. The ensuing discussion and observations of the tourism landscape and luxury consumption contain an analysis drawn from a reasonable theoretical precept about communism and luxury tourism.

Literature Review

The idea of luxury travel is linked to personalised guest service which is 'intuitive and anticipatory' (*Travel and Tourism News*, 2008). High net worth income travellers visiting a luxury destination expect an experience with efficient service provisions and special or exclusive care. They stand for their functional value, premium quality, price, the symbolic meanings of prestige and class attached to them (Han, Nunes, & Dreze, 2010; McFerran, Aquino, & Tracy, 2014). Atwal and Williams (2009) once expressed that luxury brands have to be exposed to high net worth consumers by creating innovative methods to reflect their desires and expression. This argument may explain the need to express new lifestyles as fresh, first-hand and recognised as luxury.

Tourism in the Soviet era was for public learning and ensuring affirmation from the elites of the successes of sizeable collective farm programs, major dam projects and massive industrial manufacturing activities (Gough & Roloff, 2006). In the case of Poland during the 1970s, also known as the 'decade of luxury', tourism was limited to the few state enterprises managing resorts (Szczerski & Wolańska, 2011). New hotels and infrastructure built to facilitate the leisure class' travel desires were a 'celebration of affluence and comfort' (Szczerski & Wolańska, 2011, p. 182). Hotels with stratified price from budget to premium were necessary to accommodate the much-needed foreign exchange and, politically, a representation of modernity.

Since the reign of Kim II Sung to date in North Korea, luxury hotels such as the one in the scenic Myohyang served the tourists primarily as the country's way of collecting much-needed revenues (Cornell, 2002). Dubois and Duquesne (1993) further explained that a class of people could acquire luxury at exorbitant prices for the purpose of display and creating an impression on others. This infers that to be luxurious, people must be rich. Thus, being rich is a prerequisite of the emergence of the need for luxury consumption. These scholarships on tourism, luxury and communism inform the duality of ideology and the human desire for leisure. Political systems through policies justify luxury tourism as a transaction need in some instances and, in the case of others, imbue a sense of pride as a larger construct of economic progress. This doxa runs contrary to communism. Marx and Engels (2002) carried the teachings that individuals earn wages according to their labour and skills in a capitalist society. This unit-level labour participation is part of larger global trade relations and exchange. Globalisation, in this context, displaces traditional socioeconomic organisations replacing them with an integrated market economy resembling similar characteristics and leveraged by capitalists from a more distant developed centre (Jellissen & Gottheil, 2009). Contemporary socialist economies that have adopted market reforms are seen as 'progressive and praiseworthy' (Jellissen & Gottheil, 2009, p. 35), fuelling the luxury market beyond the time of state crafted tourism by former East European communists blocs (Althuser, 2006).

Study Approach

This research study was conducted with extensive mining of local newspapers and historical records as well as contemporary scholarly publications. This heavy dependence on secondary resource was supplemented by street walk observations.

Fig. 4.1 represents luxury tourism as part of different economic systems of production. Where the state politics and economics geared towards communism

Fig. 4.1. A Conceptual Framework of Ideological Typologies and Luxury Tourism. *Source:* Author's own synthesised presentation.

and socialism, luxury tourism or travel was part of an education and promotion that the state had achieved a degree of development and people can share in that consumption. Through this, sharing experience is not all equitably done, as Cornell (2002), Gough and Roloff (2006), and Szczerski and Wolańska (2011) have highlighted the very nature of political agenda. On the other hand, with market reforms, communist states recognise the value of economic gains as well as part of a capitalist outcome for the domestic market, such that a transformation is both social and international, as well as integral and inevitable, wealth-creating and lives uplifting. China and Vietnam are paragons of this contemporary political economy, and accessibility to luxury tourism flows with an enterprising private sector. We use the above conceptual framework to understand the transition of Vietnamese luxury tourism progression, which is strongly premised on the acceptance of a market economy and global exchanges. The subsequent sections navigate the historical terrain to discuss contemporary luxury tourism developments.

The Political Lens

Vietnam is well known as one of the five contemporary communist countries for its remarkable economic progress (Rosenberg, 2017). Like most of the Southeast Asian countries, it underwent different political systems, from imperialism, feudalism and colonialism to a political system accepted by the people and, in this case, a communist regime. The recent moves towards a market-style economy is showing heartening results in social prosperity and modernity.

Communism advocates the eradication of property ownership in the hands of individuals, especially by the bourgeoisie class (Marx & Engels, 2002). However, in contemporary Vietnam, private landholding is neither ignored nor eliminated. The tacit approval of private wealth accumulation leads us to an assumption that luxury tourism is consequently an acceptable deviation from Marxists teachings. In contrast, luxury tourism emerged from a desire to have an above-average lifestyle. Luxury tourism first manifested in the form of holiday travel to hill stations and luxury hotels during the colonial period (1889–1945) and even more remarkable after the Doi Moi (economic reform) period, when a new elite class emerged from private sector stimulation. This contrasts with the post-Vietnam War (1975–1986), when the country was managed by state planning policies.

Wealth accumulation under communism is considered similar to private sector development. During the post-Vietnam War (1975–1986), the country was too ideologically driven, which stymied economic growth. On the other hand, as an emerging tourism destination, contemporary Vietnamese communism, statistics and facts on the ground indicate three key aspects: luxury tourism, private sector development and wealth accumulation as a result of privatization (Thirumaran, 2016; Thirumaran and Raghav, 2017). The original ideology of communism by Marx (1972) and the history of the post-Vietnam War (1975–1986) may repudiate the existence of these three aspects. Yet, there are still differences that keep the original nature of communism as it is, including the one-party political system, firm control over public properties and state owned companies while saliently facilitating private enterprises. In the next section, the focus is on the historical trajectory of Vietnam's economic development that eventually led to luxury tourism in the contemporary landscape.

Life in the Pre-colonial and Post-colonial Period

History informs us that the state of Vietnamese luxury during the period of kingdoms was limited to the palace dwellings though some of the exotic décor and cutlery were obtained through maritime trade (Hall, 2013). During the royal family's rule of Vietnam, many of the aristocrats and courtiers who had the excesses and access to state resources controlled vast swathes of privileges and land. Royal visits to remote villages were seen in horse carts and leisure trips by boats along rivers. Kings wore fine embroidery when inhabiting within the palace, eating rare foods.

Table 4.1 is an elaboration of the differences and give us ideas of the emergence of luxury tourism in contemporary Vietnam. As the literature mentioned, luxury life emerged from wealth accumulation, which created the desire to be seen, heard, distinguished and isolated from others (Kiessling, Balekjian, & Oehmichen, 2009; Rambourg, 2014). Such desires also drive luxury travel demands.

During 1175 to 1955, statistics showed few evidences of luxury travel except for those of royalty and aristocrats on horse carts. Like other empires, hunting was the most popular sport as well as entertainment for royals. For entertainment, Hát Cô Đầu was the favourite among the rich because of its exclusiveness and limit to small group audiences. For fashion, although there were not many types of distinctive clothes rather than the traditional dress code of 'áo dài khăn

Table 4.1. Political Rule and Luxury Tourism.

	Imperialism (1175–1955)	**Colonialism (1884–1945)**	**Post-Vietnam War (1975–1986)**	**Period of Đổi Mới (1986–Present)**
Authorities	Aristocrats	French administrator	Communist party	Communist party
Society ruling philosophy		Exploitation	Government monopoly, no private sector	Towards prosperity, innovation, equity and democracy (Vietnam 2035 report, 2016)
Social class	Aristocrats/peasants	Rulers/locals	Party leaders/workers/peasants	Party leaders/businesses/workers/farmers
Economy	Mostly locally trading	Local exploitation	Autarky (self-sufficiency, limited international trade)	Internationally trading
Amusement provision	Small singing groups/hunting for aristocrats only	Exclusive hill stations, hotels (small numbers)	No sign	Booming luxurious resorts, exclusive services, huge entertainment complexes (Ha Long, Da Nang, Nha Trang)

Source: Author's own synthesised observation.

đóng' (traditional dress of long shirt and turban), the rich tended to go understated with fine silks and colourful fabric.

When the French took control of the country in 1884, the luxury lifestyle started to emerge during colonialism. The wealthy class of people imported brand-new cars and bikes to the streets. Hunting was replaced with swimming and tennis. They also brought 'rượu vang' (foreign wines), instead of native wines, to their feasts. They visited the Opera House to savour their entertainment appetite. The woman discovered an exotic dress and new 'bob hair' style. On top of that, they made a breakthrough by building new hill stations in Đà Lạt and Sapa (Pile, 2016).

During the French colonialism, Europeans surrounded themselves with high life by creating large government bungalows, mansions, theatres, bridges and hill stations, which were meant to serve this exclusive European administrator. French-built infrastructure, in fact, set a new level of lifestyle not only for foreigners but also for the local aristocrats.

French colonial period had been long enough to leave some influences on the Vietnamese local life. During the French rule (1884–1945), the French lived separately from the locals. In the North of Vietnam, people were unaffected by the modern trend. They still played 'Mạt Chược' or Mahjong game, an exclusive game that originated from the royals, as Bell (1976) described 'for 2000 years, it was played by royalty'. Virgitti, the Mayor of Hanoi in the 1930s, described Khâm Thiên street as 'the most expensive, indulgent and crowded street' where people came to listen to 'Hát Cô Đầu', an entertainment type involving a group of rich people who invited a famous singer and a small music band to sing at a community centre (Nguyễn, 2011, transl.). During that time, most of those entertainment places were at Khâm Thiên Street, Hanoi. However, historical statistics does not provide evidence about domestic tourism. The ruling system during French colonialism did not encourage people to go beyond their residential area without a proper purpose. Thus, we can only see a few buildings left by the French colonials at that time, including the Sofitel Legend Metropole hotel, known as the Grand Hotel Metropole Palace in 1901 (Famous Hotels, n.d.). The website for this hotel states

> a luxury venue for prestigious events and a popular rendezvous for business and leisure travellers... We have the honour to welcome such famous playwrights, actors, ambassadors and heads of state from all over the world as Charlie Chaplin, Graham Greene, Somerset Maugham (Sofitel Legend Metropole, n.d.).

Together with the Hanoi Opera House, the two buildings were then refurbished for the promotion of culture and lifestyle, which served the need to 'silently communicate luxury, exclusiveness and extravagance' for the elites (McFerran et al., 2014). The emergence of luxury could be seen from the North by the Mạt Chược game and Hát Cô Đầu entertainment.

Life in the American ruled South of Vietnam during 50s–70s was more indulgent towards luxury. According to Minh Anh (2015, transl.), the people were exposed to exotic traits including fashion, vehicles and sports. Young girls followed the booming fashion with mini shorts and 'bob' hairstyle with ties. The wealthy class were identifiable by their clothing and 'Vespa', Italian motorbikes, which changed dramatically from horse carts and 'xe kéo', a kind of transportation service which the driver moved the carriage and ran with his bare feet. For sports, it changed from hunting to tennis and swimming. People came to special places like 'Lao Động' swimming pool, a favourite place where people wear 'bikini', which was a breakthrough for locals. 'Hội Xẹc' tennis courts were a favourite place for the high class. Nightlife changed remarkably from places like a group of local singers performing local songs and dances to places like 'Nữu Ước' or New York bar, a new ritzy centre where people came to have alcoholic drinks and chill out with friends. During this period, only elites can be seen with Vespa, playing tennis, swimming and come to 'Nữu Ước' bar.

In a historical account of the mystical and luxury resort in Dalat, Jennings (2019, p. 17) describes in his introduction,

> Imagine Davos, Aspen, or Chamonix in Vietnam. Nestled high in Vietnam's rugged interior, 150 miles northeast of Ho Chi Minh City (formerly Saigon), this colonial-era mountain resort features hundreds of quant French regional villas, the golf course of Vietnam's former emperor, a grand luxury hotel, colonial-era boarding schools, pagodas, and monasteries – all set against tall pine trees and artificial lakes.

Even during the colonial period, French governors and senior administrators would seek solace from the tropical heat and stay over for a period in Dalat. While European-style chalets were built during the early colonial era as part of a hill station concept, it also was used to escape from the seasonal rise in mosquito-related diseases. The Liang-Bian Palace, as Jennings highlights, remains the iconic hotel of luxury of its times to feel to be back home in France and a make-belief escape from all things Vietnamese. History informs us that the luxurious life in Vietnam has always been there, exposing with the flow of political changes.

During the post-Vietnam War from 1975 to 1986, information was limited. At this time there was national monopoly in production and distribution (Thượng, 2014), no private wealth accumulation was allowed. As a result, luxury life during this time was unknown.

In the age of Đổi Mới, the country's remarkable economic growth has fuelled the luxury market segment as evidenced in the tourism landscape. New airlines5-star cruises are on the increase. Mega entertainment complex and luxurious resorts are mushrooming. Top fashion brands are easy to find in big cities like Hanoi or Ho Chi Minh City. Golf, an exclusive sport for the rich, is now popular. Decades-old marketisation reforms have yielded rising incomes for most Vietnamese (World Bank report, 2019).

The Successful Doi Moi Reforms

Since the Doi Moi (economic restructuring) reform in 1989, the Vietnamese government has single-mindedly focused its attention on social development and economic modernisation. Remnants of French architecture and infrastructure as in Hanoi and Ho Chi Minh City remain useful in this endeavour. Thus, Opera House, Sofitel Legend Metropole Hanoi and Majestic, Rex Hotel in Ho Chi Minh City portend changes over the decades, reflecting political regime transformations from French colonialism to Vietnamese communism.

Hanoi Opera House was erected by the French colonial administration between 1901 and 1910. In its website, the theatre was described as 'venues for classical art forms such as Opera, drama … for the upper classes of French officials and a minority of wealthy Vietnamese' (Hanoi Opera House, n.d.). Over the times, the theatre has always been standing firm with its original function as the main hall for national music performance and international epic ballet performance. To every artist, it is always a dream to perform in the Hanoi Opera House. It is more than just a theatre; it is an opportunity to participate in high culture in people's mind (VOV, 2016, my translation).

In Ho Chi Minh City, Rex Hotel originally served as a French garage. Its function changed from Information Press centre to trading centre during the post-Vietnam War. Ever since classified as a five-star hotel by Vietnam Tourism Association in 2008, the hotel is one of the top 10 hotels in Vietnam. On its website, the hotel proudly stated 'the luxurious new refurbishments have been thoughtfully designed to preserve the historic integrity of the Rex Hotel while at the same time introducing sophisticated postmodern architecture' (Rex Hotel, n.d.).

Situated in the same city is another French colonial architecture called the Hotel Majestic Saigon, set up in 1925, featuring 44 guest rooms. Throughout its 92 years of operation (1925–2017), the hotel has always been an exclusive place for the wealthy in Saigon (now Ho Chi Minh City). As introduced on its official website, the hotel now features 175 deluxe rooms and 'all the rooms have the classical French colonial design with private balconies' (Majestic Saigon, n.d.). Today, it is known as a luxury hotel of heritage, hospitality and elegance. The Grand Hotel Metropole opened in 1901 and refurbished and named as Metropole Legend Hanoi in 2018 is another example of a heritage building transformed into a heritage and luxury hotel.

A number of luxury-related lifestyle tourism spaces have emerged. Fine dining restaurants that at one time served French customers in the colonial period (nineteenth century) are now similarly service guests with a French menu. There are approximately 171 fine dining restaurants in Ho Chi Minh City (Tripadvisor, 2020a) and 137 in Hanoi (Tripadvisor, 2020b). What was posted on Tripadvisor on 4 October 2018 was, out of 15 best restaurants in Hanoi with French colonial design, there are 12 restaurants following new designs and three restaurants following French design (Tripadvisor, 2018). The same website shows that out of

15 best restaurants in Hanoi, Ly Spices Garden, Ly Club Hanoi and La Terrasse carry French design themes.

There are also hotels and resorts emerging in the tourism landscape. In Phu Quoc Island three years ago, there was no hotel. As of 30 April 2017, the island of 57,400 hectares recorded the highest number of five-star hotel rooms by 8,000, which is 1.5 times more than the number in Hanoi and 1.3 times more than in Ho Chi Minh City (Khanh, 2017). Overall, Vietnam has reached the number of 32,000 rooms in five-star hotels (Khanh, 2017). Increasingly, plains and paddy fields have been replaced by luxurious hotels and resorts. Tripadvisor website on 4 October 2018 showed the record of 581 properties when we did the random search of all hotels on the island. Interestingly in 2014, there were only 15 hotel properties on the island (CBRE, 2018). After four years, the number increased into 581, most of them are private owned, which indicates the fact that the private sector is developing fast in this field. The Phu Quoc Island, which has a number of direct flights from Asian capitals, is definitely a destination for luxury travellers (Lee, 2019).

Trang Tien Plaza in Hanoi was once a restaurant named Les Magasins Reunis in 1901 for French and rich local people. The building became a commercial centre in 1960. In 2013, it became the first high-end luxury shopping centre in Hanoi. Most customers are local, but internationals do visit and shop there too. It is now home to famous international brands such as Bvlgari, Dior, Cartier and Louis Vuitton. This building is an example that remains from the Colonial period that is now reverting its function into a high-end luxury shopping centre.

Luxury Travel Ltd. (2020) in Hanoi claimed itself as the first-ever luxury tour operator in Vietnam when it opened for business in 2004. There has been a big wave of developments in entertainment complexes, luxury resorts and hotels and luxury apartments. A number of private companies have been successful from 2015 to 2017. Examples include Vingroup, Sungroup and Jetstar Pacific. Many more companies in the private sector have been active in this wave. We highlight three companies for their prominence. These three private companies have a few things in common: engaged in key luxury projects in the tourism and hospitality industry, fully supported by the government. Typically, in Vietnam, paperwork is a huge problem and takes a long time, normally 5–10 years, especially for huge projects, but these companies completed their paper procedures within 2 years. Examples of these huge projects include Vinhomes Phú Quốc Resort with an estimated investment of 11 billion USD (Sơn, 2015, transl.) and Vinhomes Central Park (hotels and complex) with an estimated investment of 1.6 billion USD by Vingroup (Trí, 2015, transl.). Sungroup joins the development waves with Bà Nà Hills Mountain Resort with an investment of 374 million USD completed in 2018 and Sun World Hạ Long complex with 342 million USD to be completed by 2019 (Sungroup, n.d.).

Vietjet Air on 11 February 2014 announced a contract with Airbus worth 9.1 billion USD as a result of the preliminary agreement on September 2013 during the official visit of Vietnamese PM to Paris. (Vietjet Air, 2014, my translation). Another contract worth 3.58 billion USD with CFM International on 31 May 2017 was made on 31 May 2017. This contract is made during the official visit of the Vietnamese PM to the United States (Vietjet Air, 2017a). During the official visit of US President Donald Trump to Vietnam on 12 November 2017, it signed a new contract with Pratt and Whitney to purchase 10 new engines and increased the total contract between the two companies to 3.7 billion USD (Vietjet Air, 2017b). Twelve years ago, back in 2007, it was rare to see many huge projects in the tourism industry in Vietnam, but now the local government is approving many projects from private sectors, especially with a focus to develop the tourism industry. The fact that these big contracts are made only during official visits of country leaders is a sign that the government is backing private entrepreneurs whenever it comes to the country's interest in developing the tourism industry. These projects are part of the recent effort from the government in making tourism a top industry by 2020 (Decree 08 NQ/TW, 2017).

Communism, Tourism and Luxury

After the Vietnamese Communist Party assumed total power of North and South Vietnam (1975), the life of luxury or exclusivity was in the quiet. Vietnamese communism was all about the ideology of fairness, shared progress and 'work by your ability, get by your demand' as stated by Marx and Engels (2002). Before Doi Moi (economic reform) in 1986, communism in Vietnam was towards ideological dogmatism. However, the enlightened Communist Party of Vietnam realised that Soviet Union era ideology has to give way to more progressive thinking, competition and market style economy was accepted as part of Doi Moi – economic restructuring.

Zhao (2014) stated 'Nobody believed China has not only become the World's Factory, but also sent their astronauts into space... China has lifted more than 500 million people out of poverty, creating the largest economic miracles in the later 20th century'. The main driving force that some scholars may attribute China's success is not the new market mechanism or the democracy, but the better governance of the Chinese government. In his article, he argued that even though with the right policies, going in the right direction, China could not be successful without their distinctive characteristic of the people and their prodigious cultures. Vietnam shares similar characteristics with China: a population with a Confucian-Buddhist philosophy, respect to the authorities, hardwork and disciplined culture. Dickson (2003) argues that one of the reasons for China's success is that the Chinese government considered the private sector as its main production force for the country to achieve its economic goals. The emergence of private entrepreneurs in China has been so important that it is converting their economic influence into political power, which was further illustrated by Xiaotong and Keith (2017).

Since Doi Moi period (1986–present), Vietnam has promulgated a slew of marketization policies. Even though some state-owned enterprises remain, they too are progressively planned for privatisation. The Decree 08 NQ/TW (2017, transl.) aims to transform the tourism industry to be a top performing economic sector in Vietnam that would be fully supported by infrastructure, quality labour force and a rich culture. Lessons are abound from neighbouring countries like Indonesia and Thailand, with similar tourism conditions: long coastal lines with many uncovered beautiful beaches, rich traditions and cultures. The absolute contribution of tourism industry to the country's GDP (Turner & Freier, 2017) in Thailand is 36.7 billion USD, Indonesia 17.0 billion, Malaysia 14.0, Singapore 12.4 and Vietnam 9.3. Clearly, Vietnam is below the average.

According to the Decree 08 NQ/TW (2017), the country will take four immediate actions: first, recondition and refresh people's mind about the tourism industry and develop infrastructure. Second, advertise on CNN and other accredited media channels. Third, build up a good environment for community involvement; and fourth, develop human resource and increase the effectiveness of the Vietnam National Association of Tourism (VNAT) through empowerment and international tourism events. The aims of this decree are, first, to change the way people do tourism business favouring stable development. Second, this decree in fact does not directly mention about the private sector involvement, but indirectly it does by saying 'develop infrastructure'. It does not simply mean roads and highways, but the whole tourism infrastructure, including entertainment complexes and luxury resorts and hotels, which needs greater private sector involvement. Second, broaden the country's tourism presence in the world. Third, the decree is a promise from the local authorities to create favourable conditions for the private sectors, a fair environment towards market competition, which deeply involves local businesses. Lastly, is the intention to prepare the labour force for an increasing need from the tourism industry and entrust VNAT to lead the private sectors towards this tourism transformation. This last purpose is more like a legacy from the contemporary Vietnamese communism, is to open the market, but always under government oversight towards the 'socialization direction'. In general, this decree is an epic mark in the country's tourism development (VNAT, 2017).

Conclusion

The Connection between Communism and Luxury Tourism

The world's understanding of communism may vary; in the post-Cold War era, communist countries are focusing on economic development and social modernisation. Vietnam appears to follow China's development footpath to prosper and modernise its economy. The five remaining communist states, North Korea, China, Vietnam, Laos and Cuba, vary in their tourism policy. Of these, Vietnam and China openly tap the market with economic reforms resulting in the speedy growth of the luxury sector in the tourism landscape. Interestingly, when it comes to luxury tourism, despite the communism philosophy, these two traits work harmoniously.

The first observation is the difference between the communism as defined by Marx and Engels (2002) and the contemporary communism that Vietnam is following. Communism supports egalitarianism, in which no private assets are allowed, while all the citizens work as much as their ability allows and get as much as they need, the fairness ensured by the government. However, the benevolent Vietnamese communist regime has been allowing and supporting privatisation as well as private wealth accumulation. New ideas and projects are accepted as long as they foster the country's goal towards modernity and prosperity while ensuring a stable environment. Eventually, this policy not only creates a new wealthy class but also a new lifestyle, which adds value to luxury tourism in Vietnam.

The second observation relates to the similarities of the development goals and strategies to support the private sector. Communism and capitalism are two economic systems, which support the same goal: to contribute to the well-being of the people. The Vietnamese experiment with Doi Moi suggests an assumption that when wealth accumulation reaches a certain level, there will be a desire for a luxury lifestyle, hence the emergence of luxury tourism, regardless of the type of political system. The market reforms are paying off to witness social development and prosperity.

- Literature suggests that communist states in the past tended to allow luxury tourism for hard foreign revenues and as a form of showcase that the country has reached a certain degree of development and modernity.
- In the contemporary, after the collapse of Cold War and a decade of market economy experiment, luxury tourism is considered a consequence of the economic transformation even if it deviates from an egalitarian society that communism ideology attempts to construct.
- Ideology is no longer imperative in the control of the way tourism generates revenue for the state and the manner it produces luxury tourism resources and behaviour, parallel to whatever the capitalist economy produces alongside or at intersections.
- The state's political structure is the backbone of the economy, supporting growth, and finding ways to mitigate any stark contrasts of social disparity within society.

Though Marxists philosophy is in favour of 'ritual obligation' and 'social duty' in participating in leisure pursuits, an unease part of it would relate to the profitability of the businesses held in the hands of the few (Bianchi, p. 21). Globalisation and the free movement of capital inform us that capitalism can coexist with a communist state regardless of the political systems.

References

Althuser, L. (2006). Marx in his limits. In *Philosophy of the encounter* (pp. 7–162). London: Verso.

Asia News Monitor. (2014, February 5). Vietnam: Economic growth still depends on capital, resources. Retrieved from https://search-proquest-com.elibrary.jcu.edu.au/docview/1494095851?accountid=16285

Atwal, G., & Williams, A. (2009). Luxury brand marketing- the experience is everything! *Journal of Brand Management*, *16*, 338–346. doi:10.1057/bm.2008.48

Bell, R. C. (1976). *Discovering mah-jong*. Aylesbury: Shire Publications Ltd.

CBRE. (2018). *Vietnam special report*. Retrieved from http://www.cbrevietnam.com/wp-content/uploads/2014/06/Vietnam_Special-Report-Phu-Quoc_June_2014_VN.pdf

Cornell, E. (2002). *North Korea under communism: Report of an envoy to paradise*. Florence: Routledge. doi:10.4324/9780203995082

De Barnier, V., & Vallete-Florence, P. (2013). Culture and luxury: An analysis of luxury perceptions across frontiers. In K.-P. Wiedmann & N. Hennigs (Eds.), *Luxury marketing: A challenge for theory and practice* (pp. 37–56). Secaucus, NJ: Westdeutscher Verlag GmbH.

Decree 08 NQTW. (2017). Making tourism a top industry in Vietnam. Retrieved from http://thuvienphapluat.vn/van-ban/Thuong-mai/Nghi-quyet-so-08-NQ-TW-phat-trien-du-lich-tro-thanh-nganh-kinh-te-mui-nhon-2017-338542.aspx

Dickson, B. J. (2003). *Red capitalists in China: The party, private entrepreneurs, and prospects for political change*. Cambridge: Cambridge University Press.

Dubois, B., & Duquesne, P. (1993). The market for luxury goods: Income versus culture. *European Journal of Marketing*, *27*(1), 35–44. doi:10.1108/03090569310024530

Famous Hotels. (n.d.). Retrieved from https://famoushotels.org/books/metropole-hanoi

Gough, M., & Roloff, M. (2006). Radical tourism: Sergei tret'iakov at the communist lighthouse. *October*, *118*, 159–178. doi:10.1162/octo.2006.118.1.159

Hall, K. R. (2013). Revisionist study of cross-cultural commercial competition on the Vietnam coastline in the fourteenth and fifteenth centuries and its wider implications. *Journal of World History*, *24*(1), 71–105. doi:10.1353/jwh.2013.0022

Han, Y. J., Nunes, J. C., & Dreze, X. (2010). Signaling status with luxury goods: The role of brand prominence. *Journal of Marketing*, *74*(4), 15–30.

Hanoi Opera House. (n.d.). History of Hanoi Opera House. Retrieved from http://hanoioperahouse.org.vn/en/hanoi-opera-house-history

Hutton, M. (2019). As Vietnam's economy booms, more luxury buyers emerge. *Mansion Global*. Retrieved from https://www.mansionglobal.com/articles/as-viet-nams-economy-booms-more-luxury-buyers-emerge-206034

Jellissen, S. M., & Gottheil, F. M. (2009). Marx and Engels: In praise of globalization. *Contributions to Political Economy*, *28*(1), 35–46. doi:10.1093/cpe/bzp003

Jennings, E. T. (2019). *Vietnamese Dalat*. Berkeley, CA: University of California Press. doi:10.1525/9780520948440-013

Kiessling, G., Balekjian, C., & Oehmichen, A. (2009). What credit crunch? More luxury for new money: European rising stars and established markets. *Journal of Retail and Leisure Property*, *8*(1), 3–23. doi:10.1057/rlp.2008.26

Khanh, N. (2017, March 5). *Việt Nam sắpchạmmốc 100.000 phòngkháchsạn 3–5 sao*. Retrieved from http://vtv.vn/trong-nuoc/viet-nam-sap-cham-moc-100000-phong-khach-san-3-5-sao-20170305233745481.htm

Knight, F. (2017). *The wealth report: The global perspective on prime property and investment*. Retrieved from http://content.knightfrank.com/research/83/documents/en/the-wealth-report-2017-4482.pdf

Lee, G. (2019). Vietnamese Island Phu Quoc is the emerging luxury destination in Asia. *Prestige*. Retrieved from https://www.prestigeonline.com/hk/travel/vietnamese-island-phu-quoc-is-the-emerging-luxury-destination-in-asia/

Luxury Travel Ltd. (2020). Who we are? Retrieved from https://luxurytravelvietnam.com/about-us

Majestic Saigon. (n.d.). History. Retrieved from http://majesticsaigon.com/history/

Marx, K. (1972). *The Marx-Engels reader* (Vol. 4). New York, NY: Norton.

Marx, K., & Engels, F. (2002). *The Communist Manifesto*. New York, NY: Penguin.

McFerran, B., Aquino, K., & Tracy, J. L. (2014). Evidence for two facets of pride in consumption: Findings from luxury brands. *Journal of Consumer Psychology*, *24*(4), 455–471.

Minh Anh. (2015). Retrieved from http://afamily.vn/choang-ngop-voi-cac-quy-co-sai-gon-xua-mac-xa-hoa-choi-thoi-thuong-20150822105142685.chn

Nguyễn. (2011). Retrieved from http://hanoimoi.com.vn/Tin-tuc/Phong-su-Ky-su/516345/tu-ca-quan-hang-giay-den-co-dau-kham-thien

Norton, S., & Solomon, J. (2000). Future policy directions for Doi Moi in Vietnam. *Global Economic Review*, *29*(2), 117–133. doi:10.1080/12265080008449791

Pile, T. (2016). Asia's nine best hill stations, and how they've stood the test of time. Retrieved from http://www.scmp.com/magazines/post-magazine/travel/article/2052522/asias-nine-best-hill-stations-and-how-theyve-stood

Rambourg, E. (2014). *The Bling Dynasty: Why the reign of Chinese luxury shoppers has only just begun*. Singapore: John Wiley and Sons Singapore.

Rex hotel. (n.d.). Rex development. Retrieved from http://www.rexhotelvietnam.com/rex-development/

Rosenberg, M. (2017). A list of current communist countries in the world. Retrieved from https://www.thoughtco.com/communist-countries-overview-1435178

Selvaraj, R., Xuan Dam, M., DuBos, L., Jang, H., & Thirumaran, K. (2021). A political economy perspective: Attracting foreign direct investments into Sri Lanka and Vietnam. *Tropical Constrained Environments and Sustainable Adaptations: Businesses and Communities*, 151–166.

Sofitel Legend Metropole. (n.d.). Retrieved from http://www.sofitel-legend-metropole-hanoi.com/

Sungroup (n.d.). Sun world Ba Na hills. Retrieved from http://www.sungroup.com.vn/detailproject/392/en

Szczerski, A., & Wolańska, J. (2011). The decade of luxury: The People's Republic of Poland and hotels in the 1970s. *Art in Translation*, *3*(2), 179–212. doi:10.2752/175613111X12988932045715

Sơn. (2015). Vingroup chính thức ra mắt Vinpearl Phú Quốc and Villas. Retrieved from http://vneconomy.vn/dia-oc/vingroup-chinh-thuc-ra-mat-vinpearl-phu-quoc-villas-20151009112443833.htm

Thirumaran, K. (2016). Challenges in emerging research on luxury destinations. *Journal of Tourism Research & Hospitality*, *5*, 1–2.

Thirumaran, K., & Raghav, M. (2017). Luxury tourism, developing destinations: Research review and trajectories. *Asian Journal of Tourism Research*, *2*(2), 137–158.

Thượng, P. C. (2014). Life in transition period: There was a time. [Đời sống thời bao cấp (bài cuối): Nhớ lại một thời]. Retrieved from https://thethaovanhoa.vn/van-hoa-giai-tri/doi-song-thoi-bao-cap-bai-cuoi-nho-lai-mot-thoi-n20140827142645444.htm

Travel and Tourism News. (2008). Oberoi notes changes in the definition of luxury.

Tripadvisor. (2018). Restaurants in Hanoi (filtered with 2 choices: French design and fine dining). Retrieved from https://www.tripadvisor.com.vn/Restaurants-g293924-Hanoi.html. Accessed on October 4, 2018.

Tripadvisor. (2020a). Restaurants in Ho Chi Minh City. Retrieved from https://www.tripadvisor.com.vn/Restaurants-g293925-Ho_Chi_Minh_City.html. Accessed on August 14, 2020.

Tripadvisor. (2020b). Restaurants in Hanoi. Retrieved from https://www.tripadvisor.com/Restaurants-g293924-Hanoi.html. Accessed on August 14, 2020.

Tsipursky, G. (2014). "Active and conscious builders of communism": State-sponsored tourism for Soviet adolescents in the early Cold war, 1945–53. *Journal of Social History*, *48*(1), 20–46.

Turner, R., & Freier, M. (2017). Travel and tourism economic impact 2017 Thailand. Retrieved from https://www.wttc.org/-/media/files/reports/economic-impact-research/countries-2017/thailand2017.pdf

Vietjet Air. (2014). Vietjet Air officially signed on Airbus deal. [VietJetAir chính thức ký hợp đồng mua máy bay Airbus]. Retrieved from https://www.vietjetair.com/Sites/Web/vi-vn/NewsDetail/tin-tuc/1162/vietjetair-chinh-thuc-ky-hop-dong-mua-may-bay-airbus

Vietjet Air. (2017a). Vietjet signs deals worth USD 4.7 billion during Prime Minister Nguyen Xuan Phuc's official trip to US. Retrieved from https://www.vietjetair.com/Sites/Web/en-US/NewsDetail/news/2768/vietjet-signs-deals-worth-usd-47-billion-during-prime-minister-nguyen-xuan-phucs-official-trip-to-us

Vietjet Air. (2017b). Vietjet and Pratt and Whitney announce selection of PurePower® Geared Turbofan™ for aircraft order. Retrieved from https://www.vietjetair.com/Sites/Web/en-US/NewsDetail/news/2964/vietjet-and-pratt-and-whitney-announce-selection-of-purepower-geared-turbofanTM-for-aircraft-order. November 12.

VNAT. (2017). A remark in the history of Vietnamese tourism industry. Retrieved from http://vietnamtourism.gov.vn/index.php/items/22955

VOV. (2016). Nhà hát lớn, giấc mơ của sân khấu đỉnh cao. Retrieved from http://vov.vn/van-hoa-giai-tri/nha-hat-lon-ha-noi-giac-mo-cua-san-khau-dinh-cao-533199.vov

World Bank. (2016). Vietnam 2035. Retrieved from https://openknowledge.worldbank.org/bitstream/handle/10986/23724/VN2035English.pdf

World Bank. (2019). GDP per capital (current US$)-Vietnam. Retrieved from https://data.worldbank.org/indicator/NY.GDP.PCAP.CD?locations=VN. Accessed on August 14, 2020.

Xiaotong, Z., & Keith, J. (2017). From wealth to power: China's new economic statecraft. *The Washington Quarterly*, *40*(1), 185–203.

Zhao, Y. (2014). What drives China's success? Retrieved from https://www.forbes.com/sites/realspin/2014/10/02/what-drives-chinas-success/#4408582b5852

Chapter 5

Challenges and Prospects for Oman in the Making of Luxury Tourism Destination

Amitabh Mishra[1] *and Mohit Kukreti*[2]

Abstract

The overall significance of tourism's role in the nation's holistic development is now a common phenomenon world over. That is why the Government of various countries are according top-most priority towards tourism development. The luxury tourism is that niche segment which is growing intense day by day and the craving of luxury traveller is uninfluenced by any economic turmoil or the crises. Tourism is now widely acknowledged as the strategic tool for economic diversification in the Sultanate of Oman. The Government is fully geared towards turning Oman into a prime luxury tourism destination by incorporating tourism in its vision 2040 strategy. The Government is actively tapping Oman's luxury tourism potential with improved tourism product development, increased projects funding, thoughtful marketing and brand promotion to reap benefits from tourism investments. Oman's unparalleled beauty, rich historic grandeur and authentic hospitality complement the desire of luxury travellers to seek unique quality and comfort, exclusivity and less ostentation. With the slogan 'Beauty has an address', the laudable effort of Oman has placed its tourism offering as an ideal upscale and luxury destination in the Middle East. The chapter explores the prospects of beautiful attractions and various services and facilities offered by Oman to qualify as a luxury destination. It also identifies the challenges faced by Oman in luxury tourism destination development.

Keywords: Luxury; hospitality; tourism; hotels; destination; Oman

[1]https://orcid.org/0000-0002-0140-5857
[2]https://orcid.org/0000-0003-0067-0103

The Emerald Handbook of Luxury Management for Hospitality and Tourism, 101–121

doi:10.1108/978-1-83982-900-020211005

Introduction

Luxury tourism is an emerging trend in tourism and is sought after while choosing tourism destinations in the current scenario. The tourists are looking for destinations which are well equipped with all the essential facilities, little above the necessary level. They are ready to pay for relaxing in a rich and splendid destination for a few days and experience rejuvenation after a much hectic and busy schedule.

With its mesmerising natural beauty, the Sultanate of Oman is clean and tidy, beautiful beaches with golden sand, tall, craggy and wild mountains, lagoons, palm-fringed oases and extensive deserts offers one of the safest, peaceful and comforting luxury tourist destinations in the world. The endless blue skies and mild temperatures between September and April make Oman a much refreshing and relaxing destination (See Fig. 5.1). Feather on the cap is the rich heritage, legendary hospitality and generous nature of its people, unique to this part of the world. All these destinations attribute combined with gastronomic delight, personalised; high-quality hospitality services provide an ideal exclusive immersive niche tourism experience to luxury travellers.

Muscat is the capital of Oman, situated on the hottest plate of the world; it surprises anyone with its lush green gardens, roads decorated with natural and colourful flowers and trees all over the main city. Muscat has also been awarded as the cleanest city in the world. With its capital city Salalah resembling the green mountains of Wales after the rainy monsoon season, the Dhofar region is a perfect destination for nature and luxury tourism.

The chapter explores the prospects of Oman to qualify as a luxury destination due to its beautiful attractions, elegant services and facilities. The

Fig. 5.1. Bird's Eye View of Shangri-La Barr Al Jissah Resort and Spa. *Source*: Authors.

chapter also highlights the challenges faced by Oman in luxury tourism destination development.

About luxury: Luxury is purely subjective and can be defined differently by people as per their perceptions (Novotná & Kunc, 2019). As per the available studies (Armitage, Roberts, & Dhillon, 2017) (Nueno & Quelch, 1998), defining and describing luxury is not a simple task. It is very subjective and influenced by the individuals' social surroundings (Correia, Kozak, & Del Chiappa, 2020).

Luxury tourism is proliferating with increasing market share. The luxury tourism is that niche segment which is growing strong day by day and the craving of luxury traveller is uninfluenced by any economic turmoil or the crises. With the increased income and earning potentials, wealthy tourists are looking for comfortable, luxurious and quality tourism and hospitality experience (Dixit, 2020; Di Donna, 2015; Nadkarni & Heyes, 2016). Considering high earning potentials and associated benefits by developing and promoting luxury tourism, several tourist destinations are trying to portray themselves as luxury tourism destination and competing to attract tourists to themselves (Correia, Kozak, & Del Chiappa, 2020). Oman is one such destination located in the Middle East region (Brenner & Aguilar, 2002). Oman is strategically positioning itself as a luxury destination. Fig. 5.1 (Bird's eye view of Shangri-La Barr Al Jissah Resort and Spa) corroborates the same luxury destination concept.

Luxury tourism highlights a specific customer segment that is very much concerned about supply and demand (Popescu & Olteanu, 2014). According to Popescu and Olteanu, *Luxury tourism is an activity constantly practised by the people with substantial material resources, even if their number may fall or rise depending on the economic impact had periods of the crisis on areas of activity in which they operate* (2014). Oman caters to travellers looking for an authentic and genuine quality Arabian experience. Oman is witnessing an increase in the number of travellers who focus on experiential travel.

Luxury tourists are wealthy people with high earning capacity. Such people don't hesitate to pay a fair amount of money for a holiday trip. Basically, by purchasing luxury tourism products, tourists satisfy their self-esteem and prestige needs (Kirkby & Alasadi, 2016). They want to increase their social reputation and differentiate themselves from their peer groups, society and relatives in terms of status. According to Popescu and Olteanu (2014), luxury tourists are between 25 and 55 years of age in general, have either intermediate or higher levels of education, have their own business or occupy middle or top management positions in large organisations. On average, they earn between 1670 USD and 2340 USD per month. They maintain a generally very high standard of living (Brenner & Aguilar, 2002). Since such people spend most of their valuable time for the well-being and betterment of the organisation, employers, and peers and subordinates, they prefer to consume luxury tourism whenever they have a chance to enjoy it. The primary purpose behind their intention is to be relaxed, recharged and to pamper themselves (Iloranta, 2019). Luxury tourists expect to get quality treatment like a celebrity with high-quality services during their holiday experience.

Luxury travel market classification: As per 'Global Opportunity Analysis and Industry Forecast (2019–2026)', luxury travel market classifies travellers based on luxury as Aspiring Luxury, Absolute Luxury and Accessible Luxury travellers. Age-wise, these travellers are segmented as Generation X, Baby Boomers, Silver Hair and Millennial. Package/product-wise, the categorisation includes Cruise/Ship Expedition, Adventure and Safari, Small Group Journey, Culinary Travel and Shopping, Celebration and Special Events, and Customised and Private Vacations (Allied Market Research, 2019).

The concept mentioned above of luxury tourism indicates that Oman has all those characteristics that qualify for a luxury tourism destination. With the tourism growth in the past decade, Oman has developed and added beautiful attractions and various services and facilities to its offering to the premium tourists. In the following text, this chapter explores the prospects of Oman to qualify as a luxury destination by reviewing its tourism potential, efforts of the Government and private sector and the challenges faced by Oman in the luxury tourism destination development.

Luxury tourism resources and activities in Oman: The Sultanate of Oman is located on the Strait of Hormuz at the south-eastern coast of the Arabian Peninsula (Crystal, 2020) (Kharusi & Salman, 2011). Owing to its strategic location and geographical closeness to Europe, Asia and Africa, Oman has become one of the most favourable and easily approachable luxury tourist destinations worldwide. The picturesque mountains, golden sand-filled deserts and dunes (see Fig. 5.2), eye-catching sceneries, historical and cultural heritage glorify the country's beauty and bewilder all the incoming visitors. The country is linked with the United Arab Emirates, Yemen and Saudi Arabia by road. Its population is around 5 million. Oman consists of eight regions: Dhofar, A'Dakhileia, Musandam, Al Sharqiya, Al Wusta, Al Batinah, Al Dahirah and Muscut (Pai, 2019). Physically, the Sultanate has three divisions: plateau, mountain ranges and coastal plain (Encyclopedia.com). Oman has a great amalgam of climate; coasts are humid, heartland Dahirah and Dakhlia regions are mostly hot, mountains are mild and Dhofar offers a wonderful monsoon season called Khareef (Hafidh,

Fig. 5.2. Desert Safari. *Source:* Authors.

Noel, Ahmad, & Jafar, 2018) etc. The Sultanate has a rich culture and heritage (Dileep & Mathew, 2017). Some popular tourism destinations of Oman are Salalah, Muscat, Jabal Al Akhdar, Jabal Shams, Musandam, Masirah Island, Sur, Sohar and Nizwa (Pai, 2019).

Oman is known for its diverse culture and tolerant people. A safe and peaceful environment, friendly behaviour of people and possession of traditional values make tourism special in Oman (Rawabi, Kimyani, Hadrami, & Malik, 2017). The nation has good ties with neighbouring countries and political stability, which attracts tourists looking for a relaxing environment. The Sultanate has several attractions and an excellent blend of modernity and traditional commitments, which are sought out by families and young people (Usubov, 2020).

Oman provides an exceptional menu of unique and memorable experiences (THR Innovative Tourism Consultants, 2016). Due to the varied activities and experiences related to them, Oman has blossomed to be an ambiguous and captivating luxury tourism destination. It offers endless activities for all age groups with a tint of innovation, exploration and exclusiveness. The visitors feel mesmerised and are influenced to visit this land of bountiful and dazzling beauty again and again. As a part of its National Tourism Strategy, Oman is focussing on developing niche tourism segments to attract luxury travellers. These niche tourism segments include sports, sailing, fishing, adventure sports, ecotourism, heritage tourism, off-roading, caves, nature reserves, etc. The following are examples of niche tourism segments in Oman which will appeal to luxury and high-end tourists.

- *Golf tourism*
  The unique natural landscape combined with sports such as golf transforms Oman into a leading destination for golfers from Denmark, Germany and United Kingdom. Muscat Hills Golf Course, Country Club, Al Mouj Golf and Ghala Valley are the three main grass golf courses apart from several sand golf courses in Oman (Oman Observer Daily, 2019). The Government, along with Oman Golf Committee, is working to promote golf.
- *Adventure sports*
  Oman offers VIA FERRATA climbing opportunities at Wadi An Nakhur (Grand Canyon) near Jebel Shams, Snake Canyon, Wadi Bani Awf, Jebel Shams (Sun Mountain), Bandar Khyaran Reserve. Wadi Ghool in Al Sharqiyah Governate, Jebel Misht, Samhan Mountain, Jebel Akhdar, Jebel Shams offer mountain climbing experience in its rocky high and steep slopes for trekkers between October and April as trekking season.
- *Diving and underwater activities*
  Oman is also a popular diving destination in the Middle East. Dimaniyat Islands, Al Makbara Bay, Al Fahil Island, Al Khayran, Al Jissa beach, Dhofar beaches, Duqm beaches, Al Sawadi beach and Isole Al Hillaniyat are the prominent destinations for diving and underwater activities in Oman. These places also offer cruises and boating experiences. Oman is an ideal destination for extreme sports such as wakeboarding, windsurfing, paragliding, kitesurfing and kiteboarding.

- *Off-roading*
  Wadi A′ Taiyeen, the Empty Quarter, Boushar Shands, Grand Canyon, Falaj Al' Jaylah are suitable destinations in Oman for off-roading by four wheel drives.
- *Wildlife watching*
  Whale watching, dolphin watching, birdwatching and turtle watching activities are equally popular in Oman, which offers a unique and exclusive experience to the tourists.
- *Caves*
  Oman offers one of the finest destinations for exploring caves of different topography and ruggedness. Muqal cave in Wadi Bani Khalid in Al Sharqiya, Majlis Al Jinn Cave in Muscat, Al Hoota Cave, Abu Habban Cave, Jernan Cave, Al Marnif Cave, Teeq Cave and Tawi Ateer Sinkhole, Al Kittan Cave offer exclusive caving experience.
- *Nature reserves*
  Oman encourages Ecotourism through the nature reserves such as Al Jabal Al Akhdar Sanctuary, Al Saleel nature park in Al Sharqiya, Al Dimaniyat Islands nature reserve in Muscat and Al Batinah region, Dhofar lagoons and Jabal Samhan in Dhofar region, Wadi Sireen Reserve in Al Hajar mountains and Turtle Reserve Ras Al Hadd in Sharqiya.
- *Valleys*
  Many valleys with wide-ranging natural attractions are known as wadis in Oman. Wadi Bani Khalid, Wadi Al Rowdha, Wadi Al Arbiyeen, Wadi Al Taiyeen, Wadi Al Shab, etc., are popular wadi of Oman.
- *Deserts*
  Al Sharqiya Sands, Ramla Tawk, Rab Al Khali, Bousher Sands are few famous deserts of Oman which offers experience of desert safari, camping and Bedouin lifestyle and culture.
- *Geo-sights*
  For geotourists, Rock Garden in Al Duqm offers naturally sculpted art forms from rocks. 30 unique geoheritage sights are available in Oman. Few famous geo-sights are available at Al Awabi Quartzite, Al Fahal Island, As Sifah Eclogite, Bousher Marble, Ripplemarks in Wadi Amdeh, Schist bedding, Mutrah Geotrail East and West, mangrove forest, etc.
- *Culture*
  Aflaj irrigation system such as Dawoodi Falaj, Ghaili Falaj, Ayni Falaj are quite popular in Oman. Falaj Daris, Falaj Al Khatmayan, Falaj Al Jaylah appear on the World Heritage List. Traditional villages such as Baled Sayd, Misfat Al Abriyeen, Wakan, Qannah, Lima, Kumzar and Al Askharah are the important Omani villages which reflect the simple life of Omanis in the past to the high-end tourists.
- *World heritage sites*
  Oman has five United Nations Educational, Scientific and Cultural Organization (UNESCO) World Heritage Sites such as Omani Falajs, Bahla Fort and Wall, Bat Tombs and settlements and Frankincense Route frequented by the

high-end tourists. These World Heritage Sites and archaeological sites, forts, castles, museums and traditional markets/souqs offer myriads of exclusive tourist attractions to the luxury tourists. The following picture (Fig. 5.3) of UNESCO World Heritage site of Bat Tomb and Settlements near Ibri provides a picturesque scenery against the blue sky backdrop ideal for photography enthusiasts.

The above-mentioned niche tourism authentic activities help luxury travellers connect to the nature, culture and local traditions of Oman. Luxury tourists are progressively becoming interested in experience-led travel. Oman has fully recognised the potential of niche tourism segments to attract luxury travellers and offer a taste of experiential tourism. This has equipped Oman with a competitive advantage in tourism over its neighbouring countries.

Public and Private Sector Efforts towards Luxury Tourism Development in Oman

The tourism industry is a relatively young and challenging yet promising and rapidly growing sector in Oman. The country opened to tourism effectively in 1987, and since then, there has been a remarkable development of the transportation system, tour operations, heritage and cultural attractions. In Oman, a tourism strategy was prepared for the first time in 1991 and reviewed in 1995, focussing mainly on the tourism products which Oman can offer and the role of the Government in achieving this plan (Parsons International Limited, 2000). The report in 2000 also predicted that leisure tourism is on the rise in Oman and will double in the coming 10 years.

Fig. 5.3. Bat Tombs and Settlements. *Source:* Authors.

In 2009, 15,85,000 inbound tourists visited Oman, whereas, in 2018, 2.3 million visitors came to Sultanate. According to the *Gulf News*, more than 3.5 million tourists visited Oman in 2019, which is 8.14% more than that of 2018 figures (Gulf-News, 2019). According to Tradingeconomics.com, 102,000 tourists visited Oman till March 2020 (Trading-Economics, 2020). It is estimated that tourist arrivals to Oman will increase at a compound annual growth rate (CAGR) of 5% between 2018 and 2023 to 3.5 million (Oman Observer Daily, 2019).

Oman is targeting responsible luxury tourists who want to experience authenticity in tourism. The Government believes in advertising Oman as an authentic and high-touch destination, believing in their culture, traditions and values. As per Oman tourism strategy, there are mainly nine business areas which Oman seeks, namely, vacations to relax, tourism and culture, nature and adventure, special interest, short breaks and stopovers, festivals, meetings, day visitors and visitors to family and friends (THR Innovative Tourism Consultants, 2016).

The Government is fully geared towards turning Oman into a prime luxury tourism destination by incorporating tourism in its vision 2040 strategy (THR Innovative Tourism Consultants, 2016). According to 'Oman Tourism Strategy', by 2040, the tourism industry is expected to offer more than 500,000 jobs in Oman and 18,877 million OMR worth of investment. Also, 6–10% increased contribution to tourism income to the Sultanate's gross domestic product (GDP), the establishment of approximately 1,200 small- and medium-scale tourism enterprises and development of local economy, and strengthened quality of life, local pride and identity and increased confidence in future (THR Innovative Tourism Consultants, 2016). The Government is actively tapping Oman's luxury tourism potential with improved tourism product development, increased projects funding, thoughtful marketing and brand promotion to reap benefits from tourism investments.

With the help of participatory events, the Government of Oman is trying to increase sporting activities to transform Oman into a global hotspot for year-round events. The Government is organising the Muscat Marathon, Oman Desert Marathon, cycling events, century ride, equestrian events, youthful ball cup and regional cultural events. These sporting events will act as a promotional tool for tourism in Oman (Destination Oman, 2019).

The Oman Tourism Development Company (OMRAN) manages many luxury tourism projects such as the Al Hoota Cave complex, Al Wafiyah Rest Area in Al Dhakhlia region, Scientific Centre in Ras Al Jinz, Al Ashkharah Rest Area in Al Sharqiyah and Salalah Rest Area in Dhofar region. OMRAN also oversees the luxury tourism projects such as Alila Jabal Akhdar hotel, the W Hotel Muscat and the Oman Convention and Exhibition Centre (OCEC), Mina Sultan Qaboos Waterfront and Madinat Al Irfan mixed-use complexes (Walichnowski, 2020).

Many private international investors are developing tourism projects ideal for luxury tourism in Oman, such as Mina Sultan Qaboos Waterfront Project (by DAMAC Properties company), Muscat Bay Project (by United Arab Emirates

(UAE)-based Eagle Hills company), Ras Al Hadd Tourism Development Project (by Qatari Diar company), Al Mouj Project (by Majid Al Futtaim company). Orascom Company from Egypt is working on Jebel Sifah in Muscat and Hawana Salalah in Dhofar governorate.

Oman airport is responsible for the development of airports in Oman. Cruise ship at Muscat and Salalah Port are managed by Marafi, which is a state-run holding company, Aysed and Salalah Port Company. National Ferries Company (NFC) established in 2008 provides luxurious seafaring transportation linking Musandam's three ports (Khasab, Lima and Dibba) with Shinas and Muscat. It also connects Masirah Island to mainland Oman. Land access to Oman from the neighbouring nations such as UAE, Saudi Arabia and Yemen is overseeing by the Royal Oman Police.

The National Tourism Strategy entails a planned investment of approximately 51.9 billion dollars to come from the private sector for tourism infrastructure development. Pursuing the Tourism Strategy 2040, the Sultanate has attracted foreign investments to support superstructures and strengthen the tourism and hospitality industry. By 2018, a total of 412 hotels were established across all the governorates inclusive of 5-star luxury hotel to hotel apartments, green and heritage lodges, whereas the number of such hotels was 248 in 2011 (OxfordBusinessGroup, 2019).

During 2019, 20 new hotels opened in Muscat, including three 5-star, three 4-star, five 3-star, six 2-star and three 1-star hotels. Accommodation supply in Muscat is dominated by the upper mid-scale segment of the market with 32% 4-star properties, 24% 5-star properties and 14% 3-star properties. The corporate demand accounts for 57% of hospitality demand in Muscat, while leisure travellers account for 32% of the total demand. By 2019, average occupancy was around 59.7% (Chaker, 2019). The accommodation growth across the luxury segment shows that Oman is progressing satisfactorily on the luxury destination development path.

A case study of a highly reputed luxury hotel, namely, 'Al Bustan Palace – a Ritz-Carlton Hotel', is presented below to showcase the luxury hospitality experience in Oman.

Case Study 5.1: Al Bustan Palace: A Ritz-Carlton Hotel

Al Bustan Palace, a Ritz-Carlton Hotel (Fig. 5.4), situated between Al Hajar Mountains and the Sea of Oman, is standing as one of the most luxurious and proud landmarks in the Sultanate of Oman. The luxury resort is methodically structured and designed to meet all the parameters of a luxury hotel stay. The resort is a miraculous amalgamation of modernity with local traditions and art of Omani hospitality and offers utmost comfort and convenience to the visitors, besides giving them a lifetime experience.

Fig. 5.4. Next Generation Luxury Experience through Renovation – Al Bustan Palace – A Ritz-Carlton Hotel. *Photo Source:* Authors.

Al Bustan Palace, a Ritz-Carlton Hotel, has redesigned and renovated 250 guest rooms and suites to meet the needs of different global travellers. Most of the renovation works were carried out by the Royal Estates at the Royal Court Affairs in Oman and renowned luxury integrated design firm WATG and affiliated design studio Wimberly Interiors. The hotel was reopened after a makeover in late 2018. The hotel rooms now showcase a magical blend of authenticity with modern design, rich furnishings, metallic works and up-to-date technology after transformation. Enriched with images of traditional Omani silverware, artefacts such as traditional jewellery and Khanjars, the hotel rooms give a feel of elegance, luxurious and refreshing stay. All hotel rooms are equipped with an electric kettle, a minibar and a flat-screen TV. There are separate seating areas in some rooms. The Grand Ballroom and meeting rooms of the hotel have also undergone a makeover.

The resort owns a 1 km long beachfront, which is the longest in Muscat. The resort has a beautiful spa named Six Senses Spa. The hotel has updated its Ritz-kids program; it is also a wonderful kids attraction, including beach and outdoor adventures. The property has introduced a new outdoor 'Family Aqua Land' to lure high-end customers with families.

Palace Hotel offers an option to ride a zip line to adventurer loving young clients. They can also spend some time in a dedicated pool (Fig. 5.5) or slide from a large traditional Omani boat.

Fig. 5.5. Refreshing Pool and Executive See View – Al Bustan Palace – A Ritz-Carlton Hotel. *Photo Source:* Authors.

One can enjoy a range of new dining experiences on the shores of the Gulf of Oman. By operating three elegant restaurants, namely Al Khiran Terrace, Beach Pavilion and China Mood restaurant, Al Bustan Palace Hotel serves a wide range of international cuisines to in-house and walk-in guests. Guests can also relax at BLU Bar or have a uniquely flavoured tea with a modern blend at The Atrium Tea Lounge.

The Al Bustan Palace, a Ritz-Carlton Hotel, is an amazing hotel property in Muscat with the highest quality of services and highly skilled staff who always keep the guests on top priority and take care of them. It offers a memorable first-class experience to the guests from arrival to departure. *Source:* Luxury-Travel_Magazine (2018) and Al_Bustan_Palace (2021).

The tourism sector in Oman continues to see significant investments in luxury tourism explicitly aimed at international visitors and permanent residents. To accommodate the increasing growth of the tourists and high-end luxury hotels, new integrated tourism complexes (ITCs) are established with a range of facilities. Such developments include high-end hotels, marinas, shops and restaurants, entertainment venues, residential villas, townhouses and apartments. These ITCs are located in attractive coastal and mountain areas and provide vacation homes for high-end expatriates.

Luxury ITC projects such as Wave Muscat, Saraya Bandar Jissa Resort, Muscat Hills and Salalah Beach Project, Shangri-La, Barr Al Jissa ITC and Jebel Sifah Boutique Hotel, Madinat Al Irfan are popular ITC projects which will facilitate the development of luxury tourism in Oman. Demand for luxury properties for holiday homes or residence of affluent expatriates has been increasing. The ITC concepts boost foreign investment in Oman while also securing a strong and steady inflow of luxury and high-end tourists in Oman (OBG, 2020). A multi-clustered luxury ITC project is also being built in Barka town. This will include integrated theme parks, wildlife attractions, water parks, equestrian centre and edutainment centre, multiple hotels, residential zones and retail areas.

Prospects and Challenges in Developing Luxury Tourism

With its rich touristic assets and ample room for growth, many prospects promise a bright future for developing luxury tourism in Oman. The following section discusses these prospects.

The first and foremost prospect is *Omani Traditional Hospitality*. Omanis are very helpful, smiling, hospitable and welcoming by nature, and this quality makes the travel experience authentic for the high-end or luxury tourists in Oman. The aromatherapy of coffee and frankincense fragrance will accompany the luxury travellers throughout their stay in Oman.

Availability of Four Emerging Luxury Tourism Trends

A 2011 report on luxury travel trends from Horwath HTL and International Luxury Travel Market highlighted four emerging trends in luxury tourism development. These trends are more privacy and less ostentation; increased demand for open range excursions through spectacular landscapes and unspoilt nature; growing interest in off beaten track destination; destination that appealed to environmentally concerned tourist (Hospitalitynet.org, 2011).

The tranquil 'Wadis' or pools in Hajar and Dhofar mountains, a walk along a vast stretch of long beaches, sand dunes and desert, rugged mountain trails offers privacy and less ostentation.

Oman offers a great variety of attractions to satisfy the demand for open rage excursions. The Sultanate's amazing tourism potential came in the form of the 'Favourite Road Trip Destination' award at the Conde Nast Traveller Middle East Readers' Choice Awards 2018–2019, the Middle East and North Africa region's premier travel industry awards held in Dubai.

Oman is home to stunning mountains, beaches, deserts and coasts, most of which remain well off the beaten track, even for locals and expatriate residents. Through the tagline 'Beauty has an address', Oman highlights its splendid natural attractions, unique heritage elements, state-of-the-art shopping facilities and high-adrenaline adventure options as the country's major tourism products to the international visitors.

Oman is committed to sustainable and responsible tourism development benefiting the guest and host community, local culture, the environment and ecosystems. In Oman, many tourism businesses have received environment certificates granted by internationally renowned institutions in alignment with the Ministry of Environment and Climate Affairs.

Strong and Robust Planning Strategy 2040

As discussed in the previous sections, Oman has a strong and robust tourism planning strategy for 2040. This strategy will develop Oman as a top-of-mind vacation destination with a sustainable competitive advantage to compete successfully in international high-end tourism markets. The Strategic Plan has identified digital marketing and promotion as an important tool to promote Oman as a high-end tourist destination. The National Tourism Strategy targets Key Performance Indicators (KPIs) in terms of foreign visitations, increasing national tourism, foreign investment, employment, etc., for reaping benefits from tourism.

'Beauty Has an Address – Oman', Branding Campaign

In 2010, the Sultanate launched its first global branding campaign under the tagline ''Beauty has an address – Oman'; this campaign marked the beginning of investment in marketing Oman as a premium luxury destination. The prevailing message of this campaign is to exhibit how Oman sees its unique hallmark program of arts, sports and cultural events as a reflection of Oman's rich cultural heritage and natural attributes. The frankincense fragrance and Arabic coffee, dates and halwa offered by the locals clad in traditional Omani dresses make tourists amazed with the feel of Omani personality, enhancing the image and the tourism brand campaign of the country. In association with Oman Air, the branding campaign was launched in Oman's major source luxury tourist markets in the United Kingdom, Germany, France, The Netherlands, Belgium, Russia, the Gulf Cooperation Council (GCC) region, India and Australia.

Laudable Promotion Efforts

Oman's Ministry of Tourism is making laudable promotion efforts to promote and place Oman on the global tourism map by participating in major travel trade shows globally to expose the country's potential as a luxury destination. Oman participates in major international tourism exhibitions such as Internationale Tourismus-Börse (ITB) Berlin, ATM, SeaTrade, IMEX and World Travel Market (WTM) to raise awareness of what the Sultanate has to offer to tourists. It will also participate in three mobile workshops to be held in the Republic of China, Saudi Arabia and the Russian Federation. The Ministry of Tourism won the regional 'Best Usage of Social Media' award in 2017 at the Arab Travel Market Expo, Dubai.

International Luxury Awards

Oman is one of the choicest tourist destinations globally. It has several awards to its kitty, such as 'Best Tourist Destination in the Arab Region Award', 'Favourite Road Trip destination Award at the Conde Nast Traveller Middle East Readers Choice Award 2018–2019'. Oman has successfully hosted the world's toughest 'Ironman 70.3 Championship'. 'Must-See' recommendations by National Geographic and Lonely Planet coveted spot on the top travel destination list for 2019 published by the *Financial Times* and the *CNN Travel* (Nambiar, 2019).

Oman hosted 26th edition of the World Awards for Travel and Tourism in November 2019 and won 16 awards during this event. Oman's Ministry of Tourism has received the 'Our Beloved Oman' award as one of the world's leading destinations. Al Bustan Palace Hotel of Oman won the 'Best Hotel Award', 'W Hotels' group won the 'Best New Hotel Award', Anantara Hotel was awarded 'World's Leading Resort Award', 'The Chedi Muscat' bagged the 'Best Luxury Honeymoon Resort'. Kempinski is Europe's oldest luxury hotel group, and its Kempinski Hotel Muscat received the award for Oman's Leading Beach Resort during 26th edition of World Awards for Travel and Tourism. The 'Oman Conference and Exhibition Centre' was awarded 'Best New Conference and Exhibition Centre in the World' and received Leadership in Energy and Environmental Design (LEED) gold certification for sustainable infrastructure in the country by the United States Green Building Council (USGBC). 'Muscat International Airport (MIA)' and 'Salalah Airports' were awarded prestigious awards. 'Oman Air' won the 'World's Leading Airport Lounge', Business Class Best International Airline's Award in the Middle East destinations, 'Best Airline Reward Program' in the world for its Sindbad program (Times-News-Service, 2019).

Safe and Transparent Tourism Investment Environment

Oman provides one of the safest and most transparent investment environments in the GCC region. The growth of Oman's tourism sector is primarily centred on high-end hotels because foreign investors can easily find avenues to invest in the tourism sector. The boost to this sector has been indicated very firmly to the world that investing in the tourism sector for the long term is determined to achieve the objectives of the 2040 National Tourism Strategy. The Government has taken inspiration from the UAE and Saudi Arabia in the establishment of Real Estate Investment Funds (REIFs), a move aimed at stimulating growth in the tourism sector as well as the real estate market in general. REIFs typically own, operate and trade in several types of commercial real estate assets such as office buildings, apartment complexes, hospitals, warehouses, shopping malls and, most importantly for the tourism sector, hotels.

The country is open for business. Oman is providing all kinds of direct and indirect opportunities for people to invest in this country's tourism sector, as well as other sectors such as the Duqm Special Economic Zone. The Government of Oman is working hard to attract foreign investment. It has cut back on bureaucratic and other administrative hurdles and provides subsidised loans and other incentives to foreign investors.

Famed Cleanliness and Safety

Oman is one of the cleanest countries in the world. Its famed cleanliness is an ideal example that other tourism destinations must follow. The cleanliness and safe destination reputation of Oman is also contributing to the increase in European visitors' arrival. The 'World Travel Trends Report' 2016/2017 by IPK International and ITB Berlin highlighted both a rise in outbound travel from Europe and a tendency for European travellers to choose destinations perceived as safe. The visitors will be surprised by the immaculate and unspoiled state of Omani businesses, streets and public spaces.

Oman is one of the safest nations in the Middle East region. The political environment is stable and peaceful, the economy is booming and there is almost no risk of terrorism. The crime rates are meagre, and foreigners are unlikely to become victims of illegal activities. Oman is also considered very safe for female solo travellers (Menafn, 2017).

Exceptional Attraction Options for the Luxury Travellers

Oman focuses on developing an exceptional list of tourism resources and experience targeting the high-end tourist to get distinctive experience truly representing the unique cultural heritage of Oman and the very essence of Arabia. Oman provides once-in-a-lifetime experiences that will lure the luxury tourist through the attractions such as forts and castles, water sports and traditional 'Dhow' boat cruise on fjords, visit to traditional villages to discover traditional architecture, cultural lifestyle and authentic Omani food, turtle nesting and egg hatching, exploring the ancient frankincense trade route in Dhofar. According to the National Tourism Strategy (NTS), the certified destination management companies and operators can offer products of 'Signature Experience' such as Glamorous Camping or glamping in the desert with stargazing, experience of ancient and authentic Bedouin traditional way of life, milking the camel, camel racing, bull fighting, tasteful Omani cuisine experience complimented with a selection of Omani coffee and dates. The beauty of Islamic architecture can be experienced by visiting the Sultan Qaboos Grand Mosque in Muscat. The high-end cultural heritage experience can be discovered at the Royal Opera House in Muscat.

New e-Visa System

The increase in visitors to Oman is also attributed to the visa application process, which has been enhanced with the implementation of a new e-visa system that is among the safest in the world. GCC passport holders such as Kuwaitis, Bahrainis, Saudis, Qataris and Emiratis do not need a visa to visit Oman. Many other countries receive the visa upon arrival, while others have to apply ahead of time (TravelDailyNews, 2020).

On the regulatory side, the Government has added more than 103 countries to the list of those whose citizens are eligible for non-sponsored tourist e-visas. The recently simplified process of obtaining visas and permits has shown the Omani Government's enthusiasm for promoting international tourism.

Enhanced Accessibility

In Oman, the majority of international tourists arrive by air. MIA and Salalah Airport were revamped with many luxury lounges to accommodate the increasing number of arrivals.

The national carrier Oman Air was rated four-star in the Major Regional Airline category by the Airline Passenger Experience Association in October 2018; it was also voted Best Airline in the Middle East, Africa and Europe at the Seven Stars Luxury Hospitality and Lifestyle Awards.

Cruise Ship Tourism

Cruise ship tourism is on the rise in Oman and contributing significantly to making Oman a luxury tourism destination. Oman is a highly liked destination in the Middle East region by luxury tourists. According to the Ministry of Tourism, Oman, Sultanate experienced 43.6% growth in cruise ship tourists from January to December 2019. During this time, a total of 2,83,488 passengers visited Oman, whereas 1,93,467 passengers visited the Sultanate during the same time in 2018 (Tourism, n.d.). Major cruise ship ports in Oman are Muscat Port, Salalah Port and Khasab Port (Rokou, 2020).

Availability of Luxury ITCs

The high-end expatriates in Oman are now permitted to buy property in specially designated developments known as ITC. This will help in the improvement of luxury tourism in Oman once more high-end expatriates purchase properties as holiday homes in Oman. Luxury ITCs projects such as Salalah Beach ITC, Al Mouj Muscat ITC, Saraya Bandar Jissah ITC, Madinat Al Irfan etc., will undoubtedly add to Oman's high-end tourism capacity.

Availability of Training Places for Human Capital Development

The high-end luxury tourism sector is now slowly inching closer towards emotional tourism experience. The very basis of emotional tourism is to take care of tourists and establish an emotional engagement between the hotel, its employees and customers to relax and enjoy the unique experience. This can be ensured by trained human capital.

The development of Tourism Human Capital is a critical element of the Oman Tourism Strategy as it assists a strong association between investment, employment creation and community development. The Strategy sets the direction for creating new education and training programs that address luxury tourism human capital needs. To improve tourism sector employees' professional knowledge and skills, national-level government and private universities have adopted new technology and tools for tourism human capital development.

Challenges for Luxury Tourism

The prospects for Oman being developed into a luxury tourism destination are pretty positive; however, few challenges might affect the country's potential as a luxury destination. These challenges are:

Limited Contribution of Tourism to the GDP

In the ninth Five-Year Development Plan, Tourism is a key priority sector, but its estimated contribution to GDP is around 2.9 only. To contribute meaningfully to Oman's economy, tourism should account for at least 10–15% of country's GDP. This can be possible by boosting the investment in the tourism sector, foreign direct investment and public–private partnership schemes. Usufruct agreements in real estate, income tax exemptions, low land rental value per m2, transferable ownership in ITCs, visa facilitation etc., are some of the incentives government has granted to overcome this challenge.

Logistics Challenge

Logistics represents a number of new opportunities and access to the tourist market within the region. The growth in demand for special transportation services requires substantial investments in logistics services. In Oman, except for Muscat, other cities like Salalah, Sur, Sohar and Nizwa do not have direct air connectivity with the generating market. The lack of direct flights to Salalah is one of the main challenges the tourism industry faces in boosting tourist numbers. To overcome this challenge, the Government should aid in increasing flights to Salalah.

Shifting Trends in the Luxury Tourism Market

The luxury tourism market is small in size, but a high paying market and is quite dynamic. Elder, wealthy couples earlier dominated the luxury industry, but now more multi-generational families are travelling for an exclusive experience. The taste and choices are continuously changing in the globalised world, so Oman should be proactive in attracting luxury tourists. To overcome the challenges posed by shifting trends in the luxury tourism market, the service providers should add life and energy to the tourism offerings and strive to offer unexpected, distinctive, personalised, individually curated luxury tourism experiences.

Week Demand and Increasing Supply of the Room

Oman's hotel room supply is expected to grow to 80,000 rooms, including 33,373 hotel rooms, 29,287 vacation home rooms and 17,262 ITC rooms. Currently, Muscat has a 53% share of the total accommodations in the country. Overcapacity in hotel rooms will lead to weekly demand and put downward pressure on hotel performance, especially in occupancy levels. This might lead to prize war. To overcome this challenge, the Government should promote tourism in other regions so as to bring the room share down.

Lack of Research in Luxury Destination Development in Oman

Oman has great offerings in niche tourism resources that can be easily capitalised to attract high-end luxury tourism segments. However, except NCT 2040, there is not enough research in luxury tourism to help develop.

Need to Break the Typed Image of Muscat and Salalah

Muscat is the commercial capital city. It has many upscale accommodation properties and facilities for MICE tourism and business tourists. Tourism promotion majorly centres on this theme. This tendency has branded Muscat as a corporate destination rather than a leisure or holiday destination. Similarly, Salalah has MICE, business and event tourism potential, whereas it is showcased only as a leisure destination. This stereotyped image poses a challenge in attracting luxury tourist. So, there is a strong need to break the image of Muscat as a corporate destination and Salalah as a leisure destination.

Seamless Experience from Arrival to Departure

Luxury tourists demand a seamless travel experience from arrival to departure, which creates a significant challenge for the destinations. The Government can ensure a seamless travel experience by embracing the whole airport experience, from a leisure journey through the airport to a clear connection to the region's tourism offering. The Government can encourage the hotels to apply features that enhance seamless operations and tourist experience on the property and after departure.

Lack of Trained Local Manpower

The luxury tourism destinations seek to balance providing a unique experience through its people and environment while ensuring that high-quality standards of services are maintained. There is a shortage of locals trained in tourism and hospitality to work in the forthcoming luxury properties because of the lack of interest shown by Omanis in joining the hospitality sector. Non-availability of skilled Omani staff for certain positions, turnover of the local workforce, training local staff, recruitment difficulties, local staff language skills, lack of professional tour guides, etc., are the main challenges in the Omanisation process. Training and quality management are, therefore, significant challenge in the development of the luxury destination.

Water Shortage and Fragile Natural Resources

Some forecasts suggest that water availability could halve across the Middle East region within the next 50 years, which may pose a challenge for the luxury tourism growth. In anticipation of this, Oman joined the Ramsar Convention Wetlands a year ago to ensure that areas attracting significant birdlife, particularly in the south of the country, are protected. Oman needs to tread carefully on the path of sustainable ecotourism to conserve its fragile natural resources.

Conclusion

The earlier concept of luxury tourism focused on tourism activities centred primarily on top-class facilities and amenities, luxury accommodation, chartered flight, private jet and yacht, and limousine service etc. However, the changing trend in luxury tourism indicates that the present-day luxury tourist seeks unique tourism experiences, hospitality comfort, personalised quality service and exclusivity. Oman's unparalleled beauty, rich historic grandeur and authentic hospitality complement this desire of luxury travellers to seek unique quality and comfort, exclusivity and less ostentation.

There is a dearth of studies exploring the potential of the Sultanate of Oman as a luxury tourism destination, but the present chapter aims to bridge this gap by highlighting that Oman has all those characteristics which enable it to qualify for the tag of a luxury tourism destination. This chapter presents the prospects of luxury tourism in Oman and addresses the challenges for Oman in the making of luxury tourism destination.

The Sultanate of Oman is abundantly blessed with spectacular tourism attractions and a blend of fascinating ancient Arabic culture with world-renowned hospitality to offer luxury tourists. Apart from these prospects, the Government has taken unprecedented steps to promote the charm of the Sultanate as a luxury tourist destination to the high net worth individuals and luxury travellers. The tourism sector is growing in line with the Oman Tourism Strategy 2040 due to sustained campaigns, initiatives and programs by the Government.

To raise the country's profile and promote investment in the sector, the Government should keep on maintaining a key strategic partnership with the private sector to develop the tourism industry by introducing new and unique attractions, ITCs, financial incentives, enhanced visa facilities, innovative services and international promotions. The hotel industry benefits from this move and expands its prize-winning offerings, particularly in the high-end segment. All these prospects are very promising in ensuring sustainable future growth; however, there are still challenges that must be overcome if Oman is to stake and maintain a claim as a leading tourist destination. To overcome the challenges in the making of luxury tourism destination, the Government should actively adopt the measures prescribed by this chapter to address the challenges such as tourism's low GDP contribution, accessibility, image barrier, weak rooms demand and increased supply, lack of trained local manpower, water scarcity etc.

Although later starter in tourism compared to its neighbouring tourism giants such as UAE and Saudi Arabia, Oman's journey to becoming an international luxury destination is quite promising and well supported by the Government. But, now the Government's main revenue-generating oil sector is quickly shrinking, so economic diversification with the help of tourism has become essential. With the slogan 'Beauty has an address', the laudable effort of Oman has begun placing its tourism offering as an ideal upscale and luxury destination internationally. Now, Oman has to aggressively realise its full tourism potential to maximise the prospects and reduce the challenges to transform itself into a world-renowned top-of-mind luxury tourism destination.

References

Al_Bustan_Palace. (2021, February 22). Retrieved from https://www.ritzcarlton.com/en/hotels/oman/al-bustan#ROOMS%20%26%20SUITES

Allied Market Research. (2019). *Luxury travel market*. [Press release]. Retrieved from https://www.alliedmarketresearch.com/press-release/luxury-travel-market.html

Armitage, J., Roberts, J., & Dhillon, Y. S. (2017). Luxury products and services and the sustainable value chain: Six management lessons from Gucci. In M. Gardetti (Ed.), *Sustainable management of luxury* (Environmental footprints and eco-design of products and processe, pp. 259–279. Singapore: Springer. doi:10.1007/978-981-10-2917-2_12

Brenner, L., & Aguilar, A. G. (2002). Luxury tourism and regional economic development in Mexico. *Professional Geographer*, *54*(4), 500–520. doi:10.1111/0033-0124.00346

Chaker, S. (2019, October 16). *GCC hospitality industry report*. Retrieved from https://www.fusion-me.com/hospitality-industry-in-oman/

Correia, A., Kozak, M., & Del Chiappa, G. (2020). Examining the meaning of luxury in tourism: A mixed-method approach. *Current Issues in Tourism*, *23*(8), 952–970. doi:10.1080/13683500.2019.1574290

Crystal, J. A. (2020, July 24). *Encyclopædia britannica*. Retrieved from https://www.britannica.com/place/Oman

DestinationOman. (2019, September 12). Retrieved from http://www.ttgmena.com/destination-oman-7/

Dileep, M. R., & Mathew, V. (2017). Attributes of destination competitiveness: A study on tourism in Oman. *Omani Journal of Applied Sciences*, *6*(1), 98–115. ISSN 1990–5149.

Dixit, S. K. (2020). *The Routledge handbook of tourism experience management and marketing*. London: Routledge.

Di Donna, G. (2015). Luxury hospitality in Croatia 2004–2014: A case study Maistra – evolution analysis of Croatian luxury tourism with a specific attention of Maistra company part of The Adris Group. *Journal of Strategic and International Studies*, *X*(4), 22–27.

Gulf-News. (2019). Gulfnews. Retrieved from https://gulfnews.com/world/gulf/oman/oman-welcomed-more-than-35m-tourists-in-2019-1.72295462

Hafidh, A., Noel, S., Ahmad, M. R., & Jafar, J. (2018). Evaluating ecotourism challenges in Oman. In M. L. Stephenson & A. Al-Hamarneh (Eds.), *International tourism development and the Gulf Cooperation Council State* (pp. 156–168). Oxford: Taylor & Francis. doi:10.4324/9781315776576-10

Iloranta, R. (2019). Luxury tourism service provision – lessons from the industry. *Tourism Management Perspectives*, *32*, 100568. doi:10.1016/j.tmp.2019.100568

Kharusi, N. S., & Salman, A. (2011). The English transliteration of place names in Oman. *Journal of Academic and Applied Studies*, *1*(3), 1–27.

Kirkby, D., & Alasadi, R. (2016). Luxury tourism as a community economic development strategy in the depleted community of Gananoque and the 1000 Islands. *International Journal of Management and Business Research*, *6*(1), 49–63.

Limited, P. I. (2000, November). *Tourism development objectives interim report*. Muscat: Ministry of Commerce and Industry, D.G. of Tourism, Sultanate of Oman.

Luxury-Travel_Magazine. (2018, October 6). Retrieved from https://www.luxury-travelmagazine.com/news-articles/al-bustan-palace-a-ritz-carlton-hotel-reopens-in-oman

Menafn. (2017, March 10). Oman sees rewards by focusing on high-end tourism. Retrieved from https://www.menafn.com/qn_news_story_s.aspx?storyid=1095917030&title=Oman-sees-rewards-by-focusing-on-high-end-tourism

Nadkarni, S., & Heyes, A. (2016). Luxury consumption in tourism: The case of Dubai. *Research in Hospitality Management*, *6*(2), 213–218. doi:10.1080/22243534.2016.1253293

Nambiar, S. D. (2019, March 8). Choicest tourist destination. Retrieved from www.omanobserver.com

Novotná, M., & Kunc, J. (2019). Luxury tourists and their preferences: Perspectives in the Czech Republic. *Tourism*, *67*(1), 90–95.

Nueno, J. L., & Quelch, J. A. (1998). The mass marketing of luxury. *Business Horizons*, *41*(6), 61–68. doi:10.1016/s0007-6813(98)90023-4

OBG. (2020, August 23). Oman's tourism officials taking multipronged approach to increase appeal of domestic offerings. Retrieved from https://oxfordbusinessgroup.com/overview/feeling-home-officials-are-taking-multi-pronged-approach-increase-appeal-domestic-offerings

OmanObserverDaily. (2019, March 6). Retrieved from https://www.omanobserver.om/oman-tourism-arrivals-to-increase-5pc-annually-to-2023/

OxfordBusinessGroup. (2019). Oman's tourism officials taking multipronged approach to increase appeal of domestic offerings. Oxford Business Group. Retrieved from: https://oxfordbusinessgroup.com/overview/feeling-home-officials-are-taking-multi-pronged-approach-increase-appeal-domestic-offerings

Pai, S. (2019). Tourism in Oman. Retrieved from http://www.srikumar.com/oman/tourism_in_oman.htm

Parsons International Limited. (2000). *Development objectives interim report prepared for ministry of commerce, director general of tourism, sultanate of Oman*. Retrieved from https://omantourism.gov.om/wps/wcm/connect/mot/1708f5d2-a8d5-4d1d-835a-eeb4865dfcd6/devolpment.pdf

Popescu, I. V., & Olteanu, V. (2014). Luxury tourism: Characteristics and trends of the behavior of purchase. *SEA – Practical Application of Science*, *II*(04), 319–324.

Rawabi, T., Kimyani, N., Hadrami, S., & Malik, M. (2017). Residents perception of tourism impacts in a Dhakhiliyah region of sultanate of Oman. *Ottoman Journal of Tourism and Management Research*, *2*, 119–134. doi:10.26465/ojtmr.2017239498

Rokou, T. (2020, February 4). Retrieved from https://www.traveldailynews.com/post/oman-posts-436-per-cent-growth-in-cruise-ship-tourists-in-2019#:~:text=Oman's%20Ministry%20of%20Tourism%20has,the%20same%20period%20last%20year

THR Innovative Tourism Consultants. (2016). Oman tourism strategy. Retrieved from https://omantourism.gov.om/wps/wcm/connect/mot/4bd8ab5a-f376-44b0-94f0-812a31bd0b99/ENGLIGH+EXECUTIVE+SUMMARY+.pdf?MOD=AJPERES&CONVERT_TO=url&CACHEID=4bd8ab5a-f376-44b0-94f0-812a31bd0b99

Times-News-Service. (2019, November 29). World travel awards concludes in Oman. *Times of Oman*. Retrieved from https://timesofoman.com/article/2302945/oman/tourism/world-travel-awards-concludes-in-oman

Trading-Economics. (2020). Trading economics. Retrieved from https://tradingeconomics.com/

TravelDailyNews. (2020, August 25). Oman welcomed more than 3 million tourists in 2018. Retrieved from https://www.traveldailynews.com/post/oman-welcomed-more-than-3-million-tourists-in-2018

Usubov, D. (2020, June 15). Oman observer. Oman guide. Retrieved from https://www.omanobserver.om/author-of-oman-guide-sultanate-favourable-tourist-destination-for-youth/

Walichnowski, P. (2020). Interview with Mr. Peter Walichnowski, CEO of OMRAN, Oman. Retrieved from https://www.global-insight.net/interview_walichnowski.html

Part 2
Managing and Marketing Luxury Experiences

Chapter 6

Consuming Luxury Tourism – Differences in Consumption Patterns in the Czech Market

Markéta Novotná[1] *and Josef Kunc*[2]

Abstract

Luxury consumption in the Czech Republic and other post-socialist countries has a shorter history than in the developed Western countries. The historical development of these countries still reflects the differences in buying behaviour. The chapter focusses on luxury travellers' behaviour and consumption patterns in the Czech Republic. In the Czech Republic, where the concept of old luxury has prevailed, it is already possible to observe a gradual shift towards a new luxury travel model. This new luxury model is associated with a high level of comfort and privacy, exclusive location and first-class services and new elements such as travelling, authenticity and sustainability.

The chapter provides different perspectives on consumption patterns. It points to consumer behaviour in luxury travel in dependence on the selected sociodemographic variables and spatial determinants. It also reveals the luxury-driven travel motivation and preferences on a luxury holiday. The results show that Czech luxury travellers are different in comparison with the traditional European markets. The Czechs fall behind the European luxury travellers, especially regarding the financial possibilities. As far as the Czech luxury traveller profile is concerned, the category of baby boomers with tertiary education and regular income is the most dominant among Czech affluent clients. The satisfaction of the needs for relaxation and the exploration of new destinations are the primary motives for taking luxury holidays. Concerning the regional income inequality, disparities between the capital city of Prague and the rest of the country are observed.

[1]https://orcid.org/0000-0002-6839-7513
[2]https://orcid.org/0000-0002-6196-0262

The Emerald Handbook of Luxury Management for Hospitality and Tourism, 125–144

doi:10.1108/978-1-83982-900-020211006

Keywords: Luxury tourism; consumer behaviour; consumption patterns; travel experiences; differences; Czech Republic; Central Europe

Introduction

Along with the development of consumer's values, social status and lifestyle, various types and forms of tourism gradually evolving and diversifying to meet demand and consumer trends. One of these forms that have come to the foreground in recent decades is luxury tourism. The combination of luxury and travel is becoming a promising industry, as potential consumers are increasingly looking for uniqueness, valuable services and rare experiences, which logically also corresponds with higher prices (Kapferer & Laurent, 2015).

An increase in purchasing power and the wealth levels are the key factors that can drive consumers to enjoy these exclusive and authentic experiences during their holidays. The growth of luxury tourism goes hand in hand with the rising demand for unique goods and personalised experiences (Page, 2009). Firstly, the research was focussed on the consumption of luxury products. Consequently, the dimension of luxury services was included. The researchers had to deal with the intangible nature of luxury services and their perceived value (Brun & Castelli, 2013).

In the Czech conditions, these activities are still in their infancy, as the understanding of aspects of luxury by Czech consumers has only recently begun to transform into a form that corresponds to affluent clients from Western European countries (Capgemini, 2019). This fact is also pointed out in this chapter, which characterises the demand segment of luxury tourism in the Czech Republic and focusses on its consumer patterns and motivations. The chapter presents the results of an extensive survey and is based on previous findings of the authors who deal with consumer behaviour in the Czech Republic (Novotná & Kunc, 2019a, 2019b).

The chapter aims to analyse the consumption patterns in the Czech market from three main perspectives. These main perspectives are shown in three case studies which provide more in-depth insight into consumer behaviour in luxury travel. (1)They point to luxury travel consumption depending on the selected sociodemographic variables and spatial determinants of luxury travel consumption in the Czech Republic. (2) They reveal the luxury-driven travel motivation and preferences on a luxury holiday. (3) The findings are compared to the traditional European luxury markets and other post-socialist Central European countries, which are in a similar trajectory of economic and social development.

Luxury Consumption in Tourism

Luxury hospitality is one of the most critical parts of the entire luxury sector. However, 'hospitality' can be defined as an industry that includes hotels, restaurants and other accommodation and catering facilities and other attractions that form the basis for tourism and the motivation to demand it (Barrows, Powers, & Reynolds, 2012). In simple terms, products and services that go beyond basic needs and exceed their usual price, quality, aesthetics and rarity can be considered luxurious (Bernstein, 1999). These goods contribute to personal satisfaction or confirm the high social status of the buyer. The desire of the consumer to be a part of the luxury market plays a vital role in deciding whether to buy a luxury product (Heine & Phan, 2011). However, the high price level of these products and services affects the number of consumers who can afford to buy those (Silverstein, Fiske, & Butman, 2003). As a result, for most people, luxury goods appear to be more distant than necessary goods. While most ordinary consumers do not have access to such products or services at all or only to a negligible extent, materiality is a common standard for the so-called affluent clients (Hansen & Wänke, 2011). Anyone with above-standard income and financial assets in a specifically determined volume can be considered an affluent client.

Theoretical Approaches to Luxury Consumption

Especially in the past, luxury was only available to high-ranking older generations and their families (Aldrich, 1989). At present, there is certain democratisation of luxury, which refers to its easier availability (Danzinger, 2005). A consumer-oriented approach with an emphasis on experiences and feelings is shaping up here. In connection with these changes, we can talk about two dimensions of luxury, which are generally recognised by many authors, led by Aldrich (1989), Silverstein et al. (2003) or Danzinger (2005). They are supporters of the division into the concept of 'old luxury' and 'new luxury'. Both approaches are still relevant. Everyone has defined their approach, target markets, strategies and other attributes. However, it is also possible to combine and intertwine them.

The old luxury approach was more frequent in the past; however, it is still applied in the present, e.g., in emerging markets (Euromonitor Passport, 2015). The contemporary 'new luxury' approach includes goods and services of high-level quality, taste and aspiration, contrary to the other products in the category. Still, they are not so expensive as to be utterly inaccessible (Silverstein et al., 2003). Today's emphasis on extraordinary travel experiences stems from the assumption that money spent on experiences can satisfy the highest level of consumers' needs (Mahika, 2011; Novotná & Kunc, 2019a; Pangea Network, 2017).

New demand trends go hand in hand with the changes in luxury travel. The model of luxury travel includes several current trends, in which the emphasis is on the intangible nature of luxury travel. The intangible nature is related to personalised services and the provision of services adapted to clients' preferences. The requirement for the complexity of services is associated with the ever-increasing value of time. All elements of the stay, from transfer planning to returning home, should be comprehensively connected so as not to unnecessarily burden passengers who want to enjoy a carefree luxury trip according to their ideas (Horwath HTL, 2015).

Digitalisation and new technologies play an important part in the entire segment of tourism (Novotná, Grajciarová, & Polehňa, 2019). Even luxury tourism is thus influenced by the use of mobile technologies, including various applications that allow on-site orientation, purchase of air tickets, accommodation and other services (PwC, 2017). Therefore, to provide rapid operational responses to potential problems, the luxury tourism service providers should not lose sight of the use of modern technology. With the expansion of smartphones, tablets, etc., the Wi-Fi connection should be a matter of course. Nowadays, moreover, iPod docking stations (sometimes with the iPod or iPad itself for communication with reception, lighting control, etc.) are increasingly a piece of standard equipment.

In addition to the requirements for a high standard of services, affluent clients are becoming increasingly interested in sustainability and social responsibility issues (Koch, 2011). Efforts to preserve the natural beauty and the environment, support local production and the local community penetrate luxury travelling. Interest in luxury or glamorous camping, known as glamping, has been rapidly increasing (Lee, Lee, & Moon, 2019). Thanks to the growing interest in this issue, participants in luxury tourism can participate in various activities such as planting trees in deforested areas or working in agriculture with local residents. Although volunteering in luxury tourism is still a new concept, it is being mentioned especially in connection with sustainable tourism (Barbieri, Santos, & Katsube, 2012). Part of the habits of contemporary luxury clients is the trend of a healthy lifestyle that is usually pursued also during their trips and holidays. Therefore, the possibility of relaxation in spa facilities and wellness centres with professional treatments and massages and leisure centres is considered an inseparable part of luxury travel (Vystoupil, Šauer, & Bobková, 2017).

Luxury Tourism Segment

While in the past, luxury was a regular consumption of exceptional people, it can be said with exaggeration that it is now an exceptional consumption of ordinary people (Roustant, 2014). The essential precondition for the growing importance of luxury tourism is to increase comfort, society's well-being and the availability of free time. According to the latest comparable Eurostat (2020) data, the wealthiest regions of the European Union (according to GDP per capita in 2016 and 2017) were typically urban regions or metropolitan areas such as London, Luxembourg, Hamburg or Brussels. It supports the claim of the concentration of affluent clients in cities and metropolitan regions. Prague, in which there is a significant share of affluent clients and is, therefore, the subject of interest in this chapter, ranked seventh out of 281 compared regions (Prague = 187%; EU28 = 100%); the whole of the Czech Republic ranked fifteenth. This fact is by no means favourable concerning the high economic disparity between Prague and the 'rest' of the republic; on the other hand, it attributes wealth and 'luxury opportunities' to the inhabitants of the Czech capital. Prague, together with the Bratislava Region (eighth place), is thus the wealthiest locality in the whole of Central and Eastern Europe.

Affluent consumers can afford to turn their dreams into reality through tailor-made experiences. Under these experiences, you can imagine stays in luxury hotels and resorts, luxury cruises, travelling including luxury shopping, unique personalised tours tailored with authentic experiences or extreme experiences associated with trips to little-explored areas (Petrick & Durko, 2015). The specifics of experiences in luxury tourism depend on the demands of travellers who usually want to experience authentic, exclusive and personalised experiences with high added value for large sums of money during their travels, e.g., individual excursions and trips, dinners in unique places (either in famous restaurants or in solitude in nature), the opportunity to experience traditional rituals with the locals in the destination, etc. (Mahika, 2011).

The information mentioned above indicates that the term 'luxury' is more or less a subjective concept, as each consumer conceives luxury differently (Novotná & Kunc, 2019b). For young generations (e.g., Generation Y or Z), a piece of luxury can be, for example, an authentic experience. According to Luxe Digital (2020), the so-called millennials seek authenticity and shareable luxury travel experiences. They are inspired by stories and their truthfulness. However, in general, the youngest generation of the demand spectrum across European countries is less motivated by the need for luxury (Larsen, Øgaard, & Brun, 2011). Despite several advantages comparing to other generations, for example, in the form of the maximum technological literacy with which they grow up and live, they often lack pragmatic things – enough financial resources, the security of a permanent and well-paid job and their own housing (Sladek & Grabinger, 2013).

Travel habits of Generation X may be more difficult. They are considered to be one of the most highly educated generations and technologically and media savvy, but also naturally sceptical, disillusioned and pragmatic (Pitta, Eastman, & Liu, 2012). This generation was widely affected by the financial crisis that occurred between 2007 and 2009. However, more than 10 years after the crisis, Generation X women look for luxury wellness travel activities to respite from their hectic life in destinations like Bali, Thailand and Bhutan (Virtuoso, 2019).

The so-called baby boomers are considered to be an important segment interested in luxury tourism. They tend to be relatively well funded and are also healthier than former seniors (Horwath HTL, 2015). According to MarTech Series (2019), boomers are responsible for 80% of all luxury travel spending. Boomers tend to be more willing to spend extra for a luxury experience. In particular, the so-called empty nest seniors aged 55–64, who are still working, can indulge in luxury goods because they have a relatively high income, and their children are financially independent of them (Collins & Tisdell, 2002).

Seniors are also more often potential clients of travel agencies, which are currently losing their competitiveness, as many things related to holiday planning are arranged online and without the help of such an intermediary (Patterson, 2018). However, despite this desire for luxury, baby boomers still actively seek out the best-perceived deal, and their budget plays a factor in the luxury travel for 57% of them (MarTech Series, 2019).

It follows from the foregoing that luxury consumption patterns differ. There are essential differences in consumer preferences for luxury travel across sociodemographic categories, which affect their financial possibilities. The following case study is used to explore the related sociodemographic issues in more details. It analysed the Czech luxury travel market by the age structure, education, economic status and consumers' place of residence.

Case Study 6.1

The case study focusses on the sociodemographic characteristics of Czech affluent clients. Moreover, it highlights the spatial point of view on their distribution. The research itself is primarily quantitative. It is methodologically based on the questionnaire survey, which is the most common method of quantitative research. The questionnaire survey conducted among clients of the major travel agency Deluxea operating in the luxury sector in the Czech Republic provides insight into the Czech luxury tourism market segment. More than 2,500 respondents, who have used the travel agency's service, were addressed in 2017 by email. Consequently, mathematical statistical and analytical methods have been applied to evaluate the results. The information obtained was updated for the purpose of this chapter based on an interview with a travel agent in 2020.

Dozens of travel agencies operate in the Czech market; however, only a few of them can be marked as true luxury. Deluxea not only belongs to the major Czech travel agencies, but also it offers general worldwide journeys and is exclusively specialised in tailor-made luxury tours and exotic destinations. Most other luxury travel agencies specialise in a selected aspect, either in terms of geography and location (North Africa and the Middle East, South and Southeast Asia, the Caribbean, the Pacific, etc.) or thematically (golf, expeditions and exploration, congress tourism, adventure trips, etc.).

As far as the sociodemographic characteristics are concerned, the category of the so-called baby boomers is the most dominant one among the affluent clients. This post-war generation was born until 1964. It represents almost 45% of the clients. By contrast, only an insignificant quantity of people under 25 years of age is found in the researched sample. Generation Z represents only 0.5% of the sample. Apart from the breakdown by cultural generations, the most frequent clients from the sample according to the sex are men between 35 and 49 years of age (21% of all respondents) and between 50 and 59 years of age (18%).

The luxury travellers are, in particular, individuals with tertiary or at least secondary education. More than 60% of the Czech luxury travellers who took part in the survey are tertiary-educated people. As far as the economic status is concerned, people with a regular income, the employed, the employers or the self-employed represent approximately 87% of the clients. Only 13% of the clients represent people who can be characterised as specific groups of the disadvantaged. This includes, for example, students, mothers on maternity leave, the unemployed or old age pensioners, who make up 67% of this category. It is also interesting that clients most often come from cities with more than 100,000 inhabitants (51.5%). The relatively large representation of municipalities in the category of up to 50,000 inhabitants is to some extent related to the considerable fragmentation of the settlement structure of the Czech Republic, which is characterised by a large number of relatively small municipalities. This confirms the fact that the distribution of respondents covers all regions of the Czech Republic, with greater representation of the largest cities with a wealthier population. Table 6.1 summarises the mentioned sociodemographic characteristics of the respondents and depicts the Czech luxury traveller generalised profile using Microsoft Office icons.

Table 6.1. Sociodemographic Characteristics of the Respondents.

Characteristic	Market Share	Luxury Traveller Profile
Age group		
Gen Z < 25 years	0.5%	
Gen Y 26–39 years	27.5%	
Gen X 40–54 years	27.1%	
Baby boomers >55 years	44.9%	
Educational stage		
Primary education	0.5%	
Secondary education	38.0%	
Tertiary education	61.5%	
Economic status		
Regular income	86.9%	
Specific groups of disadvantaged	13.1%	
Place of residence		
More than 100,000 inhabitants	51.5%	
50,000–99,999 inhabitants	11.8%	
Less than 49,999 inhabitants	36.7%	

Source: Authors.

As far as spatial determinants of the luxury travel segment in the Czech Republic regards (Fig. 6.1), more than 32% of the respondents indicated the capital city of Prague as their place of residence. Another 14% of the respondents are from the Central Bohemian Region, which has close socioeconomic links with Prague and, to a certain extent, can be considered as the capital's metropolitan hinterland. Representation of luxury clients is not negligible even in the South Moravian Region (12.9%). Its capital, Brno, is the second largest city in the Czech Republic. The finding regarding spatial concentrations of affluent clients is fundamentally affected by the city of Prague, its metropolitan hinterland, and the city of Brno.

Luxury Tourism Motivation

When considering the issue of luxury tourism, it is crucial to examine the motivations that activate the desire for experiences and understand what drives luxury travellers' behaviour. Numerous studies in the last years were focussed on examining the motivations (e.g., Baloglu & Uysal, 1996; Dixit, 2017; Novanská, Benová, & Geghamyan, 2018; Uysal & Jurowski, 1994; Yoon & Uysal, 2005) that activate the needs for experiences and subsequently satisfy the needs by participation in tourism (Lee, Chua, & Han, 2017). Based on the analysis of the motivations in tourism, the types of tourism were created (Mahika, 2011). A

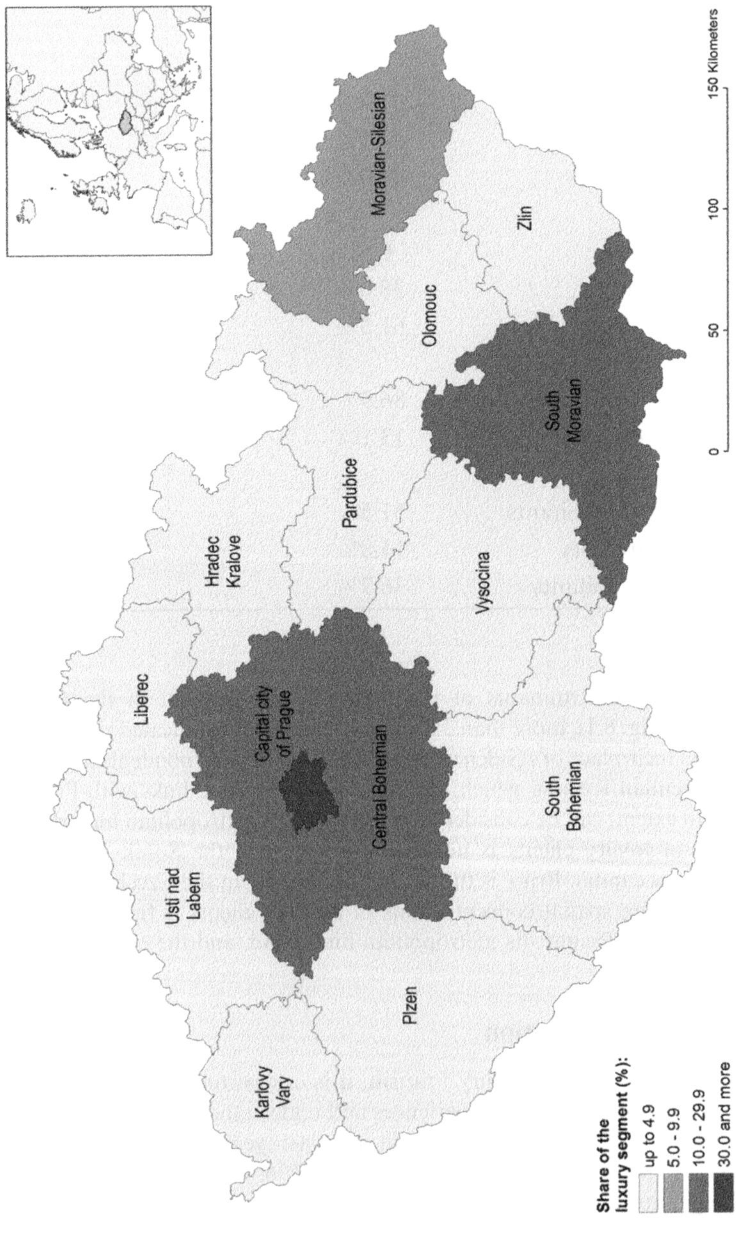

Fig. 6.1. Spatial Distribution of the Respondents in NUTS 3 Regions of the Czech Republic. *Source:* Authors.

study of demographic and socioeconomic characteristics of consumers and their psychological stimuli thus became useful not only for managers but also for businessmen who wished to comprehend the behaviour of consumers and develop successful marketing strategies (e.g., Bobková & Holešinská, 2017; Han & Hyun, 2018; Park & Yoon, 2009). The providers of tourism services had to be able to predict the changes in motivations that predetermine the tourists to buy a package holiday or select a destination. Considering this, different motivations were hugely discussed, being caused on the one hand by the real psychological motives (push factors), while on the other hand depending on specific characteristics of the destination, the so-called pull factors (Lesjak, Navrátil, Pícha, & Gilliam, 2015; Uysal & Jurowski, 1994).

The specific aspect of luxury tourism is its cross-sectional nature, spreading into other forms of tourism. Elements of luxury can thus include, for example, spa, culture and cognitive tourism, congress tourism, etc. An important unifying starting point is the motivation of individuals to satisfy their needs to form recreation, recovery, learning or organising leisure activities (Page, 2009). Hallot (2013) points out that luxury visitors and ordinary visitors have, in general, the same needs and desires. However, investigating luxury services consumption behaviour when participating in tourism activities involves additional uncertainty resulting from the lower consistency in service quality (Chen & Peng, 2018; Dixit, 2020).

Demand in luxury tourism differs from an ordinary segment, for example, by special interests and requirements. The consulting company Horwath HTL (2015) assumes the typical participant of luxury tourism to be someone who uses air transport or a ship to get to the target destination, and the purpose of their trip is predominantly to discover, relax or change the pace of life. According to Peak and Skift (2014), people nowadays want to travel under better conditions on a higher emotional and personal level. They refuse pre-prepared standardised packages with tourism products, and their travels are more authentic, adventurous, active and locally focussed. In addition to the material standard, the affluent clients begin to look for a certainly added value of an intangible nature (Koch, 2011). The intangible nature is related to personalised services from the preparation of the trip through the stay itself to the services after returning home. The basic condition is the ability to meet personal requirements and thus arrange the provision of services adapted to clients' preferences.

Case Study 6.2

The case study focusses on the preferences for a luxury holiday and the requirements for the organisation of luxury travel. In addition to this, the motivation of the Czech luxury travellers for choosing luxury holidays is investigated. The results of the survey show what factors the Czech demand segment is motivated to participate in luxury tourism and how, according to the most frequent answers of the respondents, the form of luxury holidays looks like. The data have been processed with the help of MS Excel software and SPSS statistical programme.

According to approximately 65% of answers, the Czech luxury segments prefer the personalised services during their luxury holidays. Luxury travel is, for them, also associated with a carefree stay (55%). As far as the age category is concerned, a carefree stay is a typical requirement, especially for younger representatives of luxury travellers. Luxury holidays are often tailor-made, which is confirmed by more than 48% of the clients. This aspect is preferred mainly by people over 50 years of age. At times, clients require luxury hotels in the 5-star category (26%) and absolute privacy and tactfulness (24%). At the very least, the uniqueness and attractiveness of the locality were mentioned as the requirement for luxury.

On the other hand, this factor was significant for choosing luxury travel for clients in North America (Virtuoso, 2015). Their preferences for luxury holidays are mentioned in Fig. 6.2.

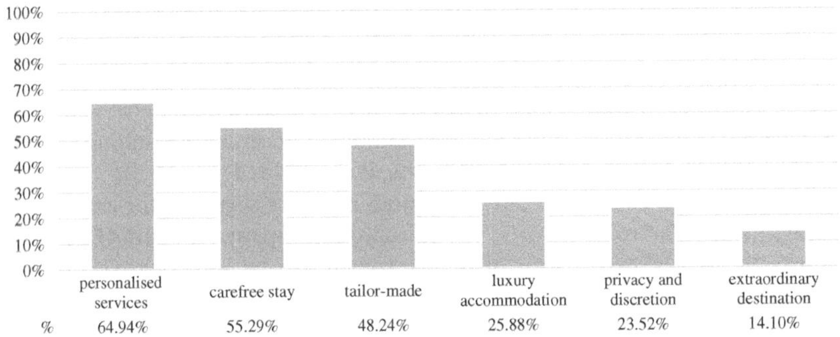

Fig. 6.2. Preferences on a Luxury Holiday. *Source:* Own processing based on Novotná and Kunc (2019a).

As the level of organisation of the luxury holidays regards, it is not surprising that the Czech affluent clients prefer all-inclusive and tailor-made luxury holidays (Fig. 6.3). Approximately 40% of them are satisfied with partly organised holidays; it means that flight and accommodation reservation is included, while the clients themselves organise other attributes of the holiday (daily programme, trips, etc.). Only 8% of the respondents prefer to organise the whole luxury travel by themselves. Standardised holiday packages without considering the individual requirements of a client are preferred by respondents from the specific groups of disadvantaged. In the case of wholly tailored luxury holidays according to the client's wishes and needs, the situation is the opposite; this level of organisation is preferred by employed people and people from Generation X and older. This is related to the fact that the more affluent the clients are, the more they tend to use tailor-made services.

For more than 90% of the participants of luxury holidays, the primary motivation for luxury travel is relaxation. The Czech luxury travellers are attracted by beaches and the sea, which is quite typical for the wealthy

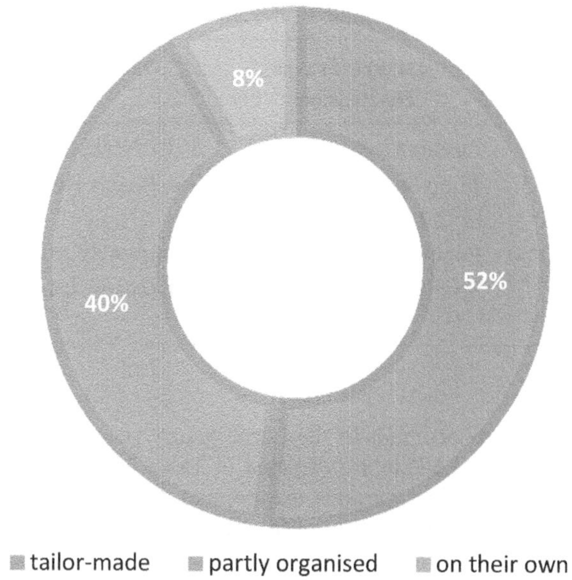

Fig. 6.3. The Level of Organisation of the Luxury Holiday. *Source:* Own processing based on Novotná and Kunc (2019a).

inhabitants of a country without the sea. Frequently, the motivation for the luxury journey emphasises the exotic places (80%) that, for many people, are determined by the distance of the chosen destination from the Czech Republic. For more than two-thirds of the respondents, the main reason is to explore the destination and its surrounding nature. This is also connected to building awareness of the destination's history and culture. For almost 35% of the respondents, entertainment (e.g., sports activities or celebration) is a crucial part of the luxury holiday. It is interesting that luxury travel, in most cases, is not connected with business trips and job-related duties.

Consumer preferences of the Czech luxury clients can also be seen in connection with specifically determined areas in the world, which are, according to the respondents, regarded to be the most luxury. The highest frequency of responses was recorded in the Indian Ocean Region. The Caribbean Sea, mentioned by about half of Czech luxury tour participants, and the Pacific Ocean Region also have a strong position. Destinations located in Southeast Asia follow with a great distance. Overall, it can be concluded that the areas of tourism, considered by Czech clients to be luxurious, are characterised by rather tropical or exotic features and a considerable distance from the Czech Republic.

Table 6.2. Travel Motivations and Preferences.

Top Motivations	**Luxury Travel Destinations**	**Main Features of Luxury Holidays**
Relaxation (94%)	Indian Ocean (77%)	8–15 Nights (82%)
Exotic (80%)	The Caribbean Sea (49%)	Sea/Beach Resort (80%)
Exploration (67%)	Pacific Ocean (38%)	Completely tailored (52%)
Entertainment (35%)	Southeast Asia (12%)	Off-season travel (41%)

Source: Authors.

In summary, consumer trends of the Czech luxury clients and the other main features of their luxury holiday are depicted in Table 6.2. The given percentage indicates how the individual variants are popular with respondents. As it was possible to record more than one answer, the percentages do not provide anywhere in the sum of the whole of 100%.

A Central European Perspective

The foreign professional literature, from Bernstein (1999), Kapferer (2008), and Han and Hyun (2018) to Iloranta (2019), has been dealing with luxury tourism to a large extent for two decades. On the contrary, the research on luxury travel and related luxury travellers' behaviour in the conditions of Central Europe has been somewhat sporadic and fragmented. From this point of view, the development of luxury tourism in post-communist Central Europe is quite different. It has a considerably shorter history than in the developed Western countries. This is due to the fact that before the transition to a market economy in the 1990s, the population of Central European countries was regulated for long decades by the centrally controlled socialistic economy.

Trade unions, state-owned companies or youth organisations sponsored numbers of holidays as part of the social development of the working class (Ivy & Copp, 1999). Economic decisions were taken by the state, which also decided on price regulation. After the transition, despite the possibility of choice, limited purchasing power led to rational buying behaviour, emphasising price as a dominant buying criterion (Czinkota & Ronkainen, 2009).

Throughout the second half of the twentieth century, the Czech Republic in Central Europe was understood as part of the so-called Eastern Bloc, and thus Eastern Europe, which has not changed during almost 30 years of the economic and social transformation after 1989. Only in 2016, Nijman, Muller, and de Blij (2016) included the Czech Republic, together with the western half of Poland, in Western Europe. After many decades, American authors have been trying to bring the Czech Republic back to its position where it belonged until 1939, in other words, to the democratic West.

Although this fact may positively affect the perception of how other countries perceive this state, the Czech Republic is far from reaching the level of the wealthy West – in terms of the level of the most robust middle class. There are, of course, the so-called 'top 10,000' (very wealthy individuals) who may afford any luxury; however, the financially stable and independent middle class is still being shaped. According to GDP per capita in purchasing power parity, in 2019, the Czech Republic was in the 27th place among the OECD countries, and of the original EU countries, it has already overtaken Greece, Portugal and Spain; however, it still lags behind neighbouring Germany and Austria by an immense 14,000 (or 16,000) USD per capita (OECD, 2020). Therefore, the purchasing power of the Czech society will have to increase even more so that their consumer behaviour concerning luxury tourism can be more emotional and, to some extent, become a standard.

These paragraphs can be concluded by recognising that historical development and common culture still reflect the differences between Central-Eastern and Western Europe in buying behaviour.

Emphasis on the Czech Republic

Despite the above-mentioned historical development, the demand segment of luxury tourism in the Czech Republic might be considered unique within Central Europe. This is also linked with the perception of luxury by the Czech consumers, which is now getting closer to the perception of luxury by the affluent clients from the traditional countries such as the United States, Japan, Germany, China, France or the United Kingdom (Capgemini, 2019). In the Czech Republic, where the concept of old luxury has so far prevailed, it is already possible, unlike other V4 countries, to observe a gradual shift towards a new luxury travel model, which is typical for the countries mentioned above. A classical luxury travel model, which is traditionally associated with a high level of comfort and privacy, exclusive location, first-class services and high social status, is beginning to be enriched with new elements such as adventure travelling, authenticity and sustainability. According to the Czech marketing agency Ogilvy (2014), only 3% of the Czech luxury segment connects luxury with a holiday.

On the contrary, they understand luxury as watches, jewellery, cars, and fashion or housing. Luxury is still a symbol of status and wealth in the Czech Republic, but the number of those who buy luxury items for their pleasure, not to impress others, is also growing. This can also be seen as a shift of the Czech luxury consumer towards the Western perception of luxury.

From a historical point of view, the beginnings of luxury tourism can be combined with spa tourism, which, in the nineteenth century, became a popular part of the life of wealthy clients, especially in western Bohemia. Later, luxury travelling was negatively affected by world wars, but the interwar period represented a development in the field of tourism. At that time, the ČEDOK (Czechoslovak Transport Office) company was founded in 1920, which began to provide Czech clients with the opportunity to travel to, in those days, genuinely exotic destinations (North Africa, Orient, Scandinavia, etc.) and to use new forms of transport (buses and planes). Interestingly, the profound difference in prices between luxury and ordinary tours outside the territory of the Czech Republic was not as noticeable as it is today. After World War II, the borders of former Czechoslovakia closed to the West. In 1948, hotels were nationalised, and exclusive travels could only be afforded by the real elite group, especially high-ranking individuals who sympathised with the then-communist regime. The possibilities for travelling abroad were considerably limited. Only trips to the so-called Eastern Bloc states were allowed, and the vast majority of visitors to Czech hotels also came from these states. After the fall of the socialist political regime in 1989, the borders opened, and inbound and outbound tourism began to revive significantly; however, it still copes with the adverse effects of the previous regime (Hajdáková, 2013).

The progress of the Czech luxury demand corresponds to the growth trend set across the world. The Czech State Agency for Tourism (CzechTourism, 2012) spoke about solvent clients with assets exceeding USD 100,000. The number of such wealthy people in the Czech Republic was estimated to be 50,000–70,000. However, these data must be taken with a pinch of salt as the affluent and solvent clients represent only a certain minimum basis for luxury tourism. By Investopedia server (2018), a better validity regarding the potential luxury clients presents the so-called concept of HNWI (high-net-worth individuals) that includes individuals with liquid assets exceeding a specific limit, while the most frequent limit is USD one million. When applying the concept of HNWI, the number of US dollar millionaires in the Czech Republic amounts to approximately 33,000 (Credit Suisse, 2017).

Compared to Poland, where 0.1% of the population was considered as belonging to the HNWI category in 2016, in the Czech Republic, it was 0.3% of the population (Credit Suisse, 2017). The value of the assets of an average inhabitant of Poland (USD 19,800) was comparable to other post-communist countries such as Slovakia and Hungary or Croatia. These countries remained behind the Czech Republic, with mean assets of USD 38,900 (KPMG, 2016).

Case Study 6.3

This case study presents the main differences between the Czech luxury market and the Western countries. It critically examines the unique dimensions of luxury from the perspective of the Czech luxury segment. The research is methodologically based on the questionnaire survey presented in the previous sections. A compilation of selected international studies and literature review on luxury travellers' behaviour research creates a conceptual framework for data evaluation and interpretation.

As shown in Table 6.3, a typical Czech participant of luxury tourism is considerably different from the Western one. The main difference can be observed in financial possibilities. On average, the Czech affluent clients are usually willing to pay less money for their luxury holidays. The annual frequency of luxury holidays is also lower. In these aspects, the Czech tourists are falling behind the European luxury tourists. The Czech luxury travel segment also lags behind the Western one in the case of travel class preference. The research showed that Czech travellers participating in luxury travelling use economy class for almost 90% of their journeys, which may also be related to lower purchasing power than Western countries.

On the other hand, the situation is similar when regarding the organisation of the holiday (the so-called 'tailor-made' packages prevail). In both segments, the Internet is also gaining prominence as a tool for finding information before the trip and sharing experiences after the trip. Apparently, digital technologies are beginning to enter the whole value chain for both luxury clients' groups. However, personal recommendations obtained from friends and experts in the travel agency are still important information for the Czech luxury traveller.

It has already been mentioned that tourism participants from developed Western countries are characterised by a new luxury travel model, which contains new elements such as experiences, adventure and authenticity. The emphasis on exoticism, relaxation, building awareness, etc., is known among Czech luxury clients. There has been a growing inclination among consumers to prioritise travel experiences over materialistic possessions, particularly among younger generations, which has made luxury travel an important choice among elite consumers. According to a 2016 Airbnb report, millennials who like to travel not only prioritise these trips over other money-oriented assets but also save up for it. Remote and unexplored destinations that allow travellers to avoid crowds and overly touristy locales are increasingly popular. The luxury travel network also witnesses significant interest in places that are rich in culture and are home to unmatched natural beauty, such as Borneo, Greenland or Oman (Virtuoso, 2019).

Conclusions

The chapter has addressed the luxury tourism behaviour and differences in consumption patterns. Theoretical knowledge about luxury behaviour was supplemented by an outline of current consumer trends, such as the emphasis placed on sustainability, social responsibility or digitalisation. Regarding the differences in consumption patterns, selected sociodemographic variables and spatial aspects were taken into account. The study pointed especially to the luxury travel segment divided by age group. The chapter also included a specific historical summary and description of the issue, especially when evaluating luxury tourism development from a Central European perspective.

The differences have been explored through three case studies focussed on the Czech luxury tourism market segment. The case studies aimed to analyse this

Table 6.3. Differences between Czech and Western Affluent Clients.

Czech Affluent Clients	Western Affluent Clients
The most frequent age group	
Baby boomers	Baby boomers
The preferences on the holiday	
Personal services	Added value
Motivations to travel	
Relaxation, exotic, awareness building	Comfort, experiences authenticity
The level of organization of the holiday	
Tailor-made	Tailor-made
The preferred season for holiday	
Spring, autumn, winter	Winter, summer
The most frequent average length of the holiday	
8 days	10 days
Travel class	
Economy class	First class
Annual frequency of holidays	
Once a year holiday	At least four holidays a year
The most frequent costs per day and person	
Up to 400 euros	Up to 1,100 euros
The character of the destination	
Beaches and sea	Unexplored destinations
Use of digital technologies	
The growing trend in the use of the Internet	Searching for information on the Internet, activity on social networks and travel websites

Source: Own processing based on Novotná and Kunc (2019a, 2019b), Pangea Network (2017), and Virtuoso (2019).

market segment and assess disparities in luxury travel consumption depending on the selected sociodemographic variables (especially age groups). Regarding spatial disparities, it provides perspectives on inequalities between the capital city of Prague and the rest of the country. Concerning the regional income inequality and a higher level of income of Prague residents, it is advisable to examine the affluent clients from Prague and their demand for luxury experiences separately.

The results show that Czech luxury travellers are tertiary-educated people with high regular income. As far as the age structure is concerned, baby boomers are the most dominant among the affluent clients. If we focus on the consumption patterns and motivation of the Czech luxury segment, the satisfaction of the needs for relaxation and exploration of new destinations are the primary motives for taking luxury holidays. The results suggest that Czech luxury travellers are different compared to the traditional West European markets, especially regarding financial possibilities. The frequency of their luxury holidays is also significantly lower. Nowadays, however, it is already possible to notice the shift from a Czech luxury consumer towards the Western way of the perception of luxury, where it is not regarded only as a symbol of social status but as own pleasure and treats.

Suppose we summarise the typical Czech luxury consumer very briefly at the end of the second decade of the twenty-first century (old luxury concept). In that case, it is more of a member of the baby boomers (over 55 years of age), in a high-income group (entrepreneur or employer/employee), looking for exotic seaside or island destinations, where all-inclusive as well as tailor-made services can be found. The length of stay, which takes place in relatively any season once a year, is between 7 and 10 days, and the consumer is willing to spend 3–4000 euro per person.

The young generation, which is beginning to have more financial resources and which will become a fundamental pillar of the luxury clients in the near future, has been increasingly turning away from the targeted visibility of its high social status associated with demanding maximum comfort of services and facilities, exclusivity and privacy, to prefer experiences, authenticity and sustainable travelling. In the future, these aspects of consumer behaviour will be typical for young travellers and consumers of the dynamically living generations Y and Z.

References

Aldrich, N. W. (1989). *Old money: The mythology of America's upper class*. New York, NY: Vintage Books.

Baloglu, S., & Uysal, M. (1996). Market segments of push and pull motivations: Canonical correlation approach. *International Journal of Contemporary Hospitality Management*, *8*(3), 32–38. doi:10.1108/09596119610115989

Barbieri, C., Santos, C. A., & Katsube, Y. (2012). Volunteer tourism: On-the-ground observations from Rwanda. *Tourism Management*, *33*(3), 509–516. doi:10.1016/j.tourman.2011.05.009

Barrows, C. W., Powers, T. F., & Reynolds, D. E. (2012). *Introduction to management in the hospitality industry*. New Jersey, NJ: Wiley.

Bernstein, L. (1999). Luxury and the hotel brand: Art, science, or fiction? *Cornell Hotel and Restaurant Administration Quarterly*, *40*(1), 47–53. doi:10.1177/001088049904000122

Bobková, M., & Holešinská, A. (2017). Networking in a destination from the perspective of virtual relationships and their spatial dimension. *Geographia Technica*, *12*(2), 10–19. doi:10.21163/GT_2017.122.02

Brun, A., & Castelli, C. (2013). The nature of luxury: A consumer perspective. *International Journal of Retail and Distribution Management*, *41*(11/12), 823–847. doi:10.1108/IJRDM-01-2013-0006

Capgemini. (2019). *World Wealth Report 2019*. Paris: Head Office, Capgemini Service.

Chen, A., & Peng, N. (2018). Examining consumers' intentions to dine at luxury restaurants while traveling. *International Journal of Hospitality Management, 71*, 59–67. doi:10.1016/j.ijhm.2017.11.009

Collins, D., & Tisdell, C. (2002). Age-related lifecycles. *Annals of Tourism Research, 29*(3), 801–818. doi:10.1016/S0160-7383(01)00081-0

Credit Suisse. (2017). *Global Wealth Report 2016*. Zurich: Board of Directors, Media Relations.

CzechTourism. (2012). *Tourism and creditworthy clientele*. Prague: The ministry of regional development of the Czech Republic, CzechTourism.

Czinkota, M., & Ronkainen, I. (2009). *Emerging trends, threats and opportunities in international marketing: What executives need to know*. New York, NY: Business Expert Press.

Danziger, P. N. (2005). *Let them eat cake: Marketing luxury to the masses-as well as the classes*. Chicago, IL: Dearborn Trade Pub.

Dixit, S. K. (2017). *The Routledge handbook of consumer behaviour in hospitality and tourism*. London: Routledge.

Dixit, S. K. (2020). *The Routledge handbook of tourism experience management and marketing*. London: Routledge.

Euromonitor Passport. (2015). *Global luxury travel trends report*. London: Euromonitor international, Headquarters.

Eurostat. (2020). *GDP at regional level*. European Commission. Luxembourg: Eurostat.

Hajdáková, I. (2013). Discreet economy: Luxury hospitality in the context of post-socialist transformation of Czech society. *Laboratorium, 5*(1), 73–97.

Hallot, A. (2013). *The future potential for developing luxury tourism and hospitality in Lapland*. Helsinki: Haaga Helia University.

Han, H., & Hyun, S. S. (2018). Role of motivations for luxury cruise traveling, satisfaction, and involvement in building traveler loyalty. *International Journal of Hospitality Management, 70*, 75–84. doi:10.1016/j.ijhm.2017.10.024

Hansen, J., & Wänke, M. (2011). The abstractness of luxury. *Journal of Economic Psychology, 32*(5), 789–796. doi:10.1016/j.joep.2011.05.005

Heine, K., & Phan, M. (2011). Trading-up mass-market goods to luxury products. *Australasian Marketing Journal, 19*(2), 108–114. doi:10.1016/j.ausmj.2011.03.001

Horwath HTL. (2015). *Tourism Megatrends. 10 things you need to know about the future of tourism*. New York, NY: Crowe Global, Horwath HTL – Hotel, Tourism and Leisure.

Iloranta, R. (2019). Luxury tourism service provision–Lessons from the industry. *Tourism Management Perspectives, 32*, 100568. doi:10.1016/j.tmp.2019.100568

Ivy, R. L., & Copp, C. B. (1999). Tourism patterns and problems in East Central Europe. *Tourism Geographies, 1*(4), 425–442. doi:10.1080/14616689908721335

Kapferer, J. N. (2008). *The new strategic brand management: Creating and sustaining brand equity long term*. Philadelphia, PA: Kogan.

Kapferer, J. N., & Laurent, G. (2015). Where do consumers think luxury begins? A study of perceived minimum price for 21 luxury goods in 7 countries. *Journal of Business Research, 69*(1), 332–340. doi:10.1016/j.jbusres.2015.08.005

Koch, K. D. (2011). Luxury tourism–Does this segment still work? In R. Conrady & M. Buck (Eds.), *Trends and issues in global tourism 2011* (pp. 179–185). Heidelberg: Springer.

KPMG. (2016). *The luxury goods market in Poland*. Warsaw: KPMG International, KPMG Sp. z o.o..

Larsen, S., Øgaard, T., & Brun, W. (2011). Backpackers and mainstreamers: Realities and myths. *Annals of Tourism Research*, *38*(2), 690–707. doi:10.1016/j.annals.2011.01.003

Lee, S., Chua, B. L., & Han, H. (2017). Role of service encounter and physical environment performances, novelty, satisfaction, and affective commitment in generating cruise passenger loyalty. *Asia Pacific Journal of Tourism Research*, *22*(2), 131–146. doi:10.1080/10941665.2016.1182039

Lee, W. S., Lee, J. K., & Moon, J. (2019). Influential attributes for the selection of luxury camping: A mixed-logit method. *Journal of Hospitality and Tourism Management*, *40*, 88–93. doi:10.1016/j.jhtm.2019.05.004

Lesjak, M., Navrátil, J., Pícha, K., & Gilliam, V. L. W. B. (2015). The predictors of the willingness to recommend a visit for diversified tourism attractions. *Czech Journal of Tourism*, *4*(2), 77–90. doi:10.1515/cjot-2015-0005

Luxe Digital. (2020). *Millennials seek authentic and shareable luxury travel experiences*. Singapore: mOOnshot digital, Luxe Digital.

Mahika, E. C. (2011). Current trends in tourist motivation. *Cactus Tourism Journal*, *2*(2), 15–24.

MarTech Series. (2019). 9 travel trends and habits of baby boomers. Marketing Technology News, California.

Nijman, J., Muller, P. O., & de Blij, H. J. (2016). *Geography: Realms, regions, and concepts*. New Jersey, NJ: Wiley.

Novanská, V., Benová, A., & Geghamyan, S. (2018). Options for recreation in the city of Bratislava based on the perception of local inhabitants. *Geographia Cassoviensis*, *8*(1), 110–128.

Novotná, M., Grajciarová, L., & Polehňa, D. (2019). Dentification of global tourism trends and their impact on destination sustainability. In V. Klímová & V. Žítek (Eds.), *XXII. International colloquium about regional sciences* (pp. 462–469). Brno: Masaryk University Press.

Novotná, M., & Kunc, J. (2019a). Experiences matter! Luxury tourism consumption patterns and motivation of the Czech affluent society. *Deturope*, *11*(2), 121–142.

Novotná, M., & Kunc, J. (2019b). Luxury tourists and their preferences: Perspectives in the Czech Republic. *Tourism: An International Interdisciplinary Journal*, *67*(1), 90–95.

OECD. (2020). *Level of GDP per capita and productivity*. Paris: The Organisation for Economic Co-operation and Development, OECD.Stat.

Ogilvy. (2014). *And how do you enjoy yourself? Luxury brands in 2014–Study*. Prague: Ogilvy, s.r.o.

Page, S. J. (2009). *Tourism management: Managing for change*. Amsterdam: Elsevier/Butterworth-Heinemann.

Pangea Network. (2017). *5th international luxury travel trends*. Utrecht: Pangaea Network Ltd.

Park, D. B., & Yoon, Y. S. (2009). Segmentation by motivation in rural tourism: A Korean case study. *Tourism Management*, *30*(1), 99–108. doi:10.1016/j.tourman.2008.03.011

Patterson, I. R. (2018). *Tourism and leisure behaviour in an ageing world*. Boston, MA: CABI.

Peak and Skift. (2014). *The rise of experiential travel–Special report*. New York, NY: Skift Team + Peak Adventure Travel Group.

Petrick, J. F., & Durko, A. M. (2015). Segmenting luxury cruise tourists based on their motivations. *Tourism in Marine Environments*, *10*(3–4), 149–157. doi:10.3727/154427315X14181438892649

Pitta, D., Eastman, J. K., & Liu, J. (2012). The impact of generational cohorts on status consumption: An exploratory look at generational cohort and demographics on status consumption. *Journal of Consumer Marketing*, *29*(2), 93–102. doi:10.1108/07363761211206348

PWC. (2017). *Global megatrends and their impact on the Middle East's travel and tourism industry*. London: Pricewaterhouse Coopers.

Roustant, J. C. (2014). *A report on green luxury*. Champagne: Groupe ESC Troyes.

Silverstein, M. J., Fiske, N., & Butman, J. (2003). *Trading up: The new American luxury*. New York, NY: Portfolio.

Sladek, S., & Grabinger, A. (2013). *Gen Z, the first generation of the 21st century has arrived!* Minneapolis, MN: XYZ University.

Uysal, M., & Jurowski, C. (1994). Testing the push and pull factors. *Annals of Tourism Research*, *21*(4), 844–846. doi:10.1016/0160-7383(94)90091-4

Virtuoso. (2015). *The 2015 Virtuoso Luxe Report. A guide to key trends and insights in luxury travel*. Fort Worth, TX: Virtuoso.

Virtuoso. (2019). *Luxury tourism market: Gen X women to drive wellness travel trend in 2020*. Fort Worth, TX: Virtuoso.

Vystoupil, J., Šauer, M., & Bobková, M. (2017). Spa, spa tourism and wellness tourism in the Czech Republic. *Czech Journal of Tourism*, *6*(1), 5–26. doi:10.1515/cjot-2017–0001

Yoon, Y., & Uysal, M. (2005). An examination of the effects of motivation and satisfaction on destination loyalty: A structural model. *Tourism Management*, *26*(1), 45–56. doi:10.1016/j.tourman.2003.08.016

Chapter 7

Marketing Luxury Experiences in an Emerging Luxury Destination of Norway

Xiang Ying Mei[1]

Abstract

Many destinations are keen to be a part of the increasing and lucrative luxury tourism market. Before targeting luxury tourists in the marketing efforts, is it essential to understand the meaning of 'luxury' and what is considered exclusive? As an emerging luxury destination, Norway certainly offers many attributes that can satisfy the needs of luxury tourists. Marketing luxury provides, nevertheless, challenges for Norway as its main attraction is nature, which is fundamentally free and accessible to anyone. Thus, it is necessary to tailor-make unique nature experiences with luxurious products, resulting in exclusivity and once-in-a-lifetime experiences beyond the masses. Luxury tourists have also been advocated as more sustainable than mass tourists as addition to the issue of carrying capacity; many are also concerned with responsible tourism. Specialist travel agents play an important role in the marketing process as they have first-hand knowledge of luxury tourists as well as acting as guardians of information communicated in marketing. In addition to mainland Norway, the Svalbard islands in the Arctic region may be considered as truly unique and exclusive destinations. While sustainable development is a major concern in Norwegian tourism in general, it is even more vital in the Arctic regions. Therefore, marketing needs to provide a balanced picture of the attractions and attributes in order to become a luxury destination that is sustainable in the long run.

Keywords: Exclusivity; uniqueness; tailor-made experiences; peripheral destinations; sustainable development; Norway

[1]https://orcid.org/0000-0002-9691-1540

The Emerald Handbook of Luxury Management for Hospitality and Tourism, 145–161

doi:10.1108/978-1-83982-900-020211007

Introduction

Trends indicate that an increasing number of tourist are seeking luxury tourism experiences in the recent years. Thus, more destinations are interested in targeting such lucrative market. While the luxury tourism market is well established in popular tourism destinations such as Spain and France, this niche market is relatively new in a country such as Norway, located on the northern periphery of Europe. Nevertheless, it is such a peripherality that attracts many luxury tourists to seek unique Norwegian experiences. According to travel agents who specialise in inbound luxury travel to Norway, Norway is an ideal destination for luxury tourists who want to experience wilderness and something extraordinary. Luxury tourists can, for instance, leave millions of dollars in one single visit (Bing, 2017). There is no doubt that it is essential to target such a market in the marketing effort with such a lucrative market. However, marketing luxury tourism may not be as straightforward, and it is too simplistic to assume that luxury tourists consist of one homogenous group. Luxury tourists may also seek various experiences depending on the destination itself and its unique attributes. The chapter explores relevant topics in understanding how marketing luxury travel may be conducted and the types of products and experiences that would appeal to contemporary luxury tourists in a destination that relies on unique experiences of nature. This is followed by two specific research objectives.

- Investigating luxury tourists and relevant topics associated with understanding 'luxury', as part of luxury tourism marketing efforts.
- Exploring the potential of Norway as an appealing luxury tourism destination based on the element of exclusivity in experiences as well as the importance of sustainability.

Although luxury tourism is not an entirely new term, there is still a scant understanding of this market, particularly as the recent trends indicate that the scope of luxury tourism is changing. Much of such trends are attributed to people's perception of what luxury means. The chapter starts with addressing some of the challenges in marketing luxury tourism, followed by a short discussion of the various types of luxury tourists, the influence of personal values and the importance of exclusiveness in the experiences. This is followed by a discussion of Norway as a potential emerging luxury tourism destination along with two specific cases focussing on the role of specialist travel agents as well as the opportunities and challenges of the Svalbard islands as a luxury destination.

Marketing Luxury

Marketing luxury tourism may be challenging as the definition of 'luxury' is debatable, and the understanding of the term varies (Atwal & Wiliams, 2009; Le & Quy, 2020). While most people might have an idea of what luxury means to them as luxury is a refinement of basic human needs, it is arguably a subjective concept that varies from person to person, much like personal experiences. Some of the past's basic items that were considered luxury have lost their luxury status due to mass consumption (Veríssimo & Loureiro, 2012). This also applies to the tourism industry, as travelling for leisure may have been a luxury in the past while cheaper airfares and boost in the economy have attributed to leisure travelling being possible for many people, leading to the development of mass tourism. Nevertheless, leisure travelling may still be considered as a luxury by some people, thus suggesting that the perception of what is a necessity and what is luxury also varies from society to society (Kemp, 1998). A similar idea may also be applied to various destination contexts as what is perceived as a luxury in one destination may not be considered as luxury in another.

Further, Swarbrooke (2018) explained that marketing luxury is about selling products and services, which are not considered everyday purchases and that is non-essential for survival. However, the issue is that the word 'luxury' is often misrepresented or misused as it has become a buzzword used by many marketers due to the positive connotation associated with the term (Swarbrooke, 2018). Thus, the notion of exclusivity becomes important in the discussion of luxury. Using exclusively as the main attribute to market luxury tourism products and services, it can be argued that a five-star hotel stay, for instance, can possibly be experienced anywhere unrelated to the destination itself. Furthermore, single dining in a luxury restaurant or stay at a luxury resort in a destination may not be adequate for those seeking unique and exclusive once-in-a-lifetime experiences. Hence, marketing efforts need to combine various exclusive products and services with the unique attributes of the destination and connect with luxury tourists by focussing on destination-related experiences (Atwal & Wiliams, 2009). The idea of exclusivity in tourism experiences is further explored in the later sections. Swarbrooke (2018) further argues that luxury should perhaps be defined by tourists themselves rather than by marketing professionals. Thus, before marketing a luxury destination, it is crucial to understand the needs and wants of the luxury tourist market.

Understanding Luxury Tourists

Although this chapter does not intend to investigate types of luxury tourists and their behaviour in depth, it is imperative to differentiate the various types of luxury tourists and types of products that appeal to them to market luxury tourism products and services that meet and even exceed their expectations. There is still much to be understood because the traditional perception of luxury tourists as a homogenous group consisting of mature consumers seeking extravagant lifestyles may be outdated (Dixit, 2017; Prayag & Hosany, 2014). Therefore, it is a misconception that luxury tourists consist of a homogeneous group. CBI Ministry of Foreign Affairs (2016) explains that it is important to understand that there are various types of luxury tourists. Some luxury tourists may be used to high-end standards in their daily lives (Prayag & Hosany, 2014), and thus it is challenging to satisfy and target such a market in the general marketing efforts.

On the other hand, another type of luxury tourist, the more significant segment, consists of individuals who have saved up travelling for a special occasion such as a honeymoon or anniversary. Such tourists seek originality and exclusivity, such as a once-in-a-lifetime experience with a wow factor rather than high-end luxury (CBI Ministry of Foreign Affairs, 2016).

Nevertheless, luxury tourists may also consist of a combination of both. Hence, marketers need to understand these elements as part of the luxury tourism marketing efforts. Evidently, some common characteristics may be found in all types of luxury tourists, such as having extremely high expectations. Moreover, such expectations must be met in all the parts of the supply chain, not just the end product, as they are willing to pay a large sum of money to enjoy top-quality products and unique experiences (Buckley & Mossaz, 2016). Other needs of luxury tourists have been identified, and some of the attributes that should be emphasised in the marketing effort are illustrated in Table 7.1.

While luxury is often associated with the high price, it is too simplistic to assume that luxury equals high price as the expensiveness may be related to the exclusivity and rarity rather than high-end luxury (CBI Ministry of Foreign Affairs, 2016). It is fundamentally about providing something that the average tourist perhaps has not even heard of, as luxury tourists often want something which is exclusive, unique and tailor-made just for them (Røeggen, 2010). Prayag and Hosany (2014) further explain that for luxury tourists, it is all about belonging to the elite within the exclusive spheres and to rise above the masses. For such reasons, it would be, for instance, unthinkable for luxury tourists to line up in a queue or to share their experiences with the mass tourists. Exclusivity and the sense of belonging to an elite group are thus some of the key attributes that marketers must focus on in order to target luxury tourists.

Table 7.1. Luxury Tourist Needs.

Needs	Attributes that Destination Marketing Should Focus On
Privacy	A calm style without ostentation is preferred.
Personalisation	Marketing must focus on personalisation and differentiation. The tourist must be treated as the only one.
Simplicity and perfection	Focus on the value chain. Services must work without mistakes, and quality must exist in all the processes.
Authenticity	There is a growing desire for exotic destinations and the searching for a different culture that is not sought and visited by the masses.
Comfort and safety	While the trend of new destinations is increasing, destinations must still emphasise on safety and comfortability.

Source: Adapted from ILTM (2010) and Veríssimo and Loureiro (2012).

Influences of Personal Values

In addition to the needs and wants, it is also imperative to understand the personal values of luxury tourists in order to market luxury tourism products and services. Although personal values in conjunction with luxury tourists, in particular, require to be further explored, personal values have been widely incorporated into tourism marketing in general (Wen & Huang, 2019). Personal values are complex terms, which will not be explored in depth in the present chapter; however, they are essential components to consider when attempting to target the luxury market. Personal values form an important part of human lives and influence greatly on emotions such as happiness, fun and a sense of security and belonging (Chattalas & Shukla, 2015; Nwankwo, Hamelin, & Khaled, 2014). Based on such a notion, luxury tourists would also evaluate luxury tourism products and services based on their personal values, which may consist of social recognition and social integration as well as living a peaceful life (Lages & Fernandes, 2005; Le & Quy, 2020). These three value dimensions are important as they are directly related to tourists' satisfaction with the products and services (Lages & Fernandes, 2005). Just like the various understanding of the concept of 'luxury', personal values would also vary according to individual tourists and their previous experiences as well as the destination attributes that match such personal values. This stresses the complex interplay between the individual and contextual influences in terms of luxury tourism products and services.

Furthermore, other values that come into play in the luxury market include experiential, symbolic and functional values (Chattalas & Shukla, 2015; Yang, Zhang, & Mattila, 2015). It can be argued that for luxury tourism, experiential values are the most sought value, although the latter two may also be sought. Thus, it is evident that a more experiential marketing approach along with key understanding of the personal values of luxury tourists is necessary in the luxury tourism industry (Swarbrooke, 2018; Veríssimo & Loureiro, 2012).

Focussing on the Exclusivity in Tourism Experiences

Although products and services are often referred to in the tourism industry, the industry is essentially an experience provider. Unique tourism experiences are conceptualised through pleasure and special feelings, which is very different compared with ordinary experiences (Dixit, 2020; Gurzki & Woisetschläger, 2017). Hence, self-fulfilment and meaningful experiences are arguable something that many tourists want to gain as travelling is more than just escaping routine and home life. As discussed, luxury tourists may have different needs that vary from person to person. A common trait for a luxury tourist, whether luxury is a part of an individuals' lifestyle (Prayag & Hosany, 2014) or someone who is seeking once-in-a-lifetime experiences, the subject of matter is exclusivity. However, exclusively does not necessarily mean having materialistic items (Bellaiche, Mei-Pochtler, & Hanisch, 2010), as many luxury consumers are increasingly interested in life experiences. Yang et al. (2015) explain that luxury consumers are seeking to enrich their happiness and enjoyment by spending more on travel to obtain unique experiences rather than to display material luxury. Many are increasingly seeking appreciation for experience-based luxury that are genuine and meaningful and thus have become more willing to pay for unique and exclusive experiences that make them happy (Bellaiche et al., 2010). This is evidently also related to personal values, as previously discussed. Such trends verify the shift in mindset from how luxury is valued from a transactional relationship to a holistic experience (Atwal & Wiliams, 2009). As stated, some common tourism products and services such as accommodation and a bed to sleep may be experienced anywhere, but it is the unique combination of destination attributes with certain attractions, products and services that are only available to a few elites which together lead to unique and exclusive experiences.

Norway as a Luxury Destination

Norwegian Tourism

Norway is in such a position that while the tourism industry is an important industry considered by many, the country's economy has traditionally not been dependent on the industry as other destinations due to the heavy dependence on the petroleum sector (Mei, 2014a). However, tourism accounts for approximately 4.6% of gross domestic product, and it brings more value to the economy than traditional sectors such as fishery and agriculture (Statistics Norway, 2020). The tourism industry is also considered to be a better alternative to the petroleum sector (Goeldner & Ritchie, 2009; Pechlaner, Fischer, & Hammann, 2005), especially when the petroleum sector phases out in the future (Jakobsen & Espelien, 2010). For such reasons, the Norwegian government has been actively marketing tourism. A long-term international marketing approach in collaboration between Innovation Norway (a governmental organisation responsible for innovation and development of Norwegian enterprises and industry) and the Norwegian tourism industry, especially from 2010 to 2015, has helped to increase interest in Norway as a destination (Norwegian Tourism Partners, 2019). While Norway is not exclusively marketed as a luxury tourism destination, the country is known for its high living cost and standard. Its peripheral location, which has been considered as a barrier in the past, is also now a pull factor for many visitors, much attributed by the uniqueness of its nature and better accessibility through increased flight capacity and more direct flights. A much weaker Norwegian krone (NOK) from 2014 has certainly also contributed to such development. The tourist numbers have steadily increased from early 2010 and until 2019 (Norwegian Tourism Partners, 2019), to the point that many stakeholders are concerned about over-tourism from the mass tourist market, particularly in the heaviest time of the year, which is the summer months. Hence, many tourism operators and locals are generally advocating for tourists who come in smaller groups and higher spenders in one visit as well as those who are more interested in the natural and cultural elements in comparison to mass tourists.

Luxury Tourism Potentials

While no recent data on luxury tourism spending in Norway exist, it was estimated earlier on that inbound luxury tourists to Norway are willing to pay more than 2000 USD per day for an exclusive Norwegian holiday experience (Røeggen, 2010). This number would be significantly higher today, whereas recent statistics indicate that an average tourist would spend around 50 to 200 USD per day (Innovation Norway, 2019). Although no official guideline exists at present on classifying the luxury tourist market based on spending, ITB (2017) defines a luxury holiday as either an international trip of three nights with a consumption of over 750 EUR per day or a more extended trip with more than four nights and consumption of more than 450 EUR per day. Such a market is therefore considered as extremely lucrative for many tourism destinations. Norway is still in its infancy when it comes to luxury tourism compared to other more established destinations such as France and Spain, at least in the traditional understanding of luxury tourism. There has generally been a lack of luxury accommodations or products that are at an international standard, which was some of the challenges to bring luxury tourists to the country. Nevertheless, recent developments such as the Thief in Oslo, Villa Åmot in Sunnfjord or Wolf Lodge in the north indicate that the destination is ready for such a market. In addition to luxury accommodations, Norway offers natural landscape, which has been classified as world-class standard with some of Europe's most dramatic, varied and unspoiled mountains and high plateau, deep valleys and fjords (Jakobsen & Espelien, 2010). It is important to point out that extravagance is not necessarily important for many luxury tourists in Norway. What they want is for everything to be connected and that the experiences are real with a good dose of the history behind the place they visit (Røeggen, 2010). Essentially, it is not necessarily about five-star hotels and private butlers, but rather about exotic adventures that only Norway can offer (Holøien, 2017). The attributes that marketing should focus on are authentic experiences, which are thought to be unspoilt by the impacts of mass tourism (Swarbrooke, 2018). For instance, tree top house hotels with limited availability, which offers unique experiences of the Norwegian nature, have become increasingly popular and something that the luxury tourist market is searching for as part of exclusive Norwegian experiences, in combination with local food of the highest class, cultural experiences, storytelling, warm hospitality and other natural attractions in a magnificent landscape (NRK, 2018). Thus, as discussed, the importance is the combination of several products and services that lead to total exclusive and unique experiences, which are only available to a group of elites. The marketing of such products and services may be difficult without the assistance of specialist travel agents, which is further discussed in case study one.

Furthermore, what is interesting about Norway is that the destination has always promoted its nature as 'free' due to an act called 'allemannsretten', which refers to the *right to roam* in 'unfenced land' and land that is not cultivated. This includes most shores, bogs, forests and mountains in Norway. Outdoor recreation is of great importance to the Norwegian national identity. Hence, the right to roam anywhere in nature by anyone, including tourists, is protected by the law (Visit Norway, 2020a). While the *right to roam act* is not unique in Norway as similar rights are found in other Scandinavian and European countries, for inbound tourists from other countries beyond the European continent, this is quite exceptional. For instance, although there is no detailed statistics on luxury tourists in Norway, as this market is still quite new, the latest available numbers indicate that North Americans are the largest group of luxury tourists in Norway at present (Innovation Norway, 2017). Hence, the idea of the *right to roam* is something very unusual for such markets.

Luxury Tourism Advocated as More Sustainable

It is difficult to omit the discussion on sustainability issues in terms of marketing luxury tourism in Norway, particularly when nature is one of its main pulling factors as well as the *right to* roam in nature. While sustainable tourism development is important in Norwegian tourism in general, it is imperative for some areas in particular such as the Svalbard islands, which is further explored in case study two. It has been long argued that mass tourism leads to several negative impacts, particularly in a destination such as Norway, where its tourism industry is heavily reliant on nature due to low population density in many areas where key natural attractions are located. For instance, many of the key natural attractions in Norway are located in small towns and areas such as Geiranger, Trolltunga, the Pulpit Rock, Flom and Lofoten where the number of tourists can triple the number of residents in peak seasons. Tourism operators and locals are thus fearing the negative consequences of the increasing number of mass tourists in the environmentally fragile areas. Some of the issues of mass tourism include littering, congested tourist trails that exceed carrying capacity and where nature is used as a toilet for tourists to relieve themselves. The debate has been going on whether tourism is a blessing or a curse for these areas (Mathisen, 2017; Sørholm, 2018). Thus, local operators seek tourists who come in small numbers, with more appreciation of the natural environment and who are more interested in engaging with the local surroundings. It has been argued that unique and memorable experiences are determined by length of stay and the degree of involvement. The more involved and engaged a tourist is, the more the experience will be considered as unique and valuable (Mei, 2014a). Arguably for some tourists in the mass tourism categories, this is quite the opposite as engagement and longer stay may not be sought.

On the other hand, luxury travel has evolved from merely high spending, consumption and self-indulgence to authenticity and responsible tourism (Swarbrooke, 2018). Luxury tourism has thus been advocated as more sustainable use of tourism resources as Norway wants tourists but not mass tourism. Small segments of tourists who have more spending power than the average tourists while being particularly conscious about nature, and at the same time counter the effect of carrying capacity in the destination, are highly welcomed by the Norwegian tourism industry (Mei, 2014a). Although it has been argued that choosing green products and services leads to a decreased level of luxury (Baker, Davis, & Weaver, 2013; Line & Hanks, 2016), travel agents targeting luxury travellers insist that luxury tourists today are interested in learning while displaying particular environmental consciousness (Larsen, 2017; Røeggen, 2010). Swarbrooke (2018) further emphasises that it may be important to make sustainable behaviour and responsible tourism more fashionable and status-enhancing in order to get even more luxury tourists onboard.

Case Study 7.1: Travel Agents in Marketing Exclusive Norway Experiences

In Norway's case, to market its unique nature, which is fundamentally free and accessible to anyone, 'luxury' is undoubtedly challenging. It is thus important that luxury Norway experiences combine nature with luxurious products that appeal to a market that is willing to spend significantly more than an average tourist. As discussed, an exclusive Norway experience may therefore consist of many products and services, which in the total lead to unique experiences beyond the masses. An example may consist of a private cruise or a helicopter ride in the famous Geiranger fjord, followed by an overnight in luxurious accommodation with food of the best quality (Bing, 2017). Furthermore, as discussed, several needs of luxury tourists, including privacy, differentiation, simplicity and perfection, authenticity, comfort and safety (ILTM, 2010; Veríssimo & Loureiro, 2012), must be considered when targeting such tourists. The unique and picture-perfect Norwegian nature and low population density certainly satisfy many of the needs and wants of such a market. Moreover, according to Ashley Rudnansky at Cox and Kings', a travel agent that specialises in the luxury market, Norway is also considered as a safe destination (Bing, 2017), which in itself, as discussed, appeals to many luxury tourists (ILTM, 2010; Veríssimo & Loureiro, 2012).

Essentially, it is imperative to provide the tourists with unforgettable, unique and exclusive experiences that leave nothing to chance. For such reasons, it is arguably difficult to offer and combine various services as well as having a great deal of understanding of the various needs of luxury tourists without considering the important role of specialist travel agents in the process. Incoming travel agents in many countries, including Norway, may have lost their important role in the mass tourism market in recent years due to the fact that more tourists are arranging and managing holidays on their own (Assaf, Barros, & Dieke, 2011; Harison & Boonstra, 2008; Mei, 2014b). Another reason is the development of the Internet and information access and flow (Mei, 2014b). However, when looking at the luxury tourism market, travel agents may have established a newfound role as time may be mostly valuable to luxury tourists in the sense that they need someone else to make all the necessary arrangements (Swarbrooke, 2018). As previously discussed, the overusage of the word 'luxury' may have diluted its meaning, which arguably makes it challenging for the low-volume luxury tourists themselves to search for relevant information and high-end tourism products and services that satisfy their needs. Furthermore, specialist travel agents' role is, in particular, relevant in Norway as the Norwegian nature has often been criticised for being 'too raw' and not facilitated for visitors as tourists are often left to fend for themselves (Mei, 2014a). Much attributed to the *right to roam act*, the idea of such an approach is for everyone to experience nature as it is. This has nevertheless been labelled as too basic. At this level where tourists expect nothing but perfection, such basicness may not be satisfactory for luxury tourists. For instance, it would be unthinkable for luxury tourists to explore nature by themselves, hoping that they would gain unique experiences. The role of a specialist travel agent would be to make luxury tourists feel special by creating the perfect experiences, which are tailor-made specifically for them (Buckley & Mossaz, 2016; Dixit, 2020; Swarbrooke, 2018) by focussing on the attributes that only Norway can offer. Recently, a dozen specialist travel agents based in the United States and the United Kingdom were invited to the Vestland County located at Western Norway, famous for its several UNESCO heritage sites and unique sceneries, to experience first-hand the attractions directed at the luxury market (NRK, 2018). Hence, it is imperative that in addition to market the unique experiences with the tailor-made combination of services and products, the agents themselves must also have sufficient knowledge of the products that they market and sell (Buckley & Mossaz, 2016). Subsequently, while there are high expectations of the expertise of specialist travel agents by the luxury market, the agents also possess a great deal of control and power in types of products and services that they believe to be good enough to market these to their highly demanding clientele. Buckley and Mossaz (2016) further argue that specialist agents would make their booking decisions based on their personal experiences of potential luxury products and services rather than external marketing materials provided by tourism operators and destination marketing organisations. Such notion highlights that destinations and their relevant tourism operators that seek to attract luxury tourists must closely align with specialist travel agents in the marketing efforts as the agents would have first-hand knowledge of the needs and wants of luxury tourists and also because they can be considered as guardians of types of information to be communicated in the marketing effort.

Case Study 7.2: Luxury Tourism and Sustainability in the Arctic Region of Svalbard

The Svalbard islands are located in the Arctic Ocean, halfway between mainland Norway and the North Pole. The islands consist of the largest island of Spitsbergen followed by Nordaustlandet, Edgeøya and Barentsøya, as well as smaller and more isolated ones including Kvitøya, Kong Karls Land of Hopen and Bjørnøya. There are 2,939 residents in total, with Longyearbyen as the administration centre (Barr & Thuesen, 2020; Visit Norway, 2020b). The location of the Svalbard islands and their distance to mainland Norway is illustrated in Fig. 7.1.

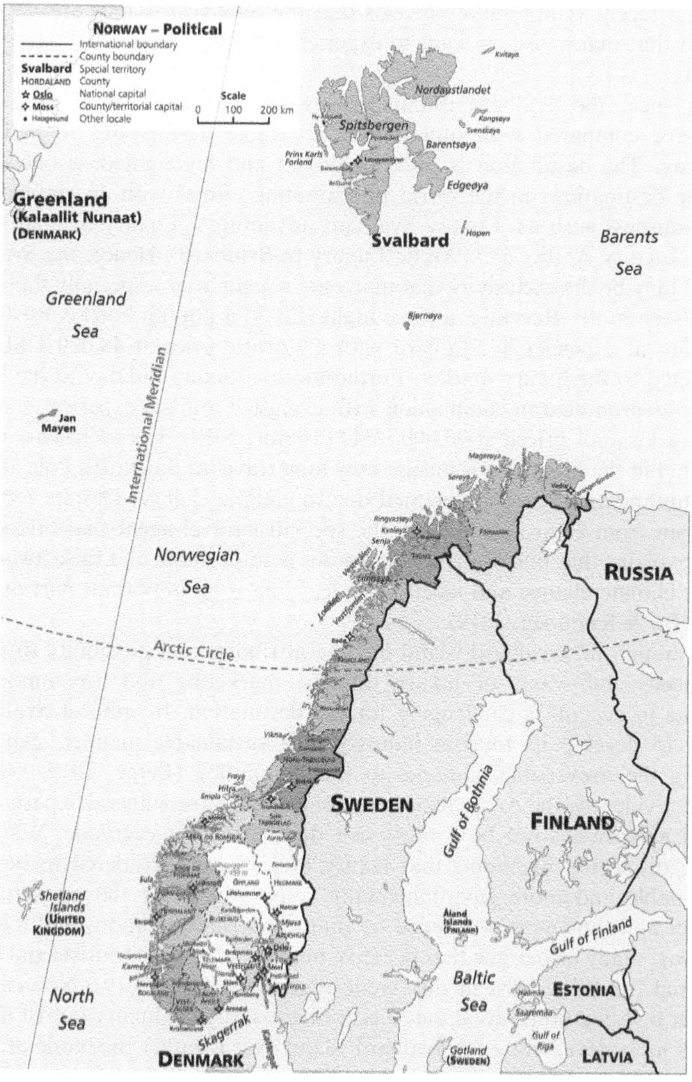

Fig. 7.1. The Location of the Svalbard Islands and their Distance to Mainland Norway. *Source:* Pluribus (2019).

The Svalbard islands may be considered as true destinations to experience once-in-a-lifetime with the key attributes of untouched Arctic nature and sceneries, in combination with polar bears, snow sledge riding, the northern light, midnight sun and the distinctive blue light in one visit (Enger, 2018). While no specific statistics are recorded on the number of luxury tourists in particular, tourist numbers in general in Svalbard have doubled in the past decade to 150,000 visitors annually last year (Øian & Kaltenborn, 2020). The destination is not as easily accessible as it involves significant cost and time to reach the islands. Besides visiting Svalbard for its key attributes and attractions, a recent visitor survey reveals that the Svalbard islands are *the* destination that many visitors want to experience 'before they die' (Enger, 2018, p. 21).

As such, the Svalbard islands are evidently considered as even more exclusive compared with other more accessible destinations in mainland Norway. The destination is often described and highlighted as one of the unique destinations in the world in marketing efforts with an emphasis on luxuriousness such as 'Luxury Svalbard adventure', 'Luxury Svalbard holiday', 'Luxury Arctic' and 'Arctic Luxury in Svalbard'. Hence, the Svalbard island may be the exclusivity that many are sought after, especially those who have 'seen it all'. Recently, a three-night stay in a pop-up hotel consisting of an igloo at a glacier in Svalbard with a starting price at 48,000 USD was marketed to the luxury market. Furthermore, a luxury holiday to the North Pole was promoted in conjunction with exclusive tourism experiences on the Svalbard islands priced at 90,000 USD (Taylor, 2019). The exclusivity is also reflected in the price as it is unsure how long travel to the North Pole and the surrounding areas can be operated due to climate change. Despite receiving criticism from environmentalists, the specialist travel agent that offers such trips explains that one of the key purposes is to promote and raise awareness about climate change and its effect as learning is an important part of these trips (Bø & Sagmoen, 2019).

Although the Svalbard island has the attributes and possibility to satisfy the needs and wants of luxury tourists, marketing and accommodating tourists in general is challenging for the destination. In spite of Svalbard's effort to develop its tourism industry in a sustainable manner, many are fearing the irreversible damages to the destination (Berge, 2019; Dvorak, 2019). Evidently, the Arctic regions are one of the most vulnerable parts of the world and should not be a must-visit destination for everyone. While the above discussion highlights that luxury tourists are considered to be more sustainable and more environmentally responsible than the mass tourism market and thus more welcomed by many tourism operations and locals in Norway, many also argue that negative impacts as long as tourists and fossil-powered travelling exist in the Arctic region (Dvorak, 2019). Nevertheless, further investigations reveal that it is not necessarily the luxury tourist market that is most devastating for Svalbard island as it is rather the mode of travel and activities. For instance, it has been estimated that greenhouse gas emissions for a cruise tourist in Svalbard are more than three times higher than a tourist travelling by aeroplane (Dvorak, 2019; Øian & Kaltenborn, 2020). Marketing Svalbard experiences to luxury tourists must also take such issues into consideration as activities such as private cruise or helicopter ride might not be as wise in Svalbard as in mainland Norway. In Svalbard's case, marketing and communication should be conducted with caution in terms of expectations. The focus should not be on attractions and activities that are fragile or create negative irreversible damages such as helicopter rides and polar bears, but rather convey the magnificent nature and the life of the Arctic wilderness (Dvorak, 2019). The discussion indicates that while destination attributes should be emphasised in the marketing efforts directed at the luxury market, not all attractions should be promoted, and sustainable and responsible tourism must be integrated in the process as something important, exclusive and even fashionable.

Conclusion

It can be quite certain of the type of product, services and experience luxury tourists are seeking as well as their expectations, although it may not always be easy to identify who luxury tourists are. From such understanding, it is possible to suggest how marketing luxury travel should be conducted and the types of products and experiences that would appeal to luxury tourists. Just like any tourism market, it is also important to keep up-to-date on the needs and wants of luxury tourists, as their preferences may change from indulging in luxury by seeking the best and most glamorous travel experiences to once-in-a-lifetime, unique and unforgettable experiences that can be pricey due to the notion of exclusivity. In Norway's case, it might not be the extravagant restaurants and accommodations that attract the most as it is rather the authenticity, unspoiled environment and exclusivity, which are only available or experienced by a few elites. In marketing destinations such as Norway as a luxury, it is important to convey a combination of various attributes, which together form exclusive once-in-a-lifetime experiences. From the case study discussions, it is stressed that specialist travel agents act both as experts of luxury tourists and the controller of communication through marketing. While it is not to suggest that all luxury tourists would rely on specialist agents as luxury tourists are essentially not a homogenous market, well-educated and informed agents certainly play an important intermediate role in marketing destinations that aim to target the luxury market. The case of the Svalbard islands further indicates that although the destinations possess great potentials to satisfy the demanding luxury tourists, such rarity also come with great responsibilities in terms of sustainable development. Although this is particularly relevant in destinations in the Arctic regions, the importance of sustainable destination development is imperative in Norwegian tourism in general due to its heavy reliance on nature. Hence, marketing luxury needs to provide a balanced picture of the attractions and attributes that are in line with what appeals to luxury tourists with sustainability to ensure a viable luxury tourism market in the long run.

References

Assaf, A. G., Barros, C. P., & Dieke, P. U. C. (2011). Portuguise tour operators: A fight for survival. *Journal of Air Transport Management*, *17*(3), 155–157. doi: 10.1016/j.jairtraman.2010.12.002

Atwal, G., & Wiliams, A. (2009). Luxury brand marketing—The experience is everything! *Journal of Brand Management*, *16*, 338–346. doi:10.1057/bm.2008.48

Baker, M. A., Davis, E. A., & Weaver, P. A. (2013). Eco-friendly attitudes, barriers to participation, and differences in behavior at green hotels. *Cornell Hospitality Quarterly*, *55*(1), 89–99. doi:10.1177/1938965513504483

Barr, S., & Thuesen, N. P. (2020). Svalbard. Retrieved from https://snl.no/Svalbard. Accessed on June 20, 2020.

Bellaiche, J.-C., Mei-Pochtler, A., & Hanisch, D. (2010). *The new world of luxury: Caught between growing momentum and lasting change*. Boston, MA: The Boston Consulting Group (BCG).

Berge, T. (2019). Luksusturisme til Nordpolen, et stort steg i feil retning [Luxury tourism to North Pole, a big step in the wrong direction]. Retrieved from https://www.highnorthnews.com/nb/luksusturisme-til-nordpolen-et-stort-steg-i-feil-retning. Accessed on June 20, 2020.

Bing, I. (2017, July 19). Luksusturisme til Norge øker: – De kan bruke flere millioner [Luxury tourism in Norway increases – They can spend many millions]. *E24*. Retrieved from https://e24.no/naeringsliv/i/qL3PnO/luksusturisme-til-norge-oeker-de-kan-bruke-flere-millioner. Accessed on July 1, 2020.

Bø, E., & Sagmoen, I. (2019). Tilbyr skreddersydde luksusreiser til Nordpolen [Offers tailored luxury trips to the North Pole]. Retrieved from https://e24.no/naeringsliv/i/50MW06/tilbyr-skreddersydde-luksusreiser-til-nordpolen. Accessed on July 20, 2020.

Buckley, R., & Mossaz, A. C. (2016). Decision making by specialist luxury travel agents. *Tourism Management*, *55*, 133–138. doi:10.1016/j.tourman.2016.02.002

CBI Ministry of Foreign Affairs. (2016). What are the opportunities for luxury tourism from Europe? Retrieved from https://www.cbi.eu/node/2597/pdf/

Chattalas, M., & Shukla, P. (2015). Impact of value perceptions on luxury purchase intentions: A developed market comparison. *Luxury Research Journal*, *1*(1), 40–57. doi:10.1504/LRJ.2015.069806

Dixit, S. K. (2017). *The Routledge handbook of consumer behaviour in hospitality and tourism*. London: Routledge.

Dixit, S. K. (2020). *The Routledge handbook of tourism experience management and marketing*. London: Routledge.

Dvorak, Z. (2019). *Bærekraftig turisme i Arktis? En studie av utviklingen på Svalbard* [*Sustainable tourism in the arctic? A study of the development in Svalbard*]. Master thesis, Nord University, Tromsø.

Enger, A. (2018). Gjesteundersøkelsefor Svalbard, 2018 [Visitor survey for Svalbard, 2018]. Retrieved from Oslo.

Goeldner, C. R., & Ritchie, J. R. B. (2009). *Tourism: Principles, practices, philosophies* (11 ed.). Hoboken, NJ: John Wiley & Sons.

Gurzki, H., & Woisetschläger, D. M. (2017). Mapping the luxury research landscape: A bibliometric citation analysis. *Journal of Business Research*, *77*, 147–166. doi: 10.1016/j.jbusres.2016.11.009

Harison, E., & Boonstra, A. (2008). Reaching new altitudes in e-commerce. Assessing the performance of airline websites. *Journal of Air Transport Management*, *14*(2), 92–98. doi:10.1016/j.jairtraman.2008.02.002

Holøien, M. (2017). Dette vil turistene vite før Norgesferien [This is what the tourists want to know before Norway holiday]. *Finansavisen*. Retrieved from https://finansavisen.no/nyheter/reise/2017/07/dette-vil-turistene-vite-foer-norgesferien. Accessed on July 1, 2020.

ILTM (International Luxury Travel Market). (2010). Global trends in luxury travel: A whitepaper. Retrieved from Cannes, France https://www.iltm.com/RXUK/RXUK_ILTM/2014/pdfs/leaders_forum_white_paper_2010.pdf?v=635264421266205336

Innovation Norway. (2017). Key figures for Norwegian tourism 2017. Retrieved from https://assets.simpleviewcms.com/simpleview/image/upload/v1/clients/norway/N_kkeltall_for_norsk_turisme_enkeltsider_NY_VERSJON_2017_5cb226fe-05b7-4e3c-87c7-56003cf87eff.pdf

Innovation Norway. (2019). Nøkkeltall for norsk turisme [Key numbers for Norwegian tourism]. Retrieved from https://assets.simpleviewcms.com/simpleview/image/upload/v1/clients/norway/N_kkeltall_om_norsk_turisme_2019_ny_c28e515f-43f7-43ea-8e7d-6fd01a6ea85b.pdf

ITB. (2017). ITB world travel trends report 2017/2018 boom or bust? Where is tourism heading? Retrieved from https://www.itb-berlin.com/media/itb/itb_dl_all/itb_presse_all/ITB_WTTR_A4_2018_interaktiv.pdf

Jakobsen, E. W., & Espelien, A. (2010). Et Kunnskapsbasert Reiseliv: Veivalg for Næringen. Retrieved from https://www.menon.no/wp-content/uploads/26et_kunnskapsbasert_reiseliv_v21.pdf

Kemp, A. (1998). Perceiving luxury and necessity. *Journal of Economic Psychology*, *19*(5), 591–606. doi:10.1016/S0167-4870(98)00026-9

Lages, L. F., & Fernandes, J. C. (2005). The SERPVAL scale: A multi-item instrument for measuring service personal values. *Journal of Business Research*, *58*(11), 1562–1572. doi:10.1016/j.jbusres.2004.10.001

Larsen, M. H. (2017, July 7). Kinesisk turistboom til Norge [Chinese tourist boom in Norway]. *E24*. Retrieved from https://e24.no/naeringsliv/i/Vbgrk1/kinesisk-turist-boom-til-norge. Accessed on June 20, 2020.

Le, N. T. C., & Quy, V. T. (2020). Personal values of luxury services consumption: A confucian culture perspective. *Journal of International Consumer Marketing*, 1–13. doi:10.1080/08961530.2020.1712292

Line, N. D., & Hanks, L. (2016). The effects of environmental and luxury beliefs on intention to patronise green hotels: The moderating effect of destination image. *Journal of Sustainable Tourism*, *24*(6), 904–925. doi:10.1080/09669582.2015.1091467

Mathisen, K. R. (2017). *Betalingsvillighet for fellesgoder blant turister i Lofoten* [-*Willingness to pay for common goods among tourists in Lofoten*]. Master thesis, Norwegian University of Life Science Ås.

Mei, X. Y. (2014a). Boring and expensive: The challenge of developing experience-based tourism in the Inland region, Norway. *Tourism Management Perspectives, 12*, 71–80. doi:10.1016/j.tmp.2014.09.003

Mei, X. Y. (2014b). Revisiting the role of incoming tour operators (ITOs): Initial findings from Eastern Norway. *European Journal of Tourism Research, 8*, 148–156.

Norwegian Tourism Partners. (2019). Norsk reiselivsåret 2019 [Norwegian tourism year 2019]. Retrieved from https://norsk-reiseliv.no/reiselivsaret-2019/. Accessed on August 1, 2020.

NRK. (2018). Slik skal dei lokke rike turistar til å legge att millionar [This is how they will attract rich tourists to leave behind millions]. Retrieved from https://www.nrk.no/vestland/slik-skal-dei-lokke-rike-turistar-til-a-legge-att-millionar-1.13936506. Accessed on August 4, 2020.

Nwankwo, S., Hamelin, N., & Khaled, M. (2014). Consumer values, motivation and purchase intention for luxury goods. *Journal of Retailing and Consumer Services, 21*(5), 735–744. doi:10.1016/j.jretconser.2014.05.003

Øian, H., & Kaltenborn, B. (2020). Turisme på Svalbard og i Arktis. Effekter på naturmiljø, kulturminner og samfunn med hovedvekt på cruiseturisme [Tourism in Svalbard and in the Arctic. Effects on the natural environment, cultural monuments and societies with an emphasis on cruise tourism]. Retrieved from https://brage.nina.no/nina-xmlui/bitstream/handle/11250/2643245/1745.pdf?sequence=3&isAllowed=y

Pechlaner, H., Fischer, E., & Hammann, E. (2005). Leadership and innovation processes – Development of products and services based on core competencies. In M. Peters & B. Pikkemaat (Eds.), *Innovation in hospitality and tourism* (pp. 31–57). New York, NY: The Haworth Hospitality Press.

Pluribus, A. E. (2019). File: Map Norway political-geo.png. Retrieved from https://commons.wikimedia.org/wiki/File:Map_Norway_political-geo.png. Accessed on July 10, 2020.

Prayag, G., & Hosany, S. (2014). When Middle East meets West: Understanding the motives and perceptions of young tourists from United Arab Emirates. *Tourism Management, 40*, 35–45. doi:10.1016/j.tourman.2013.05.003

Røeggen, A. (2010, January 21). Fikser "hemmelig" norsk luksusferie [Fixing "secret" Norwegian luxurious holiday]. Nettavisen.

Sørholm, M. (2018). *Allemannsretten møter masseturisme: Rettslige virkemidler for vern av naturgrunnlaget i områder med omfattende naturbasert turisme* [*The right to roam meets mass tourism: Legal instruments for the protection of nature in areas with extensive nature-based tourism*]. Master thesis, UiT The Arctic University, Tromsø.

Statistics Norway. (2020). Tourism Satelite accounts. Retrieved from https://www.ssb.no/en/turismesat. Accessed on July 20, 2020.

Swarbrooke, J. (2018). *The meaning of luxury in hospitality, events and tourism.* London: Goodfellow Publisers.

Taylor, C. (2019). A new pop-up hotel in the North Pole will charge guests $100,000 to stay—Take a look inside. Retrieved from https://www.cnbc.com/2019/09/18/pop-up-hotel-in-the-north-pole-will-charge-guests-100000-to-stay.html. Accessed on August 1, 2020.

Veríssimo, M., & Loureiro, S. M. C. (2012). Experience marketing and the luxury travel industry. Paper presented at the Book of Proceedings – Tourism and

Management Studies International Conference Algarve, University of the Algarve, Portugal.

VisitNorway. (2020a). The right to roam: Joys and responsibilities in Norway. Retrieved from https://www.visitnorway.com/plan-your-trip/travel-tips-a-z/right-of-access/. Accessed on June 25, 2020.

VisitNorway. (2020b). Svalbard islands. Retrieved from https://www.visitnorway.com/places-to-go/svalbard-islands/. Accessed on August 2, 2020.

Wen, J., & Huang, S. (2019). The effects of push and pull travel motivations, personal values, and destination familiarity on tourist loyalty: A study of Chinese cigar tourists to Cuba. *Asia Pacific Journal of Tourism Research*, *24*(8), 805–821. doi: 10.1080/10941665.2019.1635504

Yang, W., Zhang, L., & Mattila, A. S. (2015). Luxe for less: How do consumers react to luxury hotel price promotions? The moderating role of consumers' need for status. *Cornell Hospitality Quarterly*, *57*(1), 82–92. doi:10.1177/1938965515580133

Chapter 8

'The Butler Did It!' Luxury Accommodation Management in Tourism from the Caribbean to the Pacific Islands

Johnnel Smith[1]

Abstract

To remain competitive in an ever-changing marketplace, many accommodation managers are constantly upgrading their portfolio of service offerings for guests. Luxury suites with butler service were created as an extra component of luxury to the accommodation model. This inclusion was a welcomed innovation for guests who were willing to pay top dollar for top-of-the-line services and the ultimate in exclusivity. Butlers act as the luxury service delivery mechanisms in accommodations, going above and beyond to meet and exceed guest expectations. This chapter makes a unique contribution to the compendium of knowledge on luxury hospitality and tourism by providing insights from the supply side of luxury accommodation brands in the Caribbean and Pacific regions. A comprehensive review of the literature on the topic coupled with semi-structured interviews were conducted in the Maldives, Jamaica, St. Barts, The Bahamas, St. Lucia, Antigua, Cayman Islands, Barbados and Mexico. With an expressed need to examine the operationalisation (Miller & Mills, 2012a, 2012b) and subjectivity constructs (Godey et al., 2012a, 2012b) on Luxury Management in Hospitality and Tourism, this chapter aims to provide an excellent addition to the discourse. It also aims to address a major gap in the literature as there is a lacuna of recent research on Luxury Management in Tourism in Small Island Developing States (SIDS) from two major tourism dependent regions of the world – the Caribbean and the Pacific.

[1]https://orcid.org/0000-0001-7973-7344

The Emerald Handbook of Luxury Management for Hospitality and Tourism, 163–181

doi:10.1108/978-1-83982-900-020211008

Keywords: Luxury management; marketing; hotel butlers; accommodation; hospitality; tourism; Caribbean Islands; Pacific Islands

Introduction

From a floating breakfast in your pool to a warm, beautiful bath awaiting to pamper you when you return to your hotel room, from arranging the perfect proposal to dinners under the moonlight – the butlers did it! Oh yes, they did!

In an ultra-competitive world, tourism services must have a pulse on innovation, pushing the envelope to offer services that increase brand loyalty and customer retention to ensure repeat purchases. As a critical element in tourism services, the accommodation sector in tourism and hospitality is always on the leading edge of groundbreaking innovations. Evident, is the evolution of accommodations through the years-offering five, six and even seven-star service-catering to every need of the guest. According to Middleton and Clarke (2001), the accommodation sector plays a key role by providing the necessary infrastructure to enable convenient and comfortable travel. Hall (1995) regarded accommodation as one of the more critical components on the demand side – because accommodation has a major influence on the type of visitors who come to a destination. As information flows and guests become more exposed, the accommodation sector must meet and exceed the needs of the modern-day traveler, who may be seeking a range of experiences from basic amenities to an all-time luxury experience.

The Concept of the Luxury Hotel

Hotels – an integral feature of the accommodation model – provide facilities for business transactions, meetings, conferences, recreation and entertainment (Presbury, Fitzgerald, & Chapman, 2005). Hotels are renowned for constantly upgrading their facilities and services offering, to ensure that guests remain satisfied and that their expectations are surpassed. Hsi-Jui Wu and Liang (2005) note that the hospitality industry generally accepts hotel star ratings designated by Forbes (formerly, Mobil) Travel Guide, American Automobile Association (AAA) and other organisations (Sherman, 2007). The perception exists that a hotel that has been awarded the Forbes Five Star award and/or AAA Five Diamond is a luxury hotel, and four-star and five-star hotels generally describe themselves as luxury hotels. The luxury hotel is considered a crucial and rapidly expanding segment of the hospitality industry. The expansion of the luxury hotel segment draws considerable attention from researchers regarding the phenomenon of luxury consumption. The Smith Travel Research (STR) report shows that the United States luxury segment had the highest occupancy rate (72.3%) in the fourth quarter of 2014 (Digital Luxury Group, 2014; Mayock, 2015).

Luxury or luxe hotels are a captivating option for travelers, with their many amenities such as spas, gourmet room service, chic bars and boutiques; these can take any trip or vacation to the next level, and many brands go even further and up the ante with an ultra-exclusive benefit – butler service (Ebsworth-Goold, 2019). Many luxury/five-star hotels and resorts offer butler services for guests staying in high-end or the most luxurious/superior rooms or suites as personalised or custom-made service for the duration of a guest's stay. It is an elite, exclusive, over-the-top experience that is laser-focused on comfort and convenience (Ebsworth-Goold, 2019).

The Butler – An Added Value in Luxury Accommodation Models

An often-overlooked position in the luxury model, the butler provides exceptional services to enhance the guest experience. Tasked with going beyond excellent service delivery, butlers create 'wow' moments for their guests from orchestrating an epic wedding proposal, hassle-free check-in or not check-in at all, breakfasts at sunrise, floating-gazebo dinners, flower/rose petal baths. Tracking down long-lost relatives, planning and arranging reunions, popping champagne bottles for guests, creating a home-away-from-home experience for guests by replicating elements of a guest's home in their suite, driving long distances to deliver lost items, handling medical emergencies smoothly are just a few examples of how butlers transform a normal hotel stay into an above and beyond experience (Ferry, 2011).

Providing the ultimate in exclusivity, some resorts with butler service are only available in the very highest level of suites, providing guests with an incredible mind-boggling standard of service and supreme luxury, where every need of the guest is anticipated and provided for (Sandals Resorts International, 2020). Butlers, similar to Rolls Royces, Bentleys, Rolex watches, designer handbags, superyachts and private jets, symbolise and represent the ultimate in affluence, superior service and lifestyles (Ferry, 2011).

The primary reason some guests return is because of their butler service, with some hotels enjoying upwards of 35%–40% guest return rate (The International Institute of Modern Butlers, 2014). Ferry (2011) further shares that travelling is a lonely business, so butlers putting the mansion-away-from-the-mansion back into the equation certainly adds value to a hotel's offering.

However, one could argue that there is another component in hotels offering a similar service- the concierge. What is the difference between a concierge and a butler? Wood (2017) shares that the main difference between a concierge staff member and a butler is that concierges generally handle requests for things outside of the room, such as restaurant reservations or shopping assistance, while butlers tend to guests' needs inside the room, such as unpacking and beverage service. The role of the butler is to make the guests say 'WOW!' This is probably why guests tend to rave about their experience at these hotels, with 'nearly 100% exceptional feedback from our guests', as one General Manager raved in turn, and comments like 'the best service they have received in all their travels', as one head butler reports (Ferry, 2011).

Ferry (2011) further adds other benefits of butlers in these hotels to include:

(1) Personalised service based on a database of guest preferences;

(2) Provide a single point of contact for guests who take ownership of any problems and remove the worry from the guest experience;
(3) The development of a relationship that encourages repeat visits, with guests requesting the same butler.

With the value of butler service incontrovertible, this study sought to conduct a deep dive into the butler's world in the luxury accommodation model. There is a famous adage dubbed, 'The Butler did it!' as the chapter is aptly entitled, but how exactly do they do it, and what are the intricacies involved in this career trajectory?

Hence, the following research aims guided the study:

(1) To examine the role of the butler and the elements of butler service in luxury accommodation;
(2) To unearth the personal characteristics and service attributes of butlers and the challenges faced by butlers in job performance;
(3) To identify the process of recruitment, training and retention that enable butlers to deliver exceptional service in luxury hospitality accommodation.

Literature Review

The chapter will explore three salient components on the supply side of luxury accommodation. These encapsulate service encounters and delivery in luxury consumption and the role of employees in luxury service quality, and the process of managing talent to provide the same. The literature on the topics has been reviewed and grouped accordingly:

Theoretical Implications on Service Encounters and Delivery in Luxury Management

The first component encompasses service encounters. Butlers play a vital role in creating service encounters within the hospitality and tourism sector. Service marketing scholars usually explore the service encounter elements to divulge personal interaction as increasing the pleasure of service encounters can reduce the perceived risk associated with purchasing a service and improve the buying experience (Julian & Ramaseshan, 1994). Service encounter elements are, therefore, considered as the antecedents of experiential value.

Customer service encounters are defined as the lasting personal impressions that customers receive upon first encountering a product, service and/or company, which they hopefully take with them and communicate to others (Pine & Gilmore, 1998, 1999; Poulsson & Kale, 2004). Marketing scholars have long postulated that customer service encounters instil good customer impressions of a company (Dixit, 2017; Grove, Fisk, & Dorsch, 1998). Customer service encounters offer basic information about customer preference, and businesses can use it to improve their service. In the literature on customer service encounters, Shieh and Cheng (2007) analyzed several studies of the relationship between user experience and satisfaction. Results indicate that, compared with first-time guests, repeat guests of luxury hotels have a significantly higher perception of brand awareness and brand loyalty, resulting in higher brand attitude and purchase intention. This chapter explores how butlers create exceptional customer service encounters that lead to higher brand perception, purchase intention and repeat business.

Employees and Service Quality in Luxury Service Delivery and Management

The second component comprises service employee factors, including the kind of employees providing service to consumers. Butlers are service employees and provide critical interactive relations with guests continuously. In the past, many service brands have applied a consumer-goods approach to branding to handle a more competitive environment. This approach emphasised the role of advertising in building a strong brand but underestimated the role played by employees in strengthening the brand during their interactions with customers. More recently, the development of models that provide an understanding of the process of building a powerful service brand has emphasised the role of employees and their contribution to enhance brand equity. The role of employees in building the brand and making the brand 'come alive' is considered essential.

Hence, the interactive relationship between consumers and service employees is important to consumer evaluations. Specific behaviours, including the behaviour of service employees, are the key determinants of perceived service quality and also of consumer satisfaction (AndaleebSaad & Conway, 2006; Wu & Liang, 2005). Keng, Huang, Zheng, and Hsu (2007) found that personal interaction encounters and physical environmental interaction encounters positively influence customer experiential value. Liu et al. (2017) cite that the empirical results from their study indicated that interactions with service employees and other consumers positively influence the consumer experiential value. Additionally, only interactive relationships with service employees directly and positively affect consumer satisfaction.

Managing Talent in Luxury Service Delivery and Management

With the intimacy of exchanges and interactions between the butler and the guest, butlers are vital for perpetuating brand values. Employees must be committed to demonstrating the brand values (as expressed by top management) each time a customer interacts with the brand (Kimpakorn & Tocquer, 2009). Employees' brand commitment is defined as the extent to which employees experience a sense of identification and involvement with the brand values of the company they work for. The dimensions of employer brand (the organisation's image as seen through the eyes of associates and potential hires) are used as independent variables for explaining the level of employee's commitment to their company's brand (Kimpakorn & Tocquer, 2009).

Sherman (2005) further shares that in the luxury hotel, unequal entitlements between workers and guests – both to human attention and labour and to social and material resources generally – are prominent. Participant observation in two hotels shows that one of the myriad ways interactive workers respond to their subordinate position is to establish themselves as superior to others. They use comparative strategies of self to situate themselves as privileged on a range of symbolic hierarchies, including those of competence, authority, status, need, morality, intelligence and cultural capital. Comparative ethnographic data on these judgements show that they are not fixed categories in people's heads but rather contradictory, context-dependent orientations. The chapter further argues that even as they establish workers as symbolically superior, these strategies also constitute some people as legitimately entitled to workers' labour, thereby normalising inequality between workers and guests. This chapter explores these theories on copiability and other strategies used in job performance by butlers.

Methodology

Based on the hotel star ratings designated by Forbes Travel Guide (2009) and the 'Hotel Class Appraisal System' developed by the Taiwan Tourism Bureau (2006) for luxury-hotel restaurant classification, this study analyzed the hotel industry by sampling respondents from four-star (first class), five-star and six-star (luxury) hotels. Given the proximity and network of the researcher, the study was conducted within the Caribbean and Pacific regions. The countries sampled were: the Maldives, Jamaica, St. Barts (St. Bartholomew), The Bahamas, St. Lucia, Antigua, Cayman Islands, Barbados and Mexico.

The sample was drawn using the snowball technique and consisted of: butlers, butler supervisors, assistant head butlers and head butlers/butler managers. The researcher used the survey method of qualitative interviews (Jamshed, 2014) to collect the data. The sample was asked to complete the survey/a semi-structured interview via email return or online via Zoom and over the phone. Twenty-two interviews were conducted over a six-month period between April and September 2020. Eighteen interviews were conducted in the Caribbean and four in the Pacific. Of the 22 interviews conducted, only 20 were deemed valid for inclusion in the final results due to the challenges faced with technology and receiving incomplete responses to the survey instrument.

Findings and Discussion

Sample Profile

Table 8.1 lists the analytical results of the study/sample. Most of the respondents were male (63.6%), 26–45 years old (28.3%), currently working as butler (97%), recruited from another job in hospitality (53.6%) and university-educated (27.9%). Interviewee age and gender were relatively consistent, which validated further analysis of the study.

Table 8.1. Profile of Survey Respondents.

Characteristics	Descriptions	Statistics	(%)
Gender	Male	14	63.63%
	Female	8	36.37%
Age	Younger than 25 years	2	9.09%
	26–35	6	27.27%
	36–45	10	45.44%
	46–55	3	13.65%
	56 and older	1	4.55%
Education	Junior high school	0	0%
	Senior high school	8	36.7%
	Bachelors' degree	14	63.3%
	Master's or doctorate degree	0	0%

Your Wish Is My Command – *The Role of the Butler and the Elements of Butler Service in Luxury Accommodation*

Trained in the fine art of pouring wine, setting a table for an impressive seven-course meal, ironing a shirt and tying a tie to perfection, the main concept of butler service is to identify what an individual guest needs and make sure that they know that they are being thought of. Drawing a bath with rose petals, reserving a cabana, taking care of children and going snorkeling, each service is tailor-made directly at what would make the guests more comfortable to stay in the hotel (Travel Associate, 2019).

The work structure sees a butler working for 12 hours per day with 3–4 days off per week. Therefore, most guests will have two butlers per day. 'My duty first and foremost is to ensure the smooth operation of the butler team and to ensure they are exceeding guests' expectations. I also oversee all the guest preferences and make sure every department is briefed to deliver on their requests in a timely manner. Our duty as butlers is to ensure that every aspect of a guest vacation is taken care of from the moment, they step foot on the resort; from handling excursions to communicating any dietary restrictions, to setting up dinner reservations, to organizing chairs on the beach or reserving a private poolside cabana. Their every wish is our command! We even take things a step further. If we know it is our guest's birthday or anniversary, for instance, we will create a special moment for them' (Kickham, 2019).

As stated, prior, butlers are known for offering personalised service and act as a mini concierge for the guests. Other examples of typical butler services include ironing clothes for guests, handling meal requests and special dining requests, making drinks and snacks, scheduling treatments at the spa (sometimes in-room spa treatments) and attending to any need from the guest in a unique, one-of-a-kind, way (Sardone, 2019). Exclusivity is paramount in butler service offerings.

Respondent #1 (from the Maldives): 'We help couples with their engagement by planning the ultimate proposal. We also help with the renewal of vows. We are all about creating "WOW" moments. As butlers, it is important to listen and be savvy. We can arrange a special surprise breakfast on their patio, special dinners for honeymooners. We write messages in the sand that we can surprise the guest with when they go on their balcony. We do surprise lunches and everything with champagne'. All these surprises and delights come at no extra charge and are all a part of the butler experience. If a guest requests that we accompany them for shopping, and excursions, and even diving and having dinner, we are happy to do this with them (Corporate Maldives, 2019).

> The classic hotel butlers work around the clock to service guests' needs. Dressed in the signature three-piece suits and tails, butlers offer seamless service for those staying in the top end suites. Suites are all attended by a 24-hour butler who handles unpacking and packing, customizes the private bar with precise beverage preferences that would satisfy the crankiest headliner's contract, provides an array of personal spa amenities, coffees and herbal teas, and can create a customized guest CD music track. The Butler will also set the room's comfort level for guests, adjusting the TV, radio, drapes, temperature controls, lighting and then, like the infinity edge in the bathtub, disappear.
>
> (Sardone, 2019, para 3)

Ramani (2016) also cites that guest can expect the following from their butler service:

(1) Welcome – The relationship with the butler begins as soon as a guest checks to ensure a seamless and impeccable service for the duration of the stay. The guest is usually given a cell phone where they can call or text the butler for any assistance needed, no matter the time of day or night.
(2) Unpacking – The butler will assist a guest in unpacking. If they notice that any clothing items need pressing, special handling or shining, they will handle those for a guest, and when it's time to leave, they will assist with packing.
(3) Shopping – A butler is available to act as a personal shopper and scout out exactly what guests require with ease.

The Evolution of the Butler – *Changing Paradigms and Expectations in the Luxury Consumer*

If the term 'butler' conjures up images of *Downton Abbey*'s stone-faced Mr. Carson, think again: hotels today are trading in stuffy formality for targeted services that are personalised, useful, and unexpected – and often downright quirky (Ramani, 2016). The travel industry has moved away from the traditional definition of luxury; the high-end experiences sought after these days are experiential in nature, individually crafted to the client, and hyper-local, and the growing interest in specialised butler services is a natural progression of these trends. Themed butler services are also a way for the properties to differentiate themselves and create a surprise and delight experience for the client (Ramini, 2016).

Respondent #3 (from the Maldives): 'One morning, I brought a floating breakfast to a couple, to go above and beyond the typical in-room dining experience. This blew their mind; they were so excited'. Therefore, the mix of services designed to delight and thrill guests is ever evolving and has been even more creative as cited by Ramani (2016) and is organised below in Table 8.2.

The Challenges Faced by Butlers in Service Delivery

The respondents from the research sample noted that high customer expectations are one of the biggest impediments to service delivery. The surveys revealed that there was difficulty in meeting the increased expectations of guests. One butler mentioned that customers are more aware, travel a lot and stay in many places, and they compare us to where they have stayed before. Butler Managers also reported that guests are often unwilling to pay for the extras they expect and demand; This also occurred in the work of Presbury et al. (2005). Customer's expectations are related to their previous experiences. Unfortunately for us, a lot of our customers' previous experience is with local clubs, so they expect greater value for money, and a lot of free facilities and services … Most of our customers understand the pressure of competition, and they use it to their advantage'. Another challenge revealed by the respondents was the lack of support from other team members within the other departments in the hotel. Respondent 6 (from Jamaica) says, 'Sometimes the members of the other departments such as Food and Beverage do not respond as quickly to the demands of the guests and any time delays will reflect badly on the Butler'.

The Person Behind the Service

The sample revealed that these were critical personal characteristics that would denote someone's ability to be a great butler. 'A butler has to be exposed, knowledgeable and well-versed, to communicate effectively with guests. You want your guest to feel comfortable about sharing personal information with you. The Butler becomes your friend and a confidante', Respondent #7 (from Barbados).

The following characteristics were also cited from the sample:

- Think quickly and on your feet.
- Very caring about people.
- Dedicated to giving very personalised service.
- Attention to detail.
- Bond with guests.

Table 8.2. The New and Evolving Mix of Butler Services in Luxury Accommodation (Ramani, 2016).

Services	Description
Technology assistance	If a guest has work to complete while on the road, or perhaps they are having difficulty downloading vacation photos from their phone to their laptop or other device, this designated staff member is ready to assist with any technology trouble. The tech butler can help to access the Wi-Fi, provide the right power plugs and even resolve software compatibility issues, taking the stress out of a vacation.
Services for women travellers	More and more women are travelling abroad, both in groups and solo. To cater to these women, an all-female butler staff and chauffeur fleet exclusively for female travelers are offered at some hotels. It is an amenity that has increased in recent years, as traveller safety has become a top priority.
Pet services	Many luxury hotel chains offer butler services to four-legged friends as well. Some hotels provide pets special treats, bowls and place mats at check-in and gourmet room service options for both cats and dogs.
The Instagram butler	To ensure the ideal images are captured for social media, the Instagram butler takes guests to the most photogenic locations and advises on the best times and angles to use to capture the perfect poses in paradise. An 'InstaButler' helps guests capture 'gram-worthy moments' by taking them to spots of the property that are normally off limits.
Bonfire butler	A bonfire butler will help guests scout out a prime beach location, set up chairs and blankets and build a roaring fire over which they can roast s'mores. Guests will also learn tips on how to build a proper bonfire.
Hydrotherapy butler	Personal hydrotherapy butlers are on hand to guide each guest through the spa process, from the locker room and relaxation area using the hydrotherapy circuit, which includes scented

Table 8.2. *(Continued)*

Services	Description
	steam rooms, an ice room, hot and cold plunge pools and a water walkway lined with massaging pebbles.
Ambience butler	The ambience butler will contact a guest before their visit and will walk them through choices on aromatherapy, ambient music, healthy treats, reinvigorating elixirs and more. The ambience butler will then set the mood and stock the room accordingly in advance of the guest's arrival.
Amphibious butler	The amphibious butler is usually dressed in a logoed wetsuit; the butler wades around the pool, taking drink orders, collecting glasses, and offering cold towels and spritzes of water, helping guests maximise their in-water time and making sure they are never without a cold beverage in hand.
Energiser butler	The energiser butler will deliver extra juice to your mobile devices, whether you are on or off hotel grounds. The butler is armed with power bank, wireless chargers, which work with iPhones, androids, tablets and music players. This is especially useful if a guest forgets their phone charger.
Fragrance butler	Checked baggage fees and travel regulations limit liquids of a certain size in carry-ons; travelling with one's personal scents can be challenging. The 24-hour fragrance butler will appear at a guest's door with a silver tray laden with 10 fragrance choices; the guest can mist themselves with the scent of their choosing, then ring the butler whenever they need another spritz.
Bud-ler	The bud-ler delivers floral arrangements to guest weddings and events.
Ski butler	Ski butlers will come to a guest to outfit everyone in all the pieces needed to hit the slopes so that the guests do not have to travel with all that gear and can skip the lines at the local shops.

Table 8.2. *(Continued)*

Services	Description
Barbecue butler	The BBQ butler will help set up and man the grill, offer tips on crafting the perfect barbecue and serve a meal full of Southern flavors. You can select sides, sauces and more from three preset BBQ butler menus.
Picnic butler	The picnic butler is an experienced forest ranger who will offer guidance on foraging, making fires, spotting wildlife and outdoor survival – all topped off with a locally sourced picnic lunch. Guests can choose from a range of picnic butler experiences, including ones tailored toward couples or families.
Serenity butler	The serenity butler is dedicated to bringing physical, mental, and spiritual wellness to guests, whether it's providing one-on-one yoga instruction to a business traveler following a day of meetings or meditation tips to a bride looking to calm her nerves before the big day. The serenity butler can also create curated relaxation playlists and customised aromatherapy (including ones that help with sleep issues), and set-up in-room yoga sessions.
Oyster butler (and crawfish concierge)	The hotel's oyster butler offers tips on shucking, saucing and serving – while the crawfish concierge will teach you how to peel, cook and eat a crawfish.
Fireside butler	This is a butler who comes and fills up their outdoor hot tub and light the outdoor fireplace, so guests can relax under the stars.

The professional trajectory of a butler seems to place a lot of emphasis on soft skills, but a blend of technical and soft skills is important. 'This is a professional industry where it is not limited to anyone. Having a good attitude, discipline and integrity are always important. Stick your mind to a certain goal, keep targets to achieve and work for it. Dreams never come true without hard work and dedication. The only experience does not help when you look at higher superior posts, it requires a good academic performance too' (www.corporatemaldives.com).

In conducting one of the interviews, a butler manager from the sample shared his experience of one of the challenges faced on the job and the personal attributes that he used to overcome it. This butler has been working in this area for over 5 years and is known for creating romantic 'wow' moments for couples in love. However, his attempts to create these romantic experiences for a couple were rendered futile when he was made aware that the couple was staying at the resort was not in love or celebrating any special occasion. In fact, these romantic gestures brought up negative memories and emotions in the wife of the couple, but this butler still wanted to create an exceptional experience for the couple, and so he set to work.

Case Study 8.1: 'Spare Your Efforts...We Are Not in Love!' – An Interview with the Butler

Interview Question: Please provide an example where you went above and beyond to meet guests' needs.

Response: 'I had a couple arrive at the resort a few months ago. As this is a couple's resort and the theme is love and romance, we assume and anticipate that couples who opt for staying at this hotel are arriving deeply in love and want to keep their romance alive, but this was not the case with these guests. As is customary, I decorated their room with candles, had wine on chill and drew a beautiful bath in the love tub in their bathroom. When they entered the room and saw the decor, the wife immediately burst into tears and told me, "Please don't do anything like this ever again. Spare your efforts, we are not in Love". She later revealed that they were going through a divorce and had only decided to come to the resort as her husband's company had an alliance with the resort and received free resort passes, and they just wanted a vacation. I felt shattered and deeply disappointed that my efforts were thwarted. I was also very cautious about doing anything else for this couple for the rest of their stay, but then I remembered that my role as butler is to create superior experiences for the guest, regardless of the situation. I soon remembered something else... While I was giving this couple a tour of the property earlier, the wife deeply admired the architecture and the paintings on the walls of the property. She also shared that she was a painter too and that she used to paint with her husband when they just met and how much she missed that. I also remembered that the artist who painted these pieces visited the hotel a few months ago. I was able to get his contact, and I asked him if he would be able to come back to the property as he had a new admirer of his work. He quickly arrived at the hotel, and I was able to arrange a painting session with the artist and the couple. They were both so delighted. The wife cried again...happy tears this time. This is the true role of the butler. Going beyond the call of duty and creating unforgettable experiences. Always find ways to create an experience. Delighting a guest is possible, but it also requires listening to your guests intently and looking for connection points and moments where you can create a "WOW"!'.

Recruitment, Training and Retention

Recruitment – Finding the Best

With such high expectations for exceptional service delivery to guests, the recruitment process for the position of butler is undoubtedly intricate, but where is the talent pool, and how are amazing butlers created? Hoteliers agreed that hiring experienced butlers is ideal for hotels that want to have butler service as part of their suite of service offerings. Hiring managers and those with a passion for serving are also ideal, training them repeatedly to focus on attention to detail, especially in compiling and following guest preference databases (Ferry, 2011). The butlers in our sample revealed that they were recruited from positions such as waiter in a restaurant (4%), the food and beverage department within the hotel (64%), human resources department (4%), the housekeeping department (9%), the front office (17%) and from a career college fair at a local institution (2%).

The butler managers in the sample concurred that finding the right people with the right attitudes is not a very easy task. The pool of available talent from which to choose was perceived to be very small. Managers also reported that some people who apply for positions do not wish to pursue a career in the hospitality industry in the long term and that some persons saw this job as temporary and transient until they received the job they really wanted. When the right people are found, the training of this small talent pool is equally important.

Training – Preparing the Best

Training has as its core the development and enhancement of the skills necessary for employees to effectively handle their increased authority and responsibility and exercise good judgement (Brymer, 1991). The sample revealed that the following organisations provided training for butlers:

(1) The Guild of Professional English Butlers
(2) The British School of Etiquette
(3) SABA International Butler Academy
(4) South African Butler Academy

Kickham (2019), in her interview with a butler, shared this remark, 'The founder of the Guild of Professional English Butlers (who was contracted to train in accordance to exacting standards by the Guild of Professional English butlers, which provides butlers to nobility and celebrities), came to Grenada, so we didn't have to go to England. We trained for three months before the resort opened. We were trained on different cultures, how to fold clothes, how to communicate and above all, deliver exceptional service. The biggest thing for a butler is teamwork. We trained five days per week, eight hours a day. And we even learned how to make towel animals'!

The butlers from our sample further shared that they all received full-day workshops, where they had to practice standing on their feet for 12 hours, 'We were trained on everything, proper table setting, manners, etiquette, emotional intelligence, art and craft and the basics of child care', Respondent #20 (from the Bahamas). Wood (2017) conveys that prior hospitality experience is often a prerequisite, and having multi-cultural experience is an important factor. Some butlers undertake a two-day butler certification training, which is based on English tradition. The butlers receive special training on packing to assure that guests' clothing emerges wrinkle-free on arrival to their destination. Butlers work in different areas within the hotel, such as with the concierge, wine sommelier and guest services to ensure a complete understanding of how the hotel operates.

Klidas, Van den Berg, and Wilderom (2007) cite that in the literature, one can find a true host of skills, which, it is claimed, should be instilled in employees in the context of 'empowerment training' (Jones, Palmer, Whitehead, & Osterweil, 1996, p. 545). Emphasis is generally placed on soft skills that are aimed at enhancing employees' competence in customer care, such as interpersonal and communication skills (Ripley & Ripley, 1992). Training also needs to impart the technical and professional knowledge and skills necessary for the delivery of quality guest service (Lewis & Meadows, 1995). With such skills, employees will develop capacity to respond to customers' needs and requests with expediency.

Retention – Keeping the Best

The sample revealed that the butlers were working in their position for a time range of 6 months–12 years. Some properties face a challenge with retaining butlers, as there is a staff turnover every two years – not because of dissatisfaction, but because, having worked at the hotel, Butlers became valued commodities in the West, as well as their home countries (Ferry, 2011).

There are several solutions used to solve the problem of frequent staff turnover. These include hiring mature butlers, providing better pay and living conditions, allowing the butlers to multitask and, of course, promoting internally, so there was a career path worth pursuing (Ferry, 2011). Other strategies from the index of the solution include employee empowerment in the form of more training (ex-ante control), creating good staff morale (meta-control) and remuneration (ex-post control).

Employee Empowerment

Klidas et al. (2007) further suggest that training is probably the most frequently used 'intervention' when considering what is crucial in the implementation of empowerment (Brymer, 1991; Jones et al., 1996). Training is a means of exercising 'ex-ante control' over employee behavior, by introducing the necessary skills into the work process (Hales & Klidas, 1998). The training process further incorporates a system of intense socialisation (induction) and development designed to inculcate employees with organisational values and to instil, shape and reinforce appropriate employee attitudes (Hales & Klidas, 1998).

Effective training generally supports the goal of empowerment by enhancing morale and motivation, increasing flexibility, reducing staff turnover and allowing employees to function independently without the need for constant supervision (Hope & Mühlemann, 1997). Finally, it fosters in employees feelings of self-efficacy necessary to initiate their participation in decision-making (Conger & Kanungo, 1988) by imparting the requisite skills and knowledge needed to render such participation effective. Training enables employees to use their increased authority responsibly and effectively (Grönroos, 2000). Creating a strong customer-oriented culture is also vital in the empowerment process. Management should consistently strive to institute a culture strongly oriented to service quality and customer satisfaction. Such reinforcement requires not only the rhetoric of the company's commitment to the goal of service quality and customer satisfaction but also a more specific organisational alignment towards this goal.

In addition to a satisfactory paycheck and benefits package, butlers work for tips. There are several guidelines offered online for tipping butlers. 'There are many ways a butler can help take your travel experience to the next level!' However, should you opt for such service, it is important to remember that these men and women work extremely hard to make sure your vacation is efficient and enjoyable. In return, gratitude and gratuities are both in order if your butler has provided exemplary service. Do not tip per service (each time a meal is brought, or something is ordered or procured); instead, plan to leave your butler 5 percent of your room rate at the end of your stay (Ebsworth-Goold, 2019). The sample revealed that butlers receive tip amounts in the hundreds and thousands, ranging from $400.00USD to over $5000.00USD. One butler shared that the biggest tip she has ever received was $5700.00USD.

Conclusion

The participating hotels agreed that all five-star/diamond hotels needed to offer butler service if they expect to deliver top-level service; one even suggested that some four-star hotels should also offer butler service; as the wow factor and what it does for word of mouth promotion, repeat visits, occupancy, revenue per available room (RevPAR) and the bottom line (Ferry, 2011), but as we see from the information shared in this study, there are many intricacies involved in creating these wow factors – from recruiting the right talent to training and retention. It is also critical to note the evolution of this service in luxury accommodation management. Luxury services are always in invention and innovation mode, catering to the evolving need of the more exposed traveler. This chapter fills a major dearth in the literature on this topic and provides valuable insights into the role, training and retention on a vital component on the supply side of luxury management and practical applications for the industry. Butlers have done it in the past and will continue to do it – breaking boundaries and changing paradigms in the luxury travel market.

References

AndaleebSaad, S., & Conway, C. (2006). Customer satisfaction in the restaurant industry: An examination of the transaction-specific model. *Journal of Services Marketing*, *20*(1), 3–11. doi:10.1108/08876040610646536

Brymer, R. A. (1991). Employee empowerment: A guest-driven leadership strategy. *Cornell Hotel and Restaurant Administration Quarterly*, *32*(1), 58–68. doi:10.1177/001088049103200116

Chen, W., Wang, K. C., Luoh, H. F., Shih, J., & You, Y. (2014). Does a friendly hotel room increase senior group package tourists' satisfaction? A field experiment. *Asia Pacific Journal of Tourism Research*, *19*(8), 950–970. doi:10.1080/10941665.2013.833123

Conger, J. A., & Kanungo, R. (1988). The empowerment process: Integrating theory and practice. *Academy of Management Review*, *13*(3), 471–482.

Digital Luxury Group. (2014). The World luxury index hotels 2014: The most sought-after luxury hotels. Retrieved from http://www.digitalluxurygroup. com/intelligence/research/world-luxury-index-hotels-2014/

Dixit, S. K. (Ed.). (2017). *The Routledge handbook of consumer behavior in hospitality and tourism*. London: Routledge.

Ebsworth-Goold, E. (2019). 8 luxurious perks you might get with hotel butler service. Retrieved from https://www.travelawaits.com/. Accessed on September 10, 2020.

Ferry, S. (2011). Butler service today. Retrieved from https://www.hospitalitynet.org/opinion/4049982.html/. Accessed on September 10, 2020.

Forbes Travel Guide. (2009). World's top hotels. Retrieved from https://www.forbes.com/2009/11/09/best-luxury-hotels-lifestyle-travel-hotels.html?sh=42e014b2aa8c

Godey, B., Pederzoli, D., Aiello, G., Donvito, R., Chan., Oh, P., … Weitz, B. (2012a). Brand and country-of-origin effect on consumers' decision to purchase luxury products. *Journal of Business Research*, *65*(10), 1461–1470. ISSN 0148-2963. doi:10.1016/j.jbusres.2011.10.012

Godey, B., Pederzoli, D., Aiello, G., Donvito, R., Chan., Oh, P., … Weitz, B. (2012b). What is the value of luxury? A cross-cultural consumer perspective. *Psychology and Marketing*, *29*(12), 1018–1034. Wiley Online Library.

Grönroos, C. (2000). Service management and marketing: A customer relationship management approach. Retrieved from http://www.wiley.co.uk

Grove, S. J., Fisk, R. P., & Dorsch, M. J. (1998). Assessing the theatrical components of the service encounter: A cluster analysis examination. *Service Industries Journal*, *18*(3), 116–134.

Hales, C., & Klidas, A. (1998). Empowerment in five-star hotels: Choice, voice or rhetoric?. *International Journal of Contemporary Hospitality Management*, *10*(3), 88–95. doi:10.1108/09596119810210260

Hall, C. (1995). *Introduction to tourism: Development* (2nd ed.). Melbourne: Longman Cheshire.

Hope, C., & Mühlemann, A. (1997). *Service operations management: Strategy, design, and delivery*. Hoboken, NJ: Prentice Hall.

Jamshed, S. (2014). Qualitative research method-interviewing and observation. *Journal of Basic and Clinical Pharmacy*, *5*(4), 87–88. doi:10.4103/0976-0105.141942

Jones, P., Palmer, J., Whitehead, D., & Osterweil, C. (1996). Performance through people. *Empowerment in Organizations*, *4*(4), 23–27.

Julian, C. C., & Ramaseshan, B. (1994). The role of customer-contact personnel in the marketing of a retail bank's services. *International Journal of Retail & Distribution Management*, *22*(5), 29–34. doi:10.1108/09590559410067316

Keng, C., Huang, T., Zheng, L., & Hsu, M. (2007). Modeling service encounters and customer experiential value in retailing: An empirical investigation of shopping mall customers in Taiwan. *International Journal of Service Industry Management*, *18*(4), 349–367.

Kickham, D. (2019). Retrieved from https://www.forbes.com/sites/debbikickham/2019/10/31/why-the-butler-becomes-your-best-friend-at-sandals-grenada/#29f822c47917

Kimpakorn, N., & Tocquer, G. (2009). Employees' commitment to brands in the service sector: Luxury hotel chains in Thailand. *Journal of Brand Management*, *16*, 532–544. doi:10.1057/palgrave.bm.2550140

Klidas, A., Van den Berg, P., & Wilderom, C. (2007). Managing employee empowerment in luxury hotels in Europe. *International Journal of Service Industry Management*, *18*(1), 70–88. doi:10.1108/09564230710732902

Lewis, P. G., & Meadows, A. J. (1995). Canadian tourist information and the UK: Plans and perceptions. *Tourism Management*, *16*(1), 67–72. ISSN 0261-5177. doi: 10.1016/0261-5177(94)00009-Y

Liu, M. T., Wong, I. A., Tseng, T.-H., Chang, A. W.-Y., & Phau, I. (2017). Applying consumer-based brand equity in luxury hotel branding. *Journal of Business Research*, *81*, 192–202. doi:10.1016/j.jbusres.2017.06.014

Mayock, P. (2015). The 2015 big brands report. *Hotel News Now*. Retrieved from http://www.hotelnewsnow.com/media/File/PDFs/Misc/

Middleton, V. T. C., & Clarke, J. R. (2001). *Marketing in travel and tourism* (3rd ed.). Routledge. doi:10.4324/9780080511108

Miller, K., & Mills, M. (2012a). Contributing clarity by examining brand luxury in the fashion market. *Journal of Business Research*, *65*(10), 1471–1479. ISSN 0148-2963. doi:10.1016/j.jbusres.2011.10.013

Miller, K., & Mills, M. (2012b). Probing brand luxury: A multiple lens approach. *Journal Brand Management*, *20*, 41–51. doi:10.1057/bm.2011.64

Pine, B. J., & Gilmore, J. H. (1998). Welcome to the experience economy. *Harvard Business Review*.

Poulsson, S. H. G., & Kale, S. H. (2004). The experience economy and commercial experiences. *The Marketing Review*, *4*(3), Autumn, 267–277(11). doi:10.1362/1469347042223445

Presbury, R., Fitzgerald, A., & Chapman, R. (2005). Impediments to improvements in service quality in luxury hotels. *Managing Service Quality: International Journal*, *15*(4), 357–373. doi:10.1108/09604520510606835

Ramani, S. (2016). Retrieved from https://www.cntraveler.com/stories/2016-05-27/16-over-the-top-hotel-butler-services/. Accessed on September 1, 2020.

Ripley, R. E., & Ripley, M. J. (1992). Empowerment, the cornerstone of quality: Empowering management in innovative organizations in the 1990s. *Management Decision*, *30*(4). doi:10.1108/00251749210014743

Sardone, S. B. (2019). Butler services at hotels. Retrieved from https://www.tripsavvy.com/the-butler-does-it-1862835

Sherman, R. (2005). Producing the superior self: Strategic comparison and symbolic boundaries among luxury hotel workers. *Ethnography*, *6*(2), 131–158.

Sherman, R. (2007). *Class acts: Service and inequality in luxury hotels*. Berkeley, CA: University of California Press. doi:10.1525/9780520939608

Shieh, K. F., & Cheng, M. S. (2007). An empirical study of experiential value and lifestyle and their effects on satisfaction in adolescents: An example using online gaming. *Adolescence*, *42*(165), 199–215.

Smith, A. (2016). Smith travel research- shared, collaborative and on demand: The new digital economy. Retrieved from http://www.pewinternet.org/2016/05/19/the-new-digital-economy

Taiwan Tourism Bureau. (2006). Hotel class appraisal system. Retrieved from http://admin.taiwan.net.tw/indexc.asp

Tsai, M. T., Wu, H. L. & Liang, W. K. (2008). Fuzzy decision making for market positioning and developing strategy for improving service quality in department stores. *Qual Quant*, *42*, 303–319. doi:10.1007/s11135-006-9047-1

Wood, M. (2017). Retrieved from https://www.oyster.com/articles/what-exactly-does-a-hotel-butler-do/

Wu, C. H.-J., & Liang, R. (2005). Effect of experiential value on customer satisfaction with service encounters in luxury-hotel restaurants. *International Journal of Hospitality Management*, *28*(4), 586–593. doi:10.1016/j.ijhm.2009.03.008

The International Institute of Modern Butlers. (2014). Retrieved from http://www.modernbutlers.com/butler-service-today-five-leading-hotels-share-their-secret/. Accessed on September 3, 2020.

Corporate Maldives. (2019). Retrieved from https://corporatemaldives.com/chief-butler-at-westin-maldives-proves-the-importance-of-academic-qualification-along-with-experience/. Accessed on September 2, 2020.

Sandals Resorts International. (2020). Retrieved from https://sandal.com. Accessed on September 2, 2020.

Marriott Bonvoy. Retrieved from https://www.marriott.com/hotels/hotel-information/details-4/mlexr-the-st-regis-maldives-vommuli-resort/. Accessed on September 2, 2020.

Travel Associates. Retrieved from https://www.travelassociates.com/luxury-holidays-news/what-a-hotel-butler-can-do-for-you/. Accessed on September 2, 2020.

Chapter 9

Marketing Gastronomic Tourism Experiences as Luxury

Namita Roy[1] *and Ulrike Gretzel*[2]

Abstract

Luxury has received attention from tourism researchers as an important element of the gastronomic tourism experience. With recent research suggesting food and wine tourism being connected to luxury, it is important to explore how gastronomic tourism experiences are marketed to create such perceptions and feelings of luxury. This chapter aims to understand marketing strategies that support luxury gastronomic tourism experiences. In contrast to the definition of luxury as a performance or a value, this research conceptualises luxury as an affect which is sensed and felt in gastronomic tourism experiences. How this conceptualisation translates into marketing practice is explored for a particular gastronomic region. An in-depth analysis of the website of a destination marketing organisation in the Hunter Valley gastronomic region of Australia shows that the gastronomic tourism experience is marketed as bucolic luxury using marketing strategies of connection, congregation and repetition, all of which channel and maintain the affect of bucolic luxury. The chapter contributes to the literature on luxury marketing in the tourism context by identifying marketing strategies that can augment the affect of luxury for the gastronomy tourist.

Keywords: Food and wine tourism; gastronomic tourism; luxury; Hunter Valley; Australia; marketing

[1]https://orcid.org/0000-0003-4785-9405
[2]https://orcid.org/0000-0001-8416-1829

The Emerald Handbook of Luxury Management for Hospitality and Tourism, 183–197

doi:10.1108/978-1-83982-900-020211009

Introduction

The marketing of luxury experiences is a growing research area, especially in hospitality and tourism (Swarbrooke, 2018). Traditionally, research on luxury in tourism and marketing has defined luxury as a tangible property of the product or service measured through price, authenticity, craftsmanship, quality and exclusivity (Beverland, 2004; Correia, Kozak, & Del Chiappa, 2019; Dixit, 2020a; Taplin, 2015; Yeoman & McMahon-Beattie, 2014) based on its capacity to materially define social status. However, recent research on experiential luxury has challenged these traditional meanings of luxury and redefines the connection between social status and luxury by suggesting experiential luxury to be unconventional (Thomsen, Holmqvist, von Wallpach, Hemetsberger, & Belk, 2020), inconspicuous (Eckhardt, Belk, & Wilson, 2015) and dematerialised (Bardhi, Eckhardt, & Samsioe, 2020). In this chapter, we adopt this reconceptualisation of luxury to study the marketing of gastronomic tourism experiences.

Gastronomic tourism experiences occur as tourists travel to gastronomic regions motivated by the need to travel to savour local food and beverages (Dixit, 2019; Sormaz, Akmese, Gunes, & Aras, 2016). While much research has discussed luxury in gastronomic products, hospitality services (Chen & Peng, 2018; Otero, 2018; Yang & Mattila Anna, 2016) and wine (Beverland, 2006; Maguire, 2013; Reyneke, Berthon Pierre, Pitt Leyland, & Parent, 2011; Taplin, 2015), the connection between luxury and the gastronomic tourist experience is still underexplored (Bellini & Resnick, 2018). We conceptualise luxury as an affect in the gastronomic tourism experience that incorporates the intangible (Eckhardt & Bardhi, 2019), fluid (Holmqvist, Diaz Ruiz, & Peñaloza, 2020), sensorial (Belk, 2020; Schmitt, 2020) and personalised (Morhart, Wilcox, & Czellar, 2020) nature of luxury. We then use the case study of Hunter Valley, Australia, to understand how this affect of luxury is created through marketing communications in the digital realm.

Changing Notions of Luxury

At the core, luxury is primarily defined in the context of affluence and understood as excess and excellence (Cristini, Kauppinen-Räisänen, Barthod-Prothade, & Woodside, 2017). These meanings of luxury connect luxury and social status, which has been studied to be appropriated through conspicuous consumption (Veblen, 1899). Traditional literature on luxury follows this conceptualisation, presenting tangible ways of materially describing social status using notions of price, quality, exclusivity and authenticity. However, status and luxury have recently been studied to be disentangled or renegotiated because of democratisation of luxury or increase in access to luxury (Cristini et al., 2017) and liquidity of social structures (Bauman, 2013), making conspicuousness less reliable as a status signal (Eckhardt & Bardhi, 2019).

Recent research argues for a dematerialised experiential definition of luxury owing to growth in unconventional forms of indulgences, access-based consumption and inconspicuous consumption (Bardhi et al., 2020; Eckhardt &

Bardhi, 2019; Thomsen et al., 2020; von Wallpach, Hemetsberger, Thomsen, & Belk, 2020). While the traditional view on luxury valued material possessions making the product or brand a focus for marketing, new ways of consumption forward a more ephemeral and experiential form of luxury (Bardhi et al., 2020). Several researchers have now moved away from defining luxury in the product to luxury in the consumption experience articulating unconventional meanings of luxury felt in hedonic consumption experiences and knowledge acquisition (Holmqvist et al., 2020). Further, having extraordinary experiences has become a marker of status, challenging traditional meanings of luxury that assumed material ownership as a given (Eckhardt & Bardhi, 2019; Pantano & Stylos, 2020). Bardhi et al. (2020) argue that an experiential redefinition of luxury is also required in reference to an increase in inconspicuous consumption, which suggests experiences to be subtle and felt rather than intelligible markers of status (Morhart et al., 2020). The increasing value of experiential luxury as a status symbol owing to changing meanings of status and characteristics of luxury suggests the need to redefine luxury from an experiential perspective.

Marketing Luxury as an Experience

Defining luxury as experiential also points towards rethinking how luxury is marketed. Following traditional meanings of luxury, current luxury marketing literature focuses on branding of products and services. Marketing luxury is challenging as it requires shaping and building on the socio-cultural discourses that are specific to the product or service (Correia et al., 2019). Incorporating this specificity, most research on luxury marketing suggests using branding strategies to create and sell luxury (Atwal & Williams, 2009; Bellini & Resnick, 2018; Cristini et al., 2017; Dixit & Abraham, 2020; Kapferer, 2014; Lee & Hwang, 2011). Branding can be used to market firm-specific or consumer-perceived conceptualisations of luxury, as it allows the construction of a distinct identity using values, symbols, myths and other representations of culture (Seo & Buchanan-Oliver, 2015). However, research on branding luxury experiences is still nascent. This resonates in the tourism literature, where the focus of marketing communications cues such as stories, descriptions and imagery is on describing products and services (Silkes, Cai, & Lehto, 2013). This presents a gap in understanding the marketing of luxury in gastronomic tourism experiences from an experiential perspective.

Gastronomic Tourism Experiences

Indulging in food and beverages is one of the most important aspects of a tourism experience and often becomes the central motivation for travel. In recent years, gastronomic tourists visiting food and wine-producing regions have surged worldwide (Hall & Mitchell, 2004). Reports by the United Nations World Tourism Organization (UNWTO), (2012) argue for the economic importance of gastronomic tourism for regional development and destination marketing.

Gastronomic tourism has been referred to using different terms, including 'culinary tourism' (Bjork & Kauppinen-Räisänen, 2014; Dixit, 2019; Germann Molz, 2007; Silkes et al., 2013) and 'food tourism' (Boyne, Hall, & Williams, 2003; Hall & Gössling, 2016; Park, Kim, & Yeoman, 2019; Yeoman & McMahon-Beatte, 2016); however, it essentially highlights the idea of 'travelling for food'.

The success of gastronomic tourism is underpinned by the idea of experiencing places and cultures through their food and beverages (Dixit, 2020b; Kivela & Crotts, 2006). Food and beverages are known to be key symbols and manifestations of a destination's culture, identity, history, terroir, myths and discourses (Dixit, 2019). Tourists engage in gastronomic experiences to learn about the destination and its culture. Following this line of research, most literature on gastronomic tourism studies the phenomenon through its value in the 'locality' that is experienced by the tourist through gastronomic tourism (Bjork & Kauppinen-Räisänen, 2014; Boniface, 2017; Dixit, 2019; Okumus & Cetin, 2018).

Research on gastronomic tourism has also explored its connection with luxury. Luxurious gastronomic tourism experiences have been known to enhance destination attractiveness and destination attitudes (Batat, 2019; Chen & Peng, 2018; Peng, Chen, & Hung, 2020). Research indicates tourists travel to gastronomic regions primarily to taste distinct flavours (Chen & Peng, 2018) and savour exclusive handcrafted wines (Bellini & Resnick, 2018). In addition, tourists are drawn towards exclusive and luxurious service environments such as star-rated restaurants and personalised wine tasting when they visit a gastronomic region (Croce & Perri, 2010; Joy, Belk, Charters, Wang, & Peña, 2018; Mason & O'Mahony, 2007). However, much of this research on luxury in gastronomic tourism follows traditional notions of luxury and articulates luxury as a value in the product or the service rather than embedded in the gastronomic tourism experience (Beverland, 2004, 2006; Rokka, 2017; Taplin, 2015). For instance, Beverland (2006) suggests luxury to be defined by price and authenticity in the craftsmanship of wine. Chen and Peng (2018) and Peng et al. (2020) describe luxury for the tourists as situated in the restaurant service. Thus, despite luxury being an important driver of gastronomic tourism, it has not been explored from an experiential perspective. In line with recent conceptualisations of luxury (Bardhi et al., 2020), Bellini and Resnick (2018) argue that luxury in the gastronomic tourism experience needs to be explored beyond the product or the service but rather as embedded in the integrated tourism experience.

Luxury as Affect

Following the need to redefine luxury as 'felt' in experiences, we conceptualise luxury as an affect. Much recent literature on experiential luxury points towards luxury being sensed and felt as an affect. Hagtvedt (2020) articulates luxury in art not in the art piece but in the affect of art felt in the experience. In the same vein, Morhart and Malär (2020) point towards the notion of authenticity in luxury as emergent in the experience through employee behaviour, location, creativity, design and transparency, which transcends luxury as a materialistic value. The non-materialistic ephemeral nature of luxury is highlighted by Schmitt (2020) who discusses luxury as sensed in pleasurable extraordinary experiences. Further, Banister, Roper, and Potavanich (2020) argue for decoupling of luxury from its traditional meanings of status and wealth when consumers 'feel' luxury in everyday experiences. This literature suggests that luxury in experiences is disentangled from the product and emergent in the socio-material arrangements of the experience where luxury is sensed and felt. We articulate this felt nature of luxury as affect.

Affect is defined as a transient, fluid, sensorial, emotional yet indescribable feeling that is felt and sensed (Ahmed, 2013; Buda, d'Hauteserre, & Johnston, 2014; Germann Molz, 2015). We conceptualise luxury as felt in the experience and as constructed through the socio-material interactions people have with the environment (Everingham & Motta, 2020; Germann Molz, 2015; Holst, 2018; Olafsdottir, 2013; Silkes et al., 2013). This redefinition of luxury as a fluid momentary sensorial affect emergent in the experience is useful in understanding the manifestation of luxury in gastronomic tourism experiences and the way luxury should be communicated to prospective customers.

Through this chapter, we study how luxury can be marketed in the context of gastronomic tourism experiences following the recent conceptualisations of experiential luxury as an affect. We highlight how the affect of luxury is portrayed and sensed through digital marketing communications. Using the Hunter Valley, a food and wine region in Australia, as the context, the chapter presents an in-depth analysis of the official website of the Hunter Valley Wine and Tourism Association (HVWTA) to understand the strategies used to market gastronomic experiences as luxury. Websites present a helpful way to understand luxury marketing as they demonstrate marketing strategies as well as mediate and construct customer expectation and experience (Shen, Vel, & Khalifa, 2016). We analysed the content on the HVWTA website by studying the descriptions, images and the website design of the home page, wine, stay and dine experiences listed on the website. As affect emerges through sensing of a combination of discourses and material elements (Ahmed, 2013), websites provide a great way to understand discourses and material cues that can trigger affect. Website content analysis is an established method of research to understand luxury marketing (Baloglu & Pekcan, 2006) and is, thus, helpful in understanding how luxury is marketed as an affect. The next section gives an overview of the selected case study and then presents the analysis of the marketing strategies that are reflected in the HVWTA website.

Hunter Valley, Australia

According to Tourism Australia (2016), 52% of international tourists visiting Australia associate the country with its food and wine offerings. Hunter Valley is one of the six main gastronomic regions in Australia and recorded 13.3 million international and domestic visitors in 2019. The region is a historic wine-producing region with over 150 wineries and hosts some of Australia's most renowned fine dining restaurants. According to Tourism Australia, Hunter Valley has the highest recall rate among domestic visitors. This is not surprising as Hunter Valley is marketed as a luxurious gourmet food and wine region and, therefore, represents an aspirational destination for many Australians that is nevertheless within reach. Some of the food- and wine-related activities offered in the region include wine tasting and purchasing, having meals at wineries, eating at fine dining restaurants, boutique accommodation, staying at wineries, sampling local produce, wine classes and tours and cooking classes.

Hunter Valley features regularly in the James Halliday's book of wines, which catalogues award-winning wines of the year. The region is host to several award-winning winemakers who are one of the main attractions for the wine connoisseurs. Most wineries in the Hunter Valley are run by third-generation winemakers. These winemakers are well established and written about in wine magazines and are locally upheld as experts and celebrities. Local associations hold regular events for interactions with these winemakers. This sort of marketing of the winemaker and their wine attracts tourists through the celebrityhood of the winemaker.

Most wineries in the region have diversified their business into food production which includes cheese, olive and other local food production which they sell at the cellar door along with wine. These additional products are paired with different wines and served to customers during wine tasting sessions which provide an add-on sensorial experience to the traveller.

In addition to producing food and wine, several wineries in the region have distinguished themselves as 'commercial' wineries by adding restaurants and becoming wholesale producers of wine. According to a survey conducted by Tourism Australia in 2015, 92% of tourists considered eating at the winery as an important activity (Australia, 2015). Having combined cellar doors and restaurants, these wineries aim at providing a holistic experience to the tourist.

In addition to the wineries, the region is also host to a gastronomy-themed park with the same name called 'Hunter Valley Gardens'. These gardens not only showcase but also trade the local food and wines of the region. Hunter Valley Gardens hosts different events throughout the year, one of them being the 'Christmas Lights' which is extremely popular in the region.

Hunter Valley is marketed by the HVWTA, which is a membership-based local association of businesses in the region. They advertise member businesses through their website and magazine. Their website www.winecountry.com.au is used as one of the main information resources by the tourists and is the first link on google, while searching for keywords 'Hunter Valley'.

Marketing Bucolic Luxury

Our analysis of the HVTWA website suggests luxury in the gastronomic tourism experience is marketed as bucolic in the case of the Hunter Valley region. Bucolic is understood as country life associated with a pastoral and rural setting. Our analysis suggests an affective signalling of luxury in rustic and rural aspects of country life. While the connection between gastronomic tourism and country is established in the literature (Garrod, Wornell, & Youell, 2006; Hall, Kirkpatrick, & Mitchell, 2005; Huang, Beeco, Hallo, & Norman, 2016), we argue that the bucolic characteristics and setting of the tourism experience is marketed as a luxury experience in this particular case. The affect of luxury emerges when the traditional ideas of luxury are associated with food and wine production practices, and with material cues such as country views, landscapes and aesthetics.

The website communicates the affect or sense of bucolic luxury in the gastronomic tourism experience through its design, copywriting and use of images (Chowdhury, Olsen, & Pracejus, 2008; Kluge, Königsfeld, Fassnacht, & Mitschke, 2013; Silkes et al., 2013; Tsai & Wang, 2017). Literature suggests that words, stories and images are affective in nature, and their alignment is instrumental in creating feelings (Newell, 2018; Reilly, Klima, & Bellugi, 1990; Selting, 2010). In the Hunter Valley case, bucolic luxury is suggested by using alignments of words and imagery which affectively intertwine meanings of luxury with the country setting. In particular, our analysis suggests three strategies that create the affect of luxury, namely, connection, congregation and repetition.

Connection

Affect of bucolic luxury is sensed from the website when multiple meanings of luxury are connected to the countryside and rural setting through copywriting and material cues. Affect of luxury is communicated when multiple meanings of luxury are threaded together with food and wine production practices through words such as *'first-generation winemakers', 'boutique', 'available only at the cellar door', 'award-winning', and 'hand-picked'* and material cues such as country landscape. Affect of luxury emerges as these elements are coupled together to create the feeling of a luxurious experience that can be expected.

The website threads meanings of opulence and personalisation with countryside ideas and practices triggering affect through their positioning with material cues. The images on the website point towards the different meanings of luxury that are intentionally intertwined together with country. Phrases such as *'enjoy the view'* along with the country views depicted in some of the photographs combined with the idea of abundant space create a sense of abundance or opulence. Opulence or abundance is linked with food production and historically indicated agrarian wealth (Rokka, 2017). Thus, the idea of opulence is marketed as luxury using terroir elements such as vines, country views and vistas in the imagery.

> Enjoy a self-guided tour of our onsite museum, followed by a master class and premium tasting from our in-house wine expert. Collect your bottle of hand picked wine along with a picnic hamper to share between two and select a scenic location on the property to enjoy the view.
>
> (Listed under See and Do experience on winecountry.com.au)

The description above connects the idea of pastoral space and rural grandeur to luxury, as personalisation through use of words such as *select a scenic location* and *premium tasting from our in house wine expert*. While the sense of selecting suggests personalisation through choice, *in house wine expert* embeds meanings of exclusivity. Access to both space and expertise is suggested as personalised luxury which is consistent with emerging meanings of 'access-based' luxury in recent literature (Bardhi et al., 2020; Pantano & Stylos, 2020). What our analysis suggests is how these ideas of access-based consumption affectively connect with bucolic luxury.

Affect of bucolic luxury also emerges when meanings of luxury are suggested in the 'experiencing' of exclusivity, quality and authenticity through use of words and depictions of food and wine production practices. Using particular verbs, the website connects luxury to bucolic activities and experiences rather than the products or services. For instance, the activity of grape-picking as a wine production practice is threaded into the use of the word *'hand-picked'*, which also intertwines with the ideas of exclusivity, craftsmanship and authenticity (Beverland, 2006). Further, the use of specific nouns to describe practices and practitioners such as *'vintage'* and *'first-generation winemakers'* affectively suggests luxury through the discourse on authenticity (Taplin, 2015). Further, in the case of Hunter Valley, meanings of luxury as an experience of quality are implied in images of people enjoying the richness of terroir (Croce & Perri, 2010). For example, the pictures on the website show people enjoying the view in a way that allows the viewer to enjoy it too. The gaze of the viewer is directed to the landscape, and the experience of it is transmitted affectively to the viewer. These words and images that typically describe bucolic activities and practices convey feelings of luxury as they are intertwined with ideas of luxury through the promotion of 'experiencing'.

Congregation

The affect of luxury is also felt because of the congregation of words and phrases in the same sentence. Words and phrases with multiple underlying meanings of luxury connected to country are positioned together to create impact. This points towards the use of affective storytelling, which is an effective strategy known to trigger feelings (Selting, 2010), expectations and imaginaries (Moscardo, 2020). What our analysis suggests is the use of congregation as a strategy to produce affective stories related to bucolic luxury.

The affect of bucolic luxury is also sensed through congregation of different inconspicuous material cues such as architectural details and landscapes. Architecture and interior design of cellar doors and accommodation is used as an affective trigger of luxury on the website. The images show exterior details such as tin roofs and walls, stone and wood, country style large wooden doors that are traditionally rural and rustic materials being aesthetically congregated with trees, pots, foliage and gravel and well-dressed people to affectively signal bucolic luxury. Further, elements such as fireplaces combined with a couple enjoying wine characteristically suggest luxury through associations with personalisation and exclusivity. This use of inconspicuous material and bucolic heritage to suggest the idea of rural history and heritage being connected to luxury aligns with new ideas of luxury (Bardhi et al., 2020).

In addition to inconspicuous depictions of luxury, expensive and conspicuous ideas of wealth can also be sensed in the aesthetics of the cellar doors. High-end luxury is directly felt in the bespoke designs of cellar doors that demonstrate uniqueness and character (Iloranta, 2019; Seo & Buchanan-Oliver, 2015). While some images present designs that are either reminiscent of classic French provincial luxury, other images suggest uniqueness and craftsmanship by focussing on congregation of expensive elements such as vaulted ceilings, furniture, use of glass, wood and other details together in contemporary modern designs. Traditionally, architecture has been understood as an indicator of social status. Much architecture literature discusses the commodification of architecture in marketing hotels and restaurants (Avermaete & Massey, 2013; Sharr, 2016; Thomson, 1993). What our analysis indicates is the importance of architectural imagery in defining bucolic luxury.

An important element that congregates meanings and images of luxury and bucolic characteristics on the HVWTA website is aesthetic website design. Simplified clean navigation combined with use of neutral colours including white and grey present a neat canvas to portray the richness of bucolic colours and create a visually pleasing experience. The use of large photographs suggest focus and add to the feelings of luxury (Baloglu & Pekcan, 2006). Further, while the website is informational in providing direction towards trip planning, it is not intensely populated. Wine experiences, dine, stay and other activities are categorised into different web pages where each experience is depicted as a tile with a photo and a small description. This clarity exacerbates the affect of luxury as the clean and well-organised platform helps focus on the sensory content. While much literature discusses the importance of website aesthetics in luxury marketing (Baloglu & Pekcan, 2006; Kim, Hyun, & Park, 2020; Shen et al., 2016), the research on individual aesthetic cues in luxury marketing is only recent and needs further exploration (Kim et al., 2020). Our analysis reflects the importance of website aesthetics in creating and sustaining the affect of bucolic luxury for the user.

Repetition

The affect of luxury is sensed as users are not only barraged with phrases and imagery describing bucolic luxury, but the affect is also sensed because these congregations are encountered repeatedly in different permutations and combinations. Each web page on the website depicts these bucolic luxury congregations and connections in different ways, repeating the core messaging (Kluge et al., 2013; Schmitt, 1999).

The repetition of congregations and connections through website design results in sensorial stimulation. Literature suggests website descriptions, and images can be instrumental in triggering sensorial processes in the mind (Lee & Gretzel, 2012; Lee, Gretzel, & Law, 2010). In case of the Hunter Valley website, the affect of luxury is felt as the user is bombarded by luxury congregations and connections with every web page view. Multiple images and multiple descriptions depicting multiple experiential cues suffuse the affect of luxury for the user when browsing through the entire website. This repletion of congregations and connections amplifies affect. While repetition as a strategy is common in the marketing literature, its role in generating affect of luxury is underexplored (Reckwitz, 2016). Our analysis points towards the importance of repetition specifically in creating the affect of bucolic luxury, which is experienced and sensed by the viewer rather than thought or evaluated.

Conclusion

Literature on marketing experiential luxury is still emergent (Holmqvist et al., 2020). In this chapter, we focused on how luxury is marketed as an affect in the context of a gastronomic tourism experience. We studied the official website of Hunter Valley created by the HVWTA, Australia, to understand how the gastronomic tourism experience is affectively marketed as bucolic luxury. We identified three key strategies that have been used to create the affect of bucolic luxury. These findings are especially relevant for marketing luxury in contexts not traditionally associated with luxury, such as the regional gastronomic tourism experience.

The conceptualisation of bucolic luxury contributes towards understanding luxury as unconventional in gastronomic tourism experiences. A recent special issue of the *Journal of Business Research* points towards luxury in experiences as unconventional, fluid and affectively sensed in practice (Banister et al., 2020; Holmqvist et al., 2020; Thomsen et al., 2020). In line with this argument, we suggest that luxury can be marketed in the context of regional gastronomic tourism as an affect which connects multiple unconventional meanings of luxury with bucolic heritage, practices and cues. Further, we demonstrate how website marketing strategies of connection, congregation and repetition are used to create and sustain the affect of luxury, stimulating mental processes and triggering expectations and imaginaries that are not just cognitively sensed. Future research could explore how the affect of luxury can be facilitated for the tourist during the consumption experience by using themed marketing strategies (Roy & Gretzel, 2020).

References

Ahmed, S. (2013). *The cultural politics of emotion*. New York, NY: Routledge.

Atwal, G., & Williams, A. (2009). Luxury brand marketing – The experience is everything! *Journal of Brand Management*, *16*(5–6), 338–346. doi:10.1057/bm.2008.48

Australia, T. (2015). *Food and wine tourism in New South Wales: Executive summary*. Sydney: Tourism Research Australia.

Australia, T. (2016). *Restaurant Australia: Why its time to share our food and wine with the World*. Sydney: Tourism Australia.

Avermaete, T., & Massey, A. (2013). *Hotel lobbies and lounges: The architecture of professional hospitality*. Oxon: Routledge.

Baloglu, S., & Pekcan, Y. A. (2006). The website design and internet site marketing practices of upscale and luxury hotels in Turkey. *Tourism Management*, *27*(1), 171–176. doi:10.1016/j.tourman.2004.07.003

Banister, E., Roper, S., & Potavanich, T. (2020). Consumers' practices of everyday luxury. *Journal of Business Research*, *116*, 458–466. doi:10.1016/j.jbusres.2019.12.003

Bardhi, F., Eckhardt, G. M., & Samsioe, E. (2020). Liquid luxury. In F. Morhart, K. Wilcox, & S. Czellar (Eds.), *Research handbook on luxury branding*. Cheltenham: Edward Elgar Publishing.

Batat, W. (2019). The role of luxury gastronomy in culinary tourism: An ethnographic study of Michelin-Starred restaurants in France. *International Journal of Tourism Research*, *23*(2), 150–163. doi:10.1002/jtr.2372

Bauman, Z. (2013). *Liquid modernity*. Cornwall: John Wiley & Sons.

Belk, R. (2020). The changing notions of materialism and status in an increasingly dematerialised world. In F. Morhart, K. Wilcox, & S. Czellar (Eds.), *Research handbook on luxury branding*. Cheltenham: Edward Elgar Publishing.

Bellini, N., & Resnick, E. (2018). The luxury turn in wine tourism. In N. Bellini, C. Clergeau, & O. Etcheverria (Eds.), *Gastronomy and local development the quality of products, places and experiences*. Oxon: Routledge.

Beverland, M. (2004). Uncovering "theories-in-use": Building luxury wine brands. *European Journal of Marketing*, *38*(3/4), 446–466. doi:10.1108/03090560410518648

Beverland, M. (2006). The 'real thing': Branding authenticity in the luxury wine trade. *Journal of Business Research*, *59*(2), 251–258. doi:10.1016/j.jbusres.2005.04.007

Bjork, P., & Kauppinen-Räisänen, H. (2014). Culinary-gastronomic tourism – A search for local food experiences. *Nutrition & Food Science*, *44*(4), 294–309. doi: 10.1108/NFS-12-2013-0142

Boniface, P. (2017). *Tasting tourism: Travelling for food and drink*. London: Routledge.

Boyne, S., Hall, D., & Williams, F. (2003). Policy, support and promotion for food-related tourism initiatives. *Journal of Travel & Tourism Marketing*, *14*(3–4), 131–154. doi:10.1300/J073v14n03_08

Buda, D. M., d'Hauteserre, A.-M., & Johnston, L. (2014). Feeling and tourism studies. *Annals of Tourism Research*, *46*, 102–114. doi:10.1016/j.annals.2014.03.005

Chen, A., & Peng, N. (2018). Examining consumers' intentions to dine at luxury restaurants while traveling. *International Journal of Hospitality Management*, *71*, 59–67. doi:10.1016/j.ijhm.2017.11.009

Chowdhury, R. M. M. I., Olsen, G. D., & Pracejus, J. W. (2008). Affective responses to images in print advertising: Affect integration in a simultaneous presentation context. *Journal of Advertising*, *37*(3), 7–18. doi:10.2753/JOA0091-3367370301

Correia, A., Kozak, M., & Del Chiappa, G. (2019). Examining the meaning of luxury in tourism: A mixed-method approach. *Current Issues in Tourism*, *23*, 1–19. doi: 10.1080/13683500.2019.1574290

Cristini, H., Kauppinen-Räisänen, H., Barthod-Prothade, M., & Woodside, A. (2017). Toward a general theory of luxury: Advancing from workbench definitions and theoretical transformations. *Journal of Business Research*, *70*, 101–107. doi: 10.1016/j.jbusres.2016.07.001

Croce, E., & Perri, G. (2010). *Food and wine tourism: Integrating food, travel and territory*. Oxfordshire: CABI.

Dixit, S. K. (2019). *The Routledge handbook of gastronomic tourism*. Oxon: Routledge.

Dixit, S. K. (Ed.). (2020a). *The Routledge handbook of tourism experience management and marketing*. London: Routledge.

Dixit, S. K. (2020b). Marketing gastronomic tourism experiences. In S. K. Dixit (Ed.), *The Routledge handbook of tourism experience management and marketing* (pp. 322–335). London: Routledge.

Dixit, S. K., & Abraham, A. (2020), Brand identity and positioning in selected India chain hotel companies. In M. A. Gardini, M. C. Ottenbacher, & M. Schuckert (Eds.), *The Routledge companion to international hospitality management* (pp. 260–270). London: Routledge.

Eckhardt, G. M., & Bardhi, F. (2019). New dynamics of social status and distinction. *Marketing Theory*, *20*(1), 85–102. doi:10.1177/1470593119856650

Eckhardt, G. M., Belk, R. W., & Wilson, J. A. J. (2015). The rise of inconspicuous consumption. *Journal of Marketing Management*, *31*(7–8), 807–826. doi:10.1080/0267257X.2014.989890

Everingham, P., & Motta, S. C. (2020). Decolonising the 'autonomy of affect' in volunteer tourism encounters. *Tourism Geographies*, 1–21. doi:10.1080/14616688.2020.1713879

Garrod, B., Wornell, R., & Youell, R. (2006). Re-conceptualising rural resources as countryside capital: The case of rural tourism. *Journal of Rural Studies*, *22*(1), 117–128. doi:10.1016/j.jrurstud.2005.08.001

Germann Molz, J. (2007). Eating difference: The cosmopolitan mobilities of culinary tourism. *Space and Culture*, *10*(1), 77–93. doi:10.1177/1206331206296383

Germann Molz, J. (2015). Giving back, doing good, feeling global: The affective flows of family voluntourism. *Journal of Contemporary Ethnography*, *46*(3), 334–360. doi:10.1177/0891241615610382

Hagtvedt, H. (2020). Art and aesthetics. In F. Morhart, F. Wilcox, & S. Czellar (Eds.), *Research handbook on luxury branding*. Cheltenham: Edward Elgar Publishing.

Hall, C. M., & Gössling, S. (2016). From food tourism and regional development to food, tourism and regional development: Themes and issues in contemporary foodscapes. In C. M. Hall & S. Gössling (Eds.), *Food tourism and regional development: Networks, products and trajectories* (pp. 3–59). London: Routledge.

Hall, D. R., Kirkpatrick, I., & Mitchell, M. (2005). *Rural tourism and sustainable business*. Clevedon: Channel View Publications.

Hall, C. M., & Mitchell, R. (2004). Consuming tourists: Food tourism consumer behaviour. In *Food tourism around the world* (pp. 72–92). Burlington, MA: Routledge.

Holmqvist, J., Diaz Ruiz, C., & Peñaloza, L. (2020). Moments of luxury: Hedonic escapism as a luxury experience. *Journal of Business Research*, *116*, 503–513. doi:10.1016/j.jbusres.2019.10.015

Holst, T. (2018). *The affective negotiation of slum tourism: City walks in Delhi*. Oxon: Routledge.

Huang, W.-J., Beeco, J. A., Hallo, J. C., & Norman, W. C. (2016). Bundling attractions for rural tourism development. *Journal of Sustainable Tourism*, *24*(10), 1387–1402. doi:10.1080/09669582.2015.1115510

Iloranta, R. (2019). Luxury tourism service provision – Lessons from the industry. *Tourism Management Perspectives*, *32*, 100568. doi:10.1016/j.tmp.2019.100568

Joy, A., Belk, R. W., Charters, S., Wang, J. J. F., & Peña, C. (2018). Performance theory and consumer engagement: Wine-tourism experiences in South Africa and India. In *Consumer culture theory* (pp. 163–187). Emerald Publishing Limited.

Kapferer, J.-N. (2014). The future of luxury: Challenges and opportunities. *Journal of Brand Management*, *21*(9), 716–726. doi:10.1057/bm.2014.32

Kim, D., Hyun, H., & Park, J. (2020). The effect of interior color on customers' aesthetic perception, emotion, and behavior in the luxury service. *Journal of Retailing and Consumer Services*, *57*, 102252. doi:10.1016/j.jretconser.2020.102252

Kivela, J., & Crotts, J. C. (2006). Tourism and gastronomy: Gastronomy's influence on how tourists experience a destination. *Journal of Hospitality & Tourism Research*, *30*(3), 354–377. doi:10.1177/1096348006286797

Kluge, P. N., Königsfeld, J. A., Fassnacht, M., & Mitschke, F. (2013). Luxury web atmospherics: An examination of homepage design. *International Journal of Retail & Distribution Management*, *41*(11/12), 901–916. doi:10.1108/IJRDM-01-2013-0013

Lee, W., & Gretzel, U. (2012). Designing persuasive destination websites: A mental imagery processing perspective. *Tourism Management*, *33*(5), 1270–1280. doi: 10.1016/j.tourman.2011.10.012

Lee, W., Gretzel, U., & Law, R. (2010). Quasi-trial experiences through sensory information on destination web sites. *Journal of Travel Research*, *49*(3), 310–322. doi:10.1177/0047287509346991

Lee, J. H., & Hwang, J. (2011). Luxury marketing: The influences of psychological and demographic characteristics on attitudes toward luxury restaurants. *International Journal of Hospitality Management*, *30*(3), 658–669. doi:10.1016/j.ijhm.2010.12.001

Maguire, J. S. (2013). Provenance as a filtering and framing device in the qualification of wine AU – Smith Maguire, Jennifer. *Consumption, Markets and Culture*, *16*(4), 368–391. doi:10.1080/10253866.2012.662829

Mason, R., & O'Mahony, B. (2007). On the trail of food and wine: The tourist search for meaningful experience. *Annals of Leisure Research*, *10*(3–4), 498–517. doi: 10.1080/11745398.2007.9686778

Morhart, F., & Malär, L. (2020). Authenticity in luxury branding. In F. Morhart, F. Wilcox, & S. Czellar (Eds.), *Research handbook on luxury branding*. Cheltenham: Edward Elgar Publishing.

Morhart, F., Wilcox, K., & Czellar, S. (2020). *Research handbook on luxury branding*. Cheltenham: Edward Elgar Publishing.

Moscardo, G. (2020). Stories and design in tourism. *Annals of Tourism Research*, *83*, 102950. doi:10.1016/j.annals.2020.102950

Newell, S. (2018). The affectiveness of symbols: Materiality, magicality, and the limits of the antisemiotic turn. *Current Anthropology*, *59*(1), 1–22. doi:10.1086/696071

Okumus, B., & Cetin, G. (2018). Marketing Istanbul as a culinary destination. *Journal of Destination Marketing and Management*, *9*, 340–346. doi:10.1016/j.jdmm.2018.03.008

Olafsdottir, G. (2013). '…sometimes you've just got to get away.': On trekking holidays and their therapeutic effect. *Tourist Studies*, *13*(2), 209–231. doi:10.1177/1468797613490379

Organisation, W. T. (2012). *Global Report on food tourism*. Retrieved from Madrid.

Otero, J. (2018). High-status food is changing: New gastronomic perspectives. *International Journal of Gastronomy and Food Science*, *11*, 35–40. doi:10.1016/j.ijgfs.2017.11.003

Pantano, E., & Stylos, N. (2020). The Cinderella moment: Exploring consumers' motivations to engage with renting as collaborative luxury consumption mode. *Psychology and Marketing*, *37*(5), 740–753. doi:10.1002/mar.21345

Park, E., Kim, S., & Yeoman, I. (2019). *Food tourism in Asia* (E. Park, S. Kim, & I. Yeoman, Eds.). Singapore: Springer.

Peng, N., Chen, A., & Hung, K.-P. (2020). Dining at luxury restaurants when traveling abroad: Incorporating destination attitude into a luxury consumption value model. *Journal of Travel & Tourism Marketing*, *37*(5), 562–576. doi:10.1080/10548408.2019.1568352

Reckwitz, A. (2016). Practices and their affects. In A. Hui, T. Schatzki, & E. Shove (Eds.), *The Nexus of practices* (pp. 126–137). Oxon: Routledge.

Reilly, J., Klima, E. S., & Bellugi, U. (1990). Once more with feeling: Affect and language in atypical populations. *Development and Psychopathology*, *2*(4), 367–391. doi:10.1017/S0954579400005782

Reyneke, M., Berthon Pierre, R., Pitt Leyland, F., & Parent, M. (2011). Luxury wine brands as gifts: Ontological and aesthetic perspectives. *International Journal of Wine Business Research*, *23*(3), 258–270. doi:10.1108/17511061111163078

Rokka, J. (2017). Champagne: Marketplace icon AU – Rokka, Joonas. *Consumption, Markets and Culture*, *20*(3), 275–283. doi:10.1080/10253866.2016.1177990

Roy, N., & Gretzel, U. (2020). Themed route marketing in India. *Anatolia*, 1–12. doi:10.1080/13032917.2020.1747222.

Schmitt, B. (1999). Experiential marketing. *Journal of Marketing Management*, *15*(1–3), 53–67. doi:10.1362/026725799784870496

Schmitt, B. (2020). Sensuality and experience. In F. Morhart, K. Wilcox, & S. Czellar (Eds.), *Research handbook on luxury branding*. Cheltenham: Edward Elgar Publishing.

Selting, M. (2010). Affectivity in conversational storytelling: An analysis of displays of anger or indignation in complaint stories. *Pragmatics*, *20*(2), 229–277. doi:10.1075/prag.20.2.06sel

Seo, Y., & Buchanan-Oliver, M. (2015). Luxury branding: The industry, trends, and future conceptualisations. *Asia Pacific Journal of Marketing & Logistics*, *27*(1), 82–98. doi:10.1108/APJML-10-2014-0148

Sharr, A. (2016). Liebskind in Las Vegas: Reflections on architrecture as luxury commodity. *Critical luxury studies: Art, design, media*, 151–176.

Shen, K. N., Vel, P., & Khalifa, M. (2016). Website design: Place for luxury brands in cyberspace? *Behaviour & Information Technology*, *35*(12), 1115–1129. doi:10.1080/0144929X.2016.1188987

Silkes, C. A., Cai, L. A., & Lehto, X. Y. (2013). Marketing to the culinary tourist. *Journal of Travel & Tourism Marketing*, *30*(4), 335–349. doi:10.1080/10548408.2013.784151

Sormaz, U., Akmese, H., Gunes, E., & Aras, S. (2016). Gastronomy in tourism. *Procedia Economics and Finance*, *39*, 725–730. doi:10.1016/S2212-5671(16)30286-6

Swarbrooke, J. (2018). *The meaning of luxury in tourism, hospitality and events*. Oxford: Goodfellow Publishers Limited.

Taplin, I. M. (2015). Bottling luxury: Napa Valley and the transformation of an agricultural backwater into a world-class wine region. *Luxury*, *2*(1), 81–100. doi:10.1080/20511817.2015.11428566

Thomsen, T. U., Holmqvist, J., von Wallpach, S., Hemetsberger, A., & Belk, R. W. (2020). Conceptualising unconventional luxury. *Journal of Business Research, 116*, 441–445. doi:10.1016/j.jbusres.2020.01.058

Thomson, D. (1993). *Renaissance architecture: Critics, patrons, luxury*. Manchester: Manchester University Press.

Tsai, C.-T., & Wang, Y.-C. (2017). Experiential value in branding food tourism. *Journal of Destination Marketing and Management*, *6*(1), 56–65. doi:10.1016/j.jdmm.2016.02.003

Veblen, T. (1899/1994). *The theory of the leisure class*. New Brunswick, NJ: Transaction Publishers.

von Wallpach, S., Hemetsberger, A., Thomsen, T. U., & Belk, R. W. (2020). Moments of luxury – A qualitative account of the experiential essence of luxury. *Journal of Business Research, 116*, 491–502. doi:10.1016/j.jbusres.2019.10.060

Yang, W., & Mattila Anna, S. (2016). Why do we buy luxury experiences? Measuring value perceptions of luxury hospitality services. *International Journal of Contemporary Hospitality Management*, *28*(9), 1848–1867. doi:10.1108/IJCHM-11-2014-0579

Yeoman, I., & McMahon-Beatte, U. (2016). The future of food tourism. *Journal of Tourism Futures*, *2*(1), 95–98. doi:10.1108/JTF-12-2015-0051

Yeoman, I., & McMahon-Beattie, U. (2014). Exclusivity: The future of luxury. *Journal of Revenue and Pricing Management*, *13*(1), 12–22. doi:10.1057/rpm.2013.29

Chapter 10

Luxury Tourism and Hospitality Employees: Their Role in Service Delivery

Tracy Harkison[1]

Abstract

Delivering services that create memorable luxury accommodation experiences rely on frontline staff to engage guests on a sensory level rather than merely a functional one. This engagement includes cognitive, emotional, relational and behavioural. Hospitality and tourism industries are people-orientated – people are needed to serve people in order to create desired experiences – and it is very difficult to create satisfaction or to revisit intention in every interaction that takes place. It is this intangible characteristic of the industries, provisions and tangible cues that play an important part in enhancing the overall luxury accommodation experience. Guests are very clear as to what they expect from luxury accommodation experiences: they feel that they are paying for a service that should be personalised, and that staff should realise what they want and need. The human interaction component and the co-creation that occurs between staff and guests is an essential dimension of the industry. The influence of these interactions on guest experiences and the delivery of services will be explored in this chapter.

Keywords: Frontline staff; luxury; luxury accommodation experience; co-creation; hospitality; service

Introduction

Luxury tourism and hospitality services very much depend on employees. Not only do employees play a crucial role in the delivery of these services, but also they are central to the luxury accommodation experience that guests are now expecting. This chapter investigates and appraises the role that staff play in the delivery of luxury tourism and hospitality services and the luxury accommodation

[1]https://orcid.org/0000-0003-2409-231X

The Emerald Handbook of Luxury Management for Hospitality and Tourism, 199–219

doi:10.1108/978-1-83982-900-020211010

experience they anticipate will come with those luxury services. The role undertaken by frontline staff will be investigated and appraised in relation to their interactions with customers, as this interaction is one of the most critical elements in luxury service delivery.

To achieve an improved relationship with their customers, tourism and hospitality organisations are choreographing or developing luxury accommodation experiences via human interaction dimensions. For example, the focus is being given to their frontline staff's presentation, professional behaviour, temperament, demeanour, proactive service and how they foster the socialisation of guests in order to encourage the customer–employee relationship. In service industries such as tourism and hospitality, product quality has been replaced by service quality. This new approach is viewed as a stratagem by which luxury establishments can add value to the accommodation experiences of their guests and differentiate themselves from other offerings in their sector.

In this new approach, the product becomes less important than the interactions between the customer and the producer, and in the luxury accommodation sector, this interaction is becoming more and more critical. There is now an emphasis on creating luxury accommodation experiences in which, rather than merely being a passive observer, the guest is an active participant. It is essential for frontline staff to enable this co-creation due to the continual interactions they have with guests. In the process of co-creating unforgettable and unique luxury accommodation experiences, it is crucial to scrutinise and assess the recruitment, selection, training and development of tourism and hospitality professionals, so they are better able to anticipate and accommodate guests' requirements and desires.

The Importance of Luxury Accommodation

Accommodation is recognised as one of the most significant elements of the tourist experience and is generally one of the first services tourists demand when reaching their destination (Santoro, 2015). Furthermore, if they are happy with their accommodation, this may place them in a positive mindset, but if the accommodation does not live up to their expectations, it can ruin their entire experience (Dixit, 2020; Khoo-Lattimore & Ekiz, 2014; Kotler, Bowen, & Makens, 2010). Yet, despite the critical nature of accommodation to the tourist experience, this component has frequently been taken for granted or ignored completely (Chu, Tang, & Luo, 2016; Quan & Wang, 2004), and in terms of understanding guests' emotional triggers, the accommodation experience has remained significantly under-researched (Chu et al., 2016; Titz, 2008).

Hospitality is one of the largest and fastest-growing industries in the world. The word 'hospitality' is an umbrella term that encompasses many sectors, and when researchers are determining a target for study, it is suggested that they should focus their research on specific sectors within the industry and also on the variability within those sectors (Ottenbacher, Harrington, & Parsa, 2009). Cetin and Dincer (2014) state that luxury hotels can give researchers more insight into experiences due to their challenging and knowledgeable clientele, as well as the importance that the luxury hotel places on their own products and services. Such hotels are service-intensive and maintain a high staff-to-guest ratio in order to ensure high interpersonal guest contact within a lavish environment (Brien, Ratna, & Boddington, 2012; Harkison, Hemmington, & Hyde, 2018). Researchers argue that luxury hotels form a portion of the 'experience economy', in which employees deliver the experience as part of their relationships with both the guest and other staff (Gilmore & Pine, 2002; Sundbo & Dixit, 2020; Tosti, 2009). However, producing an experience necessitates engagement from and some type of emotional (not just functional) participation of the visitor. This emotional engagement in experience formation depends on interactions (Snel, 2013) between the visitor and the organisation, which co-creates experiences through the resulting two-way interactions (Sørensen & Jensen, 2015). Developing an understanding of the engagement behaviours of customers, such as co-creation, has developed into a key research priority for marketing, tourism and hospitality researchers (Harkison, 2018; Marketing Science Institute, 2010; Verhoef, Reinartz, & Krafft, 2010).

Luxury Hotels

In prior years, luxury hotels tended to be well known for their premium locations, imposing dimensions, superior culinary offerings, spotlessness, exquisite beauty and the provision of security, privacy and exceptionally customised services (Kucukusta, Heung, & Hui, 2014; Sherman, 2007). But then, other accommodation organisations started to believe that it was necessary that they provide further amenities in their rooms (for example, a range of familiar brands of toiletries). This trend, which was fuelled by highly competitive market environments, was called 'the amenities creep of the 80s' (Bell, 1989; Bernstein, 1999). However, by the turn of the next century, the availability of hotel facilities, for example, restaurants, bars, rooms and nightclubs, was no longer considered by guests as a luxury: for many people, these facilities had become a vital component of their lifestyle (Kandampully & Suhartanto, 2000).

This raised standard for 'average' accommodation increased competition within the luxury segment and obliged organisations to give more consideration to customer satisfaction (Lu, Berchoux, Marek, & Chen, 2015; Mattila, 1999). Examining customer satisfaction from the customer perspective highlights how they develop connections between luxury hotels as a product and the relevance the product has to their lives (Dixit, 2017; Zeithaml, Bitner, & Gremler, 2006). Luxury hotels were found to provide strong emotional experiences for their guests, who were prepared to be charged more for this (Barsky & Nash, 2002; Han & Back, 2007). That knowledge led to some hotel organisations specifically upscaling their hotels to luxury. However, luxury hotels must determine whether guests are visiting for pleasure or business because guest perceptions can be affected by the purpose of their visit (Griffin, Shea, & Weaver, 1996). According to Cetin and Dincer (2014), it is not realistic to presume that business travellers would demonstrate the same level of concern towards hotel experiences as would recreational tourists. Another factor that can affect guests' perceptions of satisfaction is the brand of the hotel.

Hotel Brands

The brand of a hotel is viewed as an implied promise of what types of amenities and levels of service guests could expect (Xu & Chan, 2010). The brand acts as the signature of a hotel chain, showcasing its services and products, and is often perceived as a mechanism for hotel organisations to distinguish and differentiate themselves in the minds of their guests (Mohsin, Hussain, & Khan, 2011; Xu & Chan, 2010). However, although a number of organisations, such as the Four Seasons and the Ritz Carlton, promoted their businesses as luxury hotels when O'Neill (2004) investigated the changing nature of luxury hotel brand values, he was unable to identify a clear definition for what 'made' a luxury hotel. In an attempt to provide some clarity, O'Neill used a retail analogy to explain the difference: 'Luxury hotels are about as different from the average hotel as a Prada bag is from a duffel bag' (O'Neill, 2004, p. 30). Presbury, Fitzgerald, and Chapman (2005) highlighted that luxury hotel managers needed to identify the distinguishing features, services, capabilities and strengths that they could deliver to guests in a consistent manner in order for their businesses to be seen as different from average hotels. Luxury hotels are also distinguished from other hotels by having well-trained and highly engaged staff (Bagnera, 2017; Cetin & Walls, 2016).

Frontline Staff

Employees, particularly those that have direct interactions with guests, are the ones who provide the service and manage the product that is the guests' experience. Given this fact, it is critical that organisations be careful to select and hire the 'correct' staff – those with personalities predisposed to delight customers. And once staff have been employed, the organisation needs to support their staff through training and development programmes and immerse them in the organisational culture (Torres & Kline, 2006).

Walls, Okumus, Wang, and Kwun (2011) suggested that, in addition to travel-related factors (for example, whether guests are travelling for pleasure or for business), human interactions can significantly affect the guest experience in luxury hotels. Given that the accommodation industry involves elevated levels of interaction between staff and guests, visitors' emotions are frequently impacted by the services and actions of the staff they encounter (Han & Back, 2007; Lewis & McCann, 2004). Visitors want staff to display professional behaviour and a positive attitude, and to provide services in a proactive manner (Walls et al., 2011).

Frontline staff interactions with visitors generally start with structured and rehearsed sequences of verbal exchanges (Scerri, Jenkins, & Lovell, 2017). Baum (2006) believed that frontline employees within luxury hotels must possess the ability to engage in conversation with visitors on any relevant matter (be it sport, music, movies, politics or any other subject), all with international awareness. However, this prerequisite relies upon employing educated frontline staff with a degree of international exposure, as well as a commitment from them to maintain up-to-date knowledge in these subjects (Baum, 2006). It has been proposed that luxury hotels must preserve high staff-to-guest ratios in order to guarantee that they can provide high levels of interpersonal guest contact in what is viewed as a lavish setting (Brien et al., 2012; Harkison et al., 2018). Bharwani and Jauhari (2013) believed that this increase in human interaction is essential for improving guests' positive experiences.

One of the major priorities in hospitality is the guest experience – hosting guests and developing the perception of hospitableness and being hosted. It has been suggested that hospitality exists within the lived experience – that the 'host' and the 'guest' partake in a 'here and now' experience that is experienced by both of them. O'Gorman (2007) described hospitality as a gift that is given and shared and that the truest gift was the act of kindness the guest experiences, turning a stranger into a friend, albeit for a short period of time (O'Gorman, 2007). The hospitality experience success rate depends on the ability of the host and guest to connect and also on how willing the guest is to engage in the experience offered by the organisation (Alcántara-Alcover, Artacho-Ramírez, Martínez-Guillamón, & Campos-Soriano, 2013; Walls et al., 2011). And, without question, visitors and staff have roles to play in the service experiences and service interactions (Carù & Cova, 2006; Lugosi, 2008).

In endeavouring to create a connection with guests, organisations choreograph or produce experiences by means of human interaction dimensions (for example, the presentation, professionalism, proactiveness and demeanour of employees; and the attitude, behaviour, socialisation and presentation of the visitor) and/or the physical environment aspects (for example, the artefacts, function, ambience, symbols, space, signs and multisensory facets) (Walls et al., 2011). However, the multidimensional impression formed by encounters with products, services and businesses is mediated by the guest's willingness to be affected by physical and/or human interaction dimensions (Walls et al., 2011).

Within the context of a luxury hotel, the dimensions of a visitor experience include the human interactions with staff and fellow guests, the personal characteristics of those visitors and the physical environment (Ariffin, Nameghi, & Zakaria, 2013; Brien et al., 2012). Walls (2013) suggested that the hotel experience for guests should be managed by a detailed strategy that starts from the pre-experience expectations and continues through to the post-experience assessments of that hotel experience. In addition, Gilmore and Pine (2002) previously noted that an orchestrated and innovative design of experiences would gradually become essential core competence of successful hotel managers. Williams (2006) recognised that hospitality was very similar to the theatre, in that both contain an element of staging the environment and that it is necessary to stage-manage the experiences. With hotels and restaurants, there is a meticulous need to prepare the front of the house as a stage, and although this can be very time-consuming, it is vital to creating the experience (Williams, 2006). Within the hospitality industry, an example of this is the employment of 'experience engineers' by Starwood

Hotels and Resorts, their principal objective being to improve the service culture at their establishments, as an improved service culture has been found to enhance guest satisfaction and loyalty by delivering better guest experiences (Walls, 2013). In an attempt to provide improved experiences for their guests, Starwood used staging in the form of Second Life software in order to obtain valuable feedback from their visitors about new hotels they were proposing to build (Penfold, 2009).

Hemmington (2007) proposed that hospitality organisations should aim to create their guest experience in a similar manner to that of good restaurants by introducing many little surprises. To develop such experiences, businesses need to rely on their employees – to utilise their ideas and creativity to repeatedly stimulate and excite their visitors throughout the entirety of their experience. However, the delivery of hospitality products can be very dependent on retaining the customers' perspective: customers do not buy service delivery, they buy experiences; they do not buy service quality, they buy memories; they do not buy food and drink, they buy meal experiences (Hemmington, 2007, p. 749).

To achieve success in the hospitality industry, it is essential for organisations to be able to design and deliver unforgettable experiences (Yuan & Wu, 2008). Cetin and Dincer (2014) refined this belief by recommending that luxury hotels with hundreds of rooms should focus more on their physical environment experiences and that smaller boutique hotels could concentrate more on the social interaction experiences within their businesses.

Hotels as a Memorable Experience

Knutson and Beck (2004) state that guests' experiences consist of all the memorable encounters throughout the entire encounter with the business. For hotels, this encompasses everything from making a reservation via various channels right through to the billing at the end of the stay. Therefore, they should concentrate on managing the visitor journey by integrating it into their positioning strategy – the journey is critical because it progresses the guest from the expectations they held prior to their visit through to appraisals they make once it is finished (Knutson & Beck, 2004; Sturken, 2018). It has been suggested that if businesses want to take control of this journey, they need to recognise two sets of clues: the first is the actual functions of the goods and services; the second is the emotions that are generated, which emanate from interactions with other things or people, for example, their perception of the perceptible aspects in the environment (Berry, Carbone, & Haeckel, 2002).

Drewer (2005) further suggested that 'rather than simply viewing the hotel as accommodation, clients are looking at the hotel to provide an experience itself' (p. 6). This concept was also investigated by McIntosh and Siggs (2005), who concentrated on specialised accommodation since they believed that there had been little consideration given to the experiential nature of accommodation, particularly in the realm of specialisation. Their research establishes that the demand for specialised accommodation businesses might be related to the experiences of luxury uniqueness, host–guest interaction, personal touches, history and comfort that these establishments can provide (McIntosh & Siggs, 2005).

It can be said that the hospitality industry possesses an intangible facet, and it needs people to deliver its product:

> Unlike most service industries, it is the manner in which the hospitality employee provides the service – as opposed to the service itself – which is critical to the customer's overall enjoyment of the product or "experience" being purchased.
> (Dawson, Abbott, & Shoemaker, 2011, p. 290)

Positive or negative encounters with other guests or with an organisation's employees can trigger a guest's emotions in various ways and impact on their overall experience (Barsky & Nash, 2002; Han, Moon, & Hyun, 2019; Sundbo & Dixit, 2020). Encounters such as chaotic arrival or departure processes, shabbily dressed staff members, or low hygiene standards might cause visitors to feel miserable with their visit (Barsky & Nash, 2002; Han et al., 2019). So it is important that hospitality organisations maintain the 'customer perspective' in mind when developing customer experiences or planning how to effectively deliver hospitality products and services (Bharwani & Jauhari, 2013).

Service Quality

In recent years, the hospitality industry, along with other service industry organisations, has seen service quality slowly replace product quality – making improvements to the service quality components of their operations is being viewed both as a way for these businesses to differentiate themselves from others in their sector and to enable them to add value to the guest experience (Knutson, Beck, Kim, & Cha, 2010). Service quality can be viewed as appropriate or personalised services – for example, hotel staff knowing a guest's name or their personal information, the outcome of which contributes to the guest feeling important and that staff are treating them as something special (Ariffin & Maghzi, 2012). Personalised service can be accomplished via the interactions between frontline staff and guests, as the host–guest relationship is a fundamental aspect of the visitor experience (Lashley, 2008; OnsØyen, Mykletun, & Steiro, 2009). From the guest's perspective, frontline staff in any service organisation *are* the brand and the service (Zeithaml et al., 2006).

Quality service is synonymous with luxury hotels (Kim, Kim, King, & Heo, 2019). For hospitality businesses to create value, they need to be able to engage their guests throughout their entire hotel experience, and both the tangible and intangible characteristics of a hotel play a critical role in enhancing that overall experience (Khoo-Lattimore & Ekiz, 2014). For example, in terms of the tangible elements, facilities always need to be in excellent condition, appropriately maintained and meticulously spotless and hygienic. In terms of the characteristics of the hotel that are intangible, staff should always attempt to highlight these to guests – for example, if a hotel is blessed with an incredible view, staff could mention this when the guest is checking in, and whenever possible, they should aim to upgrade them to a room that provides access to that incredible view.

With an experience mindset, quality service is recognised as a distinctive service (Pine & Gilmore, 2011) and is customised (Hung, 2018). This means that service is geared towards excelling at every opportunity (Walker, 2017). Luxury hotels excel in customising the services they provide to accommodate the needs of their guests (Harkison, 2016; Kimes & Ho, 2017).

Service quality is the most significant factor affecting the selection of luxury hotels (Low, 2012). A high level of service is standard for luxury hotels (Kim et al., 2019). Service, not the product, has the most impact on guests' perceptions about their experience (Cetin & Walls, 2016), and good service has a positive impact on guests' satisfaction in luxury hotels (Allan, 2016). Neuhofer, Buhalis, and Ladkin (2013) believe that hotel organisations need to stop viewing the guest experience as something that is a static part or created in isolation by the organisations themselves, but rather as something that is dynamic, highly personalised and proactively co-created by both the guest and staff at specific service encounters in real time.

Co-creating Experiences

As mentioned above, it should be the organisation that enhances the guest experience by managing it from beginning to end, but the success of the experience also depends on how much guests are willing and able to engage in their own experience (Walls et al., 2011). Organisations may be capable of creating the environment and circumstances in which it is possible for customers to have an amazing experience, but they are unable to guarantee that the outcome of a guest experience will be positive (Walls, 2013). These actions are seen as the basis of co-creation – the interaction or dialogue between the provider and customer to create value (Harkison, 2018; Shaw & Ivens, 2002).

To sincerely gain an understanding of their customers' views, businesses need to start with having a clear understanding of their customer's world. They need to know what exact services they will co-construct, precisely what they are co-constructing and how their customers perceive those co-construction processes (Andersson, 2007). An alternative way to view co-construction or co-creation is as follows: a tourism service provider and a tourist assemble resources that, together with their own time and skills, will be able to provide a consumption set that includes skills, time, goods and services, which combine to create an experience (Andersson, 2007).

Some researchers have stated that any experience could be regarded as a form of co-creation due to the customer gaining knowledge through the interaction of different components created by a service provider (Gupta & Vajic, 1999). However, Prahalad and Ramaswamy (2002) highlighted that there had been an increased impetus towards an environment where customers and companies were coming together to co-create experiences, and businesses were embracing the fact that customers were now becoming partners in co-creating their own experiences (Prahalad & Ramaswamy, 2002). Scott, Laws, and Boksberger (2009) suggested that experience had shifted from being something that is intrinsic to the guest to becoming a managerial matter, where it is developed and co-created by both the organisation and guest.

To ensure staff have the ability to interact successfully with guests, businesses now need to engage them in the process of co-creating customer value (Ramaswamy, 2009). Expanding on this point, Scott et al. (2009) suggested that the interplay between the customer and producer is not only becoming increasingly important, but it is becoming even more critical than the actual product itself, and that the emphasis must be placed on developing experiences that customers actively participate in rather than just passively observe. Echoing this, Santos-Vijande, Álvarez, and Rodríguez (2012) stated that frontline staff are becoming crucial to the enablement of co-creation activities, given the volume of interactions that they are now having with customers.

Cases Studies

Limited research has been undertaken on the topic of luxury accommodation experiences, and these have largely taken a unidimensional perspective on the issue – concentrating on either managerial or a customer perspective – while other crucial performers in the delivery of luxury accommodation experiences, particularly service staff, have generally been overlooked. Although the subject of producing experiences has been highlighted as a significant component in tourism experiences, it has also been largely ignored, particularly in relation to the accommodation sector and, more specifically, in luxury lodges and hotels. Despite conducting an extensive literature review, no previous research on developing luxury hotel or lodge experiences in New Zealand was found (Harkison, 2016).

For this reason, my research aimed to examine and evaluate the creation of the luxury accommodation experience using case studies within New Zealand. The research adopted interpretivist case study methodology, which utilised a multiple case study approach to the examination and evaluation of guests', employees' and managers' perspectives on producing a luxury hotel and lodge experience in New Zealand. In total, 6 luxury properties were used as case studies, and a total of 81 participants were interviewed while their 'experience' was in progress. The participants for this research included 27 managers, 27 employees and 27 guests. An analysis of the data produced two conceptual models that characterised what the luxury lodge and hotel experience entails within New Zealand and how it is created within New Zealand (Harkison, 2016).

The first model (case study one) clarifies the luxury hotel and lodge experience within New Zealand. This model reveals the luxury hotel and lodge experience as being characterised by the following elements:

- Being in the moment;
- Opulence;

- The essence of the luxury hotel or lodge;
- The feeling of indulgence;
- The location and setting of the properties;
- The sensations and emotions of managers, employees and guests;
- The physical surroundings of the properties;
- Delivering tailored attention;
- Hospitableness;
- Activities available at the properties and
- Guests' desires and expectations (Harkison, 2016).

Regardless of whether they were hotels or lodges, across all properties, the three strongest themes that were identified by the participants were the first three listed above: being in the moment; opulence and the essence of the luxury hotel and lodge.

The second case study introduces another model, which defines how the luxury lodge and hotel experience is developed in New Zealand. The model identifies how the luxury hotel and lodge experience is developed via the following themes:

- Setting the stage;
- The ethos of the properties;
- The actors' performances;
- Co-creation between participants;
- Guidance;
- Comparisons between properties;
- Domestic and international comparisons;
- The brand's values and beliefs and
- Projecting an image (Harkison, 2016).

The four most prominent themes highlighted by all of the participants interviewed were the ethos of the property, co-creation between participants, setting the stage and the actors' performances – and this held true across all of the luxury properties surveyed, be they hotel or lodge.

Recommendations from the Case Studies

A number of recommendations can be found in the present research outcomes that will assist managers when they are attempting to construct luxury hotel and lodge experiences. Although this chapter has been based on the role of employees, it is also the responsibility of the owners, management and the visitors to take part in the creation of the luxury lodge and hotel experience. An organisation's vision or mission statement is a form of promise and sets certain expectations, so it is critical that they work hard to deliver on those promises and expectations. Therefore, it is vital that the mission or vision statement is inculcated into the thoughts and actions of all employees and managers and that a monitoring system is created and utilised to guarantee that what has been promised is actually being

Case Study 10.1: What the Luxury Hotel and Lodge Experience Is in New Zealand?

Fig. 10.1 presents the model of what the luxury hotel and lodge experience is and summarises the empirical research findings. The central area of the model exemplifies the collective opinions of the participants of what they thought were the most crucial components required in a successful luxury hotel or lodge experience. This figure could be considered a holistic 'blueprint' required to build such an experience. In the left area are two additional components that were considered to be unique to hotels in the study, and in the right area are three extra elements that were considered to be unique to the lodges in the study (Harkison, 2016).

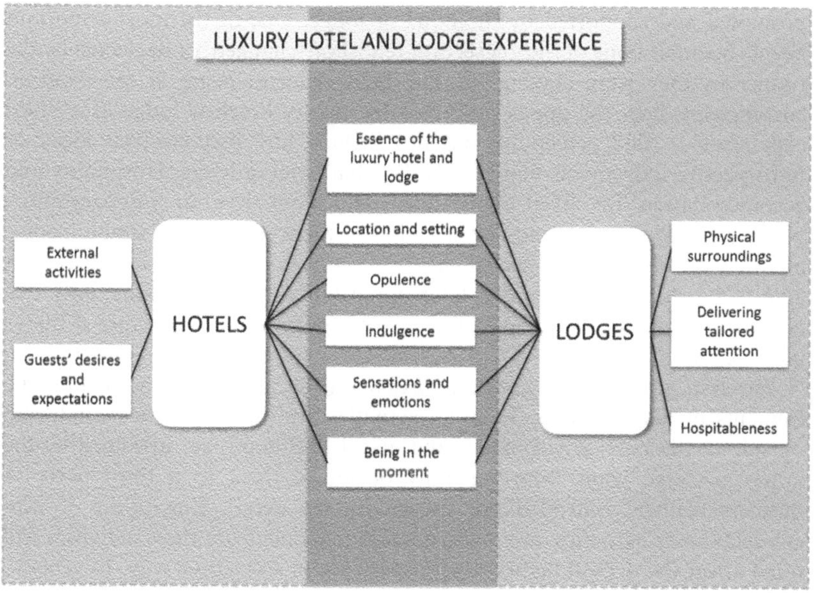

Fig. 10.1. Model of Luxury Hotel and Lodge Experience.

In the central area, the elemental *essence of the luxury hotel and lodge* is at the top, and this includes all the foundation aspects of the properties; this includes the exceptional standards that are seen as the point of difference between any 'ordinary' hotel or lodge within New Zealand and a luxury hotel or lodge. Next is *location and setting* – this relates to luxury hotels and lodges being located in either the city or countryside, but always with attractive sites with beautiful surroundings. Participants agreed that New Zealand has a distinct advantage, as it has already received global recognition as a unique holiday destination for its diversity of landscapes, scenery and locations and is seen as already having an established luxury status. The next item is *opulence*, which encompasses the various unique elements that define a luxury venue: any venues that were missing some or all of these elements were considered to be 'ordinary'. So, the different degrees of opulence, therefore, define the standards of luxury experience (Harkison, 2016).

Indulgence comes next and is seen as the convergence of the guests' desires, expectations and requirements and the ability and willingness of staff to satisfy their guests – guests were pampered, spoilt and made to feel special, with the cooperation and input of the entire staff. Both staff and manager participants explained that there was an emotional aspect associated with guest interaction and peer interactivity, and by possessing the 'right tools' and operating environment, they were able to deliver an excellent standard of service. Being able to 'get it right' for their guests also enhanced their own personal sense of fulfilment and satisfaction. The guests' *sensations and emotions* were experienced from different perspectives: here they were evaluating the property, the frontline staff and all the tangible elements that were promoted and supported by the hotel or lodge. One of the most important beliefs was that none of the guests ever felt 'short-changed' or neglected by the properties they were staying in. The final element, *being in the moment*, encompasses how the guests perceived the luxury hotel or lodge as a 'lived experience' – this perception did not include any preconceived ideas or retrospective judgements that they had of that hotel or lodge. Hospitality and accommodation are 'lived experiences', which involves an 'inseparability', because the production and consumption of these services are simultaneous, and they are also interconnected. As these services are perishable and are not able to be stored for sale at a later opportunity, properties have only one chance to impress their guests – they are not given a second chance if things go wrong (Harkison, 2016).

External activities and *guests' desires and expectations* are the two unique elements of luxury hotels. External activities are those activities conducted outside of the hotel environment and, in all likelihood, are arranged by the hotel concierge team, who generally outsource them to a third party (a business neither controlled nor owned by the hotel). Regardless of who conducts these activities, any unfavourable outcome from these activities will often affect the overall luxury hotel experience in a negative way. In general, hotels deliver more prescriptive experiences, as they are trying to deliver what they believe their guests expect and desire, in accordance with their own guidelines, operating standards and performance specifications (Harkison, 2016).

Physical surroundings, *delivering tailored attention* and *hospitableness* are the three unique elements of luxury lodges. New Zealand luxury lodges are frequently situated in fairly remote areas, with rugged topography and breathtaking scenery. In contrast, the facilities are far from rugged or spartan – in fact, most are the last word in luxury. In these lodges, staff have the ability to exercise a higher level of autonomy, and the consequence of this is the greater flexibility they have in effectively delivering tailored attention to individual visitors and thus further enhancing the luxury experience. Hospitableness, within the luxury lodge context, is a general concept that implies that all guests are being well catered to and that their well-being is the top priority of the lodge. In order to be a successful luxury property and to ensure hospitableness, it is essential that guests recognise and acknowledge these endeavours are happening in the luxury lodge (Harkison, 2016).

Case Study 10.2: How the Luxury Hotel and Lodge Experience Is Created in New Zealand?

Fig. 10.2 presents a model of how the luxury hotel and lodge experience is created in New Zealand and is based on the empirical research findings. The centre of the model represents the collective opinions of what the participants considered to be the six critical components that are required to create a successful luxury lodge or hotel experience. This model could be considered a holistic 'blueprint' for enabling the luxury accommodation experience to be developed. In the left area, there are two extra elements that are considered to be unique to the hotels that were studied, and in the right area is one extra element that is considered to be unique to the lodges studied (Harkison, 2016).

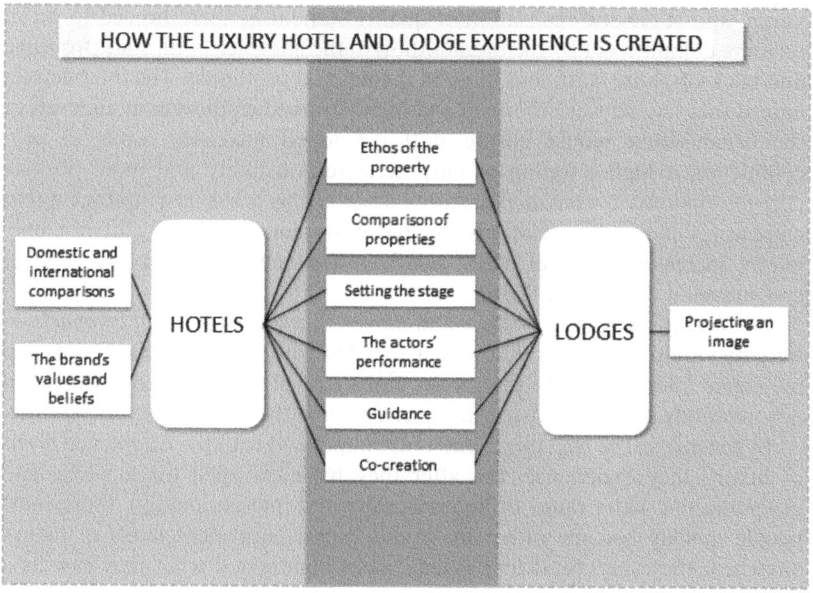

Fig. 10.2. Model of How the Luxury Hotel and Lodge Experience Is Created.

In the central block, the element *ethos of the property* is at the top – this represents the philosophies that underpinned the founding of the business. Management of the businesses all began with a vision or mission statement, which declared expectations, intentions, promises and guarantees that would impact staff and guests alike. *Comparison of properties* was agreed upon by all interviewees as essential, as they felt that luxury properties need to have the ability to positively promote their differences with comparable competing organisations. Once a vision for the property has been determined, the element of *setting the stage* is required. This stage comprises everything a luxury property offers that is tangible: those elements with which the guests will physically interact, for example, the lobby, their room, the pool, etc. These 'basic' requirements are regarded as critical to the guests' nourishment, entertainment and comfort, and therefore, there is no room for deficiencies in quantity, quality or availability (Harkison, 2016).

Once this *stage* has been set, the *play* can begin – the actors (employees and managers) must perform their parts competently, efficiently and professionally, while the audience (guests) observe *the actors' performances.* To guarantee success, the audience (guests) must become involved and actively participate in the overall performance. This active involvement is an essential ingredient of the entire luxury stay experience (Harkison, 2016).

All interviewees believed that in order to guarantee the long-term preservation and successful realisation of quality standards, there needed to be an ongoing commitment by the organisation to the *guidance* of all staff (frontline and back-of-house staff, and those in managerial positions). The interviewees agreed that having well-informed and highly trained employees at all levels of the organisation would ensure staff developed increased levels of self-confidence, a higher feeling of purposeful responsibility and more positive decision-making. *Co-creation* has been described as a business stratagem that emphasises interactive relationships and customer experience – it not only allows increased levels of active involvement from the guest but actually encourages it (Harkison, 2016).

The brand's values and beliefs and *domestic and international comparisons* were two unique elements that were associated with luxury hotels. Guests and managers interviewed felt that wider global travel had exposed a variety of new methods that could assist in the creation of the luxury hotel experience.

In general, in the beginning, guests are impressed with the heightened levels of luxury they experience, but after they have travelled for a while, and experienced a wider range of hotels (locally and internationally), those same people quickly become rather blasé and expect equivalent levels of luxury anywhere they stay. Staff and management highlighted what they saw as a definite benefit of being a member of a bigger international organisation, the fact that they were able to instil the same accepted values, beliefs and standards as the other hotels in their international group, which then ensured the creation of the same level of luxury experience (Harkison, 2016).

> *Projecting an image* was considered a unique component in the luxury lodge experience. Prior to selecting a lodge, guests frequently undertake a thorough screening of the options, commonly reviewing the full range of media available to them, including websites, social media and even any hard copy material still published by the lodges. And all participants agreed that the projection of an attractive image, accompanied with extensive information, was essential to them in their decision-making process. Indeed, they argued that the starting point for the entire luxury lodge experience was the initial contact with the lodge, which was its portrayal in that media (Harkison, 2016).

delivered. The business needs to know what their points of difference are, and it is essential that they utilise these as part of their competitive advantage – management must recognise what it is that makes their properties different from others, and then underscore and highlight these factors to all of the guests, be they regulars or potential visitors (Harkison, 2016). The businesses must *set the stage* for their visitors because this will provide the basic underpinning structure upon which employees will be able to build the luxury lodge or hotel experience for them. Ensuring that all the intangible and tangible components present at the properties are available for all visitors, with no exceptions, helps to achieve the setting of the stage – and businesses might have to develop checklists to ensure that it happens. It is very difficult, perhaps even impossible, to deliver a luxury experience if these 'building blocks' (the *stage*) have not been put in place, as they are the foundation on top of which the entire experience is constructed.

Businesses must concentrate on choosing staff with what they see as the correct appearance, behaviour, attitude, demeanour, ability to be proactive, and social skills. They also need to concentrate on attracting, developing and retaining employees who are proficient in creating and implementing luxury lodge or hotel experiences that are successful. This may require businesses to review and re-evaluate the staff-related policies and procedures that they might (or might not) currently possess. There must be clear communication of expectations from management to the staff so that they can correctly produce the promised luxury experience for guests, and this might require the development of a blueprint that details how to fashion and deliver what is required (Harkison, 2016).

Organisations will have to explore the variety of ways in which co-creation occurs within their lodges and hotels and facilitate and support increased levels of interaction and co-creation among all participants. This co-creation takes place within the properties between at least two of the participating parties (visitors, staff or management) and, frequently, external to the properties, with outside third parties (business neither owned nor controlled by the hotel or lodge). Encouraging and supporting further opportunities for co-creation to materialise is likely to impact the creation of the luxury experience in a positive way. An example of this is seen when the organisation encourages and makes it possible for employees to visit local producers of food or wine: this empowers those employees to converse confidently and knowledgeably about those matters with visitors. Providing dedicated employees (ambassadors) to interact with visitors throughout the entirety of their visit encourages improved relationships to develop, leading to greater levels of meaningful engagement, thus enabling successful co-creation. And the constant benchmarking of their own organisations with national and international counterparties will allow owners to constantly improve the experiences that they can offer their guests.

Organisations must ensure that the beliefs and values associated with their brand are adhered to and promoted by all staff, and by doing so, they will help them to filter through to their visitors. Constant employee training and support is required to ensure that the brand's beliefs and values are maintained and sustained. The image that properties portray via the various channels available to them, be they traditional or modern, needs to be carefully cultivated and managed, as it is this image and the message that it conveys that is the beginning of the creation of the luxury experience – it is sending visual and verbal signals to existing and potential visitors.

The models that have been presented through the two case studies might assist organisations to understand what a luxury experience really is and what is involved in its creation. The research demonstrates that the overall luxury lodge and hotel experience can be extremely complex, emotional and personal for all groups involved, and that if organisations are able to gain a greater appreciation and understanding of this fact, then they and their staff will more easily be able to successfully deliver positive and memorable experiences at their own luxury property (Harkison, 2016).

Conclusion

The accommodation aspect of any tourism experience is essential and can make or break the entire experience. Guests no longer want a luxury accommodation stay: they want a luxury accommodation experience. As can be seen throughout this chapter, in order to create this experience, employees are not only crucial but also central to the process.

Generally, luxury lodges and hotels possess highly trained and engaged employees, but the properties must make sure that they are attracting and recruiting the 'correct' staff. These staff members must possess personalities predisposed to delighting visitors, and they must exhibit professional behaviour, appearance, demeanour and attitudes that match the property's culture.

Given that there is an increased level of interaction with visitors at luxury hotels and lodges, businesses must make sure that frontline staff possess the confidence and ability to converse knowledgeably with visitors on a wide variety of domestic and international matters. For organisations to co-create unforgettable experiences for their guests, they need to pay attention to the guest's entire stay – and this begins even before they physically check-in at the property, and it continues through to any follow-up post the guest's departure, and encompasses everything in between. This is achieved by creating a set of 'touch points' that will enhance the overall experience, continually providing additional surprises that will not only excite but also stimulate guests during the entirety of their visit.

The luxury accommodation experience is then seen as a proactive, personal and dynamic creation, co-created by employees and guests.

References

Alcántara-Alcover, E., Artacho-Ramírez, M. Á., Martínez-Guillamón, N., & Campos-Soriano, N. (2013). Purpose of stay and willingness to stay as dimensions to identify and evaluate hotel experiences. *International Journal of Hospitality Management, 33*, 357–365. doi:10.1016/j.ijhm.2012.10.006

Allan, M. M. (2016). The relationship between service quality and customer satisfaction and retention in Ghana's luxury hotels. *IUP Journal of Marketing Management, 15*(4), 60–83.

Andersson, T. D. (2007). The tourist in the experience economy. *Scandinavian Journal of Hospitality and Tourism, 7*(1), 46–58. doi:10.1080/15022250701224035

Ariffin, A. A. M., & Maghzi, A. (2012). A preliminary study on customer expectations of hotel hospitality: Influences of personal and hotel factors. *International Journal of Hospitality Management, 31*(1), 191–198. doi:10.1016/j.ijhm.2011.04.012

Ariffin, A. A. M., Nameghi, E. N., & Zakaria, N. I. (2013). The effect of hospitableness and servicescape on guest satisfaction in the hotel industry. *Canadian Journal of Administrative Sciences/Revue canadienne des sciences de l'administration, 30*(2), 127–137. doi:10.1002/cjas.1246

Bagnera, S. M. (2017). A (diamond) cut above the rest: Improving hotel operations based on TripAdvisor rating attributes. *Boston Hospitality Review, 5*(3), 1–14. Retrieved from http://www.bu.edu/bhr

Barsky, J., & Nash, L. (2002). Evoking emotion: Affective keys to hotel loyalty. *Cornell Hotel and Restaurant Administration Quarterly, 43*(1), 39–46. doi:10.1177/0010880402431004

Baum, T. (2006). Reflections on the nature of skills in the experience economy: Challenging traditional skills models in hospitality. *Journal of Hospitality and Tourism Management, 13*(2), 124–135.

Bell, D. (1989). Finding your amenities niche. *Lodging Hospitality, 45*, 41–43.

Bernstein, L. (1999). Luxury and the hotel brand. *Cornell Hotel and Restaurant Administration Quarterly, 40*(1), 47–53. doi:10.1177/001088049904000122

Berry, L. L., Carbone, L. P., & Haeckel, S. H. (2002). Managing the total customer experience. *MIT Sloan Management Review, 43*(3), 85–89.

Bharwani, S., & Jauhari, V. (2013). An exploratory study of competencies required to co-create memorable customer experiences in the hospitality industry. *International Journal of Contemporary Hospitality Management, 25*(6), 823–843. doi:10.1108/IJCHM-05-2012-0065

Brien, A., Ratna, N., & Boddington, L. (2012). Is organisational social capital crucial for productivity growth? An exploration of 'trust' within luxury hotels in New Zealand. *Journal of Human Resources in Hospitality & Tourism, 11*(2), 123–145. doi:10.1080/15332845.2011.648840

Carù, A., & Cova, B. (2006). How to facilitate immersion in a consumption experience: Appropriation operations and service elements. *Journal of Consumer Behaviour*, *5*(1), 4–14. doi:10.1002/cb.30

Cetin, G., & Dincer, F. I. (2014). Influence of customer experience on loyalty and word-of-mouth in hospitality operations. *Anatolia*, *25*(2), 181–194. doi:10.1080/13032917.2013.841094

Cetin, G., & Walls, A. (2016). Understanding the customer experiences from the perspective of guests and hotel managers: Empirical findings from luxury hotels in Istanbul, Turkey. *Journal of Hospitality Marketing & Management*, *25*(4), 395–424. doi:10.1080/19368623.2015.1034395

Chu, Y., Tang, L., & Luo, Y. (2016). Two decades of research on luxury hotels: A review and research agenda. *Journal of Quality Assurance in Hospitality & Tourism*, *17*(2), 151–162. doi:10.1080/1528008X.2015.1047076

Dawson, M., Abbott, J., & Shoemaker, S. (2011). The hospitality culture scale: A measure organisational culture and personal attributes. *International Journal of Hospitality Management*, *30*(2), 290–300. doi:10.1016/j.ijhm.2010.10.002

Dixit, S. K. (Ed.), (2017). *The Routledge handbook of consumer behaviour in hospitality and tourism*. London: Routledge.

Dixit, S. K. (2020). *The Routledge handbook of tourism experience management and marketing*. London: Routledge.

Drewer, P. (2005). *Key note market report plus: Hotels*. Retrieved from https://www.keynote.co.uk

Gilmore, J. H., & Pine, B. J. (2002). Differentiating hospitality operations via experiences: Why selling services is not enough. *Cornell Hotel and Restaurant Administration Quarterly*, *43*(3), 87–96. doi:10.1177/0010880402433009

Griffin, R., Shea, L., & Weaver, P. (1996). How business travellers discriminate between mid priced and luxury hotels: An analysis using a longitudinal sample. *Journal of Hospitality & Leisure Marketing*, *4*(2), 63–75.

Gupta, S., & Vajic, M. (1999). The contextual and dialectical nature of experiences. In J. Fitzsimmons & M. Fitzsimmons (Eds.), *New service development* (pp. 33–51). Thousand Oaks, CA: Sage.

Han, H., & Back, K. J. (2007). Assessing customers' emotional experiences influencing their satisfaction in the lodging industry. *Journal of Travel & Tourism Marketing*, *23*(1), 43–56. Retrieved from https://doi.org/101300/J073v23n01_04

Han, H., Moon, H., & Hyun, S. (2019). Indoor and outdoor physical surroundings and guests' emotional well-being. *International Journal of Contemporary Hospitality Management*, *31*(7), 2759–2775. doi:10.1108/IJCHM-10-2018-0852

Harkison, T. (2016). *How is the luxury accommodation experience created? Case studies from New Zealand*. Doctoral thesis, Auckland University of Technology, New Zealand. Retrieved from http://hdl.handle.net/10292/9925

Harkison, T. (2018). The use of co-creation within the luxury accommodation experience – Myth or reality? *International Journal of Hospitality Management*, *71*, 11–18. doi:10.1016/j.ijhm.2017.11.006

Harkison, T., Hemmington, N., & Hyde, K. F. (2018). Luxury accommodation – Significantly different or just more expensive? *Journal of Revenue and Pricing Management*, *17*(4), 231–243. doi:10.1057/s41272-017-0085-1

Hemmington, N. (2007). From service to experience: Understanding and defining the hospitality business. *Service Industries Journal, 27*(6), 747–755. doi:10.1080/02642060701453221

Hung, K. (2018). Consumption of luxury hotel experience in contemporary China: Causality model for conspicuous consumption. *Tourism Review International, 22*(3), 171–185. doi:10.3727/154427218X15410074029607

Kandampully, J., & Suhartanto, D. (2000). Customer loyalty in the hotel industry: The role of customer satisfaction and image. *International Journal of Contemporary Hospitality Management, 12*(6), 346–351.

Khoo-Lattimore, C., & Ekiz, E. H. (2014). Power in praise: Exploring online compliments on luxury hotels in Malaysia. *Tourism and Hospitality Research, 14*(3), 152–159. doi:10.1177/1467358414539970

Kimes, S. E., & Ho, J. (2017). Revenue management in luxury hotels. *Journal of Revenue and Pricing Management, 17*(4), 291–295. doi:10.1057/s41272-017-0113-1

Kim, B., Kim, S., King, B., & Heo, C. Y. (2019). Luxurious or economical? An identification of tourists' preferred hotel attributes using best-worst scaling (BWS). *Journal of Vacation Marketing, 25*(2), 162–175. doi:10.1177/1356766718757789

Knutson, B. J., & Beck, J. A. (2004). Identifying the dimensions of the experience construct: Development of the model. *Journal of Quality Assurance in Hospitality & Tourism, 4*(3–4), 23–35. doi:10.1300/J162v04n03_03

Knutson, B. J., Beck, J. A., Kim, S., & Cha, J. (2010). Service quality as a component of the hospitality experience: Proposal of a conceptual model and framework for research. *Journal of Foodservice Business Research, 13*(1), 15–23. doi:10.1080/15378021003595889

Kotler, P., Bowen, J., & Makens, J. (2010). *Marketing for hospitality and tourism* (5th ed.). Upper Saddle River, NJ: Prentice-Hall International.

Kucukusta, D., Heung, V. C., & Hui, S. (2014). Deploying self-service technology in luxury hotel brands: Perceptions of business travellers. *Journal of Travel & Tourism Marketing, 31*(1), 55–70. doi:10.1080/10548408.2014.861707

Lashley, C. (2008). Studying hospitality: Insights from social sciences. *Scandinavian Journal of Hospitality and Tourism, 8*(1), 69–84. doi:10.1080/15022250701880745

Lewis, B. R., & McCann, P. (2004). Service failure and recovery: Evidence for the hotel sector. *International Journal of Contemporary Hospitality Management, 16*(1), 6–17. doi:10.1108/09596110410516516

Low, T. A. (2012). *Influence of consumer values and sustainable business practices on brand loyalty within luxury hotels.* Doctoral thesis, University of Bedfordshire, England. Retrieved from http://hdl.handle.net/10547/323658

Lu, C., Berchoux, C., Marek, M. W., & Chen, B. (2015). Service quality and customer satisfaction: Qualitative research implications for luxury hotels. *International Journal of Culture, Tourism and Hospitality Research, 9*(2), 168–182. doi:10.1108/IJCTHR-10-2014-0087

Lugosi, P. (2008). Hospitality spaces, hospitable moments: Consumer encounters and affective experiences in commercial settings. *Journal of Foodservice, 19*(2), 139–149.

Marketing Science Institute. (2010). *2010–2012 research priorities.* Boston, MA: Author.

Mattila, A. S. (1999). Consumers' value judgements: How business travellers evaluate luxury hotel services. *Cornell Hotel and Restaurant Administration Quarterly, 40*(1), 40–46. doi:10.1177/001088049904000121

McIntosh, A. J., & Siggs, A. (2005). An exploration of the experiential nature of boutique accommodation. *Journal of Travel Research*, *44*(1), 74–81. doi:10.1177/0047287527659 3

Mohsin, A., Hussain, I., & Khan, M. R. (2011). Exploring service quality in luxury hotels: Case of Lahore, Pakistan. *The Journal of American Academy of Business*, *16*(2), 296–303.

Neuhofer, B., Buhalis, D., & Ladkin, A. (2013). High tech for high touch experiences: A case study from the hospitality industry. In L. Cantoni & Z. Xiang (Eds.), *Information and communication technologies in tourism 2013, proceedings of the international conference in Innsbruck, Austria, January 22–25* (pp. 290–301). Heidelberg: Springer.

O'Gorman, K. D. (2007). Dimensions of hospitality: Exploring ancient and classical origins. In C. Lashley, P. Lynch, & A. Morrison (Eds.), *Hospitality: A social lens* (pp. 17–32). Amsterdam: Elsevier.

O'Neill, L. (2004). Luxury hotel values are on the rise. *Lodging Hospitality*, *60*(10), 30.

OnsØyen, L. E., Mykletun, R. J., & Steiro, T. J. (2009). Silenced and invisible: The work experience of room-attendants in Norwegian hotels. *Scandinavian Journal of Hospitality and Tourism*, *9*(1), 81–102. doi:10.1080/15022250902761462

Ottenbacher, M., Harrington, R., & Parsa, H. G. (2009). Defining the hospitality discipline: A discussion of pedagogical and research implications. *Journal of Hospitality & Tourism Research*, *33*(3), 263–283. doi:10.1177/1096348009338675

Penfold, P. (2009). Learning through the world of second life: A hospitality and tourism experience. *Journal of Teaching in Travel & Tourism*, *8*(2–3), 139–160. doi:10.1080/15313220802634224

Pine, B. J., & Gilmore, J. H. (2011). *The experience economy*. Boston, MA: Harvard Business Review Press.

Prahalad, C. K., & Ramaswamy, V. (2002). The co-creation connection. *Strategy and Business*, *27*, 50–61.

Presbury, R., Fitzgerald, A., & Chapman, R. (2005). Impediments to improvements in service quality in luxury hotels. *Managing Service Quality*, *15*(4), 357–373. doi:10.1108/09604520510606835

Quan, S., & Wang, N. (2004). Towards a structural model of the tourist experience: An illustration from food experiences in tourism. *Tourism Management*, *25*(3), 297–305. doi:10.1016/S0261-5177(03)00130-4

Ramaswamy, V. (2009). Leading the transformation to co-creation of value. *Strategy & Leadership*, *37*(2), 32–37. Retrieved from https://doi.org/101108/10878570910941208

Santoro, G. (2015). Evaluating performance in the hotel industry: An empirical analysis of Piedmont. *Journal of Investment and Management*, *4*(1–1), 17–22.

Santos-Vijande, M. L., Álvarez, B. Á., & Rodríguez, N. G. (2012). Internal marketing as a driver of market orientation and co-creation culture in the tourism sector. *African Journal of Business Management*, *6*(13), 4707–4716. doi:10.5897/AJBM11.1717

Scerri, M. A., Jenkins, J. M., & Lovell, G. (2017). A grounded theory model of service language in Australia's luxury hotels. *Journal of Hospitality and Tourism Management*, *33*, 82–92. doi:10.1016/j.jhtm.2017.09.003

Scott, N., Laws, E., & Boksberger, P. (2009). The marketing of hospitality and leisure experiences. *Journal of Hospitality Marketing & Management*, *18*(2–3), 99–110. doi:10.1080/19368620802590126

Shaw, C., & Ivens, J. (2002). *Building great customer experiences*. New York, NY: Palgrave MacMillan.

Sherman, R. (2007). *Class acts: Service and inequality in luxury hotels*. Berkeley, CA: University of California Press.

Snel, A. (2013). Experience as the DNA of a changed relationship between firms and institutions and individuals. In J. Sundbo & F. Sørensen (Eds.), *Handbook on the experience economy* (pp. 122–145). Cheltenham: Edward Elgar.

Sørensen, F., & Jensen, J. F. (2015). Value creation and knowledge development in tourism experience encounters. *Tourism Management*, *46*, 336–346. doi:10.1016/j.tourman.2014.07.009

Sturken, C. A. (2018). Luxury today: More experience, less extravagance. *Meetings and Conventions*, *53*(1), 27–30.

Sundbo, J., & Dixit, S. K. (2020). Conceptualisations in tourism experience. In S. K. Dixit (Ed.), *The Routledge handbook of tourism experience management and marketing* (pp. 15–26). London: Routledge.

Titz, K. (2008). Experiential consumption: Affect – emotions – hedonism. In H. Oh & A. Pizam (Eds.), *Handbook of hospitality marketing management* (pp. 324–352). Oxford: Butterworth-Heinemann.

Torres, E. N., & Kline, S. (2006). From satisfaction to delight: A model for the hotel industry. *International Journal of Contemporary Hospitality Management*, *18*(4), 290–301. doi:10.1108/09596110610665302

Tosti, D. C. (2009). Customer experience and value: A performance view. *Performance Improvement*, *48*(1), 37–44. doi:10.1002/pfi.20047

Verhoef, P. C., Reinartz, W. J., & Krafft, M. (2010). Customer engagement as a new perspective in customer management. *Journal of Service Research*, *13*(3), 247–252. doi:10.1177/1094670510375461

Walker, J. R. (2017). *Introduction to hospitality* (7th ed.). Boston, MA: Pearson.

Walls, A. R. (2013). A cross-sectional examination of hotel consumer experience and relative effects on consumer values. *International Journal of Hospitality Management*, *32*, 179–192. doi:10.1016/j.ijhm.2012.04.09

Walls, A., Okumus, F., Wang, Y., & Kwun, D. J. W. (2011). Understanding the consumer experience: An exploratory study of luxury hotels. *Journal of Hospitality Marketing & Management*, *20*(2), 166–197. doi:10.1080/1936823.2011.536074

Williams, A. (2006). Tourism and hospitality marketing: Fantasy, feeling and fun. *International Journal of Contemporary Hospitality Management*, *18*(6), 482–495. doi:10.1108/09596110681520

Xu, J. B., & Chan, A. (2010). A conceptual framework of hotel experience and customer-based brand equity: Some research questions and implications. *International Journal of Contemporary Hospitality Management*, *22*(2), 174–193. doi:10.1108/09596111011018179

Yuan, Y.-H., & Wu, C. (2008). Relationships among experiential marketing, experiential value, and customer satisfaction. *Journal of Hospitality & Tourism Research*, *32*(3), 387–410. doi:10.1177/1096348008317392

Zeithaml, V. A., Bitner, M. J., & Gremler, D. D. (2006). *Services marketing: Integrating customer focus across the firm* (4th ed.). New York, NY: McGraw-Hill.

Chapter 11

The Language of Luxury: Decoding the Luxury Hotel Brand's Marketing Communication

Senthilkumaran Piramanayagam[1] *and Partho Pratim Seal*[2]

Abstract

The market for luxury products and services plays a significant role in the world economy. The luxury hotel market is a crucial segment within the global market for luxury products and services. Luxury hospitality recorded a 5% growth rate along with demand for luxury cruises recording the growth of 7%, the highest among all luxury segments. The remarkable performance of luxury products and services over a period is attributed to the laudable marketing communication strategies of luxury marketers. In this research, we aimed to analyse how a multi-brand hospitality firm differentiates its luxury brand with other luxury brands in the portfolio, using textual messages aimed to communicate the uniqueness of the brands in its official websites. The case study method and content analysis are adapted to achieve the research objective. The study results show that different luxury brands under the brand portfolio of Marriott International can communicate the differences through the textual contents. Most of the differentiation relies on brand-specific features, traditions, services, location of the hotel and metadata on the Internet, followed by branded differentiators. It may be concluded that Marriott has been successful to an extent in using text contents in the website to differentiate its luxury brands.

Keywords: Luxury hotels; brand portfolio; brand differentiation; metadata; Marriott hotels; content analysis; case study method

[1]https://orcid.org/0000-0001-6746-0421
[2]https://orcid.org/0000-0001-6867-8713

The Emerald Handbook of Luxury Management for Hospitality and Tourism, 221–242

doi:10.1108/978-1-83982-900-020211011

Introduction

Currently, luxury is not any more a small sector and restricted to a happy few. The luxury offerings, along with their omnipresence, brands are presently part of globalisation and consumption societies. Luxury traditionally is defined as something which is not ordinary in considering the daily living needs (Vickers & Renand, 2003). Both luxury products and services have been defined, but the definitions many a time become vague and challenging to manage as luxury is continuously evolving and it is quite subjective (Catry, 2003; Dubois, Czellar, & Laurent, 2005; Kapferer, 1998; Tynan, McKechnie, & Chhuon, 2010). The luxury meaning also changes with the place, person and time (Michman & Mazze, 2006; Yeoman & McMahon-Beattie, 2006). Some researchers have characterised luxury as non-necessity (De Barnier, Rodina, & Valette-Florence, 2006). Luxury is connected with several characteristics such as aesthetics, exclusivity, excellent quality, premium price, scarcity and uniqueness (Chevalier & Mazzalovo, 2012; Kapferer, 1998; Nueno & Quelch, 1998). In this chapter, we define a luxury hotel as 'any hotel or resort which is distinctive in terms of surroundings, quality of service, choice of amenities, priced high, and highly rated by the grading agencies that brings esteem for the guest and satisfies both functional and psychological needs' (Kapferer, 1998).

The market for luxury products and services plays a significant role in the world economy. The role of the luxury industry is unique; even during the economic depression, the luxury market continued stable growth (D'Arpizio, Levato, Zito, & Montgolifer, 2015). In 2019, the revenue generated by the top 100 companies in luxury was \$247 billion, with an annual growth of 10.8% (Deloitte, 2019). Eighty percent of the total luxury market accounted for luxury cars, hospitality and personal luxury goods (D'Arpizio, Levato, Prete, Fabbro, & Montgolfier, 2019). The luxury hotel market is a key segment within the global market for luxury products and services. Luxury hospitality recorded a 5% growth rate along with demand for luxury cruises recording the growth of 7%, the highest among all luxury segments. The luxury hotel market size is expected to reach US\$222.9 billion by 2024, with a compounded annual growth rate of 4.5% (ARC, 2019).

Various factors are driving the demand for luxury hotels. From the demand side, increased number of high net worth individuals (Timetric, 2013), growing tourism, business travellers, international events, opulent corporate culture, disposable income rise in China and India and rising desire of customers for luxurious lifestyle have powered the demand for luxury hotels (Global Data, 2016). Other factors that increased the demand for luxury hotels are the supply side, where hoteliers' keen effort to bring technology, superior service to manage competition and attract wealthy clients results in an implicit hierarchy in the hotel product offerings (Feng, Wang, & Ryan, 2018). Continuous marketing innovation, attractive membership programmes and technology-based customer relationship management initiatives also fuel the demand for luxury hotels.

'*Luxury sells dreams*' (Kapferer, 2015, p. 1). Constructing and selling dreams is the core mission of luxury brands. What makes a luxury product and service different from others are rarity, exclusivity, prestige and creativity. While rarity can be achieved through heritage and technical know-how, exclusivity can be obtained using perennity and selectivity in terms of distribution and targeting. Branding is an important process that makes a product to be perceived as a luxury. 'Perception drives behaviour' (Chew, 2016, p. 56) and the perceptions are created by brand messages that individuals see and hear. The success of a brand relies on how it establishes the perceptual association between its products and the target customers. When the association is strong, it is able to distinguish from others, more on when there is an information overload (Chew, 2016).

Managing the image of exclusivity and prestige of luxury products and service through branding is critical for its continued success. The successful branding not only focusses on niche markets but also helps the brand to extend its reach to broader markets which obviously lead to more profit. A brand differentiates itself by building individual identity and makes an enduring long-term cognitive link between organisation, customer and the brand. Brand differentiation significantly contributes to the success of brands (Aaker, 2004; Porter, 1985). Differentiation of brands becomes vital when a different market segment exists (Lancaster, 1984). However, differentiation is very complex when a firm has to manage brands with similar product or service characteristics under its brand portfolio. The different brands owned by a parent company are called brand portfolios.

The reason for a firm to have multiple brands under the same product category is to enhance the market coverage and higher combined sales. Various brands help marketing firms to pursue different customer market segments in terms of price, channels of distribution and geographical dispersion. Other benefits for the firms are to increase internal efficiency through competition within the firm, restrict the brand switching of customers who seek variety, spread risk, achieve economies of scale, physical distribution, merchandising and advertising. Each brand, under its portfolio, contributes to the firm to maximise its brand equity. The brand portfolio strategy of a firm specifies the scope, role and interrelationship between the brands (Aaker, 2004). The main goal of the portfolio strategy is to differentiate, create synergy and energise the brands. However, one of the challenges in branding is to create differentiation; without it, a brand offering of the firms will be vulnerable (Aaker, 2004; Dixit, 2017; Keller, Parmeswaran, & Jacop, 2015). The absence of meaningful differentiation is a cause for concern (Verma, 2011). Communication media like websites, point of sale and advertising are manifestations of brands essential for the brand identity (Chevalier & Mazzalovo, 2012). For hospitality service firms, unlike other luxury products, differentiation of brands becomes more complicated due to its inherent nature of intangibility (Rao, 2004; Verma, 2011).

The emergence of the Internet has radically changed the way marketers communicate with target customers (Cheng, 2016). Websites are currently considered as the primary tool for information, communication, reputation and interaction and a strategic channel to disseminate information. Websites have a significant role in a firm's effort to sell its services as an organisations' virtual marketing representative. A hotel brand's website has paramount importance as it deciphers the feelings, values and experiences associated with the brand among the prospective customer (Wei, 2020). The correct and appropriate use of language has a significant influence on attracting customers' attention and effect on customers' decisions (Kelly-Holmes, 2019). Although a great deal of research literature found on how hotel brands use social media communication, to date, studies on how multi-brand luxury hotel company differentiates one luxury brand among its other luxury brands in the portfolio are rare, specific to websites. Furthermore, there are no studies on brand communication or messages as cue or signal to distinguish a luxury hotel brand from its counterparts. This chapter aimed to extend existing knowledge on how a multi-brand hospitality firm differentiates its luxury brand from other luxury brands in the portfolio using textual messages on its official websites.

Case Study 11.1: Dreams of Luxury

The Skylark, an international hotel chain, grew from the chain of restaurants in the capital city of Delhi in early 1940 by Sunderlal and Shyamlal Aggarwal. The restaurant *Zayka* initially was in Delhi and then later it spread across the country. As the number of Zayka restaurants expanded, the proprietor moved on to steel and other heavy industries post-independence as the government was keen to promote those industries. The company moved on to hotels, and the first hotel in Mumbai was built in 1970 and later, when India was hosting Asian Games in 1982, the second property was opened in 1981. After the initial success of two hotels, the hospitality chain expanded by strategic acquisition of properties mostly the heritage properties, which were palaces of the Maharajas with a '*Majestic*' brand to allow a royal experience for the customers.

With the continuous pursuit for growth, the Skylark hotel diversified its business. It acquired an international brand in 1990, which was a top fine dining restaurant chain in London and also had some properties across Europe. The next five years were very crucial for the company as they purchased two more hotel brands that had a presence in America and Europe. The Skylark hotel brand was not able to penetrate the market and also showed significant potential growth opportunity, so the company came up with a moderate brand. The moderate brand *Skylark Universal* was created to attract a large segment of customers who were most dissatisfied with similar competitive brands. Considering it, the Hotels of Skylark Universal started providing additional facilities like loyalty programmes and airport pick-up for business clientele and spa facilities.

The initial success of the brand Skylark led to an increase in its presence from hotels to resorts and into timeshare properties. Within a span of two more years, it purchased a well-known brand *Ebony* in the hospitality. It also added its own luxury brand *Emerald* to its collection. With the expansion in the hospitality business, the steel and heavy industries which were not doing well, the stakes for the same were sold. Later in 2005, Skylark broadened its horizon and added two more brands to enhance its market segment. It added a full suite hotel, full-service apartments for long-stay guests and a new brand Skylark Inn for the economy class. The reason for the segmentation being done was to satisfy the needs of both the travellers for business and the travellers for leisure as their needs are varied. The other reason for the brand extension was necessary so as to find new avenues for growth that might have been affected by the traditional business restrictions.

As the brand expanded, they branched into two: the franchisee properties and the properties which were directly being owned by the Skylark group. The franchisee properties were enhanced by two more acquisitions of hotel chains. In the year 2009, Skylark again added a new brand *Mahogany* to its all-suite rooms, which were 15% larger than the regular hotel rooms. In 2011, the Skylark brand came up with a new contemporary boutique hotel, Gold Collections, for the luxury guest, which was designed by famous Architect Mr K. Sundaram and the interiors were done by Ms B. Chitnis. It continued to add a new brand to its portfolio Cedar, for their strong personality independent hotels in the last decade targeting stylish young guest to their properties.

In 2015, the Zayka restaurant chain stakeholdership was shared by Mr B. Satyapaul who later on in 2018 took over the restaurant chain as the Vice President and Owner of Zayka. The Skylark brand though started with its humble beginning with the restaurant would not fit the current state of being part of a multinational hospitality brand. Currently, the Skylark International hotel has 1,200 properties across 60 countries worldwide and in the year 2018 had $4 billion in revenue.

Reviews of Literature

Branding Luxury

Branding played a crucial role in the transformation of luxury. A brand is a sort of 'beyond the actual product' (Chevalier & Mazzalovo, 2012). American Marketing Association (AMA) defines a brand as a 'name, term, sign, symbol, or design, or a combination of them, intended to identify the goods and services of one seller or group of sellers and to differentiate them from those of competition'. Branding is considered both a science and art. It is a broad impression management measure undertaken by marketers. A brand has great value than a non-brand product as it carries a unique promise. A brand with a promise helps the users to simplify decision-making, reducing the risk.

For marketers, brands are the most important intangible asset. For many firms, the value of this intangible asset is more than its tangible assets. The increased significance of branding has emerged from various factors. Commodification and technological advancements in the last decade enabled most producers to develop and deliver quality products that meet customers' expectations.

This made the customers view competing products functionally identical. Differentiating through branding becomes essential when the organisation has limited power to differentiate its product through its functional features. Brands become the new frontier for competitive differentiation in the highly crowded and homogenous market. A strong brand has multiple management implications. Strong brands can attract, develop and retain a talented pool of human resources, which drives stakeholders' growth and value. A strong brand makes the selling process easier in the market, generating good market share and mindshare among customers. Strong brands are significantly associated with quality, innovation, novelty, intellectual rights, efficient supply chain and logistics (Chew, 2016).

Creating value is the key for brands to succeed. Strategy and tactics are two essential components to create brand value. At the same time, the strategic component of brand management involves identifying market and defining values, the tactics components involved in designing a brand and communicating its value to the target market. Therefore, a brand's success is defined by the viability of its strategy and the effectiveness of its communication tactics (Alexander, 2018).

Kapferer (2008) divided the brands into four layers of a brand pyramid: brands, upper range brands, luxury brands and 'griffe'. A regular brand or a private label is a branded good made on a mass scale. Upper range brands consist of expensive and premium quality products known as mass prestige goods. Luxury brands are characterised by high-quality products manufactured in small quantities. Griffe is of unique luxury specimens known as 'materialised perfections'. The physiological and psychological needs of individuals are fulfilled by regular brands. Premium brands tend to fulfil social and belonging needs, also helping individuals to distinguish themselves from others. The esteem and self-actualisation needs of individuals are fulfilled by luxury brands by offering ideational and sentimental values. Luxury products act as a symbol of an individual's achievement by which they have been respected and recognised by others.

Brand Differentiation

The core purpose of a brand is to identify the market offering, differentiate from other competing brands and create unique functional, psychological and monetary values for the customer. A brand must be able to articulate its value in the form of brand communication to build a meaningful brand image in the mind of the target customers. While the functional value is created by the functional superiority of the products, psychological value is created by enhanced customer experience in product usage. Monetary value is created by emphasising the monetary benefits of the market offerings. By creating these values, a brand can influence the customers' behaviour in terms of purchase, the pattern of use, frequency of use and increased likelihood of sharing of positive experience, which subsequently benefits the marketing firm. Brand differentiation resulted in multiple benefits to marketers. Shares of more differentiated brands in the market outperformed others and valued as premium than firms with low relative brand value. To ensure this advantage, managers must be better at communicating it (Chew, 2016).

Dimensions of Brand Differentiation

Brand differentiation is the engine if we consider the brand as a train. Brand fails if brand differentiation is not approached as an essential marketing strategy (Agris, 2001). Brands must have their unique or different selling point, a reason for customer to purchase the products (Aaker, 2004). Brand differentiation has been done through diverse forms such as marketing, positioning, quality, durability, customer service, distribution or the product itself (Porter, 1985). However, creating and maintaining a point of differentiation is difficult and also challenging, as aggressive competitors often copy the point of difference quickly. Aaker (2003) proposed a solution to manage and create an everlasting point of difference, 'branding the differentiator'. He suggested four forms of branded differentiators: branded features, branded service, branded ingredients and branded programmes. Aaker cautions that the 'differentiator' must be worthy, meaningful, matters most, pertinent and more than just a name.

Branded features insist consumers the specific features which are valued and perceived visually by the consumers. Branded ingredient, a component of technological innovation specific to brand through explicit and implied claims, helps to create credibility. Branded services are another form of brand differentiation by providing additional services by bundling existing products and services. Branded service helps a brand to develop a new product or extend the market, which the rivals cannot. Branded programmes are a great source of differentiation. Loyalty, product or service upgrade and website support programmes are well-known forms of branded programmes. The branded programme helps marketers to build strong relationships with their customers (Aaker, 2003).

Aaker (2003) has also advocated for the use of external brand differentiators: branded alliance and co-brands, for firms who find it difficult and costly to build branded differentiators internally. Alliance with another firm's brand, its credibility and exclusive use through agreement can be used as an external brand differentiator. The exclusive criterion is essential for an external brand differentiator as it should not be available for competitors. The association for differentiation may be for external brands' features, services, programmes and ingredients. For a luxury hospitality firm, incorporating all branded differentiators is very complex, especially branded features and branded ingredients. The complexity arises due to its inherent nature of intangibility and perishability. However, other than the above-branded differentiators specific to products, service firms do have a list of brand differentiators: firms name, logos, tagline, look and feel or the collection of above all called as 'collection of distinctions' (Schultz, Doerr, & Frederiksen, 2013). Verbal and visual elements in the advertising communication are also used as cue or symbol of product and brand differentiation. Verbal cue consists of name, tagline or lexicon – a set of keywords and attributes related to the brand. These cues, when communicated frequently, do reinforce the brand in the minds of customers. The verbal and visual cues are significant for brands that are intangible (Evans & Hastings, 2008).

In many cases, marketers utilise website metadata descriptors to announce their uniqueness specific to particular brands. The effective use of metadata also increases the search engine optimisation and definitive advantage over competitors (Burgess, Sellitto, & Karanasios, 2009; Mortimer, Brooks, Smith, & Hiam, 2012). Franzen and Moriarty (2008) listed other few strategies of differentiation: design or style, channel or location, price, salience, symbolic, customer service, customer intimacy and loyalty package.

Communication of Differentiation

Once the 'collection of distinction' has been identified, it is the firm's responsibility to make a claim, announce the value and announce the position of the firms in the market in a distinct and compelling manner. The messages should act as 'signals' or 'cues' of differentiation (Sadler, 2003). The firms often believe that their messages are potent and unique, but fail to understand the dynamics of communication (Schultz et al., 2013). The communication or message architecture of a brand should have three elements: key message, key objections and a point of proof. The key message should be consistent with brand positioning with greater level details, key objections are points to deals with the objection, or resistance arises out of your key message, and the evidence or point of proof makes your messages believable through reviews, awards, testimonials of experience and so on (Schultz et al., 2013).

In the past, researchers have studied organisations' websites for varied reasons: luxury brands communication strategies in social media (Wang, Huang, & Pérez-Ríos, 2020), multimodal analysis of luxury hotel home pages (Cheng, 2016), the online image of multinational hotel brands on different social media platforms (Huang & Cai, 2010), semantic comparison of values communicated in social media (Jabreel, Morene, & Huertas, 2017), latent class segmentation of luxury brands website (Díaz, Martín-Consuegra, & Estelami, 2016), move and linguistic characteristics of hotel home pages (Sitthanakul & Dhanesschaiyakupta, 2017) and the effectiveness of hedonic versus utilitarian message appeal in luxury product communication (Amatulli, De Angelis, & Donato, 2019). Despite its significance, the strategies or communication message appeals to differentiate the luxury hotel brands through online text contents in the official websites have not received much attention from the researchers.

Methodology

This research aimed to analyse how a multi-brand hospitality firm differentiates its luxury market offerings amongst its luxury brands in its portfolio through communication signals or cues through text messages. The case study method is followed to fulfil the objective of the current study. The case study method is very appropriate for this type of research as it allows to focus on a single service provider who holds a dynamic and diversified brand portfolio (Eisenhardt, 1989). Marriott International is a multi-brand, multinational and diversified hospitality company that is selected as a study unit. Marriott is the largest hotel operator in the world and owns 30 different brands, 7,000 plus properties with 1.33 million hotel rooms spread across 131 countries. Marriott International is cited by various researchers for its comprehensive and complex hybrid brand management known for its 'branded house', 'house of brands', 'sub-brands' and 'endorsed brands' strategies (Connell, 1999; Dixit & Abraham, 2020; Dev, Thomas, Buschman, & Anderson, 2010; Edwards, Rosenbaum, & Cheng, 2018; Morritt & Weinstein, 2012; Wong & Wickham, 2015).

The brand portfolio of Marriott features 30 distinct brands, fulfilling different market segments, varying from classical luxury hotels to branded serviced apartments. The 30 unique brands in the portfolio are classified into three segments: Luxury, Premium and Select, further differentiated as classic and distinctive in each segment. The branding portfolio structure of Marriott is presented in Fig. 11.1.

In this research, a content analysis approach has been adopted to analyse the marketing communication messages presented on the website of Marriott International. Content analysis is defined as 'a technique for gathering and analysing the content of a text; the content refers to words, meanings, pictures, symbols, ideas, themes, or any message that can be communicated' (Neuman, 2014, p. 219). Content analysis is an extensively used method to analyse verbal communication messages and the contents of websites in specific, owing to its flexibility. The aim of content analysis is to analyse how the master brand 'Marriott International' differentiates its individual luxury brand within other luxury brands in the portfolio through textual content or messages used to inform

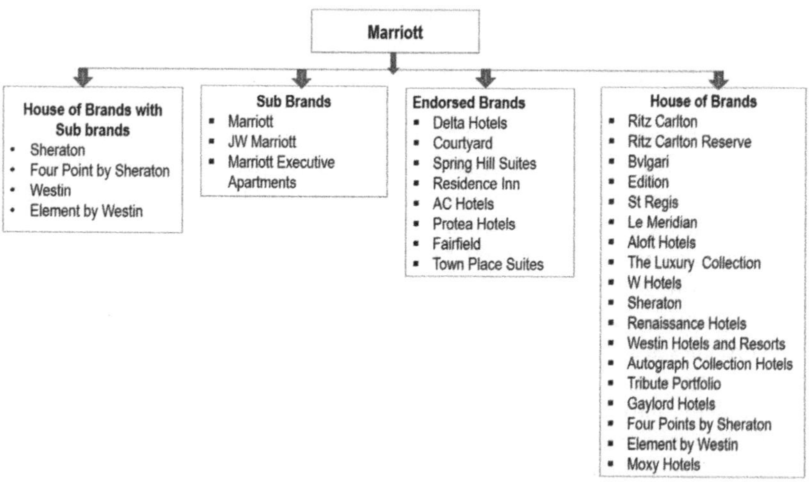

Fig. 11.1. Brand Portfolio Structure of Marriott International Inc.

and persuade the customers through websites. An eight-stage protocol for content analysis is followed, as suggested by Neuendorf (2002).

Stage 1: *Identification of research question:* Communication strategies for luxury brands are essential to keep their luxury appeal (Wang et al., 2020). Unlike other luxury product brands, hospitality brands share two unique challenges in formulating marketing communication strategies. The first challenge for all the luxury hotel brands is to differentiate its market offerings from other brands, and the second challenge is to differentiate its luxury brand from other luxury brands in its brand portfolio. In this research, we aimed to understand the communication cues used to differentiate luxury brands in the portfolio.

Stage 2: *Identification of variables:* Through the review of literature, six brand differentiators were identified: taglines, branded differentiators, external brand associations including co-branding, metadata descriptors in the Internet, verbal cues of luxury and point of proof.

Stage 3: *Defining the category and unit of measurement:* All the home pages and other web pages of each hotel brands of Marriott listed under the luxury and premium segments are considered as categories. The unit of analysis is words, phrases, sentences, trademarks, taglines, internal and external brands and other textual cues appearing on the website that signals the brand differentiation. The verbal cues on luxury are identified through the existing lexicon related to design, luxury, location, customer service and salience.

Stage 4: *Coding and creation of theme:* A codebook with categories and measurement for four dimensions of brand differentiators is created. All the words, phrases and sentences were categorised into four dimensions.

Stage 5: *Data Collection:* The data were collected from home pages, and the other linking web pages specific to each brand of Marriott under the luxury and premium segments. The period of data collection was from January to February 2020.

Stage 6: *Training coders:* Two trainers were trained and recruited to code and reconcile the differences in coding among the coders.

Stage 7: *Coding:* The text content collected from the web pages coded in the codebook.

Stage 8: *Data analysis:* This study adapts lexicon-based content analysis to the textual data available in the websites of various luxury brands of Marriott International Inc. in order to understand how the textual data are used to communicate the difference between individual luxury brands under the portfolio. The reference corpus of terms related to 'luxury' is compiled using *WordSmith Tool 8.0* (Scott, 2020). The top five concordances, a word or word phrases which frequently appear in the web pages of each brand, were considered as 'word cues'. The word cues are divided into four subcorpora: luxury, customer service, brand-specific features and location of the hotels. After following the proposed eight-stage content analysis protocol, the content was classified as per the six types of brand differentiators. The differentiators are identified based on the existing reviews. The list of brand differentiators and categories analysed in the text analysis is presented in Table 11.1.

Table 11.1. List of Brand Differentiators.

Content	Categories
Tagline	Specific tagline for each brand
Branded differentiators	Branded features, branded services, branded ingredients, branded programmes
External brand associations	Brand alliance with external brands and co-branding
Metadata description	Text content available in the Internet on the brand
Verbal cues on brand differentiation	Design or style, channel or location, price, salience, symbolic, customer service, customer segment
Point of proof	Awards, reviews, testimonials, customer intimacy

Analysis

The text data appear on the home page, and the subsequent linking pages of various luxury hotel brands of Marriott International Inc. are collected to analyse how individual luxury brands are able to differentiate themselves from other luxury brands under the firms' brand portfolio. The branded differentiation in terms of the tagline, branded differentiator, external brand associations and metadata description of all the luxury brands of Marriott is presented in Table 11.2. The first element of analysis is a tagline or motto statement that appears on the website. The taglines of individual luxury brands are unique. In terms of branded differentiators, W Hotels brands have seven, while The Ritz-Carlton have five. Most branded differentiators owned by the Marriott are membership or loyalty programmes, such as Marriott Bonvoy®, Marriott Rewards®, The Ritz-Carlton Rewards®, SPG®, Bonvoy Boundless™ and Marriott Bonvoy Moments™. Some of the branded differentiators are unique events or vacation packages hosted by individual luxury brands. The branded loyalty programme 'Marriott Bonvoy™' is associated commonly with all the individual luxury brands of Marriott except 'Ritz-Carlton Reserve', 'Bvlgari' and 'Edition'.

These three brands have their unbranded loyalty and membership programmes which are unique to individual hotels under the brand spread over different geographical locations. All the luxury hotel brands of Marriott have multiple external branded associations to differentiate its luxury offerings. Major motives for external brand associations are related to design and architecture of the hotel premise, furniture and fittings used in the room, leisure and recreational services, events, car and limousine rentals, private flights and yachts, membership credit card, spa and health services, limited edition shoes, toys, perfumes, and jewellery.

Table 11.2. Brand Differentiators – Tagline, Branded Differentiator, External Brand Associations and Meta Description.

	Luxury Brands of Marriott International Inc.							
Content	**The Ritz-Carlton®**	**St. Regis®**	**JW Marriott®**	**W Hotels®**	**Ritz-Carlton Reserve®**	**Bvlgari®**	**Edition®**	**The Luxury Collection®**
Tagline/ motto	'We are ladies and gentlemen serving ladies and gentlemen'	'The Best Address'	'Inspired by the principles of mindfulness'	'Bold meets beautiful'	'Your home at the ends of the earth'	'B Good, B Playful, B Joyful'	Extraordinary style and exceptional service	Be transformed A destination can change you forever
Branded differentiators	Ritz Kids®, Marriott Bonvoy®, Marriott Rewards®, The Ritz-Carlton Rewards® and SPG® The Ritz-Carlton Club®	Marriott Bonvoy™	Marriott Bonvoy™ Family By JW™ Spa by JW®	AWAY® Spa WET® BAR, W Hotels Music Festival® Wake Up Call® WET Deck® Bonvoy Boundless™ Marriott Bonvoy Moments™	Not specified	Not specified	Not specified	Marriott Bonvoy™
External brand associations	Six Senses, La Prairie and ESPA, Sub-Zero®, Wolf®, Gaggenau®, Miele® and Kohler® Greg Norman, Jack Nicklaus and Tom Fazio	Beijing Staycation Package with LEGO Bricks	The Joffrey Ballet, Marriott Bonvoy Tours and Activities Hertz Rentals	Bowler James, O², Brindley, Ricardo Cavolo, Muroexe Barcelona's Sound Coca-Cola Arena Chicago Cubs, FC Bayern. Mercedes-Benz, American Express, Chase	Jean-Michel Cousteau	Antonio Citterio Enzo degli Angiuoni Höbepergh	Mykita Comme Des Garcons and Byredo Miansai J. Hannah and M. Cohen Blamo Lucien Clergue	Lalique Patrick Hellmann Unique Elixir Paris Delisle Frank Lefebvre

Metadata description – brand tone	Experience The Ritz-Carlton luxury hotels and resorts that craft unforgettable travel experiences in the world's top destinations.	Book your luxury hotel stay with St. Regis Hotels. We offer 5-star luxury accommodation along with signature services, exquisite experiences and timeless traditions to create an unforgettable stay.	Discover classic luxury in JW Marriott hotels, with mindful experiences and inspiring spaces to improve your well-being and enrich your mind, body and soul.	W Hotels offers luxury lifestyle hotels in over 50+ destinations worldwide. Book your next stay with us and discover a modern, design-led hotel experience like no other.	An exploration of an unusual place, the excitement of an unexpected adventure, an indulgence in uncommon luxury. It is Ritz-Carlton Reserve – a rare place set aside for those who appreciate an exotic sanctuary tucked away in the most exquisite corners of the world.	Enter and discover our luxury hotels and resorts in the some of the most exclusive places in the world. Grant yourself a Bvlgari experience.	Stay in sophisticated style at EDITION Hotels, a unique concept in the lifestyle hotel space conceived by Ian Schrager and Marriott International. The New Generation of Luxury.	The Luxury Collection brings you extraordinary boutique hotels across the globe.

Other non-brand differentiators were frequently used by individual luxury brands to differentiate their services. The Ritz-Carlton often emphasised their 'Gold Standard of hospitality' of service and '100 years of service excellence' in communication and St. Regis frequently articulated traditional 'Rituals' and 'Bloody Mary', a cocktail 'invented' at the St. Regis New York's King Cole Bar. The luxury brand Bvlgari uses the term 'Il Ristorante Romito' and 'Il-Ristorante-Luca-Fantin' to emphasise its Italian cuisine and contemporary luxury.

All the luxury brand websites have unique metadata description of a text content presented along with the website name on the Internet to differentiate the essence and uniqueness of each luxury brand. Meta descriptions like '*luxury hotels and resorts that craft unforgettable travel experiences in the world's top destinations*', '*5-star luxury accommodation along with signature services, exquisite experiences and timeless…*', '*luxury lifestyle hotels*', '*an exploration of an unusual place, … an unexpected adventure, and uncommon luxury*', '*discover our luxury hotels and resorts in … the most exclusive places in the world*' and '*Stay in sophisticated style, a unique concept in the lifestyle hotel*' are able to communicate the uniqueness of brand in terms of location, experience, style, quality of service, design and aesthetic aspects of different brands under the portfolio. The websites of different brands under the Marriott's portfolio have used verbal cues frequently to differentiate their brands. Most of the cues are related to four aspects which are specific or unique to the luxury brands: defining the luxury experience, quality of customer service, brand-specific themed features or rituals or location.

The branded differentiation in terms of target segment, tagline, branded differentiator, external brand associations and meta description of all the luxury brands of Marriott is presented in Table 11.3.

To distinguish their luxury offering, specific phrases are used and these specific words are not repeated or used by other luxury brands. For example, specific phrases like 'time-honoured traditions', 'beyond expectations', 'finest luxury experience' and 'Gold Standard of hospitality' are used only by The Ritz-Carlton, throughout its individual hotel brands and same way, 'Legacy of luxury', 'Timeless Elegance', 'Midnight supper', 'connoisseurs' and 'Bloody Mary' are used by St. Regis. The most commonly used 'point of proof' to differentiate the luxury brand are guest reviews, awards and testimonials. Customer intimacy is also used as a point of proof by certain brands through customer stories narrated in the web pages and brand-owned magazines. This core theme of the story is how hotels fulfil unique customer needs on different occasions.

Discussion

The brand value is the fundamental asset which has to be cautiously managed in a website by the chains so that it does not lose its identity and reputation. The spectacular performance of luxury brands over a period is attributed to the laudable marketing communication strategies of luxury marketers (Amatulli, De Angelis, Pichierri, & Guido, 2018). Communication strategies for luxury brands are essential to continue their growth (Wang et al., 2020). The success of the

Table 11.3. Brand Differentiators – Target Segment, Verbal Cues and Point of Proof.

Content	Luxury Brands of Marriott International Inc.							
	The Ritz-Carlton®	**St. Regis®**	**JW Marriott®**	**W Hotels®**	**Ritz-Carlton Reserve®**	**Bvlgari®**	**Edition®**	**The Luxury Collection®**
Target segment	Everyone – luxury	A new-generation luxury travellers	Family, luxury, wellness	Adult, luxury, lifestyle hotels	Adult, luxury, wellness	Luxury, family, Italian	Family, luxury	Family, luxury
Verbal cues on brand	*Luxury* 'Time-honoured traditions' 'Beyond expectations' *Service* 'Finest luxury experience' 'Gold Standard of hospitality' *Brand-specific* '100 years of history' 'The Ritz-Carlton Club' 'World-class golf' *Location* 'World's top	*Luxury* 'Legacy of luxury' 'Timeless elegance' *Service* 'Butler service' 'Midnight supper' *Brand-specific* Traditional rituals Connoisseurs Bloody Mary Polo Antiques Scent Sabrage St. Regis	*Luxury* 'Anticipate every need' 'Relieve stressors' *Service* 'Pamper yourself' 'Impeccable service' *Brand-specific* 'Mind, body and spirit' 'All is well' *Location* 90 distinctive properties 25 countries	*Luxury* 'Music, fashion, design, fuel' 'Mystery, magic adventure' *Service* Detox, retox and repeat *Brand-specific* Adult, LGBTQ+ *Location* Stunning destinations Most lust-worthy resort	*Luxury* 'Indulgence in uncommon luxury' 'Luxury with a playful nature' *Service* 'Farm-to-table cuisine' *Brand-specific* Wellness programmes 'distinctly local touch' 'Ingredients from area	*Luxury* 'Timeless glamour' 'Heritage of magnificent Roman jeweller' *Service* 'Truly remarkable' *Brand-specific* Bvlgari Gifts Zimbabwe opaque marble Il Ristorante Chef Niko Romito Il-Ristorante-	*Luxury* 'An intimate, seductive environment' 'Personal and unique' 'Luxury is being heard before you even speak' *Service* 'Best in service, food, beverage and entertainment' 'A feeling of being somewhere so special'	*Luxury* 'Extraordinary boutique hotels across the globe' 'Unique, authentic experiences' 'Lasting, treasured memories' *Service* 'Epicurean experiences' 'Exceptional tastes' *Brand-specific* 'Heritage' 'Local cuisine' *Location* Close proximity to railway station, shopping center and

Table 11.3. *(Continued)*

Content	Luxury Brands of Marriott International Inc.							
	The Ritz-Carlton®	**St. Regis®**	**JW Marriott®**	**W Hotels®**	**Ritz-Carlton Reserve®**	**Bvlgari®**	**Edition®**	**The Luxury Collection®**
	destinations' 'City centre'	Boutique, John Jacob Astor *Location* 'Finest addresses' 'World's most inspiring destinations'			farms' *Location* 'A rare place', 'most exquisite corners of the world' 'Unusual place'	Luca-Fantin *Location* 'Major cosmopolitan cities' "most prestigious area"	*Brand-specific* Market Matador room *Location* 'Stunning microcosms' 'World's top cities'	places of historic/heritage interest
Point of proof	Malcolm Baldrige National Quality Award, guest stories – let us stay with you	Forbes Travel Guide: Four-Star Award Guest review	Forbes Travel Guide – Five-Star Award (2020), guest reviews	TripAdvisor – top 25 luxury hotels – China (No. 11), guest reviews	AAA Five Diamond Award (2016, 2017, 2019) Guest reviews	Forbes Travel Guide 2020 – 5-Star Award Guest reviews	Testimonials Guest reviews	Guest reviews

brands under a brand portfolio depends on how each brand is differentiated, and the uniqueness of each brand is presented to the customers using marketing communication. In this research, we aimed to extend existing knowledge on how a multi-brand luxury hospitality firm differentiates its individual luxury brand from other luxury brands in the brand portfolio. A content analysis approach has been adopted to analyse the marketing communication messages presented on the website of seven luxury hospitality brands of Marriott International Inc.

The major differentiators identified through the review of literature have been extensively used by the luxury brands of Marriott. Each luxury brand under the portfolio has a unique tagline. Almost all the hotel brands tried to differentiate their product offering by highlighting their location, customer service, aesthetics of hotels, brand-specific service features, traditions and rituals. The finding is similar to the studies of Schultz et al. (2013). Branded differentiators are another major tool used for brand differentiation. Except for the common loyalty programme at 'Marriott Bonvoy', many hotels do not possess any branded differentiators. The Ritz-Carlton and W Hotels brands have a maximum number of branded differentiators. The strategies specified are in line with the findings of Aaker (2003, 2004); Agris (2001) and Porter (1985).

Other luxury brands in the Marriott portfolio have tried to have non-branded differentiation in terms of the cuisine, architecture, location and theme-based service offerings. Another feature that is commonly used to communicate the uniqueness of each luxury brand under the Marriott portfolio is meta descriptions. Through the meta description, individual luxury brands are able to communicate what type of luxury is offered, types of clientele and the location (Burgess et al., 2009; Dixit & Abraham, 2020; Mortimer et al., 2012). Verbal cues are another commonly used element for brand differentiation by the study units. The findings of this study are in coherence with the studies of Evans and Hastings (2008). The brand managers are able to communicate the differences through the texts. Most of the differentiation relies on location, followed by aesthetics of hotels, customer services, traditions and segment-specific services. It may be concluded that Marriott has been successful to an extent in using text contents to differentiate its luxury brand. However, there is a greater need and scope for brand differentiation through communication message as more luxury brands compete to win the minds of the same customer and also sustain the benefits of luxury brands.

Despite the pioneering attempt to understand the role of language in luxury brands communication to differentiate its brand under the portfolio, this research study has several limitations which may be addressed in future studies. The web pages of luxury hotels analysed in the study have multiple layers of communication. Most of the home pages have limited information on brand-related specific information. The ultimate aim of the home page of the website is to initiate reservation. Much of the spaces in the home page of individual brands are occupied with images and very few text content that describes the brand-related features. The subsequent links are directed towards the web pages of different properties under the brand, which makes the difficulty in understanding the essence of each brand. The vast number of hotels under the same brand made the data collection and analysis very complex; for example, under The Ritz-Carlton, there are 97 hotels and resorts, where each property has some unique feature. However, in this research, the text data are collected only from the home pages of brands and six pages of two old individual hotels based on the year of establishment. The content analysis, with its inherent limitations, posed challenges in comprehensive analysis and measurement of the link between different communication message strategies used by luxury brands and their impact on brand differentiation. Although words are used frequently to express the brand offerings, images and other visual elements are more powerful than words.

The brand differentiation tactics adopted in this study are more relevant to product-related brands rather than to service brands due to the unique characteristics of services, for example, branded ingredient, a commonly used brand differentiator rarely applicable to services like hotels, where it sells intangible service and experiences. There is an absence or rarity of research on brand differentiators in service firms which holds a complex brand portfolio. Research works in this direction will benefit the luxury service brands as many economies move towards service as a dominant contributor to the overall economic growth. It is also observed that human brands are also portrayed to differentiate brand offerings, for example, golf professionals associated in designing the golf course of hotels, Michelin-starred chefs with restaurants, professional architects for facility design and so on. This human brand association extends the scope for future research on the significance of human brands in brand differentiation. The current research work may be extended to understand how hotel brands, whether it is a chain of hotels or an independent hotel, are differentiating their market offerings from the competing luxury brands.

Conclusion and Implications

Currently, luxury is no longer a small sector aimed to satisfy a few. Globalisation and increased trend of consumption-oriented society has given a thrust to luxury offering and brands. The market for luxury products and services is successful and proliferating during the last three decades. Luxury hotels are a vital market segment within the global luxury market. The luxury hotel market was growing much faster than other hotel segments in terms of its market value. Luxury hotels represent the third-largest market within the global market of luxury products and services. The word 'luxury' is the most overused word in marketing; almost all the companies and brands call themselves luxury service providers. The challenge for a brand that offers luxury is to sustain the image of luxury as it is critical to differentiate its products and services from other look-alike. Luxury brands compete in the market, not on their quality but on their symbolic capital. This research demonstrated how a multi-brand luxury hospitality firm has tried to differentiate its luxury brands from other brands in the portfolio. The result of the study has major implication for marketers.

As the website is the prime source of information, the web page's messages must differentiate its luxury offerings from other brands. It is challenging for the hospitality marketers to differentiate the brand only by describing the features, location, services and amenities, mostly tangible in nature. They must identify some intangible elements which go beyond to achieve the differentiation. Developing a strong unique rituals associated with a brand, building an innovative and sustainable branded differentiators, powerful association with other luxury brands and stories that reflect the symbolic value that fulfils the customers' psychological needs should be presented in the website. This study shows that brands such as St. Regis have adapted the storytelling and unique brand rituals in their communication to the target clientele. A story may be on brand heritage, its founder, the brand's evolution and how the brand associate with luxury can be portrayed. Ceaser Ritz's adaption of Gold Standard of Service is a living evidence of how a service brand can differentiate its intangible services by communication from its personal touch and service quality. A brand's exclusivity may be projected by using the difference in its geographical location and limited availability. Another time-consuming and expensive differentiation by a brand is to develop a branded differentiator. Continuous process and product innovation are central to develop branded differentiators which are very difficult to copy by the competitors. The symbolic values that are unique to the brand must be incorporated in every medium of communication that helps the firms to immerse their customer so that customers can create an emotional connection with the brand. It is concluded from this research study that brand should create a notable distinction to match the promise and unique brand core in their marketing communication. After all, luxury is all about the unequivocal distinction.

References

Aaker, D. A. (2003, October). The power of the brand differentiator. *MIT Sloan Management Review*.

Aaker, D. A. (2004). *Brand portfolio strategy*. New York, NY: Free Press.

Agris, S. (2001, March). Presentation to Stanford University.

Alexander, C. (2018). *Strategic brand management* (2nd ed.). Chicago: Cerebellum Press.

Amatulli, C., De Angelis, M., & Donato, C. (2019). An investigation on the effectiveness of hedonic versus utilitarian message appeals in luxury product communication. *Psychology and Marketing*, *37*(4), 523–534. doi:10.1002/mar.21320

Amatulli, C., De Angelis, M., Pichierri, M., & Guido, G. (2018). The importance of dream in advertising: Luxury versus Mass market. *International Journal of Marketing Studies*, *10*(1), 71. doi:10.5539/ijms.v10n1p71

ARC. (2019). Global luxury hotel market growth (status and outlook) 2019–2024. Retrieved from http://www.arcognizance.com/report/global-luxury-hotel-market-growth-status-and-outlook-2019-2024#Desc

Burgess, S., Sellitto, C., & Karanasios, S. (2009). *Effective web presence solutions for small businesses: Strategies for successful implementation: Strategies for successful implementation*. London: IGI Global.

Catry, B. (2003). The great pretenders: The magic of luxury goods. *Business Strategy Review*, *14*(3), 10–17. doi:10.1111/1467-8616.00267

Cheng, F. W. (2016). Constructing hotel brands: A multimodal analysis of luxury hotel homepages. *Iberica, 31*, 83–107.

Chevalier, M., & Mazzalovo, G. (Eds.). (2012). *Luxury brand management: A world of privilege*. Singapore: John Wiley & Sons. doi:10.1002/9781119199168

Chew, W. (2016). *Differentiation: How to win in a disruptive market*. Singapore: Write Editions.

Connell, J. (1999). Diversity in large firm international franchise strategy. *Journal of Consumer Marketing, 16*(1), 86–95. doi:10.1108/07363769910250796

D'Arpizio, C., Levato, F., Daniele Zito, D., & Montgolfier, J. (2015). Luxury goods worldwide market study fall-winter 2019: Eight themes that are rewriting the future of luxury goods. Retrieved from https://www.bain.com/globalassets/noindex/2020/bain_digest_eight_themes_that_are_rewriting_the_future_of_luxury-goods.pdf

D'Arpizio, C., Levato, F., Daniele Zito, D., & Montgolfier, J. (2019). Luxury goods worldwide market study fall-winter 2015: A time to act—How luxury brands can rebuild to win. Retrieved from https://www.bain.com/contentassets/bc52b47699884108900c65dcb73f6a91/bain_report_global_luxury_2015.pdf

De Barnier, V., Rodina, I., & Valette-Florence, P. (2006). Which luxury perceptions affect most consumer purchase behavior? A cross-cultural exploratory study in France, the United Kingdom and Russia proceedings of the international marketing trends conference.

Deloitte. (2019). *Global powers of luxury goods 2019 bridging the gap between the old and the new*. Retreived from https://www2.deloitte.com/content/dam/Deloitte/ar/Documents/Consumer_and_Industrial_Products/Global-Powers-of-Luxury-Goods-abril-2019.pdf

Dev, C. S., Thomas, J. H., Buschman, J., & Anderson, E. (2010). Brand rights and hotel management agreements. *Cornell Hospitality Quarterly, 51*(2), 215–230. doi:10.1177/1938965510363389

Díaz, E., Martín-Consuegra, D., & Estelami, H. (2016). A persuasive-based latent class segmentation analysis of luxury brand websites. *Electronic Commerce Research, 16*(3), 401–424. doi:10.1007/s10660-016-9212-0

Dixit, S. K. (2017). *The Routledge handbook of consumer behaviour in hospitality and tourism*. London: Routledge.

Dixit, S. K., & Abraham, A. (2020). Brand identity and positioning in selected India chain hotel companies. In M. A. Gardini, M. C. Ottenbacher, & M. Schuckert (Eds.), *The Routledge companion to international hospitality management* (pp. 260–270). London: Routledge.

Dubois, B., Czellar, S., & Laurent, G. (2005). Consumer segments based on attitudes toward luxury: Empirical evidence from twenty countries. *Marketing Letters, 16*(2), 115–128. doi:10.1007/s11002-005-2172-0

Edwards, K. J. D., Rosenbaum, M. S., & Cheng, M. (2018). Research note: Legal implications of soundalike hotels in China: Haiyatt and Marvelot. *Tourism Management, 68*, 162–167. doi:10.1016/j.tourman.2018.03.010

Eisenhardt, K. M. (1989). Building theories from case study research. *Academy of Management Review, 14*, 532–550.

Evans, W. D., & Hastings, G. (2008). *Public health branding: Applying marketing for social change*. Oxford: Oxford University Press.

Feng, R., Wang, Y.-C., & Ryan, B. (2018). Service experiences at luxury hotels: Business Tourist's perspectives. *Quality Services and Experiences in Hospitality and Tourism*, 181–193. doi:10.1108/s2042-144320180000009013

Franzen, G., & Moriarty, S. (2008). *The science and art of branding*. New York, NY: M.E. Sharpe.

GlobalData. (2016). *The global luxury hotels market to 2020*. London. Retrieved from https://store.globaldata.com/report/tt0338mr–the-global-luxury-hotels-market-to-2020/

Huang, J. Z., & Cai, L. A. (2010). Online image of multinational hotel brands on different language platforms. *Journal of China Tourism Research*, *6*(3), 279–295. doi:10.1080/19388160.2010.505134

Jabreel, M., Moreno, A., & Huertas, A. (2017). Semantic comparison of the emotional values communicated by destinations and tourists on social media. *Journal of Destination Marketing and Management*, *6*(3), 170–183. doi:10.1016/j.jdmm.2016.03.004

Kapferer, J. N. (1998). Why are we seduced by luxury brands? *Journal of Brand Management*, *6*(1), 44–49. doi:10.1057/bm.1998.43

Kapferer, J. N. (2008). *The new strategic brand management*. London: Kogan Page.

Kapferer, J. N. (2015). *Kapferer on luxury: How luxury brands can grow yet remain rare*. London: Kogan Page Publishers.

Keller, K. L., Parameswaran, M. G. ., & Jacob, I. (2015). *Strategic brand management* (4th ed.). New Delhi: Pearson Education.

Kelly-Holmes, K. (2019). The linguistic business of marketing. In C. Thurlow (Ed.), *The business of words: Wordsmiths, Linguists, and other language workers* (pp. 36–50). London: Routledge.

Lancaster, K. (1984). Brand advertising competition and industry demand. *Journal of Advertising*, *13*, 19–30.

Michman, R. D., & Mazze, E. M. (2006). *The affluent consumer: Marketing and selling the luxury lifestyle*. Westport, CT: Praeger Publishers.

Morritt, R., & Weinstein, A. (2012). *Segmentation strategies for hospitality managers*. New York, NY: Routledge. doi:10.4324/9780203050095

Mortimer, R., Brooks, G., Smith, C., & Hiam, A. (2012). *Marketing for dummies* (3rd ed.). London: John Wiley & Sons.

Neuendorf, K. A. (2002). *The content analysis guidebook*. Los Angeles, CA: Sage Publications.

Neuman, W. L. (2014). *Social research methods: Qualitative and quantitative approaches*. Essex: Pearson Education Limited.

Nueno, J. L., & Quelch, J. A. (1998). The mass marketing of luxury. *Business Horizons*, *41*(6), 61–68. doi:10.1016/s0007-6813(98)90023-4

Porter, M. E. (1985). *Competitive advantage*. New York, NY: Free Press.

Rao, K. R. (2004). *Services marketing*. New Delhi: Pearson Education India.

Sadler, P. (2003). *Strategic management*. London: Kogan Page Publishers.

Schultz, M., Doerr, J. E., & Frederiksen, L. (2013). *Professional services marketing* (2nd ed.). London: John Wiley & Sons.

Scott, M. (2020). *WordSmith Tools version 8*. Stroud: Lexical Analysis Software.

Sitthanakul, K., & Dhanesschaiyakupta, U. (2017). An analysis of move and linguistic characteristics of hotel homepages. *People: International Journal of Social Sciences*, *3*(2), 1057–1076. doi:10.20319/pijss.2017.32.10571076

Timetric. (2013). *The global luxury hotels market – Key trends and opportunities to 2017*. London: Timetric. Retrieved from https://www.marketresearch.com/product/sample-7788288.pdf

Tynan, C., McKechnie, S., & Chhuon, C. (2010). Co-creating value for luxury brands. *Journal of Business Research*, *63*(11), 1156–1163. doi:10.1016/j.jbusres.2009.10.012

Verma, H. (2011). *Services marketing: Text and cases*. New Delhi: Pearson Education.

Vickers, J. S., & Renand, F. (2003). The marketing of luxury goods: An exploratory study – Three conceptual dimensions. *The Marketing Review*, *3*(4), 459–478. doi: 10.1362/146934703771910071

Wang, R., Huang, S., & Pérez-Ríos, N. G. (2020). Multinational luxury brands' communication strategies on international and local social media: Comparing Twitter and Weibo. *Journal of International Consumer Marketing*, *32*(4), 313–323. doi:10.1080/08961530.2019.1710736

Wei, L. (2020). Examining corporate communications of environmental responsibility on corporate websites: Main themes, linguistic features, and text reuse. *Journal of Promotion Management*, *26*(7), 1013–1037. doi:10.1080/10496491.2020.1746467

Wong, T., & Wickham, M. (2015). An examination of Marriott's entry into the Chinese hospitality industry: A brand equity perspective. *Tourism Management*, *48*, 439–454.

Yeoman, I., & McMahon-Beattie, U. (2006). Luxury markets and premium pricing. *Journal of Revenue and Pricing Management*, *4*(4), 319–328. doi:10.1057/palgrave.rpm.5170155

Chapter 12

Branding Luxury Travel

Arup Kumar Baksi[1]

Abstract

Luxury branding, in the context of tangible luxury consumables, has received academic attention. But the notion remained inconclusive in the context of consumption of luxury intangibles. The travel setting provides an excellent backdrop to explore the complex cognitive process of assigning meaning to the relationship between travellers and luxury travel brands. The shifting image of luxury consumption from elitism to mass aspirational, too, needs to be studied for its transformative implications. The chapter focused on developing a brand relationship scale, namely, TraveLux, in the context of luxury travel consumption and tested its robustness to explain the shared sentiments and emotions of travellers, engaged in luxury travel, across social media. The chapter identifies a four construct instruments capturing the essence of immersive experience, ethnocultural acculturation, passion and excitement and self-congruence as a seedbed of luxury brand affinity for travellers. TraveLux was also found to capture the shared experience of travellers consuming luxury travel brands, thereby establishing a synch between the instrument constructs and manifested human cognition in real-life situations. The study expanded on the volume of literature pertaining to luxury branding in the context of product-oriented industry and addresses the existing void in understanding traveller–brand relationships in luxury travel contexts. The study implicates a theoretical change in branding concept in perceiving luxury brands as price-based exclusivity to a transformative cultural experience. Further extrapolations of the study could be made by incorporating subtle behavioural patterns of travellers in perceiving luxury and subsequent evocation and predisposition towards decision-making.

Keywords: Luxury travel; TraveLux; relationship; brand; experience; cultural

[1]https://orcid.org/0000-0002-9206-3994

The Emerald Handbook of Luxury Management for Hospitality and Tourism, 243–270

doi:10.1108/978-1-83982-900-020211012

Introduction

The luxury industry, in general, has been witnessing a sustainable growth trend and is expected to remain booming over the next few years. Primarily this growth has been attributed to the Millennials, likely to represent approximately 45% of the global luxury products market by 2025 (Shin, Eastman, & Mothersbaugh, 2017) with a projected share of wallet of approximately $65 billion per year. However, the luxury marketers are re-evaluating the swings in consumer preferences and aspirational drives towards luxury brands (Kapferer & Laurent, 2016) and are re-strategising their brand initiatives by embedding 'sense of exclusivity' and 'perceived co-creation of value'. Simultaneously, luxury marketers address the evolving and shifting challenges of sustainability by franchising a brand balance effort to propagate sustainable luxury. These shifts require an augmented branding effort by the luxury product and service marketers to create cultural capital for the luxury consumers who have evolved from an aspirational entity to an experiential self, seeking self-actualisation rather than prestige.

The travel industry, too, has been experiencing these sweeping changes as the global travel market, which was valued at $65 billion in 2018, has been projected to grow at a CAGR of 7.9% during 2019–2026. Unlike the manufacturing and/or product-oriented industry, the travel and hospitality industry does not have a standalone structure; rather, it resembles a spider web of networked industries. In addition, the segmental variation is also proliferative in terms of attributional orientations, as the markets are segregated into personalised and private vacations, adventure and safari, cruise expedition, micro-group journey, celebration and special events, and culinary travel and shopping. Multi-attribute segmentation or geo-clustering also requires a careful understanding of product/service positioning using potent differentiators, which subsequently transcribes into brand elements. Therefore, a standardised branding technique will not communicate the inherent heterogeneity of the industry and may fail to evoke a sense of preference. Ambidextrous branding techniques are being exploited by luxury travel marketers, whereby a luxury brand may be mounted on dual dimensions having both exploitative and exploratory features. Luxury service brands demand more personalisation, and therefore, the marketers engaged in travel and allied services need to mine travellers' psyche to have a deeper understanding of the values that they ascribe to a luxury brand. With the growing practice of data-driven consumer profiling based on psychographics, luxury brands are attempting a more customer-centric engagement strategy to stimulate intense loyalty and ensure lifetime customer value. The neo-luxury hospitality branding, for example, has been mounted on core brand values driven by personalised services and enabled by digital logistics. Branding luxury travel is a much more fluid concept, depending on the context surrounding the individual traveller's circumstances.

Review of Literature

Luxury has a perennial fascination for mankind as the term has continued to evoke shifting aspirations (Novotna & Kunc, 2019) from zipped offers based on exclusivity to bespoke designs. The term challenges the very notion of product or service design as one has to balance the cognition of rarity with desirability. The ambivalence in the semantic roots of the term is omnipresent across the geospatial existence and through the cultures and their hybrids. Conventionally, the term epitomised 'wealth and status'. Over the last two decades, the emerging globalised and digitised market forces (Okonkwo, 2007) ensured that luxury products and services reach emerging economies and the consumption guided by symbolic and non-utilitarian entities (Kapferer & Laurent, 2016; Sreejesh, Sarkar, & Roy, 2016). These changing paradigms induced challenges for luxury brand builders to embed, in the luxury brand architecture, the 'aspirational quotient' in the context of mass consumption (Nobre & Simoes, 2019; Sreejesh et al., 2016) without compromising the value metaphor of premium pricing associated with luxury consumption (Kapferer & Laurent, 2016). For example, Louis Vuitton, presently mass-consumed as a neo-luxury item compared to Rolls Royce, grounded on the conventional luxury platform, reveals this luxury paradox (Cristini, Kauppinen-Raisanen, Barthod-Prothade, & Woodside, 2017). Conventional marketing instruments to foster branded posits for luxury consumption, therefore, also became redundant as they fail to capture the emerging brand essence of co-creation (Kim & Ko, 2012; Merz, Zarantonello, & Grappi, 2018) to resonate with the consumer psyche. With aristocracy being replaced by meritocracy, the role of luxury has changed. Luxury brands no longer compete on 'quality' but on a latent social hierarchy to accrue symbolic capital. Therefore, luxury needs brands for propagation.

Theories governing luxury branding span across a wide range of approaches. Tynan, McKechnie, and Chhuon (2010) emphasised the 'phenomenological' nature of the value co-creation of the luxury brands, while Roper, Caruana, Medway, and Murphy (2013) highlighted the 'interpretivist' perspective to explain the construction and internalisation of luxury value creation. The contemporary theory governing luxury branding rejected the age-old managerial view of a brand which looks to translate the brand-identity into a perceivable brand-image (Schroeder, 2009) and adopted a polysemic approach that focused on contextuality and established a brand on strong cultural footing (Bengtsson, Bardhi, & Venkatraman, 2010; Seo & Buchanan-Oliver, 2015). Theories propagating frameworks for luxury branding have considered a number of inputs over time, namely luxury disposition behaviour (Lee, 2013), ethnocultural infusion (Liao & Wang, 2009), utilitarian benefits and symbolic meanings (Sung, Choi, Ahn, & Song, 2015), value co-creation (Tynan et al., 2010), culture-driven brand personification metaphors/motifs (Aguirre-Rodriguez, 2014), aesthetics and ideologies (Chitturi, Chitturi, & Raghavarao, 2010; Townsend & Sood, 2012), emotions of exclusivity and superiority (Dion & Arnould, 2011), etc. Becker, Lee, and Nobre (2018) proposed the BECKER model of luxury branding, which essentially focused on synchronising luxury product qualities with the consumer psyche and forwarded the concept of proxy quality. Proxy quality, associated with perceived value-driven luxury, is likely to promote equity judgements and socio-emotional justice and help in calibrating the relationship between consumer and luxury brands. Theories revealing consumers' attitudes towards luxury brands were also studied, and three major clusters were identified, namely the elitist, the democratic and the distant (Dubois, Czellar, & Laurent, 2005) with varying degrees of consumption motives.

Academic researchers have also delved into the psychic roots to identify the psychosomatic drivers that activate participation in luxury travel (Han & Hyun, 2018; Novotna & Kunc, 2019). Tutek, Roth, and Simatic (2018) posited the neo-luxury travel as a 'transformational and soul-searching experience' and identified a set of five perspectives for transformational impact, namely purpose, personalisation, experiential, balanced disconnection and seamless journey, that may be embodied as focal brand elements to connect with the cultural layout. The enactment of 'travelling' has to interact intensely with culture, ethnicity, artefacts, heritage and prevailing art forms embedded in the host communities. The structural proximity of luxury brands with the world of art offers a metaphorical bridge to relate consumers (with luxury consumption mindset) with luxury brands (Joy, Sherry, Venkatesh, Wang, & Chan, 2012). That art and culture can be a joinder in consumer luxury brands dyad, which have also been studied extensively by Koronaki, Kyrousi, and Panigyrakis (2017), whereby they found that art-based initiatives as a relationship-strengthening tool transcend geographical barriers. Artification of travel brands, therefore, embarked on philanthropic and experimental collaboration (Baumgarth, Lohrisch, & Kastner, 2014; Kapferer, 2014) for luxury embeddedness in culture. Luxury travel brands, therefore, were viewed to possess ethnocultural meanings (Wilcox, Kim, & Sen, 2009) targeted to gratify the inherent social appetite of consumers (Bian & Forsythe, 2012; Kim & Ko, 2012) and luxury consumption, conceptually, shifted from the notion of 'having-to-being' to 'owning-to-experiencing' (Cristini et al., 2017). The traveller–brand relationship, in the context of luxury travel consumption, therefore, was explained with brand experiences (Chun, Diehl, & MacInnis, 2017; Dixit, 2020; Payne, Storbacka, Frow, & Knox, 2009) evoked by a bundle of (travel) brand stimuli (Brakus, Schmitt, & Zarantonello, 2009) responsible for the construction of the brand image. These relational constructs framed a brand architecture for luxury travel, which essentially infused cultural consumption with the fundamental brand offer (Baksi, 2016, 2017). Prahalad and Ramaswamy (2004) observed that the value co-creation process was also an evolving dimension in experiential travelling, which assigned meanings to the luxury travel brands (Vargo & Lush, 2004). The notion of value co-creation as luxury branded content in travel context reflected in the observations of Lee and Hwang (2011) as they postulated a triangulation of functionalism, experiential symbolism and symbolic interaction to understand the derived sensorial pleasure and gratification of actualisation needs.

Research initiatives could also be tracked to assess the implication of relational constructs in branding luxury travel. As services shifted from transactional mode to relationship-driven operations, the consumer-based brand equity (CBBE) was used to explain the traveller–brand affinity (Huang & Cai, 2015; Liu, Wong, Tseng, Chang, & Phau, 2017; Yang & Lau, 2016) in the context of luxury consumption. One of the major issues in branding luxury travel is the range of consumptions (from high-tangible to high-intangible) associated with the travel itself, and hence a pooled-brand concept might bridge the abstract and asymmetric nature of the relationship between the traveller and the luxury travel brand. Sumbly and Siraj (2019) talked about brand clues that could take care of this asymmetry and propagate a homogeneous perception of the brand. Earlier, Vigneron and Johnson (2004) adopted a bifurcated proposition in their luxury brand framework and posited non-personal perceptions (conspicuousness, uniqueness and quality) and personal perceptions (hedonic and extended-self) as drivers of luxury consumption. It was reinforced in the studies made by Yang and Mattila (2016) and Jain, Roy, and Ranchhod (2015). The inward view of luxury consumption explained the dynamism of the relationship between the brand and the consumer as self-indulgence in luxury consumption manifested the aspirational instinct of actualisation and esteem seekers (Canziani, Watchravesringkan, & Yurchisin, 2016; Liang, Ghosh, & Oe, 2017). While developing an instrument to capture the nature of psychosomatic luxury alignment of a traveller, it is only but imperative to understand these relational realms that trigger luxury embracement (Escalas, 2013). This would provide insights into the gradual shift from nuclear-luxury travel (constrained to elitism) (Kapferer, 2014) to peripheral luxury travel (unconstrained aspirational affinity) (Bizarrias et al., 2017; Seo & Buchanan-Oliver, 2015) and would serve as a building block to construct a framework for luxury travel brands.

The luxury concept in the travel sector emerges from luxury-specific experiences, which are likely to form the scaffold of luxury travel brands. However, these experiences in shaping traveller–luxury brand relationship remain largely unexplored. Literature also remained inconclusive about the process by which travellers ascribe meaning to luxury brands which demands the construction of invigorating brand narratives in communication and marketing (Beverland, 2004; Kapferer, 2006). One of the significant shortcomings of the emerging theories of branding luxury travel services, which includes a bouquet of personalised services, namely beauty salons, spas, private travel services (e.g. private jets, cruises, yacht, etc.), tour operators, event designing, etc., is its predominant focus on either corporate or consumer-based brand equity. Luxury brands are not limited to value-added assets; they also possess socio-cultural, ideological and political identities that infuse consumption with meaning (Holt, Quelch, & Taylor, 2004; Schroeder, 2009). An understanding of the luxury travel brands requires insights into the culture and ideologies that encapsulate their consumption, in addition to conventional branding concepts, such as strategy, equity and value (Schroeder, 2009). Additionally, in the context of luxury travel consumption, the brand–consumer relationship is expected to assume a critical role in a transformational experience. This study, therefore, focused on designing an instrument to capture the relational notion of luxury travel brands and test the same to explain the perceived value of luxury travel brands as opined by travellers in social media.

Conceptualising TraveLux Instrument

The neo-luxury consumption posited travellers as value co-creators (Tynan et al., 2010), and in the process, they relived the luxury experience (Merz et al., 2018).

The absorptive and immersive experience a traveller derives from the value co-creation process embeds a sense of exclusivity, excellence, novelty and creativity in configuring a luxury brand (Cristini et al., 2017). Self-brand connection Brandão, Pinho, and Rodrigues (2019) was found to explain how the complexity of interactions, perceived image and aspirational experiences were governing consumer-luxury brand relationship (Kaufman, Loureiro, & Manarioti, 2016). Social identification theory was also used to explain the consumers' inherent desire to construct their social identity based on the commonality of luxury brand attributes (Berrozpe, Campo, & Yagüe, 2017), and the hedonic nature of luxury consumption captured the complexity of intimacy of the relationship (Nobre & Simoes, 2019).

The study postulated TraveLux as an instrument to measure the intensity of the relationship between the traveller and the brand in the context of luxury travel. Aaker, Fournier, and Brasel (2004) distinguished 'sincere brands' from 'excitement brands' on the ground of the duration of the brand–consumer relationship and observed that 'sincere brands' tend to have stronger bonds compared to 'excitement brands'. One of the reasons for such differential brand engagement might be attributed to the volatility of sentiments and feelings associated with 'excitement brands'. Aaker et al.'s (2004) findings resonated with the proposition of brand romanticism of Fletcher, Simpson, Thomas, and Giles (1999). The theory placed 'intimacy-loyalty' as an expression for 'sincere brands' and 'passion' as the bridging element for 'excitement brands'. TraveLux thematically incorporated the socio-individual facets (Vickers & Renand, 2003) to capture the essence of traveller–luxury brand relationship based on metaphorical expressions (Fog, Budtz, Munch, & Blanchette, 2010), ethnocultural acculturation (Baksi, 2016, 2017) and brand meaning (Nobre & Simoes, 2019). Nobre and Simoes (2019) took a similar kind of disposition in the context of pure product market by pooling the features of some renowned global luxury brands to examine the consumers' affinity with it. This study mounted TraveLux for the more complicated (in terms of heterogeneity and intangibility) travel market. Therefore, in a way, the study expanded the realm of luxury consumption and forayed into the psychosomatic bonding between travellers and luxury travel brands. The TraveLux instrument intended to map the perceived socio-emotional benefits of the travellers and hence, used the established scale items dimensionalised into intimacy-loyalty, passion (Fletcher et al., 1999), commitment and self-connection (Aaker et al., 2004). While commitment represented the behavioural bonds (Fournier, 1998), self-connection explained the manner in which travellers' personal identity relates to luxury consumption. Intimacy-loyalty depicted travellers' cognitive beliefs and passion (Fournier, 1998) to explain the degree of excitement in the relationship. Apart from the scale items which were already studied empirically, this study considered the variables studied by Nobre and Simoes (2019), de Kerviler and Rodriguez (2019), Liu et al. (2017), and Atwal and Williams (2009). Appropriate items to fit the Indian landscape of luxury travel consumption were also scaled. The TraveLux instrument drew thematic content and positioning platform of some of the globally renowned luxury travel brands.

Methodology

The study was bifurcated into two phases. The first phase focused on the development of the TraveLux instrument and testing the same for internal reliability, dimensionality and validation of the scale. The second phase tested the scale in explaining the perceived shared sentiments associated with luxury travel experience and interaction with luxury travel brands from the user-generated content (UGC) of the travellers in social media.

The TravelLux scale was developed with scale items derived from the review of the literature. Thematically the scale infused the essence (positioning, personality, projected identity etc.) of twenty (20) luxury travel brands that travellers had the opportunity to interact with during their luxury trips. The luxury travel brands were chosen from a wide range of luxury travel and hospitality services, namely accommodation (Leela Palace, Taj Falaknuma, Oberoi Villas, Rambagh Palace, ITC); dining (Wasabi, Bukhara, Yuuka, Olive Qutub); travel accessories (Luis Vuitton, Globe Trotter); train safari (Palace-on-wheels. Maharaja Express); road travel (Olevia, Volvo B11R); airliner (Emirates, Etihad); tour operator (LuxoIndia, Abercrombie and Kent) and spa and wellness (Park Hyatt). The respondents were asked to assess their relationships with the luxury travel brands that they had interacted with on a set of behavioural and attitudinal patterns. The 'commitment' and the 'self-connection' constructs were adopted from Aaker et al. (2004) and Nobre and Simoes (2019). The constructs of 'intimacy-loyalty' and 'passion' were adopted from the study of Fletcher et al. (1999), Nobre and Simoes (2019). The study postulated 'art and culture' as a possible construct of consumer–brand relationship, specifically in the context of luxury travel. However, this construct remained untested for luxury travel. The scale items for this construct were generated from the studies of Schroeder (2005), Joy et al. (2012), Koronaki, Kyrousi, and Panigyrakis (2017), Baumgarth et al. (2014), and Kapferer (2014). A seven-point interval scale was used for the study. The survey instrument also used elements of the Brand Relationship Quality Scale (Ekinci, Yoon, & Oppewal, 2005) and the Customer Brand Engagement Scale (Bijmolt et al., 2010; So, King, & Sparks, 2014; Hollebeek, 2009; Macey & Schneider, 2008; Patterson, Yu, & de Ruyter, 2006) to make the measurement construct (TraveLux) more robust and to be externally validated. An initial Focus Group was used to cross-check the content validity and ambiguity of questions. The final layout of the survey instrument was finalised with 25 scale items. The study was carried out in three prominent tourist destinations of India, namely Goa, Kerala and Rajasthan, based on inbound tourists and luxury travel engagements and consumptions. The study was cross-sectional in nature spanning from October 2018 to December 2019.

Following a hermeneutic phenomenology (Kafle, 2013) research design to capture the experiential nature (subjective) of the study, a convenience sampling technique was deployed targeting a specific traveller group. With the support from an appropriate intervening field agency to administer the questionnaire and to conduct the survey, initially, 434 travellers were identified based on their preferential alignment and engagement in luxury consumption while travelling.

However, the final survey result yielded 367 legitimate responses. Non-response bias was tested by assessing the differences between early and late respondents with regard to the mean of all the variables (Armstrong & Overton, 1977). No significant differences were found between the two groups of respondents. This suggested that response bias was not a problem in the study. The sample comprised 58.85% male (216) and 41.15% female (151) with an average age of 49 years. The sample consisted of 248 overseas and 119 domestic tourists. The sample also predominantly had a high percentage of respondents having higher academic/professional qualifications (85% graduation and above).

The second phase of the study tested the TraveLux brand relationship scale for its explaining and predictive capacity of shared sentiments and emotions, expressed as UGC over social media, in the context of consumption of luxury travel brands. The purposive sampling approach was used in accessing social media as it exposes information-rich cases that demonstrate the phenomenon of interest intensely (Patton, 2002). Three online luxury traveller forums were identified based on their blog followings over social media, namely thesqua.re (Twitter follower-62.98K; Facebook followers-165.20K; Alexa Rank-28.56K), FlyerTalk (Twitter follower-58.73K; Facebook followers-35.57K; Alexa Rank-29.8K) and Nomadic Matt (Twitter follower-120K; Facebook followers-255.70K; Alexa Rank-27.21K). Tweets posted on the Twitter account of the three selected luxury travel forums were collected for a period of six months, starting from June 2019 to December 2019. The units of analysis were the micro-blogs depicting luxury travel experience, contextual brand interactions and aspirational fulfilment shared over the Twitter platform.

IBM-SPSS 25, AMOS and NVivo 12 were used in data analysis.

Data Analysis

Phase-I Study

The data analysis was carried out in two stages: the first stage entailed Exploratory Factor Analysis (EFA) and the calculation of Cronbach's alpha to test internal consistency of the measures and to appraise scale items' performance within their respective dimension (Churchill, 1995); and, the second stage deployed the Confirmatory Factor Analysis (CFA) to refine the scale and check the validation issues.

EFA results revealed five factors for the TraveLux instrument based on significant factor loading (factors loaded with >0.650 were retained) nomenclated as: 'immersive experience' (6 items); 'ethnocultural acculturation' (4 items); 'passion and excitement' (4 items) and 'self-congruence' (4 items). The Cronbach's alpha was found to be 0.899, indicating good reliability levels of the scale (Nunnally, 1978). The KMO value is 0.879, which indicated sample adequacy for the study. The Bartlett's test of sphericity exhibited significant value (chi-square = 721.325, df = 325, sig. = 0.000). This suggested that we may reject the null hypothesis that our correlation matrix is an identity matrix and will conclude that the variables are correlated highly enough to provide a reasonable basis for EFA.

The total variance explained by the five factors was found to be 71.29 %, which was significant enough. Table 12.1 summarises the results for EFA and Cronbach's alpha.

Table 12.1. EFA Results (Phase-I Study): Rotated Component Matrix.[a]

Scale Items	Factors			
	Immersive Experience	**Ethno-cultural Acculturation**	**Passion & Excitement**	**Self-congruence**
Luxury travel brands explain my preferences in life (IE1)	0.889			
When I interact with my luxury travel brand, I forget everything else around me (IE2)	0.878			
Time flies when I am interacting with my luxury travel brand of choice (IE3)	0.851			
I feel satisfied while interacting intensely with my luxury travel brand (IE4)	0.832			
When interacting with my luxury travel brand, it is difficult to detach myself (IE5)	0.801			
My luxury travel brands galvanise mood change pleasantly (IE6)	0.791			
I prefer my luxury travel brands as they are reliable and consistent with my values (ECA1)		0.882		

Table 12.1. *(Continued)*

Scale Items	Factors			
	Immersive Experience	**Ethno-cultural Acculturation**	**Passion & Excitement**	**Self-congruence**
I opt for particular luxury travel brands which embed and reflect local culture, traditions, festivals and ethnicity (ECA2)		0.879		
I prefer my luxury travel brands as they promote social interaction (ECA3)		0.848		
I prefer my luxury travel brands as they aim to induce consciousness and change beliefs, attitudes and behaviours (ECA4)		0.827		
My relationship with my luxury travel brand has been thrilling (PE1)			0.861	
My relationship with my luxury travel brand has been exciting (PE2)			0.844	
I am passionate about my luxury travel brands (PE3)			0.819	
I spend a lot of my discretionary time thinking about my luxury travel brands (PE4)			0.787	
My luxury travel brand says a lot about the kind of person I would like to be (SA1)				0.907

Table 12.1. *(Continued)*

Scale Items		Factors			
		Immersive Experience	**Ethno-cultural Acculturation**	**Passion & Excitement**	**Self-congruence**
Association with my luxury travel brands lets me be a part of a shared community of like-minded consumers (SA2)					0.891
My luxury travel brands fit well with my current stage/status of life (SA3)					0.870
My luxury travel brands connect with the part of me that really makes me going (SA4)					0.835
Determinant of correlation			0.008		
Total variance explained			71.29		
KMO			0.879		
Barlett's Test of Sphericity	*Chi-square*		721.325		
	Df		325		
	Sig.		0.000		
Cronbach's α			0.899		

Extraction method: Principal component analysis; Rotation method: Varimax with Kaiser normalisation.
[a]Rotation converged in five iterations.

Common Method Bias

As the study was rationally grounded on existing theories with contextual realism, EFA was also used to assess Common Method Bias (CMB) (Arya, Sethi, & Paul, 2019). Considering the fact that validity issues might yield a potentially misleading conclusion (Campbell & Fiske, 1959), the assessment of variance attributable to the measurement method (Podsakoff, MacKenzie, Lee, & Podsakoff, 2003) was applied. Testing the biasness was critical for the study as the data were obtained through a self-administered questionnaire (using the online platform), and response to both the predictor and criterion variables were generated from the same respondent. The results revealed that the first factor explained the total variance of 23.129% (<50%), which confirmed the absence of CMB is not a problem in this study. Further, it was established that the measures were free from common method variance (<50%) and hence concluded to be an insensitive test which did not support the fundamental assumption of Herman's single-factor test (Podsakoff et al., 2003).

Full-information Maximum Likelihood (FIML) was used for CFA. The items were limited to load on its pre-specified factors, and the factors were allowed to correlate freely. All 18 scale items were retained in CFA. Fig. 12.1 depicts the measurement model with standardised regression weights of each item on the

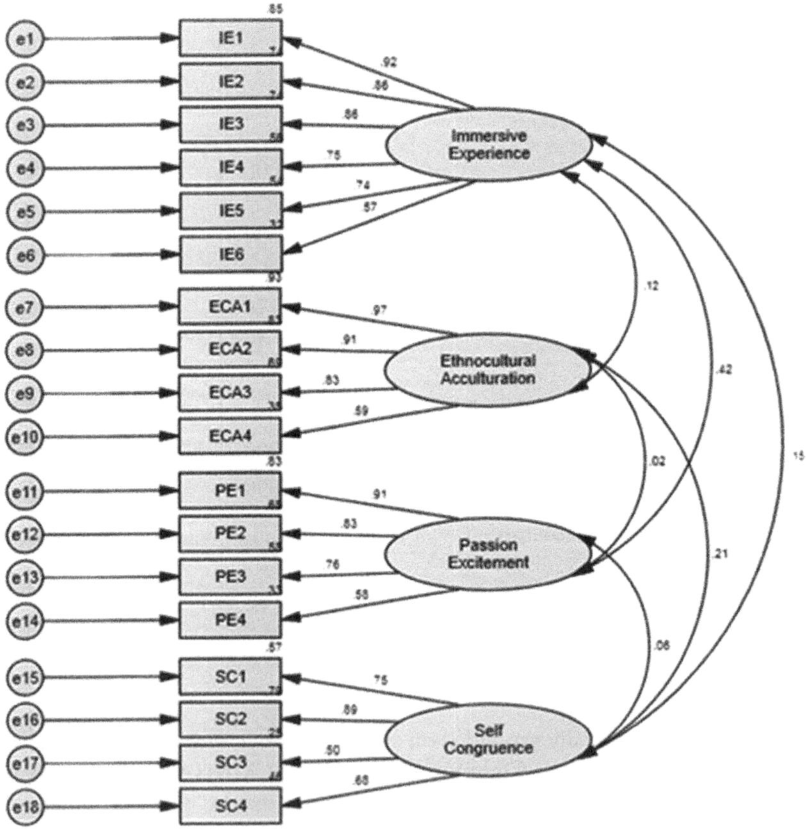

Fig. 12.1. Standardised Model (Zeroorder from Phase-I Study).

respective construct. Results reflect the unidimensionality and internal consistency of the measures.

Adequate fit was achieved with the data as the CMIN/DF was found to be 1.728 (ref. value: <3, Arya et al., 2019) (Table 12.2a). The chi-square for this model was 534.009 (df = 309, p = 0.000). The chi-square statistic is sensitive to sample size; therefore, the following fit indices (Table 12.2b) were also considered: Normed Fit Index (NFI) (0.931), Comparative Fit Index (CFI) (0.934), Tucker–Lewis Fit Index (TLI) (0.929), Goodness-of-fit Index (GFI) (0.937) and RMSEA (<0.05). All these indexes presented acceptable values (Hair, Black, Babin, & Anderson, 2010).

The reliability issue had been addressed adequately as the composite reliability (CR) was found to be >0.7 for all constructs (Table 12.3). Convergent validity was established as the average variance extracted (AVE) was found to be greater than 0.5 for all constructs (Fornell & Larcker, 1981) and CR > AVE (Bagozzi, 1980). The maximum shared variance (MSV) was found consistently less than AVE, which established discriminant validity (Hair et al., 2010). The analysis also reported Hancock's H [MaxR (H)] measure of reliability, which required fewer assumptions than Cronbach's α (Hancock & Mueller, 2006). Hancock's H reflected the minimum achievable or anticipated construct reliability associated with the measurement model of each factor (Hancock & Mueller, 2006). It ranged from 0.863 (self-congruence) to 0.981 (ethnocultural acculturation) and hence exceeded the recommended minimum value of H = 0.70 (Hancock & Mueller, 2006) (Table 12.3).

The TraveLux brand relationship instrument was further tested to understand the performance of the instrument at a higher order level (Dabholkar, Thorpe, & Rentz, 1996; Nobre & Simoes, 2019). The measurement model (Fig. 12.2) retained the four zero-order factors, with all the scale items having significant loading value.

The higher-order model achieved adequate fit as the CMIN/DF was found to be 2.314 (ref. value: <3, Arya et al., 2019) (Table 12.4) and considered to be significant. The chi-square for this model was 689.719 (df = 298, p = 0.001). The baseline comparisons were also presented in Table 12.5. The Goodness-of-Fit Index (GFI) (0.932), Normed Fit Index (NFI) (0.928), Comparative Fit Index (CFI) (0.923) and the Tucker–Lewis Fit Index (TLI) (0.924) were found significant (Hair et al., 2010). The RMSEA value (0.043) was also found <0.04 and hence significant (Hair et al., 2010). The immersive experience, the zero-order factor, had a factor loading of 0.94 in the higher-order model. Similarly, ethnocultural acculturation scored 0.82, passion and excitement scored 0.91 and self-congruence had a factor loading of 0.85. The results indicated that the TraveLux brand relationship instrument might

Table 12.2a. Model Fit Analysis (CMIN/DF).

Model	**NPAR**	**CMIN**	**DF**	***P***	**CMIN/DF**
Default model	78	534.009	309	0.000	1.728

Note: NPAR, Number of distinct parameters; DF, Degrees of freedom; *P*, Significance value; CMIN/DF: minimum discrepancy divided by degree of freedom.

Table 12.2b. Model Fit Analysis (NFI, GFI, CFI, TLI, RMSEA).

NFI	GFI	TLI	CFI	RMSEA
0.931	0.937	0.929	0.934	0.039

Note: NFI, Normed fit index; GFI, Goodness-of-fit index; TLI, Tucker–Lewis index; CFI, Comparative fit index; RMSEA, Root mean square error of approximation.

Table 12.3. Construct Validity (CR/AVE/MSV/MaxR(H)).

	SC	IE	PE	ECA	CR	AVE	MSV	MaxR(H)
SC	**0.723**				0.806	0.518	0.044	0.863
IE	0.004	**0.794**			0.908	0.628	0.172	0.954
PE	0.064	0.415	**0.781**		0.858	0.608	0.172	0.968
ECA	0.210	−0.001	0.025	**0.838**	0.900	0.698	0.044	0.981

Note: SC, Self-congruence; IE, Immersive experience; PE, Passion and excitement; ECA, Ethno-cultural acculturation, Diagonal elements show square roots of AVEs (average variance explained). Establishment of discriminant validity. The principal diagonal (values in bold) shows the Average Variance Extracted is greater than the sum of the Squared Correlations between the variables.

be represented with four dimensions, which are used by the travellers to explain their relationships with luxury travel brands. The dimensions are indicative of hedonic relationship based on experience, acculturation, engaged passion, the scope of excitement and self-congruence.

CR was found to be >0.7 for all constructs and hence confirmed significant levels. Convergent validity established as the average variance extracted (AVE) was found to be greater than 0.5 for all constructs and CR > AVE. The maximum shared variance (MSV) and the average shared variance (ASV) were found consistently less than AVE, which established discriminant validity (Hair et al., 2010) (Table 12.5).

Phase-II Study

Phase II of the study was designed to taste whether the TraveLux brand relationship instrument had incorporated adequate dimensional spread to capture perceived and experienced luxury travel consumptions. For this, the study planned to use verbal and visual content, shared on social media, in association with luxury travel and allied consumptions and compared it with the TraveLux dimensional scores obtained from the Phase-I study. Tweets were captured using hashtags 'experience with luxury travel brands' and 'luxury travel experience' from the three identified luxury travel platforms, namely thesqua.re, FlyerTalk and Nomadic Matt. Tweets and retweets were captured using the NCapture feature of NVivo-12 for a period of six months (June 2019 to December 2019). A

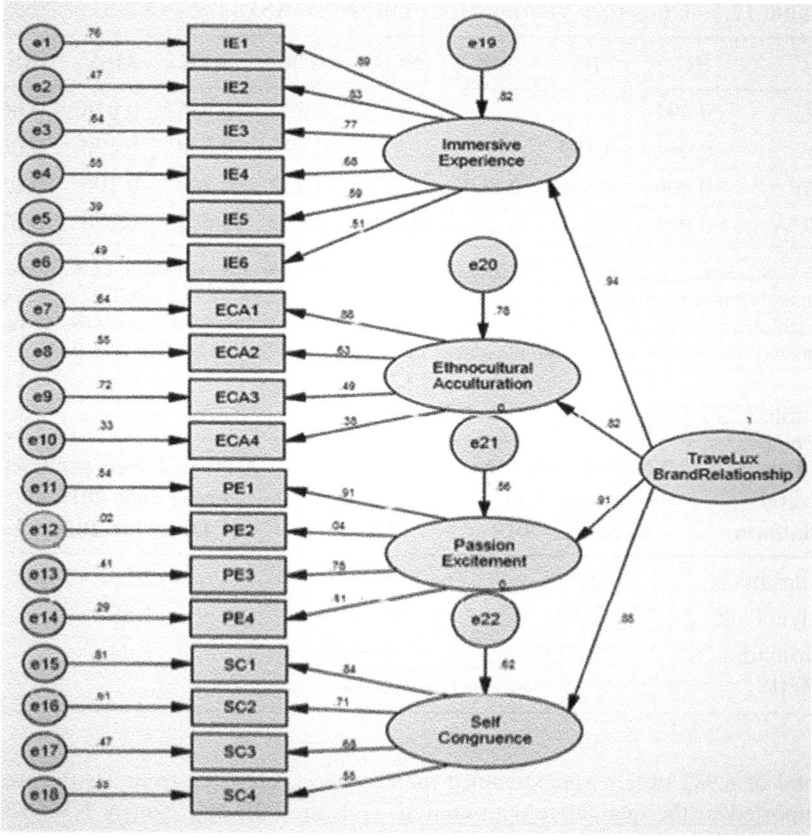

Fig. 12.2. TraveLux Brand Relationship – Higher Order Model.

Table 12.4. Goodness-of-fit Indices for the Proposed Model and RMSEA Value.

	CMIN					Baseline Comparisons				
Model	**NPAR**	**CMIN**	**DF**	***P***	**CMIN/ DF**	**NFI**	**GFI**	**TLI**	**CFI**	**RMSEA**
Default model	121	689.719	298	0.001	2.314	0.928	0.932	0.924	0.923	0.043

Note: NPAR, Number of distinct parameters; DF, Degrees of freedom; *P*, Significance value; CMIN/DF: minimum discrepancy divided by degree of freedom; NFI, Normed fit index; GFI, Goodness-of-fit index; TLI, Tucker–Lewis index; CFI, Comparative fit index; RMSEA, Root mean square error of approximation.

Table 12.5. Construct Validity (CR/AVE/MSV/ASV).

	SC	IE	PE	ECA	CR	AVE	MSV	ASV
SC	**0.797**				0.837	0.635	0.016	0.005
IE	−0.051	**0.799**			0.897	0.638	0.120	0.060
PE	0.006	0.326	**0.864**		0.922	0.746	0.106	0.054
ECA	0.021	0.271	0.267	**0.819**	0.859	0.671	0.276	0.077

Note: SC, Self-congruence; IE, Immersive experience; PE, Passion and excitement; ECA, Ethno-cultural acculturation, Diagonal elements show square roots of AVEs (average variances explained). Establishment of discriminant validity. The principal diagonal (values in bold) shows the Average Variance Extracted is greater than the sum of the Squared Correlations between the variables.

Table 12.6. Details of Extracted Twitter Data.

Luxury Travel Platform	No. of Tweets from June 2019 to December 2019	No. of Users (with Unique ID)	Average Tweets per User during June 2019 to December 2019
Thesqua.re	2,348	1,023	2.29
FlyerTalk	1,879	998	1.88
Nomadic Matt	2,775	1,429	1.94

total of 6,982 tweets were streamed in. The tweets collected from Twitter were imported to the qualitative data analysis software (QDAS), namely NVivo-12. Table 12.6 computed the details of Twitter data extracted.

Following the analytic process outlined by Tian et al. (2019) and Fan and Niu (2016), analysis on the texts and graphics, used in the tweets, was conducted using open, axial and selective coding (Corbin & Strauss, 1990). The tweets were read and contextual themes were open-coded followed by assigning interpretive labels in graphical elements of the tweets to frame the nodes. Axial coding was used to classify and cluster nodes based on homogeneity of thematic content. Finally, selective coding was done to summarise the identified nodes based on thematic expressions. Sentiment analysis was deployed to assess the cognitive alignment of the users with respect to their experiential manifestations of luxury consumption while travelling. Dictionary-based (Lexicon) method was used to compute the sentiment scores which ranged from '+1' (extremely positive) to '−1' (extremely negative). Machine algorithm used the following equation to calculate the sentiment score centred around expressed positivity/negativity via text's (T) and cross-referring it with a dictionary (D):

$$S_D^T = \frac{\sum_{W \in D} h_D(W) \cdot f^T(W)}{\sum_{W \in D} f^T(W)} = \sum_{W \in D} h_D(W) . p^T(W)$$

Where W denotes each of the word cross-referred with dictionary D; word-sentiment scores as $h_D(w)$; frequency of word as $f^T(w)$ and normalised frequency of w in T as $p^T(w) = f^T(w)/\sum$.

The world classification and clustering were done using the Affective Norms of English (ANEW) dictionary, and two major clusters were identified as measures of emotions: (1) valence and (2) arousal (Russell, 1980). While 'valence' represented the direction of the emotion, 'arousal' captured the quantum of physical response. Machine algorithm (supervised classification process using Naïve Bayes theorem) assigned a score for each tweet for valence and arousal. Higher 'valence score' implied greater mark-ups of tweets in the positive direction and a higher 'arousal score' denoted enhanced intensity in expressed emotions. Summation of valence and arousal scores for each word of a tweet was done to obtain the sentiment scores. Four clusters, based on the text-clustering process (Meyer, Hornik, & Feinerer, 2008), could be identified by the algorithm and were nomenclated based on thematic proximity of the tweets: experience, self-reflection and cultural absorption. Fig. 12.3 represented the clustered heat map.

The dimensional scores for TraveLux (immersive experience, ethnocultural acculturation, passion and excitement and self-congruence) were imputed from the CFA process conducted in the Phase-I, which served as an input to further impute the TraveLux brand relationship score. Multiple regression analysis was used to assess the capability of TraveLux brand relationship instrument to predict shared sentiments on Twitter platform in the context of luxury travel experience and interaction with luxury travel brands during travel. Therefore, the null-hypothesis was posited as:

H_0 = TraveLux does not influence sentiments and emotions associated with interaction and consumption of luxury travel brands.

Hence the regression equation takes the form as:

$$Y(\text{TraveLux}) = \alpha + X_1(\text{experience})\beta_1 + X_2(\text{cultural\_absorption})\beta_2 + X_3(\text{self\_concept})\beta_3$$

Before applying the regression procedure, the TraveLux and the sentiment scores were converted to their natural logarithm values. The descriptive statistics for all variables were summarised in Table 12.7.

The first model included all control variables. However, the control variable, namely 'twitter followers' was found intolerant with this model on account of multicollinearity issue as the variance inflation factor (VIF) was found to be >10 and did not comply with the threshold limit (Gefen, Straub, & Boudreau, 2000) and hence eliminated from the second model. In the third model, the study introduced the average score for clustered sentiments (experience, cultural_absorption and self_concept) to assess the notion of sentiments (with regard to interaction and consumption of luxury travel brands) in shaping the brand relationships with luxury travel brands. The results of the multiple regression analysis were tabulated in Table 12.8. The R-square value for the third model was found to be 0.487 with p-value 0.001 (significant at

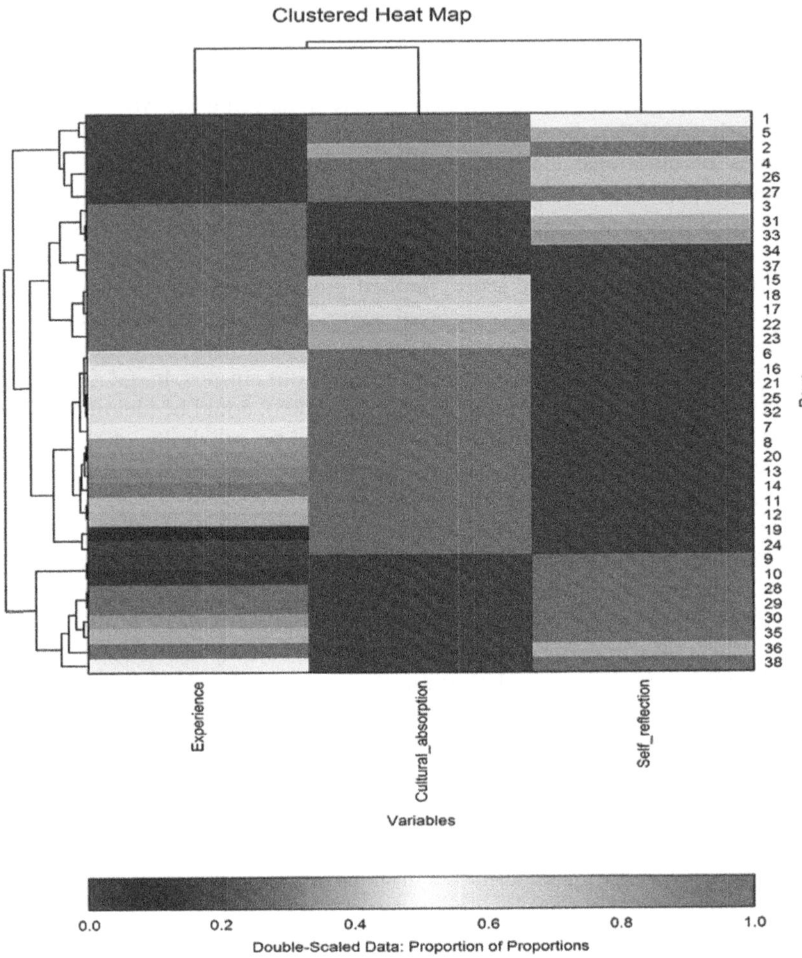

Fig. 12.3. Clustered Heat Map of Sentiments.

level of 5%). The standardised regression coefficients for the average sentiment scores were:

(1) for experience: 5.231 ($p < 0.01$)
(2) for cultural_absorption: 9.870 ($p < 0.01$)
(3) for self_concept: 3.287 ($p < 0.05$)

Conclusion and Discussion

The chapter focused on unravelling the mystery and myth associated with luxury travel and branding issues. Contemporary travellers perceive luxury travel to be

Table 12.7. Descriptive Statistics of the Variables under Study.

Variables		**Min.**	**Max.**	**Mean**	**Std. Deviation**
TraveLux		2.33	6.58	5.0457	0.46866
Sentiment clusters	Experience	21	3,976	471.61	523.964
	Cultural_absorption	14	4,561	507.76	689.901
	Self_concept	8	1,271	132.32	192.341
Control variables		**Min.**	**Max.**	**Mean**	**Std. Deviation**
Volume of Tweets/month (in the study time)		318	4,712	1764	987.032
Time engagement with Twitter (in months)		63	121	99.62	18.692
Twitter followers		2.8	1,987	682.17	601.267
Retweets (dummy variable)		0	21	4.78	7.982

Table 12.8. Regression Models.

Dependent Variable		**TraveLux (Model-1)**	**TraveLux (Model-2)**	**TraveLux (Model-3)**	**Reject Null Hypothesis**
Constant		2.124	2.098	2.207	
Twitter followers		0.45 (0.000)			
Time engagement with Twitter (in months)		0.13 (0.009)	−0.07 (0.003)	1.29*(0.012)	
Log (Volume of Tweets/ month) (in the study time)		−1.76 (0.128)	2.27*(0.109)		
Log (Average Sentiment Scores)	Experience			6.78**(1.72)	Yes
	Cultural absorption			12.92***(3.31)	Yes
	Self-concept			2.76*(0.005)	Yes
Retweets		0.117**(0.089)	0.365**(0.142)	0.442**(0.197)	
R-square		0.268	0.326	0.487	
Average *R*-square		0.242	0.307	0.471	
F		12.79	18.67	33.73	

$^{*}p < 0.10$; $^{**}p < 0.05$; $^{***}p < 0.01$.

an aspirational experience encapsulated within the realm of emotional exclusivity and embedded uniqueness. Travellers interact with luxury travel brands more frequently as the luxury travel brands continued to be affordable and accessible. The shifting dynamism of traveller relationship with luxury travel brands, therefore, needs to be explored. The existing body of literature postulated luxury consumption in the context of travel motives (Berrozpe et al., 2017; Brandao et al., 2019; Kaufman et al., 2016; Nobre & Simoes, 2019). This study expanded on the findings of Nobre and Simoes (2019), which extrapolated luxury consumption in the context of tangible consumption to a hypothesised luxury brand in the field of luxury travel with intangible consumption patterns. The major contribution of this study could be attributed to the construction of the TraveLux brand relationship instrument, conceptualised to capture the relational issues between a traveller and a luxury travel brand. The study drew inspiration from the conventional theory of consumer–brand relationship (Fletcher et al., 1999; Fournier, 1998), brand relationship quality (Aaker et al., 2004) and luxury brand relationship (Heine & Phan, 2011). It reinforced the theory of subjective and hedonic consumption associated with experiential travelling (Bian & Forsythe, 2012; Kapferer, 2014; Vickers & Renand, 2003) and the self-connection with the luxury-travel brand. The study revealed the subjective nature of luxury consumption with ethnocultural symbolism explaining the brand meaning and the acculturation process emphasising brand affinity. The hedonic manifestations were noted as travellers expressed their passion and excitement to be associated with the luxury-travel brands, which, they believed, reflected their inherent urge to enter into a relationship on the basis of perceived utility-thrill and snob value. The self-connection nature of the instrument linked the spread of perceived socio-emotional gratification, self-identification and self-actualisation. The study revealed a four-dimensional construct for the TraveLux instrument, namely, 'immersive experience', 'ethnocultural acculturation', 'passion and excitement' and 'self-congruence'.

Immersive experience entailed the uninhibited cognitive attachment of the traveller with the luxury travel brands. The intense relationship between the traveller and the luxury travel brands was reflected in the scale items ('Time flies when I am interacting with my luxury travel brand of choice'; 'When interacting with my luxury travel brand, it is difficult to detach myself' and 'My luxury-travel brands galvanise mood change pleasantly') and emphasised the nature of aspirational values that drove this relationship. Earlier studies had empirically identified cultural symbolism to impart brand meaning (Torelli, Keh, & Chiu, 2010) and conceptualised as perceived consensus about the degree to which the brand symbolises the abstract image of a certain cultural group. The construct of 'ethnocultural acculturation' captured the concept of cultural symbolism and the inherent intent of the traveller, with the motive of luxury consumption, to be an integral part of the brand's cultural and ethnic fabric. The absorption of ethnocultural essence was reflected in the scale items ('I opt for particular luxury travel brands which embed and reflect local culture, traditions, festivals and ethnicity', and 'I prefer my luxury travel brands as they aim to induce consciousness and change beliefs, attitudes and behaviours'). Passion instils excitement when it comes to luxury consumption in the context of travel behaviour ('My relationship with my luxury travel brand has been exciting' and 'I am passionate about my luxury travel brands'). The findings supported the aspect of mass-consumed luxury brands possessing the ability to induce passion and excitement, to bond with consumers (Nobre, 2015; Nobre & Simoes, 2019). The retained passion and excitement scale items linked luxury travel to the more aspirational experience and subjective evocation that luxury travel brands might propagate. The dimension of self-congruence reflected the theory that posited that consumers responded more positively to brands that were aligned with their self-concept (Sirgy, 1982), thus predicting consumers' attitudes and purchase intentions (Aaker, 1999; Hong & Zinkhan, 1995). Retained scale items, namely 'Association with my luxury travel brands lets me be a part of a shared community of like-minded consumers', conveyed the meaning of self-expression, while statements, namely 'My luxury travel brand says a lot about the kind of person I would like to be' and 'My luxury travel brands fit well with my current stage/status of life' underlined the aspirational self of the travellers that evoked a sense of relationship with the luxury travel brands. This consolidated the proposition of value co-creation (Merz et al., 2018) on the basis of perceived brand value (Payne et al., 2009) predisposed from cherished relationships and abstractions (Heine & Phan, 2011). The findings of the study further re-established the transition of the luxury concept based on perceived social images driven by exclusivity to aspirational immersive experience grounded on intimate and intensive subjective encounters. The travel sector might just provide the ideal kind of platform to experiment with luxury travel brands with scope for personalisation to infuse a greater sense of self-congruence.

The study tested the TraveLux instrument for its capacity to impact traveller behaviour on social media platforms. Studies conducted on netnography by Tian et al. (2019), Ling, Ariff, Zakuan, and Tajudin (2014), Munezero, Montero, Sutinen, and Pajunen (2014), and Adjei (2016), etc. reflected on the power of social media to amplify perceived notions by sharing text and graphical content with regard to consumptions, interactions and experience. The study randomly selected three online luxury travel platforms to stream in tweets reflecting sentiments and emotions seeding out of the luxury travel experience and checked for its dependency with the TraveLux scores. Three clustered sentiments, namely 'experience', 'cultural_
absorption' and 'self-concept' were empirically found to be shaped by the TraveLux instrument. Theoretically, it consolidated the TraveLux instrument from a validation point of view and supported the theory of consumer engagement on social media to share motives for luxury brand consumption.

The study had implications for the brand designers and the brand marketers too. In the travel, tourism and hospitality market, the emergence of 'experience designers' conveys the emphasis that goes into brand creation. Contemporary luxury travel brands have widened their scope of consumption, but the perennial essence of cultural symbolism, self-reflection, aristocracy and elitism have remained embedded. The TraveLux Brand Relationship instrument offers a framework for the brand managers to assess the nature and intensity of the relationship between traveller and luxury brands. This instrument will also assist in conceptualising brand differentiators and design brand communications. The scale can also identify the specific brand connectors, namely cultural symbolism, immersive experience, psycho-somatic behaviours, brand meaning, etc., given the nature of travel and tourism products to be marketed. The TraveLux instrument was found evoking consistent results across two zones considered for the study. To be very precise, the instrument was tested for luxury travellers across the major tourist destinations of India, where overseas travellers prefer to travel. Therefore, the luxury brand designers may ground their luxury travel brand architecture on uniform cognition and appeal.

The study was limited to three specific tourist destinations of India and was cross-sectional in nature. Non-probabilistic sampling technique was used to capture the experiential nature of the study. Therefore, the study cannot be generalised over a wider geographical and socio-demographical spread. Consumer–brand relationships are expected to evolve over time, and hence the dimensional re-orientations are required in the travel context. Further extrapolations of the study could be made by incorporating variables in the model that would explain behavioural patterns (cognitive, affective and conative) in relation to luxury travel brands. The study used only the Twitter platform for sentiment analysis. Facebook and Instagram platforms might be considered for future study for gaining deeper insights. The study was restricted to text mining only, and therefore, analysis of graphics (still and motion) might provide additional information.

References

Aaker, J. L. (1999). The malleable self: The role of self-expression in persuasion. *Journal of Marketing Research*, *36*(1), 45–57. doi:10.2307/3151914

Aaker, J., Fournier, S., & Brasel, S. A. (2004). When good brands do bad. *Journal of Consumer Research*, *31*(1), 1–16. doi:10.1086/383419

Adjei, J. K. (2016). The role of social media in national discourse and mobilisation of citizens. In Proceedings of International Conference on Collaboration Technologies and System, Orando, FL (*1*, pp. 559–566).

Aguirre-Rodriguez, A. (2014). Cultural factors that affect brand personification strategy effectiveness. *Psychology and Marketing*, *31*(1), 70–83. doi:10.1002/mar.20676

Armstrong, J. S., & Overton, T. S. (1977). Estimating nonresponse bias in mail surveys. *Journal of Marketing Research*, *14*(3), 396–402. doi:10.1177/002224377701400320

Arya, V., Sethi, D., & Paul, J. (2019). Does digital footprint act as a digital asset? – Enhancing brand experience through remarketing. *International Journal of Information Management*, *49*, 142–156. doi:10.1016/j.ijinfomgt.2019.03.013

Atwal, G., & Williams, A. (2009). Luxury brand marketing: The experience is everything! *Journal of Brand Management*, *16*, 338–346. doi:10.1057/bm.2008.48

Bagozzi, R. P. (1980). *Causal models in marketing*. New York, NY: John Wiley.

Baksi, A. K. (2016). Destination bonding: Hybrid cognition using Instagram. *Management Science Letters*, *6*(1), 31–46. doi:10.5267/j.msl.2015.12.001

Baksi, A. K. (2017). Craftourism: Development and validation of a scale to measure visitors' behavioural pattern based on identified travel motives. *Tourismos, 12*(1), 101–125. Retrieved from etem.aegean.gr/images/TourismosVOL12NO1.pdf

Baumgarth, C., Lohrisch, N., & Kastner, O. L. (2014). Arts meet luxury brands. In B. Berghaus, G. Müller-Stewens, & S. Reinecke (Eds.), *The management of luxury* (pp. 127–142). London: Kogan Page Publishers.

Becker, K., Lee, J. W., & Nobre, H. (2018). The concept of luxury brands and relationship between consumer and luxury brands. *Journal of Asian Finance, Economics and Business, 5*(3), 51–63. doi:10.13106/jafeb.2018.vol5.no3.51

Bengtsson, A., Bardhi, F., & Venkatraman, M. (2010). How global brands travel with consumers: An examination of the relationship between brand consistency and meaning across national boundaries. *International Marketing Review, 27*(5), 519–540. doi:10.1108/02651331011076572

Berrozpe, A., Campo, S., & Yagüe, M. J. (2017). Understanding the identity of Ibiza, Spain. *Journal of Travel and Tourism Marketing, 34*(8), 1033–1046. doi:10.1080/10548408.2016.1272525

Beverland, M. (2004). Uncovering theories-in-use: Building luxury wine brands. *European Journal of Marketing, 38*(3/4), 446–466. doi:10.1108/03090560410518648

Bian, Q., & Forsythe, S. (2012). Purchase intention for luxury brands: A cross cultural comparison. *Journal of Business Research, 63*, 1443–1451. doi:10.1016/j.jbusres.2011.10.010

Bijmolt, T. H. A., Leeflang, P. S. H., Block, F., Eisenbeiss, M., Hardie, B. G. S., Lemmens, A., & Saffert, P. (2010). Analytics for customer engagement. *Journal of Service Research, 13*, 341–356. doi:10.1177/1094670510375603

Bizarris, F. S., Stehlau, S., & Brandao, M. M. (2017). The moderating role of brands for low income luxury consumers. *Independent Journal of Management and Production, 8*(3), 936–954. doi:10.14807/ijmp.v8i3.639

Brakus, J., Schmitt, B. H., & Zarantonello, L. (2009). Brand experience: What is it? How is it measured? Does it affect loyalty? *Journal of Marketing, 73*(May), 52–68. doi:10.1509/jmkg.73.3.52

Brandão, A., Pinho, E., & Rodrigues, P. (2019). Antecedents and consequences of luxury brand engagement in social media. *Spanish Journal of Marketing - ESIC, 23*(2), 163–183. doi:10.1108/SJME-11-2018-0052

Campbell, D. T., & Fiske, D. W. (1959). Convergent and discriminant validation by the multitrait-multimethod matrix. *Psychological Bulletin, 56*(2), 81–105. 10.1037/H0046016.

Canziani, B., Watchravesringkan, K., & Yurchisin, J. (2016). A model for managing service encounters for neo-luxury consumers. *Worldwide Hospitality and Tourism Themes, 8*(1), 41–52. doi:10.1108/WHATT-10-2015-0036

Chitturi, R., Chitturi, P., & Raghavarao, D. (2010). Design for synergy with brand or price information. *Psychology and Marketing, 27*(7), 679–697. doi:10.1002/mar.20351

Chun, H., Diehl, K., & MacInnis, D. (2017). Savoring an upcoming experience affects ongoing and remembered consumption enjoyment. *Journal of Marketing, 81*(3), 96–110. doi:10.1509/jm.15.0267

Churchill, G. A., Jr (1995). *Marketing research: Methodological foundations*. Chicago, IL: The Dryden Press.

Corbin, J., & Strauss, A. (1990). Grounded theory research: Procedures, canons, and evaluative criteria. *Qualitative Sociology*, *13*(1), 3–21. doi:10.1007/BF00988593

Cristini, H., Kauppinen-Raisanen, H., Barthod-Prothade, M., & Woodside, A. (2017). Toward a general theory of luxury: Advancing from workbench definitions and theoretical transformations. *Journal of Business Research*, *10*(1), 101–107. doi: 10.1016/j.jbusres.2016.07.001

Dabholkar, P. A., Thorpe, D. I., & Rentz, J. (1996). A measure of service quality for retail stores: Scale development and validation. *Journal of the Academy of Marketing Science*, *24*(1), 3–16. doi:10.1007/BF02893933

Dion, D., & Arnould, E. (2011). Retail luxury strategy: Assembling charisma through art and magic. *Journal of Retailing*, *87*(4), 502–520. doi:10.1016/j.jretai.2011.09.001

Dixit, S. K. (2020). *The Routledge handbook of tourism experience management and marketing*. London: Routledge.

Dubois, B., Czellar, S., & Laurent, G. (2005). Consumer segments based on attitudes toward luxury: Empirical evidence from twenty countries. *Marketing Letters*, *16*(2), 115–128.

Ekinci, Y., Yoon, T. H., & Oppewal, H. (2005). An examination of the brand relationship quality scale in the evaluation of restaurant brands. *Advances in Hospitality and Leisure*, *1*, 189–198. doi:10.1016/S1745-3542(04)01012-4

Escalas, J. (2013). Self-identity and consumer behavior. *Journal of Consumer Research*, *39*(1), 1–6. doi:10.1086/669165

Fan, Y., & Niu, R. H. (2016). To tweet or not to tweet? Exploring the effectiveness of service recovery strategies using social media. *International Journal of Operations and Production Management*, *36*(9), 1014–1036. doi:10.1108/IJOPM-10-2013-0461

Fletcher, G. J. O., Simpson, J. A., Thomas, G., & Giles, L. (1999). Ideals in intimate relationships. *Journal of Personality and Social Psychology*, *76*(1), 72–89. doi: 10.1037/0022-3514.76.1.72

Fog, K., Budtz, C., Munch, P., & Blanchette, S. (2010). *Storytelling: Branding in practice* (2nd ed.). Copenhagen: Springer.

Fornell, C., & Larcker, D. F. (1981). Evaluating structural equation models with unobservable variables and measurement error. *Journal of Marketing Research*, *18*(February), 39–50. doi:10.1177/002224378101800104

Fournier, S. (1998). Consumers and their brands: Developing relationship theory in consumer research. *Journal of Consumer Research*, *2*(4), 343–373. doi:10.1086/209515

Gefen, D., Straub, D., & Boudreau, M.-C. (2000). Structural equation modeling and regression: Guidelines for research practice. *Communications of the Association for Information Systems*, *4*(1), 7. doi:10.1086/209515

Hair, J., Black, W., Babin, B., & Anderson, R. (2010). *Multivariate data analysis* (7th ed.). Upper Saddle River, NJ: Prentice-Hall.

Hancock, G. R., & Mueller, R. O. (2006). Structural equation modeling. In G. R. Hancock, S. L. Stapleton, & R. O. Mueller (Eds.), *The reviewer's guide to quantitative methods in the social sciences* (pp. 371–383). London: Taylor & Francis.

Han, H., & Hyun, S. S. (2018). Role of motivations for luxury cruise traveling, satisfaction, and involvement in building traveler loyalty. *International Journal of Hospitality Management*, *70*, 75–84. doi:10.1016/j.ijhm.2017.10.024

Heine, K., & Phan, M. (2011). Trading-up mass-market goods to luxury products. *Australasian Marketing Journal*, *19*(2), 108–114. doi:10.1016/j.ausmj.2011.03.001

Hollebeek, L. D. (2009, November–December). Demystifying customer engagement: Toward the development of a conceptual model. Paper presented at the ANZMAC 2009 conference, Melbourne, VIC, Monash University.

Holt, D. B., Quelch, J. A., & Taylor, E. L. (2004). How global brands compete. *Harvard Business Review*, *82*(9), 68–75.

Hong, J. W., & Zinkhan, G. M. (1995). Self-concept and advertising effectiveness: The influence of congruency, conspicuousness, and response mode. *Psychology and Marketing*, *12*(1), 53–77. doi:10.1002/mar.4220120105

Huang, Z. J., & Cai, L. A. (2015). Modeling consumer-based brand equity for multinational hotel brands – When hosts become guests. *Tourism Management*, *46*(2), 431–443. doi:10.1016/j.tourman.2014.07.013

Jain, V., Roy, S., & Ranchhod, A. (2015). Conceptualising luxury buying behavior: The Indian perspective. *Journal of Product and Brand management*, *24*(3), 211–228. doi:10.1108/JPBM-07-2014-0655

Joy, A., Sherry, J., Jr, Venkatesh, A., Wang, J., & Chan, R. (2012). Fast fashion, sustainability, and the ethical appeal of luxury brands. *Fashion Theory*, *16*(3), 273–295. September.

Kafle, N. P. (2013). Hermeneutic phenomenological research method simplified. *Bodhi: An Interdisciplinary Journal*, *5*, 181–200. doi:10.3126/bodhi.v5i1.8053

Kapferer, J. N. (2006). The two business cultures of luxury brands. In J. Schroeder & M. Salzer-Mörling (Eds.), *Brand culture* (pp. 67–76). London: Taylor & Francis.

Kapferer, J. N. (2014). The artification of luxury: From artisans to artists. *Business Horizons*, *57*(3), 371–380. doi:10.1016/J.BUSHOR.2013.12.007

Kapferer, J.-N., & Laurent, G. (2016). Where do consumers think luxury begins? A study of perceived minimum price for 21 luxury goods in 7 countries. *Journal of Business Research*, *69*(1), 332–340. doi:10.1016/j.jbusres.2015.08.005

Kaufman, H. R., Loureiro, S. M. C., & Manarioti, A. (2016). Exploring behavioural branding, brand love and brand co-creation. *Journal of Product and Brand Management*, *25*(6), 516–526. doi:10.1108/JPBM-06-2015-0919

de Kerviler, G., & Rodriguez, C. M. (2019). Luxury brand experiences and relationship quality for Millennials: The role of self-expansion. *Journal of Business Research*, *102*, 250–262. doi:10.1016/j.jbusres.2019.01.046

Kim, A. J., & Ko, E. (2012). Do social media marketing activities enhance customer equity? An empirical study of luxury fashion brand. *Journal of Business Research*, *65*, 1480–1486. doi:10.1016/j.jbusres.2011.10.014

Koronaki, E., Kyrousi, A. G., & Panigyrakis, G. G. (2017). The emotional value of arts-based initiatives: Strengthening the luxury brand–consumer relationship. *Journal of Business Research*, *85*, 406–413. doi:10.1016/j.jbusres.2017.10.018

Lee, E. J. (2013). A prototype of multicomponent brand personality structure: A consumption symbolism approach. *Psychology and Marketing*, *30*(2), 173–186. doi:10.1002/mar.20596

Lee, J., & Hwang, J. (2011). Luxury marketing: The influences of psychological and demographic characteristics on attitudes toward luxury restaurants. *International Journal of Hospitality Management*, *30*(3), 658–669. doi:10.1016/j.ijhm.2010.12.001

Liang, Y., Ghosh, S., & Oe, H. (2017). Chinese consumers' luxury value perceptions–A conceptual model. *Qualitative Market Research: An International Journal*, *20*(2), 247–262. doi:10.1108/QMR-01-2017-0010

Liao, J., & Wang, L. (2009). Face as a mediator of the relationship between material value and brand consciousness. *Psychology and Marketing*, *26*(11), 987–1001. doi: 10.1002/mar.20309

Ling, Y. J., Ariff, M. S. B. M., Zakuan, N., & Tajudin, M. N. M. (2014). Brand personality, brand loyalty and brand quality rating in the contact lens perspective. *Review of Integrative Business and Economics*, *3*(2), 433–448. doi:10.22124/jsmd.2019.3880

Liu, M. T., Wong, I. A., Tseng, T., Chang, A. W., & Phau, I. (2017). Applying consumer-based brand equity in luxury hotel branding. *Journal of Business Research*, *81*, 192–202. doi:10.1016/j.jbusres.2017.06.014

Macey, W. H., & Schneider, B. (2008). The meaning of employee engagement. *Industrial and Organisational Psychology*, *1*, 3–30. doi:10.1111/j.1754-9434.2007.0002.x

Merz, M. A., Zarantonello, L., & Grappi, S. (2018). How valuable are your customers in the brand value co-creation process? The development of a customer co-creation value (CCCV) scale. *Journal of Business Research*, *82*(January), 79–89. doi: 10.1016/j.jbusres.2017.08.018

Meyer, D., Hornik, K., & Feinerer, I. (2008). Text mining infrastructure in R. *Journal of Statistical Software*, *25*(5), 1–54. doi:10.18637/jss.v025.i05

Munezero, M. D., Montero, C. S., Sutinen, E., & Pajunen, J. (2014). Are they different? Affect, feeling, emotion, sentiment, and opinion detection in text. *IEEE Transactions on Affective Computing*, *5*(2), 101–111. doi:10.1109/TAFFC.2014.2317187

Nobre, H. (2015). *The new-luxury brand relationship scale: Measuring relationships with mass-consumed luxury brands* (working paper). Aveiro: University of Aveiro. doi:10.13106/jafeb.2018.vol5.no3.51

Nobre, H., & Simoes, C. (2019). NewLux brand relationship scale: Capturing the scope of mass-consumed luxury brand relationships. *Journal of Business Research*, *102*, 328–338. doi:10.1016/j.jbusres.2019.01.047

Novotna, M., & Kunc, J. (2019). Luxury tourists and their preferences: Perspectives in the Czech Republic. *Tourism*, *67*(1), 90–96.

Nunnally, J. C. (1978). *Psychometric theory*. New York, NY: McGraw-Hill.

Okonkwo, U. (2007). *Luxury fashion branding: Trends, tactics, techniques*. Basingstoke: Palgrave Macmillan.

Patterson, P., Yu, T., & de Ruyter, K. (2006, December 4–6). Understanding customer engagement in services. Paper presented at the ANZMAC 2006: Advancing theory, maintaining relevance, Brisbane, QLD.

Patton, M. Q. (2002). Two decades of developments in qualitative inquiry: A personal experiential perspective. *Qualitative Social Work*, *1*(3), 263–283. doi:10.1177/1473325002001003636

Payne, A., Storbacka, K., Frow, P., & Knox, S. (2009). Co-creating brands: Diagnosing and designing the brand experience relationship. *Journal of Business Research*, *69*(3), 379–789. doi:10.1016/j.jbusres.2008.05.013

Podsakoff, P. M., MacKenzie, S. B., Lee, J.-Y., & Podsakoff, N. P. (2003). Common method biases in behavioral research: A critical review of the literature and recommended remedies. *The Journal of Applied Psychology*, *88*, 879–903. doi:10.1037/0021-9010.88.5.879

Prahalad, C. K., & Ramaswamy, V. (2004). Co-creation experiences: The next practice in value creation. *Journal of Interactive Marketing*, *18*(3), 4–9. doi:10.1002/dir.20015

Roper, S., Caruana, R., Medway, D., & Murphy, P. (2013). Constructing luxury brands: Exploring the role of consumer discourse. *European Journal of Marketing*, *47*(3/4), 375–400. doi:10.1108/03090561311297382

Russell, J. A. (1980). A circumplex model of affect. *Journal of Personality and Social Psychology*, *39*(6), 1161–1178. doi:10.1037/h0077714

Schroeder, J. E. (2005). The artist and the brand. *European Journal of Marketing*, *39*, 1291–1305. doi:10.1108/03090560510623262

Schroeder, J. E. (2009). The cultural codes of branding. *Marketing Theory*, *9*(1), 123–126. doi:10.1177/1470593108100067

Seo, Y., & Buchanan-Oliver, M. (2015). Luxury branding: The industry, trends, and future conceptualisations. *Asia Pacific Journal of Marketing and Logistics*, *27*(1), 82–98. doi:10.1108/APJML-10-2014-0148

Shin, H., Eastman, J. K., & Mothersbaugh, D. (2017). The effect of a limited-edition offer following brand dilution on consumer attitudes toward a luxury brand. *Journal of Retail and Consumer Services*, *38*, 59–70. doi:10.1016/j.jretconser.2017.05.009

Sirgy, M. J. (1982). Self-concept in consumer behavior: A critical review. *Journal of Consumer Research*, *9*(3), 287–300. doi:10.1086/208924

So, K. K. F., King, C., & Sparks, B. (2014). Customer engagement with tourism brands: Scale development and validation. *Journal of Hospitality & Tourism Research*, *38*(3), 304–329. doi:10.1177/1096348012451456

Sreejesh, S. P., Sarkar, A., & Roy, S. (2016). Validating a scale to measure consumer's luxury brand aspiration. *Journal of Product and Brand Management*, *25*(5), 48–65. doi:10.1108/JPBM-06-2014-0647

Sumbly, Y., & Siraj, S. (2019). A conceptual framework of brand experience for luxury brands. *Journal of Management Research and Analysis*, *6*(1), 60–64. doi:10.18231/j.jmra.2019.011

Sung, Y., Choi, S. M., Ahn, H., & Song, Y. A. (2015). Dimensions of luxury brand personality: Scale development and validation. *Psychology and Marketing*, *32*(1), 121–132. doi:10.1002/mar.20767

Tian, X., He, W., Tang, C., Li, L., Xu, H., & Selover, D. (2019). A new approach of social media analytics to predict service quality: Evidence from the airline industry. *Journal of Enterprise Information Management*, *33*(1), 51–70. doi:10.1108/JEIM-03-2019-0086

Torelli, C., Keh, H. T., & Chiu, C. Y. (2010). Cultural symbolism of brands. In B. Loken & R. Ahluwalia (Eds.), *Brands and brand management: Contemporary research perspectives* (pp. 113–132). New York, NY: Routledge.

Townsend, C., & Sood, S. (2012). Self-affirmation through the choice of highly aesthetic products. *Journal of Consumer Research*, *39*(2), 415–428. doi:10.1086/663775

Tutek, E., Roth, C., & Simatic, A. (2018). *Rethinking luxury in hospitality. Industry report*. Retrieved from https://www.howarthhtl.com.on.16/07/2020

Tynan, C., McKechnie, S., & Chhuon, C. (2010). Co-creating value for luxury brands. *Journal of Business Research*, *63*(11), 1156–1163. doi:10.1016/j.jbusres.2009.10.012

Vargo, S. L., & Lusch, R. F. (2004). Evolving to a new dominant logic for marketing. *Journal of Marketing*, *68*(1), 1–17. doi:10.1509/jmkg.68.1.1.24036

Vickers, J. S., & Renand, F. (2003). The marketing of luxury goods: An exploratory study–Three conceptual dimensions. *The Marketing Review*, *3*, 459–478. doi: 10.1362/146934703771910071

Vigneron, F., & Johnson, L. W. (2004). Measuring perceptions of brand luxury. *Brand management*, *6*(2), 484–506. doi:10.1057/PALGRAVE.BM.2540194

Wilcox, K., Kim, H. M., & Sen, S. (2009). Why do consumers buy counterfeit luxury brands? *Journal of Marketing Research*, *46*(2), 247–259. doi:10.1509/jmkr.46.2.247

Yang, F. X., & Lau, M. (2016). LuXurY hotel loyalty—A comparison of Chinese gen X and Y tourists to Macau. *International Journal of Contemporary Hospitality Management*, *27*(2), 1685–1706. doi:10.1108/IJMCH-06-2014-0275

Yang, W., & Mattila, A. S. (2016). Why do we buy luxury experiences? *International Journal of Contemporary Hospitality Management*, *28*(9), 1848–1867. doi:10.1108/IJCHM-11-2014-0579

Part 3
Technology and Other Contemporary Facets in Luxury

Chapter 13

Technology and Luxury in Tourism and Hospitality

Sandra Maria Correia Loureiro[1]

Abstract

This chapter intends to give insights into the use of technologies such as virtual reality (VR), augmented reality (AR) and artificial intelligence (AI) in tourism and hospitality. Thus, the objectives are to: (1) give an overview about the meaning of luxury the use of technologies in tourism and hospitality, (2) provide examples about the use of such technologies (e.g., VR, AR, AI in practice) and (3) point out suggestions about further research in the interface of luxury and technologies in tourism and hospitality. The chapter is of interest to both academics and practitioners since it presents luxury, VR, AR and AI concepts, gives examples of technologies applications and points out critical areas for future research.

Keywords: Virtual reality; augmented reality; artificial intelligence; luxury tourism; luxury hospitality; luxury consumption

Introduction

Consumption of luxury goods/services in different shapes has been a prominent feature of societies since the beginning of civilisation (Kapferer & Bastien, 2012). There is, however, no clear consensus on the definition of luxury because of its ambivalent nature (Wiedmann, Hennigs, & Siebels, 2009). This is based on the fact that luxury implies different things to different groups of people across time and cultures (Wiedmann et al., 2009). Recently, Ko, Costello, and Taylor (2019) attempt to consolidate previous definitions claiming that luxury is associated with high quality and offers authentic value, an image of prestige, premium price and inspiration connection with the luxury product/brand.

[1]https://orcid.org/0000-0001-8362-4430

The Emerald Handbook of Luxury Management for Hospitality and Tourism, 273–284

doi:10.1108/978-1-83982-900-020211013

Tourism (e.g., destinations, attractions, museums) and hospitality (e.g., hotel, lodging, restaurants, cruise) may also have at least two or three of these five aspects. New technologies, such as virtual reality (VR), augmented reality (AR) and artificial intelligence (AI), have been more often used in both tourism and hospitality to improve the service provided and enhance the experience (Cheong, 1995). These technologies may be designed as engagement-facilitating technologies because they have the capacity to affect humans at leisure and work. These technologies may contribute positively to perform the tasks more easily at work and to expand and extend the tourist experiences or even consumption experiences (Kumar, Dixit, Javalgi, & Das, 2016; Ng & Wakenshaw, 2017; Verhoef et al., 2017).

In this vein, this chapter pretends to go further in understanding the use of technologies in luxury tourism and hospitality contexts. Therefore, the objectives of the current chapter are to: (1) give an overview about the meaning of luxury the use of technologies in tourism and hospitality, (2) provide examples about the use of such technologies (e.g., VR, AR, AI in practice) and (3) point out suggestions about further research in the interface of luxury and technologies in tourism and hospitality.

Background

Luxury Consumption

Luxury consumption has been a prominent society characteristic since the beginning of civilisation (Dixit, 2017; Ghosh & Varshney, 2013; Turunen, 2018). Yet, a clear consensus about the definition of luxury has been hard to achieve (Wiedmann et al., 2009). Luxury may imply different things to different groups of people across time and cultures (Ghosh & Varshney, 2013; Turunen, 2018; Vigneron & Johnson, 2004; Wiedmann et al., 2009). An underlying term in generally understanding luxury consumption is *conspicuous consumption* invented by Veblen (2012), that is, the utility of certain good/service by dint of which the social status is exposed.

Luxury has been approached by a variety of contexts (Turunen, 2018), for instance, in economics (e.g., Leibenstein, 1950), sociology (e.g., Bourdie, 1994) and marketing (e.g., Dubois, Laurent, & Czellar, 2001; Kapferer & Bastien, 2009, 2012; Vigneron & Johnson, 2004; Wiedmann, Hennigs, & Siebels, 2007, 2009) or tourism and hospitality (e.g., Loureiro, et al., 2019a; Loureiro, et al., 2019b; Zeferino, Loureiro, & Rita, 2014).

Within the scope of marketing, tourism and hospitality, conceptualisations of luxury are commonly regarded in the light of three perspectives: product/brand management perspective, consumption perspective and consumer perspective (Fionda & Moore, 2009; Ghosh & Varshney, 2013; Ko et al., 2019; Turunen, 2018). The first advocates that luxury brands are typically defined in terms of premium brand characteristics (Fionda & Moore, 2009; Ghosh & Varshney, 2013; Turunen, 2018; Vigneron & Johnson, 2004). Thus, products (goods, services, experiences) are addressed as ‘physical manifestations’ of luxury, and unique product attributes are addressed as the root of perceptions of a brands' luxuriousness (Turunen, 2018, p. 7). Thus, high price and premium quality are commonly identified within conceptual frameworks of luxury (Fionda & Moore, 2009; Turunen, 2018). However, luxury is not simply a composition of superior product attributes (Heine & Phan, 2011; Ko et al., 2019; Turunen, 2018). As Turunen (2018, p. 7) suggests: 'Exclusive characteristics may help to differentiate one brand from another, but the attributes alone do not constitute the experience of luxury (…)'. This conducts us to the consumption perspective, stressing that not only physical characteristics and attributes are important, but symbolic characteristics are also requested aspects of brands (Turunen, 2018). However, the experience of the symbolic side of luxury good/service cannot be achieved unless it is socially recognised and appreciated (Kastanakis & Balabanis, 2011; Leibenstein, 1950; Wiedmann et al., 2007). At the core of social orientation is the theory of conspicuous consumption, which implies that 'expensive possessions are solely for external reasons' (Daswani & Jain, 2011, p. 133). Conventionally, luxury goods have been positioned according to symbolic aspects, focussing on ostentatious signalling of one's wealth and social strata (Turunen, 2018). However, it has been argued that the luxury market has progressed from a traditional conspicuous consumption model to an experiential, consumer self-orientation model, which becomes even more complex to pursue, especially on an international level (Turunen, 2018; Wiedmann et al., 2007, 2009).

In this vein, literature on luxury consumption has recently emphasised not exclusively on the social orientation of luxury symbolism but also on personal orientation (Tsai, 2005). For example, Wiedmann et al. (2007, 2009) refer to the social function as 'social value' as opposed to 'individual value', Daswani and Jain (2011) call 'outer/public self' as opposed to 'private/inner self', Kastanakis and Balabanis (2011) claim as 'inter-dependent self-concept' as opposed to 'independent self-concept' and Turunen (2018) entitles as 'symbolic to others' as opposed to 'symbolic to self'. Effectively, the personal orientation of luxury symbolism is at the very heart of contemporary consumer society (Tsai, 2005; Daswani & Jain, 2011; Turunen, 2018). In sum, social orientation and self-orientation of luxury goods/services should work in tandem to achieve desirable results.

The last perspective – called the consumer perspective – somehow combines the two previous. First, luxury brands have traditionally been communicated in terms of its symbolic meanings, where a great deal of attention is devoted to the conspicuousness of luxury goods (Chevalier & Mazzalovo, 2012). Second, the expansion of the luxury market has resulted in a shift in marketing management of luxury brands from a conventional conspicuous consumption model and to a new consumer's self-oriented model, thus changing the way consumers interpret luxury (Wiedmann et al., 2007). Hence, from a consumer standpoint, luxury is an endeavour to identify the motivations behind their consumption both within and across cultures (Vigneron & Johnson, 2004). This is on the grounds that the concept of luxury is to some extent based upon consumers' perceptions that subsequently assist in creating the image of luxury (Vigneron & Johnson, 2004). The consumer perspective entails the combination of aforementioned perspectives, thus offering a holistic picture of luxury consumption and is sought to be explained through perceived luxury value (Wiedmann et al., 2007).

As a summary, Ko et al. (2019) conduct a literature review attempting to analyze and consolidate the previous definition in five main aspects, namely luxury good/services should: (1) be high quality, (2) offer authentic value via desired benefits, whether functional or emotional, (3) have a prestigious image within the market built on qualities such as artisanship, craftsmanship or service quality, (4) be worthy of commanding a premium price and (5) be capable of inspiring a deep connection, or resonance, with the consumer/tourist/guest.

Emerging Technologies of VR, AR and AI

VR, AR and AI are technologies and algorithms that have deserved the attention of researchers and practitioners in recent times (Guttentag, 2010; Williams & Hobson, 1995). A VR environment means a completely synthetic and immersive world that could be a completely unknown environment or a real-life mimicry (e.g., Huang, Backman, Backman, & Chang, 2016; Huang, Backman, Backman, & Moore, 2013). AR is a collaborative experience of a real-world environment where tourists can see the real-world 'augmented' by computer-generated perceptual information (Loureiro et al., 2019a; Milgram, Takemura, Utsumi, & Kishino, 1994). Thus, in AR, the experience combines the real world with virtual objects overlaid, usually by wearing see-through displays (e.g., Bimber & Raskar, 2005; Buhalis & Yovcheva, 2013; Milgram et al., 1994).

VR and AR have had an important influence on the tourism industry (Guttentag, 2010; Tussyadiah, Jung, & tom Dieck, 2017). For instance, tourists can learn more about iconic landmarks without a tourist guide, or a museum or an art gallery through the smartphone applications. Other tourist contexts have also employed these technologies, such as hotels, restaurants, museums and attractions (e.g., Han, Hwang, & Woods, 2014; Jung, Chung, & Leue, 2015; Tussyadiah et al., 2017; Tussyadiah, Wang, Jung, & tom Dieck, 2018).

AI has its origins in ancient cultures of Greek (e.g., the mythological robot Talos), Chinese (e.g., Yueying Huang's dogs) and other mythologies (Nahodil & Vitku, 2013), where automatons were believed to be imbued with real minds, capable of wisdom and emotion. However, the term emerged in a workshop at Dartmouth College (United States) in 1956 (Nilsson, 2010), which is dubbed the birth of AI.

Since then, research on AI has stemmed from different fields of knowledge. Social scientists have been discussing ethical and legal implications of AI (Cath, 2018), computer scientists have developed advanced deep learning algorithms (LeCun, Bengio, & Hinton, 2015), while researchers in business management have studied the impacts of AI on customers, firms and stakeholders in an increasingly automated and interrelated business world (Huang & Rust, 2018). Yet, such advances in AI research have mainly been done in isolated silos with few interdisciplinary exchanges. Similarly, a unique and consensual definition of AI has been hard to get. Recently, Russell and Norvig (2016) summarise the various definitions of AI systems into four categories along two dimensions: reasoning–behaviour dimension and human performance–rationality dimension. These are:

(1) systems that think like humans, (2) systems that act like humans, (3) systems that think rationally and (4) systems that act rationally. AI systems should have the following capabilities: *natural language processing* to communicate in a natural language, *knowledge representation* to store information, *automated reasoning* be used as the stored information to answer questions and to draw new conclusions and *machine learning* to adapt to new circumstances and to detect and extrapolate patterns (e.g., Huang & Rust, 2018; Russell & Norvig, 2016). Yet, this lack of a solid definition does not prevent the spread of research of applications of these systems in the world.

Four types of intelligence may be considered when talking about AI: mechanical, analytical, intuitive and empathetic (Huang & Rust, 2018) claim for four types of AI. Mechanical AI is the ability to perform routines and tasks automatically (e.g., Kim, 2007). Analytical intelligence refers to the ability to process information for problem-solving and learn from that process (Sternberg, 2005). Nevertheless, these machines do not have a mind, conscious states or subjective awareness (Azarian, 2016).

Intuitive intelligence means the ability to think creatively and be able to adjust to novel situations (Sternberg, 2005). AI with intuitive intelligence may perform function demanding for more flexibility than the previous types of intelligence. Intuitive AI aggregates self-awareness, sentience and consciousness (Azarian, 2016). For instance, Google's DeepMind AlphaGo can simulate instinct (BBC News, 2016), the poker player Libratus can do strategic thinking with incomplete information (The Wall Street Journal, 2017) and Watson from IBM can understand, learn and interact (IBM, 2017) into the business-to-business context. Empathetic intelligence refers to the ability to understand other humans' emotions and respond emotionally (Goleman, 1996; Johnson, 2014). Empathetic skill considers communication, leadership, advocating, teamwork, charisma or the ability to negotiate (Caprino, 2012). Jobs related to these abilities are negotiators, politicians or psychiatrists. Yet, a lengthy ongoing discussion has been done about if AI systems can feel in the same way as humans do (Minsky, 2006).

Examples of Applications of Technologies in Luxury Tourism and Hospitality

VR enables real-time sharing of adventure and luxury among friends or family who are spread around the world. It will mean that guests and their loved ones can enjoy simulated adventures in real time, whether it be a walk through the Serengeti or an ocean swim with extinct marine life. Simon Scoot, VP of global brand strategy at InterContinental Hotels and Resorts, said the group's collaboration with Popcorn and partnership with Bernerd was to ensure the brand remains at the forefront of luxury travel (Scoot, 2020; VIT, 2020).

Luxury hotels are using VR as a new form to advertise and communicate the architecture, facilities and service. For instance, Shangri-La Hotels and Resorts' hotels and destinations allow an immersive experience, where potential guests may have an initial experience before booking (Shangri-la, 2020). Luxury attractions, such as important museums, are following the same process of advertising the facilities and exhibition through VR. This is, for instance, the case of 'Visit famous museums around the world from home' in Italia Living, what has been also particularly useful in a time of confinement due to the coronavirus (Italia Living, 2020). Likewise, destinations are using VR, sometimes connecting hotels and their surroundings for aspiration (Luxury Travel Advisor, 2020).

Regarding AR, this technology can introduce guests to the hotel's facilities by guiding them about the directions and amenities with images overlapped with the real world. Indeed, guest can point their smart device and get the information they want in their native language, call for laundry service or order food at the restaurant and even show directions and routes to guests. This is the case of one of the first hotels using AR called Holiday Inn. In the Holiday Inn London Kensington Forum, guests were offered the experience of seeing some of Britain's 2012 Olympic and Paralympic athletes in action in the hotel's reception, bedrooms and halls through AR (Stylus, 2020).

AI systems are promising to transform the tourism and hospitality industry, changing processes, services and facilities through AI-powered robotics. Having an intelligent hotel that surpasses guest expectations is the strategy for hotels. There are certain elements that make hotels intelligent, such as concierge robots, digital assistance, voice-activated services, travel experience enhancers, automatic data processing (machine learning). All this transformative process leads to the use of AI robots not only to lessen the human participation with its voice-activated assistants but also to smarten the way hotel guest service is delivered to guest's satisfaction. From guest's personal choices to their smallest of needs, AI-enabled hotel concierge bots can offer minute service and careful assistance. For instance, Apple's Siri starts providing voice-activated assistance to its mobile users, Amazon Echo and Alexa have also joined the race of forming a richer, more delightful guest experience using the power of machine learning of AI software.

The luxury hospitality industry can capture a large amount of data though AI software, guest purchases, travel choices, journey patterns and itinerary, location preferences, hotel rating enquiries and payment methods. These data can be used to provide memorable and meaningful experiences to luxury hotel guests as they travel, enquire, stay and enjoy the luxurious hotel amenities. Therefore, in-depth guest service with properly harnessed guest insight may increase the perception of value for guests, adding to the wonderful luxury atmosphere of luxury hotels (e.g., architecture, design, layout).

For instance, the Hilton Worldwide Hotel is using a new AI-based concierge Connie who is transforming the hotel guests' experience. Connie is an AI-enabled robot with two feet of height performing just like a human concierge. Connie can assist arriving hotel guests at the front desk standing on its bipedal support having the attributes of being relies on IBM's Watson AI for computing intelligence, uses travel database WayBlazer, gives real-time recommendations for visit-worthy sites and attractions, answers guest queries on the spot, learns from frequent guest interactions and provides care and support (Hilton, 2020) as a robotic concierge.

Another example is the 5-star Cosmopolitan of Las Vegas. This hotel uses the AI concierge Rose. In the Rose calling card, which guests receive on check-in, guests may read: 'Know my secrets. Text me', and 'I am the answer to the question that you never asked' (Cosmopolitan, 2020). Radisson Blu–owned Edwardian Hotel also has an AI assistant. Guests staying at Radisson Blu can request the assistance of AI called Edward by text message. Edward is developed to deliver experiences for guests who prefer digital brand interaction. Thus, Edward can deliver information on local bars and restaurants, as well as deal with complaints (Radisson, 2020).

Hotel guests can use the hotel's special mobile device to intermingle with AI-driven chatbots to perform several tasks: order meals or drinks, suggest dishes, unlock doors, control room temperature and lighting, schedule dinners/lunch, handling evening reservations, taxi booking and itinerary planning, recommendation of interesting sightseeing and visits, GPS capabilities, gamification and amusement, and answer critical queries. An example of a luxury hotel that provides digital travel companions offering intelligent voice-activated apps is the Signature Lux Hotels in South Africa (Signature, 2020).

Future Directions and Suggestions

With the shift in the logic of living the luxury tourism experience by an extensive incorporation of technologies such as VR, AR and AI systems, several research questions emerge. First, researchers and managers should be focused on understanding how technology creates value within luxury tourism. So, what are the benefits and risks for tourists and guests in using VR, AR and AI technologies? How can managers and other stakeholders join to communicate the use of such technologies in a way that can create value for luxury guests and tourists? How to standardise and personalise goods/services that emerge from these technologies?

Second, the transformation operated by the technologies will demand for new models and frameworks that could explain the approach/avoidance and acceptance. Thus, what other models and frameworks will emerge? And what about to understand the tourist/guest consumption in an area of high technology?

Third, the amount of data that it is possible to collect and analyze may expose problems of data access and protection. Therefore, laws about data protection should be investigated and created. Understand how long luxury tourists do not mind giving personal information to develop the research on this field. Researchers also need to investigate how different cultures, tourist personality traits and religions see the use and evolution of engagement-facilitating technologies.

Finally, the evolution of AI algorithms will bring the possibility to interact between different equipment in a network that could integrate, for example, luxury hotels, luxury restaurants, tours and tourist attractions. AI agents at hotels and other places are taking repetitive tasks and maximise efficiency. Yet, AI systems may demonstrate the possibility to be creative due to their own learning process (Christies, 2018; Guardian, 2018). This may enhance the interaction between them and hotel guests and suggest innovative concepts for those who manage luxury hotels or other tourist attractions. In this vein, (1) how to segment luxury tourists to understand what type of technology they prefer and the level of interaction between them and AI systems they desire?, (2) how the luxury service can be fully automated using different levels of AI intelligence?, (3) in a stage that AI system evolve to a more emotional and creative way, may tourists and guest will prefer them as employees, instead of human beings?, (4) what theories will emerge to explain the relationship between tourists and AI systems, and (5) how to prepare new laws of rights and duties of AI machines?

Conclusions

Technologies such as VR, AR, AI are transforming the way luxury tourism is selected, provided and evaluated by tourists/guests. These technologies contribute to delivering hyper-connected luxury tourism experiences. The main challenge for a successful VR adoption is to know how to create the right content to engage luxury tourists, usually employed during the process of promoting the hotel, destination or other luxury attractions or after the experience to be used as a way to enhance the memories created through the experience in the real world, real place. AR may have an important role in incrementing the luxury experience by displaying suggestions to enhance the experience, where to go, whom to ask for some information, among other utilities. AI has numerous applications that aim to improve tourist experience on several levels, and many of them are not yet widespread. If businesses do not prepare now for the near future, they may not be able to compete over the next decades. Managers should explicitly attempt to conduct short- to medium-term plans to start thinking in this transition, and given the broad scope of AI, it is fundamental to start by prioritising problems to solve (for instance, engage tourists, the creation of innovative services or optimising operations).

The AI and related technologies are still in the early stage of development. These new types of projects and strategies clearly need further inputs, either from academics that are pursuing to enlarge new boundaries to knowledge, or practitioners, such as those who manage luxury destinations, hotels or attractions, who want to understand how organisations can benefit from this admirable new world. First, AI systems will inevitably lead to a new set of organisations and suppliers that will try to offer goods/services in this specific environment with the highest quality as possible. Second, the type of AI – mechanical, analytical, intuitive and empathetic (Huang & Rust, 2018) – will be developed in distinct phases, as technology developments will respond to these different forms of AI distinctively. For example, the path to an empathetic AI seems to be more distant to achieve than the other forms. Moreover, these types of AIs are not equally accepted, effective and efficient by all tourists (for instance, some tourists will free afraid or doubt in the presence of creative AI systems). Thus, the introduction of AI in distinct social environments should include different forms to access its success. This assessment should include essential elements, such as appropriateness to the target (we may consider segmentation among luxury tourists, depending on age and personal profile) or the use of the most recent technology advancements.

As technology evolves, the degree of reliability will undoubtedly grow, leading to an adaptation either for tourists and organizations' requirements, which will fulfil all stakeholders' needs and approval. Therefore, using AI systems demands a large sum of knowledge, which can be achieved through already existing environments, or by emerging developed forms of technology. It is time for all stakeholders, from organisation to employees or tourists, to start focussing on improving the experience within the available technology, introducing real-life experiences using AI systems, which may lead to changes in the experience and acceptance of AI.

Living in a pandemic time where behaviours tend to change and also considering the sustainability of the planet, it is expected that in the near future, tourists will tend to physically travel less (e.g., due to new habits, job loss, hygienic safety conditions, rising prices). Thus, a new business model may emerge using VR, AR and even AI for those who do not want or have the possibility to travel physically to the destinations, and also these technologies may contribute to add value to tourist good/services by reducing costs and creating a more personalised experience at the destination or at the home of the 'new' tourists who do not travel physically.

References

Azarian, B. (2016). A neuroscientist explains why artificially intelligent robots will never have consciousness like humans. Raw Story, March 31. Retrieved from http://www.rawstory.com/2016/03/aneuroscientistexplainswhyartificiallyintelligent robotswillneverhaveconsciousnesslikehumans/. Accessed on January 19, 2018.

BBC News. (2016). Artificial intelligence: Google's AlphaGo beats go master lee Se-Dol. Retrieved from http://www.bbc.com/news/technology35785875. Accessed on December 4, 2018.

Bimber, O., & Raskar, R. (2005). *Spatial augmented reality merging real and virtual worlds*. Cambridge, MA: A K Peters.

Bourdieu, P. (1994). *Distinction: A social critique of the judgement of taste*. London: Routledge.

Buhalis, D., and Yovcheva, Z. (2013). Augmented reality in tourism. 10 unique applications explained. *Digital Tourism Think Thank – thinkdigital.travel*. Retrieved from https://thinkdigital.travel/wp-content/uploads/2013/04/10-AR-Best-Practices-in-Tourism.pdf. Accessed on September 3, 2019.

Caprino, K. (2012). What you don't know will hurt you: The top 8 skils professionals need to master. *Forbes*, April 27. Retrieved from http://www.forbes.com/sites/kathycaprino/2012/04/27/whatyoudontknowwillhurtyouthetop8skillsprofessionals needtomaster/print/. Accessed on January 24, 2018.

Cath, C. (2018). Governing artificial intelligence: Ethical, legal and technical opportunities and challenges. *Philosophical Transactions A: Mathematical Physical and Engineering Sciences*, *376*(2133), 20180080. doi:10.1098/rsta.2018.0080

Cheong, R. (1995). The virtual threat to travel and tourism. *Tourism Management*, *16*(6), 417–422. doi:10.1016/0261-5177(95)00049-T

Chevalier, M., & Mazzalovo, G. (2012). *Luxury brand management*. Singapore: John Wiley and Sons Singapore Pte. Ltd.

Christies. (2018). Retrieved from https://www.christies.com/features/A-collaboration-between-two-artists-one-human-one-a-machine-9332-1.aspx. Accessed on January 18, 2020.

Cosmopolitan. (2020). Retrieved from https://www.guestreservations.com/the-cosmopolitan-of-las-vegas/booking?gclid=EAIaIQobChMIu6WS5r2u6QIVgrLVCh0LUAh7EAAYASAAEgLibfD_BwE

Daswani, A., & Jain, V. (2011). Understanding luxury branding: A conceptual framework. *Media Asia*, *38*(3), 131–139. doi:10.1080/01296612.2011.11726892

Dixit, S. K. (2017). *The Routledge handbook of consumer behaviour in hospitality and tourism*. London: Routledge.

Dubois, B., Laurent, G., & Czellar, S. (2001). *Consumer rapport to luxury: Analyzing complex and ambivalent attitudes* (No. 736). Jouy-en-Josas: HEC Paris.

Fionda-Douglas, A., & Moore, C. M. (2009). The anatomy of the luxury fashion brand. *Journal of Brand Management*, *16*(5), 347–363. doi: 10.1057/bm.2008.45

Ghosh, A., & Varshney, S. (2013). Luxury goods consumption: A conceptual framework based on literature review. *South Asian Journal of Management*, *20*(2), 146–159.

Goleman, D. (1996). *Emotional intelligence: Why it can matter more than IQ*. London: Bloomsbury Publishing.

Guardian. (2018). Retrieved from https://www.theguardian.com/music/2018/oct/22/ai-artificial-intelligence-composing. Accessed on January 19, 2020.

Guttentag, D. A. (2010). Virtual reality: Applications and implications for tourism. *Tourism Management*, *31*(5), 637–651. doi:10.1016/j.tourman.2009.07.003

Han, H., Hwang, J., & Woods, D. P. (2014). Choosing virtual–rather than real–leisure activities: An examination of the decision–making process in screen-golf participants. *Asia Pacific Journal of Tourism Research*, *19*(4), 428–450. doi:10.1080/10941665.2013.764333

Heine, K., & Phan, M. (2011). Trading-up mass-market goods to luxury products. *Australasian Marketing Journal*, *19*(2), 108–114.

Hilton. (2020). Retrieved from https://newsroom.hilton.com/corporate/news/hilton-and-ibm-pilot-connie-the-worlds-first-watsonenabled-hotel-concierge. Accessed on January 20, 2020.

Huang, Y. C., Backman, K. F., Backman, S. J., & Chang, L. L. (2016). Exploring the implications of virtual reality technology in tourism marketing: An integrated research framework. *International Journal of Tourism Research*, *18*(2), 116–128. doi:10.1002/jtr.2038

Huang, Y. C., Backman, S. J., Backman, K. F., & Moore, D. (2013). Exploring user acceptance of 3D virtual worlds in travel and tourism marketing. *Tourism Management*, *36*, 490–501. doi:10.1016/j.tourman.2012.09.009

Huang, M.-H., & Rust, R. T. (2018). Artificial intelligence in service. *Journal of Service Research*, *21*(2), 155–172.

IBM. (2017). *IBM's your learning and watson—together a game changer in learning*. Retrieved from https://www.ibm.com/downloads/cas/G8DVQ9MB. Accessed on August 14, 2021.

Italia Living. (2020). Retrieved from https://italialiving.com/articles/travel/visit-famous-museums-around-the-world-from-home/. Accessed on January 19, 2020.

Johnson, H. (2014). 6 soft skills every professional needs. OnlineDegrees.com. Retrieved from https://futurism.com/majorfirmannouncesitsreplacingitsemployeeswithai/. Accessed on January 24, 2018.

Jung, T., Chung, N., & Leue, M. C. (2015). The determinants of recommendations to use augmented reality technologies: The case of a Korean theme park. *Tourism Management*, *49*, 75–86.

Kapferer, J. N., & Bastien, V. (2009). The specificity of luxury management: Turning marketing upside down. *Journal of Brand Management*, *16*(5–6), 311–322.

Kapferer, J. N., & Bastien, V. (2012). *The luxury strategy: Break the rules of marketing to build luxury brands*. London: Kogan Page publishers.

Kastanakis, M., & Balabanis, G. (2011). Signalling effects in luxury consumption. In A. Bradshaw, C. Hackley, & P. Maclaran (Eds.), *E – European advances in consumer research* (Vol. 9, pp. 537–539). Duluth, MN: Association for Consumer Research.

Kim, M. (2007). Challenges on the development of robotic intelligence. In 16th IEEE international conference on robot and human interactive communication, Jeju, South Korea.

Ko, E., Costello, J. P., & Taylor, C. R. (2019). What is a luxury brand? A new definition and review of the literature. *Journal of Business Research*, *99*, 405–413. doi:10.1016/j.jbusres.2017.08.023

Kumar, V., Dixit, A., Javalgi, A., & Das, M. (2016). Research framework, strategies, and applications of intelligent agent technologies (IATs) in marketing. *Journal of the Academy of Marketing Science*, *44*(1), 24–45.

LeCun, Y., Bengio, Y., & Hinton, G. (2015). Deep learning. *Nature*, *521*, 436–444.

Leibenstein, H. (1950). Bandwagon, Snob and Veblen effects in the theory of consumers' demand. *Quarterly Journal of Economics*, *64*(2), 183–207.

Loureiro, S. M. C., Guerreiro, J., Eloy, S., Langaro, D., & Panchapakesan, P. (2019a). Understanding the use of virtual reality in marketing: A text mining-based review. *Journal of Business Research*, *100*, 514–530. doi:10.1016/j.jbusres.2018.10.055

Loureiro, S. M. C., Japutra, A., & Kwun, D. (2019b). Signalling effects on symbolic status and travellers' well-being in the luxury cruise industry. *International Journal of Tourism Research*, *21*(5), 639–654. doi:10.1002/jtr.2287

Luxury Travel Advisor. (2020). Retrieved from https://www.luxurytraveladvisor.com/aspirational/virtual-hotel-and-destination-experiences. Accessed on March 8, 2019.

Milgram, P., Takemura, H., Utsumi, A., & Kishino, F. (1994). Augmented reality: A class of displays on the reality-virtuality continuum. *Systems research*. In H. Das (Ed.), *Telemanipulator and telepresence technologies* (Vol. 2351, pp. 282–292). Boston, MA: Photonics for Industrial Applications. doi:10.1117/12.197321

Minsky, M. (2006). *The emotion machine*. New York, NY: Simon & Schuster.

Nahodil, P., & Vitku, J. (2013). How to design an autonomous creature based on original life approaches. In J. Kelemen, J. Romportl, & E. Zackova (Eds.), *Beyond artificial intelligence: Contemplations, expectations, applications* (pp. 161–180). Berlin: Springer.

Ng, I., & Wakenshaw, S. (2017). The internet-of-things: Review and research directions. *International Journal of Research in Marketing*, *34*(1), 3–21.

Nilsson, N. J. (2010). *The quest for artificial intelligence: A history of ideas and achievements*. Cambridge: Cambridge University Press.

Radisson. (2020). Retrieved from https://www.edwardian.com/brands/radisson-bluedwardian-en.html. Accessed on January 19, 2020.

Russell, S. J., & Norvig, P. (2016). *Artificial intelligence: A modern approach* (3rd edn.). Essex: Pearson.

Scoot, S. (2020). Retrieved from https://www.hospitalitynet.org/appointment/79004247.html. Accessed on February 16, 2020.

Shangri-la. (2020). Retrieved from http://www.shangri-la.com/vr/en/index.htm. Accessed on January 21, 2020.

Signature. (2020). Retrieved from https://signatureluxhotels.com/artificial-intelligence-in-the-hotel-industry/. Accessed on March 2, 2020.

Sternberg, R. J. (2005). The theory of successful intelligence. *Interamerican Journal of Psychology*, *39*(2), 189–202.

Stylus. (2020). Retrieved from https://www.stylus.com/qqsqrx. Accessed on March 8, 2019.

The Wall Street Journal. (2017). How artificial intelligence will change everything. Retrieved from https://www.wsj.com/articles/howartificialintelligencewillchangeeverything148885632. Accessed on March 8, 2019.

Tsai, S.-P. (2005). Impact of personal orientation on luxury-brand purchase value: An international investigation. *International Journal of Market Research*, *47*(4), 427–452. doi:10.1177/147078530504700403

Turunen, L. L. M. (2018). *Interpretations of luxury: Exploring the consumer perspective*. Cham: Palgrave.

Tussyadiah, I. P., Jung, T. H., & tom Dieck, M. C. (2017). Embodiment of wearable augmented reality technology in tourism experiences. *Journal of Travel Research*, *57*(5), 597–611.

Tussyadiah, I. P., Wang, D., Jung, T. H., & tom Dieck, M. C. (2018). Virtual reality, presence, and attitude change: Empirical evidence from tourism. *Tourism Management*, *66*, 140–154. doi:10.1016/j.tourman.2017.12.003

Veblen, T. (Ed.). (2012). *The theory of leisure class*. New York, NY: Dover Publications Inc.

Verhoef, P. C., Stephen, A., Kannan, P. K., Luo, X., Abhishe, V., Andrews, M., … Zhang, Y. (2017). Consumer connectivity in a complex, technology-enabled, and mobile-oriented world with smart products. *Journal of Interactive Marketing*, *40*, 1–8.

Vigneron, F., & Johnson, L. W. (2004). Measuring perceptions of brand luxury. *Journal of Brand Management*, *11*(6), 484–506. doi:10.1057/palgrave.bm.2540194

VIT. (2020). Retrieved from https://www.webintravel.com/vr-to-play-key-role-in-shaping-the-future-of-luxury-travel/

Wiedmann, K. P., Hennigs, N., & Siebels, A. (2007). Measuring consumers' luxury value perception: A cross-cultural framework. *Academy of Marketing Science Review*, *7*(7), 1–21.

Wiedmann, K. P., Hennigs, N., & Siebels, A. (2009). Value-based segmentation of luxury consumption behavior. *Psychology and Marketing*, *26*(7), 625–651.

Williams, P., & Hobson, J. S. (1995). Virtual reality and tourism: Fact or fantasy? *Tourism Management*, *16*(6), 423–427. doi:10.1016/0261-5177(95)00050-X

Zeferino, E. C., Loureiro, S. M. C., & Rita, P. (2014). How visitors of luxury cruises perceive a tourist destination: Internal determinants, satisfaction and intention to return. *Journal of Tourism and Development*, *2*(N° 21/22), 49–58.

Chapter 14

Luxury Hospitality and the Adoption of Voice Assistants: The Role of Openness to Change and Status Consumption

Cesare Amatulli,[1] Andrea Sestino,[2] Alessandro M. Peluso,[3] and Gianluigi Guido[4]

Abstract

New technologies represent an important challenge for all sectors across the world, becoming a relevant opportunity for the luxury hospitality industry as well. This study empirically investigates the effects of openness to change and status consumption orientation in influencing luxury hotel guests' perceived usefulness of voice assistants when integrated with their staying. Specifically, we shed light on the potential interplay between these two constructs in shaping guests' perceived usefulness of these devices. Finding suggests that openness to changes and status consumption represents two potentially alternative aspects that managers could leverage to encourage hotel guests' perceived usefulness and vocal assistant devices. This implies that openness to change on the part of luxury hotel guests may be a necessary but not sufficient condition for these devices to be accepted. Rather, luxury hotel managers should also consider the status consumption orientation of their guests.

Keywords: Internet of things; voice assistant; smart object; luxury hotel; openness to change; status consumption; luxury experience

[1]https://orcid.org/0000-0002-6296-2569
[2]https://orcid.org/0000-0003-2648-4093
[3]https://orcid.org/0000-0003-4018-9429
[4]https://orcid.org/0000-0002-3578-4795

The Emerald Handbook of Luxury Management for Hospitality and Tourism, 285–303

doi:10.1108/978-1-83982-900-020211014

Introduction

The luxury market is becoming increasingly complex (Chan, Okumus, & Chan, 2017; Chan, To, & Chu, 2015), coherently with new emerging needs of consumers, who are looking for new luxury stimulus and experiences (Oe, Sunpakit, Yamaoka, & Liang, 2018), and new meanings that luxury consumption is assuming (Seo & Buchanan–Oliver, 2019). Recent data indicate that the luxury market will continuously grow at a pace of 3–5% per year until 2025, reaching a total value of between EUR 320 and 365 billion (Bain and Company, 2018). In this backdrop, there is a need for luxury literature to investigate the recent changes in the sector in order to understand how to internalise the rapidly evolving trends, such as population growth, technological progress and dynamic changes in tourists' behaviour (Dixit, 2017; Kapferer, 2014). In the last few decades, tourism has witnessed constant growth and profound diversification, which have led this industry to be one of the fastest-growing economic sectors. This growth has been qualified by phenomena like luxury and sustainability (Hennings, Wiedmann, Karmann & Behrens, 2013; Kapferer, 2014), experiential luxury (von Wallpach, Hemetsberger, Thomsen, & Belk, 2020), the redesign of services in luxury and Generation Z consumers (i.e., those who born between the 90s and the first decade of 2000s; Jain, Vatsa, & Jagani, 2014).

Technological advances have profoundly changed the way people travel by enabling more interactive and engaging experiences. In the hospitality industry, the introduction of high-quality accessories, sophisticated technologies and growing attention to details in hotels have enhanced luxury experiences by considerably improving the quality of hotel stay (Atwal & Williams, 2017). Considering innovation, sustainable hospitality and technological evolution made of chatbot and artificial intelligence (AI), travellers are looking for luxury and endurable guest experiences. Hotels and other accommodation facilities will no longer be able to build their success solely on brand names but should be able to innovate while remaining closely related to their traditions. This means that they should merge their traditional distinctive features with the most cutting-edge technological innovations.

In the hospitality industry, technology has been one of the strongest trends for years due to the ability to redesign travel experiences, from the booking phase to checkout, bringing benefits to both the guests and the hosts. This has offered relevant opportunities for penetrating and revitalising service design and delivery (Stamboulis & Skayannis, 2003). Indeed, new technologies represent an important challenge for all sectors of the world (Dwivedi et al., 2019; Sheng, Amankwah-Amoah, & Wang, 2019). New technology adoption, especially in the luxury industry, could show different opportunities (i.e., tom Dieck, Jung, Kim, & Moon, 2017). Thus, considering Italy as a relevant research setting for the luxury hospitality industry (Giusti, 2018), we conducted a structured survey study in order to shed light on the factors that can help shape the utility that guests may perceive in the technological devices introduced in luxury hotels. The chapter explored the role of two important consumer-related constructs: *openness to change*, as a personal degree of changes' acceptance and management (Wanberg & Banas, 2000), and *status consumption orientation*, as a construct particularly relevant in the luxury field that refers to the extent to which consumers buy luxury goods and services to express their status (O'Cass & McEwen, 2004). Our study highlighted that *openness to change* and *status consumption orientation* may play a critical role in increasing the perceived usefulness of the voice assistants (VAs) in luxury hotels, so that luxury hotel managers might leverage these two factors to induce guests to perceive these devices as more useful and their stay experiences as more luxurious. In such a way, we contribute to the literature stream on luxury hospitality, as our findings suggest how to revitalise hospitality services, even in light of the contingent challenges posed by the recent COVID-19 emergency and the related social distancing measures (Fisher & Wilder–Smith, 2020). Second, we contribute to the luxury consumption literature, as our findings indicate that status consumption orientation may favour technology acceptance in luxury hotels. From a practical perspective, our work carries managerial implications for luxury hotel managers, who may leverage our results to introduce new technologies in an attempt to increase the perceived quality of their services.

Theoretical Framework

Technology Adoption in Tourism and Hospitality

Technology aroused interest in a large number of sectors, including tourism and hospitality (Petrović, Milićević, & Djeri, 2017). New technologies can support consumers throughout the whole experience, from pre-consumption to post-consumption, by helping them make the more accurate choice and share either positive or negative experiences (Liu, Shin, & Burns, 2019). These technologies can become an integral part of products or services, thus offering augmented experiences (Benckendorff, Xiang, & Sheldon, 2019; Buhalis, 2019; Dixit, 2020). In tourism and hospitality, technological devices like AVs might contribute to breaking down the barriers between the real and virtual world, thus helping managers offer immersive tourist experiences (tom Diek et al., 2017).

Some studies (Benckendorff et al., 2019; Xiang, 2018) have investigated how technology can support the tourism industry by providing tools for understanding human behaviour (Tung & Ritchie, 2011). For instance, exploiting Internet of things (IoT) technologies, the contents of tourism experience and culture have been maximum exploited (Sestino, Prete, Piper, & Guido, 2020), thereby contributing to the diffusion of new concepts, such as intelligent tourism services, intelligent tourism ecosystems (Gretzel, Koo, Sigala, & Xiang, 2015) and 'smart' tourist destinations (Matos, Pinto, Barros, Martins & Au-Young, 2019). Furthermore, the emergence of social media has contributed to the promotion of tourist destinations (Kirǎlová & Pavlíčeka, 2015) and increased the relevance of word of mouth as a primary information source for actual and prospective travellers. In addition to helping travellers in the booking phase, technology has also improved the hotel stay experience.

Over the last few years, the concept of experience has become central to tourism and hospitality, and the term 'experience' has been used in literature to indicate the process of using tourist services from a consumer perspective (Tung & Ritchie, 2011). Tourism research has long acknowledged the importance of tourists' experiences, qualifying them as a situation in which tourists escape from ordinary and everyday life (Larsen, 2007). Therefore, the concept of consumption in tourism is more tied to a desire for an extraordinary experience than a mere need for satisfaction, even during a hotel stay experience (Cinotti, 2015). Luxury hotels have tried to direct their efforts toward the creation of memorable and unique stay experiences in an attempt to generate in their guests' involvement and loyalty (Hosany & Witham, 2010; Prayag, Hosany, & Odeh, 2013; Tung & Ritchie, 2011; Walls, Okumus, Wang, & Kwun, 2011).

From a business perspective, the current technological level offers exciting opportunities for observing human behaviour, beginning from the data generated by the interactions (Erevelles, Fukawa, & Swayne, 2016). Consumers adopt these emerging technologies in different ways, and, in general, with the purpose of simplifying their daily life supported by smartphones, Fitbit, tablets and voice recognition devices (Hsu & Lin, 2016; Sestino, 2019). These new technologies, enabled by the IoT and AI, could increase the ability of luxury companies to communicate and provide personalised services to their consumers (Deloitte, 2019; Liu et al., 2019). The integration of technological components has become a central challenge for luxury hotels (Pham, Nguyen, & Luse, 2018). However, for luxury hotels, compared to mass-market ones, the use of new technologies is still a misty issue; only a few hotel chains (e.g., Marriott; CPP Luxury, 2018) have adopted technological tools, such as VAs devices. This reluctance to adopt new technological devices may lie in the scepticism that traditionally characterises luxury brand managers, who may believe that the incorporation of technological components in their products or services may undermine the human and warm relationships they typically build with customers (Athwal, Istanbulluoglu & McCormack, 2019a; Li, Rong, & Thatcher, 2012; Rhee, 2018).

Voice Assistants and Self-service Technologies

In the wide range of technological devices enabled by IoT as smart object based on the continuous interaction via the Internet (Sestino et al., 2020), VAs, as 'software agents that run on purpose-built speaker devices or smartphones' (Hoy, 2018, p. 82), could help guests enjoy the hotel's services and facilities. VAs typically use AI-based algorithms to offer hotel recommendations as well as to suggest experiences and places to visit. Through VAs, guests can easily interact with software to adjust the room's lighting, the temperature, the humidity, the opening and closing of the curtains and so on, simply using their voice. Around the world, about 74% of consumers use tools based on voice recognition such as VAs (Capgemini, 2019) to search or purchase services and products, to issue orders that typically required manual interaction (i.e., lighting bulbs, opening windows, heating activations) and to carry out hands-free activities, thanks to this maximum integration of systems in home automation. Given their characteristics, VAs quickly acquired significant importance in the information supply and acquisition process and are based on the paradigm of self-service technology. Regarding self-service technologies, intended as technological interfaces that allow customers to benefit from services and carry out activities independently of the involvement of employees' indirect services (Zeithaml, Bitner, & Gremler, 2010), VAs are replacing many vis-à-vis interactions to make service transactions more accurate, convenient and faster, without human involvement.

However, it is necessary to investigate the potential of these VAs and how their use in luxury hotel rooms is perceived by guests. At first glance, one may reason that the concept of technology, when applied to luxury goods, could dilute the charm of a luxury brand (Connolly, 2000). However, luxury marketers need to understand how to embrace new technological devices and incorporate them into their offerings to keep pace with technological progress and be competitive (Athwal, Wells, Carrigan, & Henninger, 2019b). Indeed, VAs are technological interfaces that can enable consumers to benefit services independently of the involvement of direct service employees (Meuter, Ostrom, Bitner, & Roundtree, 2003). VAs are revolutionising the whole experience and, particularly, the service encounter, which traditionally entails a face-to-face interaction between a customer and a supplier (Solomon, Surprenant, Czepiel, & Gutman, 1985). Integrating devices, such as VAs, in hotels could positively influence the hotel stay experience by combining human interaction and technology. With a varying level of intrusiveness, technology can totally or partially replace most traditional services, such as in tourism or shopping. Specifically, according to Collier and Mellier (1998), we could imagine that VAs can evolve both in co-routed services, in which a large way of managing the interaction is accompanied by an equally moderate number of 'paths' through which the customer uses the (hybrid) service, or routed-services provider, in which the delivery process is fully automated and standardised. However, in the luxury field, the human component, in the form of receptionists and other hotel staff members, remains fundamental in service delivery; they do not communicate only with words but also with their bodies (i.e., gestures, smiles, elegant habit, tone of voice) (Bonini, 2015). Consequently, the quality of a stay experience derives not only from the physical environment but also from the quality of social interaction between the hotel personnel and guests (Walls et al., 2011). Indeed, both the physical environment and social interaction have been recognised as important antecedents of guests' satisfaction and loyalty (Gupta & Vajic, 1999). Social interaction, in particular, is a critical factor that can influence guests' behaviours by means of staff care toward guests, technical knowledge, willingness to serve, friendly relationships and providing all the assistance needed to help guests (Cetin & Dincer, 2014). Considering their nature, hotel services require a cultural approach based on quality and a guest-centred approach. The implementation of such an orientation requires the involvement of employees at all levels, from senior management to auxiliary staff, which leads to an improvement of terms of service quality and guest satisfaction (Gronroos, 1994). In this backdrop, in which human contact is still central, the introduction of VAs could have a terrific effect if only hotel managers were able to identify the tourist segments that might most appreciate such technological devices.

Literature has partially explored the role of VAs as a guest's support tool by focussing on the quality of the interaction as a key determinant of VAs' perceived quality and intention to use. Further studies investigated the relative advantage of using service robots compared to their human correspondents (Liu et al., 2019). Moriuchi (2019) conducted an empirical study on the effect of VAs on consumer involvement and loyalty by contributing to the technology acceptance model (TAM; Davis, Bagozzi, & Warshaw, 1989). Differently from the TAM, the model proposed by Moriuchi (2019) posits that an individual's behavioural intention is determined by purchase intentions and perceived usefulness, considering the moderating role of localisation. Specifically, localisation refers to the expectations of online users to receive suggestions consistently with the external environment (Singh, 2011); localisation involves adapting the Web content from which the VAs draws in terms of the local language, accent, jargon, cultural values and other specific preferences of the venue in order to guarantee hyper-personalised suggestions and interactions (Gollnhofer & Schüller, 2018). Despite the growing demand for technologies to be integrated into marketing strategies such as VAs, the context still appears to be little known, especially in terms of interaction with consumers. In this scenario, no studies explored the role of guests' openness to change and status consumption orientation. Moreover, recent studies have highlighted two interesting main factors such as personalisation and non-invasion. In this regards, today's consumers seem less concerned with privacy issues as they are aware that, to obtain certain benefits, they should share some personal data (Gutierrez, O'Leary, Rana, Dwivedi, & Call, 2019).

Innovativeness and Openness to Change

Scholarly research continuously tried to explain the mechanisms underlying consumers' technology adoption. This topic has long been discussed in the literature since the initial version of the TAM (see for a review, Mortenson & Vidgen, 2016). The barriers to consumers' acceptance have been studied in recent years and are generally traced back to 'extrinsic' and 'intrinsic' factors (Ertmer, 2005; Rogers, 2009). Existing research has mainly revolved around models and theories, such as the unified theory of acceptance and use of technology (UTAUT) (Venkatesh, Morris, Davis, & Davis, 2003), Rogers' (2003) innovation diffusion theory and the theory of planned behaviour (TPB) (Ajzen, 1991). The TAM (Davis, Bagozzi, & Warshaw, 1989) focuses on intrinsic factors and highlights that perceived usefulness and perceived ease of use of technological tools could influence the consumers' openness and availability to use new technologies. On the contrary, the lower the control and management of the benefits produced by innovation, the greater the difficulty that the individual will accept and adopt innovation.

Consumers may, therefore, be inclined to give away part of their privacy in exchange for a positive outcome in terms of customised services and assistance.

Specifically, Rogers (2003) proposed five characteristics/attributes that innovation should have to be adopted in terms of *relative advantage, compatibility, complexity, trialability* and *observability*. Among these five attributes, considering the characteristics of VAs' integration in hotel rooms, we felt to mainly consider the following ones: *relative advantage*, which refers to how innovation is perceived compared to available solutions; *compatibility*, or the extent to which the innovation is perceived as consistent with existing values and previous user experiences and needs; *complexity*, or the extent to which the innovation is perceived as complex to be understanding and used. According to these studies (Rogers, 2003), innovation adoption takes the form of a process characterised by five phases: *awareness*, whereby an individual is exposed to innovation without holding information about it; *interest*, whereby the individual considers innovation as potentially relevant and wishes to obtain additional information about it; *evaluation*, whereby the individual tries to imagine the future utility of innovation; *trial*, whereby the individual makes a first experience with the innovation; and *adoption*, whereby the individual applies and uses that innovation stably.

One of the main reasons that could hinder the adoption of technologies has been discovered in the *resistance to change*, which refers to an individual's reluctance to incorporate new systems or ways to do things in his or her life (Ellen, Wiener, & Cobb-Walgren, 1991). Wamber and Banas (2000) investigated this phenomenon from the opposite perspective by analysing *openness to change*, which refers to an individual's propensity to accept changes (Wanberg & Banas, 2000). *Openness to change* represents a necessary condition for an innovation to be adopted (Miller, 2006, p. 60) and, thus, could be fundamental to understanding consumer behaviour concerning new, innovative and disruptive technological devices (Rogers, 2003) such as VAs, when integrated into a traditional setting such as hotel luxury rooms.

Luxury Consumption and Status Consumption Orientation

Luxury consumption literature has underlined that there are two different prevailing approaches to luxury buying. More specifically, past studies have shed light on the so-called dichotomy of luxury consumption, which distinguishes 'externalised' versus 'internalised' luxury on the basis of buying motivations (Amatulli & Guido, 2011, 2012; Guido et al., 2020). Externalised luxury is mainly related to external motives and is associated with ostentation and desire to belong to a social group, thus by status. Whereas internalised luxury is mainly related to internal motives and is more associated to individual style than to social image and social status. Consistent with such a dichotomy, Han, Nunes, and Drèze (2010) introduced the difference between 'parvenus' and 'patrician' consumers. Parvenus consumers are characterised by a high need for status and buy luxury goods for prestige and social status, whereas patrician consumers are characterised by a low need for status and are less motivated by the social position–signalling function of luxury purchases. Importantly, such dichotomy underlines how status-signalling is important in luxury and, therefore, the central role played by status consumption orientation (Dixit, Lee, & Loo, 2019; Vigneron & Johnson, 2004; Wiedmann, Hennigs, & Siebels, 2009). Status consumption orientation refers to 'the behavioural tendency to value status and acquire and consume products that provide status to the individual' (O'Cass & McEwen, 2004, p. 34). Status consumption orientation is affected by consumers' self-monitoring, whereby consumers aim to enhance their overall image in social contexts (O'Cass & Frost, 2002). In line with this line of reasoning, O'Cass and McEwen (2004) emphasised that status consumption orientation is 'a matter of consumers' desires to gain prestige from the acquisition of status-laden products and brands' (p. 27). Importantly, Eastman, Goldsmithg, and Flynn (1999) have widely investigated status consumption orientation, underlining that it is related to 'an interest in consuming for status, which involves a desire for status and conspicuous consumption' (p. 43). In particular, consumers characterised by high-status consumption orientation get satisfaction mainly from the audience reaction to the possession of a specific product and the displaying of a certain purchasing power. Typically, this orientation is related to an opportunity to buy higher-priced products and services in order to inflate one's ego and to show wealth. Specifically, consumers who tend to buy luxury mainly for status instead of individual style (i.e., externalised luxury consumers) are more willing to buy luxury products and services because they can convey status. Such consumers are also more prone to pay a premium price for status-signalling offerings and value more luxury brands and luxury products/services that include a snob appeal (Eastman et al., 1999).

In sum, the extent to which consumers desire luxury because of its status-signalling function may significantly influence their perceptions and their choices. Thus, such consumers' orientation is relevant to luxury marketing, as it may drive consumers toward specific products. Furthermore, consumers characterised by higher levels of status consumption may be attracted by different communication messages and offerings. In the same way, status consumption orientation can play a central role in the context of luxury hospitality, even in terms of the tendency to adopt technology when integrated into the hospitality offering. On the basis of the theories discussed above, in this chapter, we reason that guests' openness to change may have a positive effect on the perceived usefulness of VAs. However, we also expect that such effect is moderated by guests' status consumption orientation. Specifically, we hypothesise that luxury hotel guests' openness to change may lead them to perceive the use of VAs in the room as something useful. Nevertheless, we also hypothesise that such perceived usefulness is diluted when guests are highly oriented to status consumption. Formally:

H. Luxury hotel guests' openness to change has a positive effect on the perceived usefulness of VAs in their rooms. However, the magnitude of this effect decreases in the presence of higher (vs. lower) levels of status consumption orientation.

In sum, we think that openness to change may have a positive effect on guests' perceived usefulness of VAs in the rooms of luxury hotels, but this effect is diluted when guests are mainly oriented to a luxury consumption approach based on status-signalling (see Fig. 14.1 for a representation of the conceptual model).

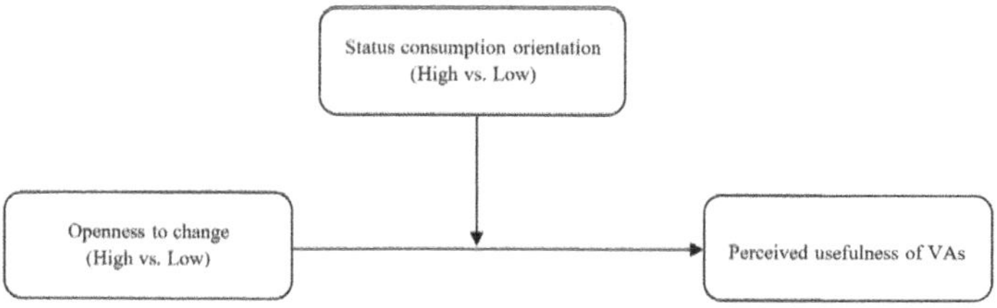

Fig. 14.1. Conceptual Model.

Methodology

To provide empirical support to our hypothesis, we conducted a structured survey study. The questionnaire was divided into three sections. The first section briefly described VAs for luxury hotels, providing a scenario in which these technological devices were integrated into hotel rooms. The second section featured measures of the relevant constructs, which were assessed on seven-point Likert scale (1 = 'completely disagree' and 7 = 'completely agree'). The measured constructs were: *openness to change* (3 items based on Wamberg & Banas, 2000); status consumption orientation (5 items based on O'Cass & McEwen, 2004). The final section collected insight toward travel frequency ('each week'; 'each month'; 'less than once a year'; 'twice a year') and some socio-demographic data (i.e., gender and age).

Results

We collected 108 questionnaires. However, two questionnaires were removed as respondents failed an attention check item ('If you are reading this question, please select answer 5'). The respondents' age ranged between 18 and 78 years ($M = 40.01$, $SD = 14.26$). About their travel frequency, 10.4% of the respondents travelled each week, 36.8% travelled each month, while 52.8% travelled few times per year (particularly 39.6% twice a year, and 13.2% less than once a year). To test our hypothesis, we conducted a moderation analysis. We first mean-centred the measures of openness to change and status consumption orientation, which served as the independent variable and the moderating variable, respectively. Thus, we employed the PROCESS SPSS macro, which was developed by Hayes (2017) to support scholars in running moderation and other regression-based analyses. When employed for conducting a moderation analysis, the PROCESS macro (Model 1) allows one to specify the independent variable, the moderator and the dependent variable. Thus, it automatically computes the main effects of both the independent and moderating variables on the dependent variable, along with the effect of an interaction term, which is obtained by multiplying the independent variable by the moderating one. If this effect is statistically significant, then moderation is established. More importantly, the macro allows one to qualify such a moderation by estimating conditional effects or the effects of the independent variables on the dependent variable at different levels of the moderator.

In our case, openness to change served as the independent variable, status consumption orientation was the moderator, while perceived usefulness of VAs served as the dependent variable.

The obtained results show that guests' openness to change ($b = 0.42$; $t = 5.70$; $p < 0.001$) and status consumption orientation ($b = 0.33$; $t = 5.47$; $p < 0.001$) exerted a significantly positive effect on perceived usefulness of VAs. More importantly, there was also a significantly negative effect of the interaction between openness to change and status consumption orientation ($b = -0.13$; $t = -2.73$; $p < 0.01$). To probe the nature of this interaction, we looked at the conditional effects of openness to change on the dependent variable at different levels of status consumption orientation. Consistent with our hypothesis, the obtained results showed that the effect of openness to change on perceived usefulness was higher at a lower level (M – 1SD) of status consumption orientation ($b = 0.59$; $t = 7.66$; $p < 0.001$) than at a higher level (M + 1SD) of this variable ($b = 0.24$; $t = 2.11$; $p < 0.05$) (see Table 14.1 for summary results). Therefore, status consumption orientation influences the effect of openness to change on the perceived usefulness of VAs in luxury hospitality. This means that guests with different levels of status consumption orientation may exhibit different attitudes toward VAs in response to different levels of openness to change. Specifically, guests with a greater, as compared to smaller, openness to change may be much more inclined to consider VAs as useful devices in the presence of lower, rather than higher, levels of status consumption orientation.

To confer robustness to our results, we explored the interaction effect by using the Johnson–Neyman 'floodlight' technique (Spiller, Fitzsimons, Lynch, &

Table 14.1. Summary of Results Related to the Moderated Model.

Variables	Main Effects	Conditional Effects
Independent Variable (IV): Openness to Change (continuous and mean-centred variable)	Effect of the IV on the DV: $b < 0.42$, $p < 0.001$	Conditional effect of the IV on the DV at a low level (M – 1SD) of MOD: $b = 0.59$, $p < 0.001$
Moderator (MOD): Status Consumption Orientation (continuous and mean-centred variable)	Effect of the MOD on the DV: $b = 0.33$, $p < 0.001$	Conditional effect of the IV on the DV at the average level of MOD: $b = 0.42$, $p < 0.001$
Dependent Variable (DV): Perceived Usefulness of the VA (continuous variable)	Effect of the interaction (IV × MOD) on the DV: $b = -0.13$, $p < 0.01$	Conditional effect of the IV on the DV at a high level (M + 1SD) of MOD: $b = 0.24$, $p < 0.05$

Note: $N = 106$.

McClelland, 2013). This technique, based on Johnson and Neyman's (1936) test, allows one to plot the effects of the independent variable on the dependent variable (on the *x*-axis) across all possible values of the moderator (on the *y*-axis). The test identifies the region(s) of the significance of the plotted effect. The obtained results clearly show how the magnitude of the positive effect of openness to change on the perceived usefulness of VAs decreased at increasing levels of status consumption orientation, thus supporting our hypothesis (see Fig. 14.2). In the figure below, the shaded area between confidence bands indicates the region within which the effect is statistically significant.

General Discussion

Our results showed that both openness to change and status consumption orientation positively influence the perceived usefulness of VAs introduced in luxury hotels. Interestingly, a significant negative interaction between these two variables emerged from the analysis, which suggested that the positive effect of openness to change on perceived usefulness decreases in magnitude as the level of status consumption orientation increases.

This finding suggests that openness to change and status consumption orientation are two potentially alternative factors on which managers could intervene to

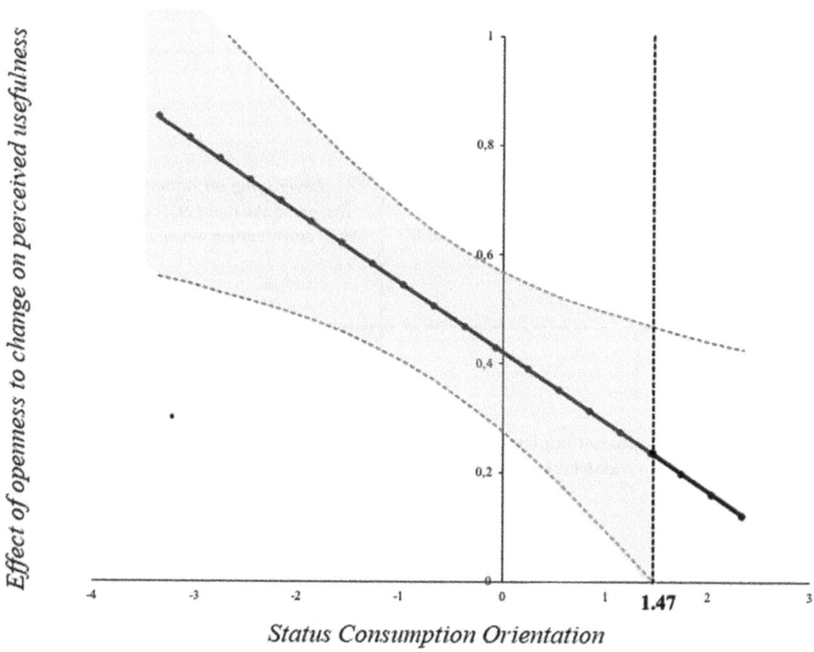

Fig. 14.2. Effect of Openness to Change on Perceived Usefulness as a Function of Status Consumption Orientation.

encourage the perceived usefulness of VAs among luxury hotel guests. This implies that luxury hotel guests' openness to change may be a necessary but not sufficient condition for acceptance of these devices. Therefore, luxury hotel managers should also consider the level of status consumption orientation of their guests. Interestingly, the results have been integrated in a coherent manner by developing an useful matrix to guide marketers' and managers' actions (see Fig. 14.3).

A further element of the hotel service delivery process consists of the physical support and the environment within which the technology is implemented. The environment in which the service is provided greatly influences the customers' choices, satisfaction and many other behaviours. The characteristics of the physical support of numerous services are evolving rapidly as technological progress and automation penetrate the service sector; the evolution of physical support, in order to be used as an effective marketing lever, should be clearly illustrated to consumers so that they can become aware of the advantages they can derive. This further confirms that to guide customers' attention toward the use of technologies such as self-service technologies SSTs and VAs, it is essential for hoteliers to promote the relative advantages compared to the traditional staff service and its ease of use (Meuter, Ostrom, Roundtree, & Bitner, 2000). Furthermore, our study is in line with the literature suggesting that a key to competitiveness for luxury hotels is to focus on both personalised service and SSTs (Kucukusta, Heung, & Hui, 2014), contributing but to highlight the role of the status consumption orientation. Indeed, our study's results show that hotel

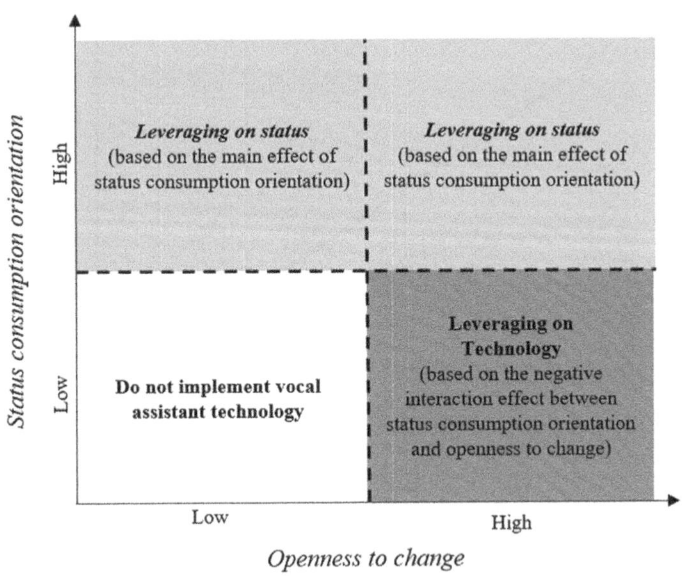

Fig. 14.3. The Strategic Focus at Different Levels of Status Consumption Orientation and Openness to Change.

guests would commonly accept these technologies. Moreover, considering the current proliferation and purchasing trend of home VAs, together with increased connectivity with smart appliances, one may reasonably imagine a consequent adaptation of customers to accomplish activities by using only voice commands such as in their daily lives. Moreover, our findings indicate that at higher levels of openness to change and status consumption, strategies that focus on emphasising the status consumption orientation obtained from experiences of staying in luxury hotels, should be encouraged.

The present study has both theoretical and managerial implications. Theoretically, it advances the current understanding of the drivers that may influence the use of new technologies in the field of luxury hospitality. Although past studies have examined the role of technology in hospitality, limited research has so far addressed how the perceived usefulness of new technological devices may depend on guests' openness to change. The current study also extends research on status consumption orientation in luxury hospitality. To date, no research has analysed the role of this individual characteristic with regard to the adoption of technological devices in luxury hospitality.

Managerially, this research suggests luxury hotel managers how to deal with new technological devices. Indeed, before investing in new technological tools, like VAs, luxury hotels should understand, on the one hand, to what extent their clientele is open to changes and, on the other hand, what kind of approach to luxury characterises such guests. Furthermore, considering that a typical characteristic of luxury markets is that many consumers buy luxury goods and services to signal their status (Amatulli & Guido, 2012; Guido et al., 2020; Vigneron & Johnson, 2004; Wiedmann et al., 2009), emphasising the role of status consumption making luxury hotels as more luxurious integrating new technologies in their services could contribute to increase both guests' satisfaction, and acting on their intention to use new technologies as they perceive benefits for their status. Furthermore, considering the current contingencies and the pandemic due to the COVID-19 diffusion and the acquired central role of digital technologies (Wang, Ng, & Brook, 2020), the use of such devices as VAs could guarantee the resumption of tourism activities by limiting the interaction between people in compliance with the policies of social distancing imposed by governments.

Conclusions

The introduction in luxury hotels of new and increasingly cutting-edge technologies has significantly changed the way guests and tourists perceive holidays. To keep up with the needs of travellers, luxury hotels and other accommodation facilities begin to evolve by leveraging above all new technologies. Our results show how technology is expressed above all through luxury. The progress of hi-tech tourism has not been few, but the real innovation will be evident only in the next decades. It is likely that, in the future, each hotel will equip rooms with virtual reality services, games, AI; the integration of IoT technologies, such as VAs, is only the first step toward a radical transformation. Thus, guests no longer have to interface with a receptionist, but maybe with a robot with the most varied features to have access to some information on the best restaurants in the area or the best itineraries to follow, and some services (such as Wi-Fi), will be nothing but one of the basic services one can expect to find. In a period of transition and change as the present one, it is not always natural that tourists are willing to accept these new technologies during their stay experience. Thus, the purpose of this research was to study the role of openness to change and status consumption orientation as two important consumer-related constructs toward the perceived usefulness of VAs when introduced into luxury hotel rooms.

The results of our research, conducted in a relevant geographical context like Italy, shed light on these constructs as fundamental in guests' technology acceptance. The hypotheses formulated were empirically supported. At high levels of openness to change and low levels of status consumption orientation, leveraging technologies and their capabilities could be an interesting marketing strategy. Marketing efforts should be directed to communicate benefits deriving from VAs use as a part of revitalised luxury hotel service. Therefore, it is important that luxury hotel managers give the guests messages that stimulate the perception of a more luxurious hotel, emphasising the role of their status consumption orientation. In this way, the results deriving from the introduction of these new technologies would be maximised.

Our research improves the current understanding of drivers that can influence the use of new technologies in the field of luxury hospitality, as in the case of the VA considered. Although previous studies examined the role of technology in hospitality (such as in Chan et al., 2017; Chan, Okumus, & Chan, 2018), our contribution has shed light on how the perceived usefulness of new technological devices may depend on the guests' openness to change, combined with the status consumption orientation analysis. Indeed, this study also extends the search for status consumption orientation in luxury hotels; no research analysed the role of this characteristic in the context of the technology adoption in the tourism field, and specifically hoteliering. Before investing in new technological tools, managers and marketers should understand, on the one hand, to what extent their customers are open to change and, on the other, what kind of approach to luxury characterises them. Since this study focuses on the relationship between the VAs integration in luxury hotels and the role of openness to change and status consumption orientation influencing guests' intention to use, future studies could analyse the effects of arousal through post-experience behaviour. Furthermore, additional dependent variables such as experience satisfaction, personal involvement, memory and attachment to the experience lived in the hotels could be considered. Moreover, the study was narrowed to the Italian context. Although some results could be extended as a general outcome, further research could examine properly the European Union (EU) market as a whole, thus enlarging the analysis sample. To conclude, understanding of luxury travel has revolved around exclusive and authentic experiences, nuanced by finer elements with a focus on value rather than price: New technologies could, thus, contribute to toughening guests' perception of their experiences, providing a more integrated and luxurious environment revolutionising their traditional experiences.

References

Ajzen, I. (1991). The theory of planned behavior. *Organisational Behavior and Human Decision Processes*, *50*(2), 179–211. doi:10.1016/0749-5978(91)90020-T

Amatulli, C., & Guido, G. (2011). Determinants of purchasing intentions of fashion luxury goods in the Italian market: A laddering approach. *Journal of Fashion Marketing and Management*, *15*(1), 123–136. doi:10.1108/13612021111112386

Amatulli, C., & Guido, G. (2012). Externalised vs. internalised consumption of luxury goods: Propositions and implications for luxury retail marketing. *The International Review of Retail, Distribution and Consumer Research*, *22*(2), 189–207. doi:10.1080/09593969.2011.652647

Athwal, N., Istanbulluoglu, D., & McCormack, S. E. (2019a). The allure of luxury brands' social media activities: A uses and gratifications perspective. *Information Technology & People*, *32*(3), 603–626. doi:10.1108/ITP-01-2018-0017

Athwal, N., Wells, V. K., Carrigan, M., & Henninger, C. E. (2019b). Sustainable luxury marketing: A synthesis and research agenda. *International Journal of Management Reviews*, *21*(4), 405–426. doi:10.1111/ijmr.12195

Atwal, G., & Williams, A. (2017). Luxury brand marketing–The experience is everything! In J. N. Kapferer, J. Kernstock, T. Brexendorf, & S. Powell (Eds.), *Advances in luxury brand management* (pp. 43–57). Cham: Palgrave Macmillan. doi:10.1007/978-3-319-51127-6_3

Bain and Company. (2018). Luxury goods worldwide market study: Fall–Winter 2018. Retrieved from https://www.bain.com/contentassets/8df501b9f8d6442e-ba00040246c6b4f9/bain_digest_goods_worldwide_market_study_fall_winter_2018.pdf. Accessed on April 30, 2020.

Benckendorff, P. J., Xiang, Z., & Sheldon, P. J. (2019). *Tourism information technology*. Boston, MA: CABI.

Buhalis, D. (2019). Technology in tourism-from information communication technologies to eTourism and smart tourism towards ambient intelligence tourism: A perspective article. *Tourism Review*, *75*(1), 267–272.

Capgemini. (2019). *Smart Talk: How organizations and consumers are embracing voice and chat assistants*. Retrieved from https://www.capgemini.com/it-it/news/smart-talk-how-organizations-and-consumers-are-embracing-voice-and-chat-assistants/

Cetin, G., & Dincer, F. I. (2014). Influence of customer experience on loyalty and word-of-mouth in hospitality operations. *Anatolia*, *25*(2), 181–194. doi:10.1080/13032917.2013.841094

Chan, E. S., Okumus, F., & Chan, W. (2017). The applications of environmental technologies in hotels. *Journal of Hospitality Marketing & Management*, *26*(1), 23–47. doi:10.1080/19368623.2016.1176975

Chan, E. S., Okumus, F., & Chan, W. (2018). Barriers to environmental technology adoption in hotels. *Journal of Hospitality & Tourism Research*, *42*(5), 829–852. doi:10.1016/j.tourman.2015.09.021

Chan, W. Y., To, C. K. M., & Chu, W. C. (2015). Materialistic consumers who seek unique products: How does their need for status and their affective response facilitate the repurchase intention of luxury goods? *Journal of Retailing and Consumer Services*, *27*, 1–10. doi:10.1016/j.jretconser.2015.07.001

Cinotti, Y. (2015). From hospitality to experience: Comparison between tourist experience in guest houses and hotels. HAL.

Connolly, D. J. (2000). Shifting paradigms: Using information technology to enhance service dyads in luxury hotels. *Journal of Hospitality & Leisure Marketing*, *7*(2), 3–38. doi:10.1300/J150v07n02_02

CPP Luxury. (2018). Marriott International to introduce innovative voice technology. Retrieved from https://cpp–luxury.com/marriott–international–to–introduce–innovative–voice–technology. Accessed on April 24, 2020.

Davis, F. D., Bagozzi, R. P., & Warshaw, P. R. (1989). User acceptance of computer technology: A comparison of two theoretical models. *Management Science*, *35*(8), 982–1003.

Deloitte. (2019). *Global powers of luxury goods 2019: Bridging the gap between the old and the new*. Retrieved from https://www2.deloitte.com/content/dam/Deloitte/global/Documents/ConsumerBusiness/gx–consumer–business–gplg2019–report.pdf. Accessed on April 30, 2020.

Dixit, S. K. (Ed.). (2017). *The Routledge handbook of consumer behaviour in hospitality and tourism*. London: Routledge.

Dixit, S. K. (Ed.). (2020). *The Routledge handbook of tourism experience management and marketing*. London: Routledge.

Dixit, S. K., Lee, K.-H., & Loo, P. T. (2019). Consumer behaviour in hospitality and tourism. *Journal of Global Scholars of Marketing Science*, *29*(2), 151–161. doi:10.1080/21639159.2019.1577159

Dwivedi, Y. K., Hughes, L., Ismagilova, E., Aarts, G., Coombs, C., Crick, T., & Galanos, V. (2019). Artificial Intelligence (AI): Multidisciplinary perspectives on emerging challenges, opportunities, and agenda for research, practice and policy. *International Journal of Information Management*, *57*, 1–47. in press. doi:10.1016/j.ijinfomgt.2019.08.002

Eastman, J. K., Goldsmith, R. E., & Flynn, L. R. (1999). Status consumption in consumer behavior: Scale development and validation. *Journal of Marketing Theory and Practice*, *7*(3), 41–52. doi:10.1080/10696679.1999.11501839

Ellen, P. S., Wiener, J. L., & Cobb-Walgren, C. (1991). The role of perceived consumer effectiveness in motivating environmentally conscious behaviors. *Journal of Public Policy and Marketing*, *10*(2), 102–117.

Erevelles, S., Fukawa, N., & Swayne, L. (2016). Big data consumer analytics and the transformation of marketing. *Journal of Business Research*, *69*(2), 897–904.

Ertmer, P. A. (2005). Teacher pedagogical beliefs: The final frontier in our quest for technology integration? *Educational Technology Research and Development*, *53*(4), 25–39.

Fisher, D., & Wilder–Smith, A. (2020). The global community needs to swiftly ramp up the response to contain COVID–19. *The Lancet*, *395*(10230), 1109–1110. doi:10.1016/s0140–6736(20)30679–6

Giusti, C. (2018). Reforming of the European Union financial supervisory and regulatory architecture: The Italian system as an example of a virtuous strategy between the judiciary and the authority of financial control. Global Conference on Business, Hospitality, and Tourism Research (GLOSEARCH 2018), Ho Chi Minh.

Gollnhofer, J. F., & Schüller, S. (2018). Sensing the vocal age: Managing voice touchpoints on alexa. Marketing review St. *Gallen*, *35*(4), 22–29.

Gretzel, U., Koo, C., Sigala, M., & Xiang, Z. (2015). Special issue on smart tourism: Convergence of information technologies, experiences, and theories. *Electronic Markets*, *25*(3), 175–177.

Gronroos, C. (1994). From marketing mix to relationship marketing: Towards a paradigm shift in marketing. *Asia-Australia Marketing Journal*, *2*(1), 9–29.

Guido, G., Amatulli, C., Peluso, A. M., De Matteis, C., Piper, L., & Pino, G. (2020). Measuring internalised versus externalised luxury consumption motivations and consumers' segmentation. *Italian Journal of Marketing*, *2020*(1), 25–47. doi: 10.1007/s43039–020–00002–9

Gupta, S. V., & Vajic, M. (1999). The contextual and dialectical nature of experiences. In J. Fitzsimmons & M. Fitzsimmons (Eds.), *New service development: Creating memorable experiences* (pp. 33–35). Thousand Oaks: Sage.

Gutierrez, A., O'Leary, S., Rana, N. P., Dwivedi, Y. K., & Calle, T. (2019). Using privacy calculus theory to explore entrepreneurial directions in mobile location-based advertising: Identifying intrusiveness as the critical risk factor. *Computers in Human Behavior*, *95*, 295–306.

Han, Y. J., Nunes, J. C., & Drèze, X. (2010). Signaling status with luxury goods: The role of brand prominence. *Journal of Marketing*, *74*(4), 15–30. doi:10.1509/jmkg.74.4.15

Hayes, A. F. (2017). *Introduction to mediation, moderation, and conditional process analysis: A regression–based approach*. New York, NY: Guilford publications.

Hennigs, N., Wiedmann, K. -P., Klarmann, C., & Behrens, S. (2013). Sustainability as part of the luxury essence: Delivering value through social and environmental excellence. *The Journal of Corporate Citizenship*, *52*, 25–35.

Hosany, S., & Witham, M. (2010). Dimensions of cruisers' experiences, satisfaction, and intention to recommend. *Journal of Travel Research*, *49*(3), 351–364.

Hoy, M. B. (2018). Alexa, Siri, Cortana, and more: An introduction to voice assistants. *Medical Reference Services Quarterly*, *37*(1), 81–88.

Hsu, C. L., & Lin, J. C. C. (2016). An empirical examination of consumer adoption of Internet of Things services: Network externalities and concern for information privacy perspectives. *Computers in Human Behavior*, *62*, 516–527. doi:10.1016/j.chb.2016.04.023

Jain, V., Vatsa, R., & Jagani, K. (2014). Exploring generation Z's purchase behavior towards luxury apparel: A conceptual framework. *Romanian Journal of Marketing*, *2*, 18–29.

Johnson, P. O., & Neyman, J. (1936). Tests of certain linear hypotheses and their application to some educational problems. *Statistical Research Memoirs*, *1*, 57–93.

Kapferer, J. N. (2014). The future of luxury: Challenges and opportunities. *Journal of Brand Management*, *21*(9), 716–726. doi:10.1057/bm.2014.32

Királová, A., & Pavlíčeka, A. (2015). Development of social media strategies in tourism destination. *Procedia-Social and Behavioral Sciences*, *175*, 358–366.

Kucukusta, D., Heung, V. C., & Hui, S. (2014). Deploying self–service technology in luxury hotel brands: Perceptions of business travelers. *Journal of Travel & Tourism Marketing*, *31*(1), 55–70. doi:10.1080/10548408.2014.861707

Larsen, S. (2007). Aspects of a psychology of the tourist experience. *Scandinavian Journal of Hospitality and Tourism*, *7*(1), 7–18.

Li, X., Rong, G., & Thatcher, J. B. (2012). Does technology trust substitute interpersonal trust? Examining technology trust's influence on individual decision–making. *Journal of Organizational and End User Computing*, *24*(2), 18–38. doi: 10.4018/joeuc.2012040102

Liu, X., Shin, H., & Burns, A. C. (2019). Examining the impact of luxury brand's social media marketing on customer engagement: Using big data analytics and natural language processing. *Journal of Business Research, 125*, 815–826. doi:10.1016/j.jbusres.2019.04.042

Matos, A., Pinto, B., Barros, F., Martins, S., Martins, J., & Au-Yong, M. (2019). Smart cities and smart tourism: What future do they bring? In *World conference on information systems and technologies* (pp. 358–370). Cham: Springer.

Meuter, M. L., Ostrom, A. L., Bitner, M. J., & Roundtree, R. (2003). The influence of technology anxiety on consumer use and experiences with self–service technologies. *Journal of Business Research, 56*(11), 899–906. doi:10.1016/S0148–2963(01)00276–4

Meuter, M. L., Ostrom, A. L., Roundtree, R. I., & Bitner, M. J. (2000). Self-service technologies: Understanding customer satisfaction with technology-based service encounters. *Journal of Marketing, 64*(3), 50–64.

Miller, J. G. (2006). Insights into moral development from cultural psychology. In *Handbook of moral development* (pp. 393–416). Psychology Press.

Moriuchi, E. (2019). Okay, Google!: An empirical study on voice assistants on consumer engagement and loyalty. *Psychology & Marketing, 36*(5), 489–501.

Mortenson, M. J., & Vidgen, R. (2016). A computational literature review of the technology acceptance model. *International Journal of Information Management, 36*(6), 1248–1259.

O'Cass, A., & Frost, H. (2002). Status brands: Examining the effects of non–product–related brand associations on status and conspicuous consumption. *Journal of Product & Brand Management, 11*(2), 67–88. doi:10.1108/10610420210423455

O'Cass, A., & McEwen, H. (2004). Exploring consumer status and conspicuous consumption. *Journal of Consumer Behaviour, 4*(1), 25–39. doi:10.1002/cb.155

Oe, H., Sunpakit, P., Yamaoka, Y., & Liang, Y. (2018). An exploratory study of Thai consumers' perceptions of "conspicuousness": A case of luxury handbags. *Journal of Consumer Marketing, 35*(6), 601–612. doi:10.1108/JCM–01–2017–2058

Petrović, J., Milićević, S., & Djeri, L. (2017). The information and communications technology as a factor of destination competitiveness in transition countries in European Union. *Tourism Economics, 23*(6), 1353–1361. doi:10.1177%2F1354816616653529

Pham, L., Nguyen, P. T. H., & Luse, D. (2018). Technology readiness and customer satisfaction in luxury hotels: A case study of Vietnam. *International Journal of Entrepreneurship, 18*(2), 183–208. doi:10.1504/IJMDM.2019.098648

Prayag, G., Hosany, S., & Odeh, K. (2013). The role of tourists' emotional experiences and satisfaction in understanding behavioral intentions. *Journal of Destination Marketing & Management, 2*(2), 118–127.

Rhee, J. (2018). *The robotic imaginary: The human and the price of dehumanized labor*. Minneapolis, MN: University of Minnesota Press.

Rogers, E. (2003). *Diffusion of innovations* (5th ed.). Free Press: New York.

Rogers, R. (2009). *The end of the virtual: Digital methods* (Vol. 339). Amsterdam: Amsterdam University Press.

Seo, Y., & Buchanan–Oliver, M. (2019). Constructing a typology of luxury brand practices. *Journal of Business Research, 99*, 414–421. doi:10.1016/j.jbusres.2017.09.019

Sestino, A. (2019). Business development, marketing automation and predictive analysis: An integration perspective–An overview towards new opportunities for studying consumer behavior and business integration. *SSRN Electronic Journal*. Retrieved from https://ssrn.com/abstract=3316759https://dx.doi.org/10.2139/ssrn.3316759

Sestino, A., Prete, M. I., Piper, L., & Guido, G. (2020). Internet of Things and Big Data as enablers for business digitalisation strategies. *Technovation*, *98*, 102173.

Sheng, J., Amankwah–Amoah, J., & Wang, X. (2019). Technology in the 21st century: New challenges and opportunities. *Technological Forecasting and Social Change*, *143*, 321–335. doi:10.1016/j.techfore.2018.06.009

Singh, N. (2011). *Localisation strategies for global e-business*. Cambridge: Cambridge University Press.

Solomon, M. R., Surprenant, C., Czepiel, J. A., & Gutman, E. G. (1985). A role theory perspective on dyadic interactions: The service encounter. *Journal of Marketing*, *49*(1), 99–111.

Spiller, S. A., Fitzsimons, G. J., Lynch, J. G., Jr, & McClelland, G. H. (2013). Spotlights, floodlights, and the magic number zero: Simple effects tests in moderated regression. *Journal of Marketing Research*, *50*(2), 277–288. doi:10.1509%2Fjmr.12.0420

Stamboulis, Y., & Skayannis, P. (2003). Innovation strategies and technology for experience–based tourism. *Tourism Management*, *24*(1), 35–43. doi:10.1016/S0261–5177(02)00047–X

tom Dieck, M. C., Jung, T. H., Kim, W. G., & Moon, Y. (2017). Hotel guests' social media acceptance in luxury hotels. *International Journal of Contemporary Hospitality Management*, *29*(1), 530–550. doi:10.1108/IJCHM–10–2015–0552

Tung, V. W. S., & Ritchie, J. B. (2011). Exploring the essence of memorable tourism experiences. *Annals of Tourism Research*, *38*(4), 1367–1386.

Venkatesh, V., Morris, M. G., Davis, G. B., & Davis, F. D. (2003). User acceptance of information technology: Toward a unified view. *MIS Quarterly*, *27*(3), 425–478.

Vigneron, F., & Johnson, L. W. (2004). Measuring perceptions of brand luxury. *Journal of Brand Management*, *11*(6), 484–506.

von Wallpach, S., Hemetsberger, A., Thomsen, T. U., & Belk, R. W. (2020). Moments of luxury–A qualitative account of the experiential essence of luxury. *Journal of Business Research*, *116*, 491–502. In press. doi:10.1016/j.jbusres.2019.10.060

Walls, A., Okumus, F., Wang, Y., & Kwun, D. J. W. (2011). Understanding the consumer experience: An exploratory study of luxury hotels. *Journal of Hospitality Marketing & Management*, *20*(2), 166–197. doi:10.1080/19368623.2011.536074

Wanberg, C. R., & Banas, J. T. (2000). Predictors and outcomes of openness to changes in a reorganizing workplace. *Journal of Applied Psychology*, *85*, 132–142.

Wang, C. J., Ng, C. Y., & Brook, R. H. (2020). Response to COVID–19 in Taiwan: Big data analytics, new technology, and proactive testing. *Jama*, *323*(14), 1341–1342. doi:10.1001/jama.2020.3151

Wiedmann, K. P., Hennigs, N., & Siebels, A. (2009). Value-based segmentation of luxury consumption behavior. *Psychology and Marketing*, *26*(7), 625–651.

Xiang, Z. (2018). From digitisation to the age of acceleration: On information technology and tourism. *Tourism Management Perspectives*, *25*, 147–150. doi:10.1016/j.tmp.2017.11.023

Zeithaml, V. A., Bitner, M. J., & Gremler, D. D. (2010). Services marketing strategy. *Wiley International Encyclopedia of Marketing*, 12–23. doi:10.1002/9781444316568.wiem01055

Chapter 15

Influence of Branded Luxury Guestroom Amenities on Guests' Hotel Buying Decisions: A Case of Five-star Hotels in Delhi

Vikas Gupta[1] and Saurabh Kumar Dixit[2]

Abstract

This study aims to determine whether the branded luxury guestroom amenities provided in five-star hotels of Delhi influence the guest's hotel purchase decisions. It also identified the amenities which are found to be most and least influential in affecting the guest's hotel selection and purchase behaviour. The study was conducted in the three upscale five-star hotels of Delhi. The selection of amenities and brands to be considered as luxury was based upon three focus group interviews with the room's division manager of the hotels. A structured questionnaire was drafted to identify the most and least useful hotel amenities among the respondents, influencing their hotel selection and purchase behaviours. Wi-Fi in the guestroom was found to be the most valuable amenity, with stationery items regarded as least valuable. It was also found that the guest's hotel selection and purchase decisions were significantly influenced when luxury branded amenities were placed in the guestroom. Guests were even found to pay extra when the hotel provided access to luxury branded amenities in the guestrooms. This is a novel attempt to find how the guests' hotel selection and purchase intentions are influenced by the placement of branded luxury amenities in guestrooms.

[1]https://orcid.org/0000-0003-0975-336X
[2]https://orcid.org/0000-0002-9478-084X

The Emerald Handbook of Luxury Management for Hospitality and Tourism, 305–319

doi:10.1108/978-1-83982-900-020211015

Keywords: Luxury; hotels; buying behaviour; guest amenities; Delhi; brand

Introduction

Most of the five-star hotels in India offer an extensive array of guest amenities (including branded or luxury) grounded on the guest's perceptions (mostly precise) of what they desire or need at a five-star category hotel and claim to use them frequently. Although for most guests, the type of amenities remains mostly the same, a considerable fraction also 'over forecast' what kind and types of room amenities they might use in the guestroom. Provision of guestroom amenities and their wide varieties in five-star hotels is a multi-billion dollar trade worldwide as the kind and varieties of amenities provided usually define the hotel quality and overall guest experiences (Heo & Hyun, 2015). Careful and sensible curation of guestroom amenities into a convincing worth proposition to entice hotel guests is the principal obligation and duty of the hotel brand administrators, proprietors and workers. The main challenge for the hotel managers (brand) and marketers is to keep track of not only the most preferred and highly valued amenities in the hotel guestrooms but also the changing nature of these preferences and guest's needs and demands which are constantly evolving with time. This chapter will, therefore, assist the hotel brand managers, marketers and administrators to determine the guestroom amenities (especially branded or luxury) which are presently trendy (in-fashion) and make sense to be included in their guestroom amenities list as per the preference of guests in the five-star hotels of Delhi.

When they are to be provided, the type of amenities to be provided and how many are to be provided in a hotel guestroom has been a long debate and discussion among the hotel proprietors, brand managers and marketers. A significant problem for them is to regulate what kind of guest amenities must be provided in guestrooms as they are an important element of a hotel brand's distinctiveness. Provision of guestroom amenities in five-star category hotels is significant as these amenities evidently influence a guest's preliminary decision regarding booking a hotel room and subsequently influences their intention to re-book the hotel and re-visit intentions (Kim, Li, Han, & Kim, 2015). Considering these aspects, the hotel management needs to find out the kind of guestroom amenities that visitors essentially use throughout their stay and assess the amenities which are least used or utilised. As the room amenities are a significant part of a hotel guestroom, the guests' preferences should be considered before their inclusion as they are an important element of the guest's inclusive experience. For instance, around eight decades ago, the provision of an air conditioner in the guestroom was a new guest amenity and the inclusion of television in the early 1960s. However, considering the present scenario, both these amenities have become very common.

The hotel business's amenities contest and thrust to provide the best amenities to the guests continue till the current date. For example, in the year 1983, luxury hotels in Europe tested the idea of concierge service in the lobby. This later became a big success and expanded globally. Lately, the provision of in-room amenities was emphasised by the hotel's management, which was a neglected area. Two problems mount the question of which guestroom amenities the five-star hotels must consider providing to the guest. The primary matter is regarding what guestroom amenities guests say that they want to have in the rooms. However, this study will discuss those amenities that the guests actually use based on their previous experiences regarding the hotel. Distinctive responses to these questions are vital, as hotel management needs to be aware of the amenities (especially branded or luxury) which significantly influences guests' hotel purchase decisions and their re-visit intentions which will be the prime focus of this chapter.

As the hotel guests have become more urbane with the diversification of the hospitality market, the prospects of providing quality service, especially with respect to offering branded or luxury guestroom amenities, are estimated to increase proportionally. It is important for the hotels to be steadily conscious of their patrons' wants and requirements. To meet these expectations, the five-star hotels should provide diverse varieties of guestroom amenities and services to their patrons (Kim et al., 2015). Therefore, this study will try to find out whether luxury or branded guestroom amenities may be utilised as indications for patrons to appraise a hotel. The hotel managers usually depend upon their instinctive wisdom regarding luxury or branded guestroom amenities to improve guest's overall gratification and enhance their brand performance. But every guestroom amenity comes with a price, and accumulating more of these amenities necessitates more outlay by the hotel. The number of guestroom amenities and the related expenses has an unswerving influence on the overall room maintenance bills and must also be well-thought-out much in advance.

The effectiveness or appeal of guestroom amenities upsurges only when novel or dissimilar varieties are appreciated and used by the patrons. Therefore, this study emphasises the number or their varieties in the hotel and their luxury branding. This will create significant insinuations for the promotion and pricing approach of the hotels. Promotional approaches and pricing assembly may be planned more efficiently if the hotel managers comprehend upon the influence of luxury or branded guestroom amenities on patrons' hotel assessments and willingness to pay a higher price for such rooms. The prime objective of this chapter is to identify the guestroom amenities guests consider most or least valuable. Specifically, it tries to inspect whether patrons trust or believe that luxury or branded guestroom amenities give the five-star hotels some sort of a premium inclusive image and whether guest's perceptions regarding the hotel influence their willingness to pay a price premium and re-visit intentions.

Literature Review

Guestroom Pricing Approaches in the Global Hospitality Industry

Guestroom pricing has obtained significant consideration from researchers and scholars in hospitality, as efficient pricing is important for the enduring accomplishment of a hospitality trade. The majority of the researches conducted to date (Israeli, 2002; Israeli & Reichel, 2001) on the guestroom pricing approaches in the hospitality sectors are done from the consumer demand perspective. Numerous scholars discovered the elements that significantly impact the hotel's room rates from several viewpoints. A study conducted by Israeli, Adler, Mehrez, and Sundali (2001) suggested that establishments may indicate tactical resources to prevailing target markets to rationalise the price premium. The influence of other significant factors such as hotel affiliation, brand image and size to forecast the guestroom pricing is also explained in the study (Israeli et al., 2001; Israeli & Uriely, 2000). Hotel star ratings were also a significant factor that can be a predictor of guestroom price (Israeli, 2002).

Though numerous methods, for example, latent growth curves (e.g., Coenders, Espinet, & Saez, 2003), conjoint analysis (e.g., Goldberg, Green, & Wind, 1984) and quantile regression approach (e.g., Hung, Shang, & Wang, 2010) have been used, and the hedonic pricing approach was found to be the utmost commonly used method for guestroom pricing especially in the five-star hotels from the perspective of consumer demand. This method assesses products and amenities as assemblies of traits or features. It suggests that the experiential price of any product or service is the sum of unobserved prices of the package of characteristics linked with it. This approach was also used by Chen and Rothschild (2010) in Taiwan, and it was instituted that the location of the hotel, the convenience of LED television and the existence of meetings, incentives, conferences and exhibitions (MICE) amenities have substantial impacts on guestroom prices of the hotel.

Though most studies on guestroom pricing in hotels have utilised the perspective from the supply end, however, very few scholars have debated this problem from the perspective of demand. For instance, a study conducted by Monty and Skidmore (2003) projected the customers' readiness to pay for hotel amenities utilising the data gathered from guests. It was found that the location of the hotel and duration of the year are significant predictors of pricing for guestrooms. Further, Danziger, Israeli, and Bekerman (2006) examined the influence of tactical hotel assets, for example, hotel brand, price, star categorisation, size of the hotel, quantity of in-house restaurants, hotel location and size of the swimming pool, in defining the consumer perceptions regarding the guestroom price in the Israeli hospitality industry.

Although earlier research results have contributed significantly towards the extant literature on hospitality, however, very restricted implications have emerged out for the five-star hotels as tactical assets and peripheral issues are rather problematic for the five-star hotels to regulate or modify in a small span of time. Room amenities may be imagined as software for five-star hotels as modifying these amenities, or their brand is comparatively easy. Consequently, this chapter emphasised the impact of guestroom amenity (especially luxury or branded) on the patrons' willingness to pay a price premium from the perspective of demand to deliver hands-on insinuations for the hotel managers. Through the application of focus group interviews/discussions with the previous hotel guests and current room division managers of a few high-end hotels in New Delhi, this study will identify the amenities which are found to be most influential and least influential in affecting the guest's hotel selection and purchase behaviour. It will also determine whether the branded or luxury guestroom amenities provided in the guestrooms of five-star hotels in Delhi influence the guests' hotel purchase decisions and their willingness to pay a price premium.

Influence of Luxury or Branded Guestroom Amenities on Hotel Guestroom Pricing

According to the AHLA (*American Hotel and Lodging Association*), the *Four Seasons* hotel group started the concept of offering guestroom amenities in the year 1975. A guestroom amenity may be demarcated as an additional facility or merchandise provided to the hotel guests in the guestrooms, usually free of cost (although it is already included in the room tariff) (Vallen & Vallen, 2005). Hotel amenities, however, may be defined as an additional service or merchandise present in the hotel such as gym, concierge, spa, salon and so on. As the hotel patrons have become more cultured and urbane and the hospitality market has become more varied, five-star hotels are taking utmost care by curating guest-room amenities as per their needs and desires. The inclusion of branded and luxury amenities to redefine contemporary hotel experiences has now become a new trend apart from providing standard amenities like the LED televisions and other such amenities. This inclusion is both in terms of the provision of branded toiletries as well as upgraded premium room linen, customised mattress, sofa cushions and pillows, latest tea and coffee dispensers, integrated lighting system, massage chairs, among many others.

Almost all five-star hotel chains have upgraded their meaning of luxury by including modern, contemporary and more user- and environment-friendly luxury branded guestrooms amenities in their room amenities checklist. Few examples of this are upgradation in the line of toiletries by *The Oberois* hotel chain from *Biotique* to the luxury brand *Forrest Essentials* which has a branded line of items from natural perfumes to shampoos, conditioners and fragrant body washes. Similarly, hotel *ITC Maurya* also provides a selected range of their house-made luxury toiletry brand *Essenzia Di Wills* for its guests to redefine their inclusive hotel experience for the better. Five-star category of hotels are not only looking into modifying their toiletries, but they are also taking all the essential efforts to make the guestroom experience more comfortable and refined for their guest by including services which are state of the art and highly consumed by guests during their stay. The traditional tea kettle with Nespresso machine in the guestrooms of luxury hotels like *The Oberoi, Taj Hotels and Resorts, ITC Hotels*, etc., is an example of the same.

Now, hotel managers believe that improving the line of luxury or branded guestroom amenities significantly enhances the inclusive appeal of the hotel to the guests. Though it is a relatively important aspect of hotel and guest experiences, scholars have drawn minimal attention in this area. Therefore, the purpose of this chapter will be to deliberate the amenities provided in the hotel guestrooms. It will also discuss how the branded and luxury guestroom amenities may augment the guests' assessment and appraisal of a hotel before making a purchase decision and significantly increases their willingness to pay a price premium for the same room.

Luxury branding is a significant subject in hospitality and largely influences the room pricing for a five-star hotel. Simoes and Dibb (2001) advocate that in the service industry like hotel business, famous luxury brands help build consumer trust, thus allowing them to envisage and appreciate the intangible services in an improved way and significantly decrease their perceived risks. Previous researches (Johnson & Russo, 1984; Park & Lessing, 1981) on data recollection and usage recommend that previous information regarding merchandise and service (or acquaintance with the product brand name) affect the degree to which patrons recollect and use the knowledge for the assessment of the quality of service and preference of services being offered. Investigators and scholars have also inspected the persons' propensity to utilise merchandise or product-linked knowledge such as brand and value to attribute quality. For instance, Rao and Monroe (1988) demonstrated in their study that previous information of a product or its acquaintance has a moderating influence on the level to which price and merchandise knowledge is utilised to assess the quality. This infers that when persons' brand acquaintance upsurges, the usage of inherent indications for quality of service valuations becomes sturdier.

Luxury may be defined as 'a situation of plentifulness or excessive comfort and cosiness' or 'anything substituting to the desire or ease but not unconditionally essential'. Further, as defined by Nueno and Quelch (1998), luxury merchandises are those items 'whose proportion of functionality to worth is little, though the ratio of imperceptible and situational efficacy to value is high'. Likewise, Hagtvedt and Patrick (2009) theorise a luxury product as the one that has the finest quality goods, delivers liking as an essential benefit and links with patrons on a psychological and emotional front. The supremacy of luxury brands to interconnect multifaceted communication rapidly with emotive psychological influence may be elucidated by the concept of a positivity bias. Positivity regarding a particular luxury brand signifies the propensity for remarkable valuations of merchandise and products to result in excessively optimistic assessments. Therefore, it is suggested that luxury branded amenities may be utilised to augment the inclusive image of the hotel and consequently increase the patron's willingness to pay a price premium and enhance their intention to return by increasing the appeal of the hotel.

Methodology

A qualitative research approach was applied in which focus group interviews were conducted with the room division managers of three luxury hotels in Delhi. This was done through online videoconferencing platforms to find out their criterion of selection of luxury amenities in the five-star hotels. This study also comprised 14 non-random, in-depth interviews with the former guests of these luxury hotels in Delhi. Their contact information was taken from TripAdvisor India, and they were approached through an invitation email to be a part of this study. Initially, a

Case Study 15.1: Responsible Luxury – A Case Study of ITC Maurya, New Delhi

About the Hotel – Accommodations That Define Luxury

ITC Maurya is a signature hotel of ITC (Indian Tobacco Company) Hotels Group and is named after the famous Maurya dynasty, which ruled India in the past. This hotel had utilised the contemporary style of luxury, which combines the historical era with the latest state-of-the art infrastructure to provide an unparalleled hotel experience to its guests. It provides 438 spacious rooms that offer an unmatched luxury to its exquisite guests in the heart of the National Capital of India – New Delhi. This hotel has now become an identity of luxury in Delhi and a favoured destination for business leaders, heads of states, celebrities and elites, worldwide.

This hotel has become a new name for luxury in Delhi. The luxury accommodations combine the best of class space, standards of service, interiors and unmatched hospitality. This hotel has set up new standards of excellence by the revolutionary concept of exclusive accommodation. The hotel rooms are divided into three sections: *ITC One, Executive Club* and *The Towers* so as to cater to the diverse needs of numerous guests and offer high-end services. Each room category is uniquely designed and appointed luxuriously to comply with the wishes and desires of the guests. The rooms are named after various famous rulers and personalities of the Mauryan era and include presidential suites such as The Chanakya and The Chandragupta, among others. The interiors of these rooms are also designed to portray the luxurious living of emperors of that era. The hotel also accommodates the invaluable and attractive art collections by famous Indian artists like M. F. Hussain, Krishen Khanna, Tyeb to Akbar Padamsee, among many others. It is also famous for its very famous and pioneer restaurant in Asia – "*The Bukhara*" which offers sumptuous northwest frontier cuisine with its elite patrons being the famous leaders and national heads from around the world.

Amenities Provided in the Guestrooms

Most guestrooms in ITC Maurya are named and inspired by the famous leaders or rulers of the Mauryan dynasty, and this is also visible in the themes, artistry and royal richness of the rooms. From state-of-the-art security systems, personalised butler service to private room entrances, all the luxuries of the rooms are beyond imagination. The same is also reflected in the guestroom amenities provided to the guests during their stay in these spaces. ITC Maurya feels proud to provide a blended luxurious comfort with a dash of its well-appointed suites and heritage guestrooms. From the walk in the spacious closets to the luxurious bath area, the level of experience offered is beyond comparison.

The privileges of the guests (usually for presidential suites) start from limousine pickup from the airport to a private entry of the guest to the hotel. The range of guestroom amenities includes: massage chair, cribs (on request), seasonal fruit platter, complimentary chocolate and night mint box, FHD television, Nespresso coffee maker, customised mini-bar, iPod docking station, Essenza Di Wills Homme and Femme bath toiletries, ceramic bathtub with jacuzzi, magnifying shaving mirror, pillow top bed, pillow menu, Evian water bottles, branded tea/coffee replenishments, choice of mattress, guestroom slippers, hairdryers, bathrobe or gown, bluetooth-enabled speakers, integrated lighting system, personalised wake-up call, Chanel No. 5 perfume, personalised butler service, high-speed Internet access, among many others. Considering the luxury standards set up by ITC Maurya, it has now become a part of Delhi's contemporary culture and a major destination for every visitor looking for delicate and personalised service with luxury at heart.

request was sent to 34 guests, but only 14 accepted the proposal. For those guests who accepted the proposal, individual interviews were scheduled and conducted through the Zoom videoconferencing platform. The sample was found to be adequate in similar researches conducted on customer decision-making in service trade situations (Dixit, 2017; Gupta, Khanna & Gupta, 2018, 2019). The selection of participants was grounded on the following criteria, including having stayed in any luxury five-star hotel in Delhi in the past one year for at least 24 hours and have used the luxury branded amenities (or having a fair idea about them) in the guestrooms. The survey instrument utilised a pre-defined set of questions based on earlier literature on branded luxury guestroom amenities and their influence on guests' intentions to re-visit and pay a price premium (Gupta, 2019; Kaplan & Haenlein, 2010).

The interview was conducted in English, and when required, they were translated into Hindi. The complete data collection process took three weeks to take responses from 14 guests and three managers. The majority of the respondents were males (64.28%) and between 35 and 45 years (57.14%). Though the results for this study are pinched from a very limited sample, they can be utilised for individuals of a wider population and fragment of a group driven by luxury brands to choose a particular hotel property (Lennon, Rentfro, & Curran, 2012) for their trip purposes. Miles and Huberman's (1994) factor method was used in which the aspects affecting the guests' hotel purchase decision were recognised via previous literature, and discrete factors were inspected and elucidated in the outcomes of the chapter laterally with quotes from the participants. The interview questionnaire included constructs such as participants age, gender, type of travel on the subsequent themes: influence of luxury and branded guestroom amenities on their hotel selection and decision behaviours; the impact of these amenities on their willingness to pay a price premium and intention to re-visit the hotel; and their most and least preferred guestroom amenities based on their usage during the stay. Participants were then requested to provide a list of all the branded amenities they remember, estimate the room tariff and specify the highest price they may pay for the guestroom. The survey instrument also requested participants to recognise anyone guest amenity they would upgrade in the guestroom and would be willing to pay a price premium.

Results and Data Interpretation

Results of the interviews suggest that among all the guestroom amenities, toiletry items such as bath gel, shampoo and moisturiser were found to be most valuable and important for the guests as it was mentioned by almost all the respondents. The influence of toiletry brands was also found to be significantly influencing the decision-making regarding the booking of the room by the guests and defining luxury perceptions about the hotel:

> Before making any final decision on the booking of a hotel, I usually check the hotel website or other hotel booking platforms to have a look at the amenities provided in the rooms through photos. After being completely satisfied by the photos, I make a final hotel booking.
>
> For me, guestroom amenities, especially toiletries, play a significant role in defining luxury. If I found through the online photos that the toiletry brands are sub-standards, I consider some other hotel options.
>
> I have stayed in many luxury hotels in the past. Based on my experience, I can state that the ones who provided the luxury guestroom amenities, including branded toiletries, customised room linen counting pillows and mattresses, integrated lighting, etc. define better luxury and more sophisticated brand value.

The next most influential luxury guestrooms amenity after toiletries was revealed as the bed and bath linens. Almost all the guests interviewed mentioned that the quality and comfortability of bed and bath linen was also a major influencer in their decision-making regarding the purchase of the hotel guestroom:

> Being an orthopaedic patient, my first priority while booking a guestroom is to find the availability of suitable mattress and quality bed linen. I usually inform the hotel much in advance about my requirements by raising a special request.
>
> For me the cleanliness and quality of bed and bath linen is one of the top-most factors before selecting a hotel room. I can compromise on price but cannot compromise on my comfort and good sleep during the stay.

Table 15.1. List of Most Preferred and Least Preferred Guestrooms Amenities.

Most Preferred Luxury Guestroom Amenities		**Least Preferred Luxury Guestroom Amenities**	
Rank 1	Toiletries	Rank 1	Business stationery
Rank 2	Bed, mattresses and bath linen	Rank 2	Room slippers
Rank 3	Bathtub or jacuzzi	Rank 3	In-room safe
Rank 4	FHD television	Rank 4	Iron
Rank 5	Nespresso coffee machine	Rank 5	Hairdryer
Rank 6	Complimentary high-speed Wi-Fi	Rank 6	Magnifying shaving mirrors
Rank 7	Personalised mini-bar	Rank 7	Bluetooth-enabled speakers
Rank 8	Cologne/perfume	Rank 8	PlayStation/Xbox console
Rank 9	Integrated room lighting	Rank 9	Radio alarm clock
Rank 10	Pillow menu	Rank 10	Study table

Source: Authors.

A list of 10 most preferred (useful) and 10 least preferred (useful) guestroom amenities are provided in Table 15.1 based on the interviews conducted with the guests and by conducting focus group interviews among the room division managers of three five-star luxury hotels in Delhi. Based on the inputs provided both by the guests as well as through the previous experiences of the managers regarding the guest's preference stationary regarding the luxury amenities, this list was prepared.

Findings of the one-to-one interview conducted with the respondents also revealed that in the luxury five-star hotels, guests don't look for standard amenities like telephone, stationery items or hairdryers but always demands something which is exclusive and branded. Most of the interviews revealed that personalised service in the guestroom along with the exclusivity of guestroom amenities is a noteworthy factor that influences the inclusive experiences of the guests. If the services provided in the guestrooms do not match with the price paid for the room, then the overall guest experience is diminished, and their intention to re-book the same hotel or their re-visit intentions are negatively influenced. Therefore, the placement of adequate guestroom supplies keeping the perceptions of the consumers in mind is important for the hotel managers to decide upon.

> Being a frequent traveller, I have a habit of comparing the facilities provided by one hotel with another. If I found that the facilities (especially the room amenities) provided in one hotel is better

> worth for my money, then I will definitely consider that hotel for my future hotel bookings.
>
> Whenever I book a luxury hotel for my stay, I usually consider experiencing the exquisite offerings in the room. I don't remember whether I have ever used the standard items like hairdryers or stationery.

This study also compiled a list of 12 branded luxury items that the respondents remembered about their guestroom experiences in the five-star hotels in Delhi, and it is presented in Table 15.2. Among these amenities, Bose headphones, Nespresso coffee machine, Evian water and Osim massagers were frequently remembered by almost all the respondents. The same was also confirmed by the room division managers in their focus group discussions.

One of the main objectives of this study was to determine whether the guests would pay a price premium if they are provided with access to branded luxury guestroom amenities. It was revealed through the survey interview that the majority of the respondents were willing to pay a premium price for the same room if they were provided with branded luxury amenities. One of the respondents also provided an example where for just US$ 30 per night, they were upgraded to a more luxurious room with better space and luxurious amenities. However, it was also found that the willingness to pay a premium price for branded or luxury amenities was stronger during a leisure trip. For business

Table 15.2. List of Branded Luxury Guestroom Amenity Items Remembered by the Respondents.

S.No.	Name of the Luxury/Branded Guestroom Amenities
1.	Bose headphones
2.	Evian water
3.	Osim Massagers
4.	Hermes Perfume
5.	iPad
6.	Samsung FHD Television
7.	Nespresso dispenser
8.	Jo Malone room spray
9.	Perrier sparkling water
10.	Bollinger Champagne
11.	Forrest essentials' toiletries
12.	Simons mattresses

Source: Authors.

travellers, though, the intention to pay a price premium for luxury guestroom amenities was found to be low.

> Luxury always comes with a price. If I have an option of enjoying better services and branded guestroom amenities during a leisure trip with an additional cost, I will definitely choose the option.
>
> During my business travel, I don't really care about the type of amenities being offered in the guestroom as most of my time is spent outside the room doing other activities. Therefore, for these trips, I usually just need a cosy bed with adequate standard amenities.

Conclusion and Future Implications

Suitable selection of guestroom amenities in a five-star hotel may improve patron gratification and helps in minimising outlays both with respect to the room amenity and correspondingly towards the costs incurred in overhauling the guestroom, therefore, assisting in the organisation's inclusive effectiveness (Jones, 2005). This chapter endeavoured to recognise the least preferred and most preferred luxury guestroom amenities from the perspective of hotel guests and also examined whether branded and luxury guestroom amenities influence the consumer's willingness to pay a price premium for the same room. The results of this study have numerous managerial and practical implications. The results sustain the statement that patrons' willingness to pay is influenced if luxury or branded amenities are provided in the guestroom. It was revealed that this also employs a positive influence on the guests' perceptions towards the guestroom. When branded or luxury amenities are positioned in the guestroom, guests' assessment of room tariff and their willingness to pay a price premium gets amplified. These findings provide a number of practical implications for curating guestroom amenities and deciding guestroom tariffs.

Hotel managers may offer branded and luxury guestroom amenities to rationalise and justify the room tariffs, especially during the peak seasons. Five-star hotels like The Oberoi, Taj Hotels and ITC Hotels are constantly keeping contemporary amenities in their guestrooms by upgrading them as per the market trends and preferences of their consumers. While branded and luxury guestroom amenities seem to be perceived as a competitive advantage, however, their enduring outlays are a point of apprehension. The results of this chapter specify that few guests are, in fact, willing to pay a price premium for the branded or luxury amenities in their guestroom. Out of three respondents in this study, two exposed that they are willing to pay extra if luxury or branded amenities are provided in the guestrooms. Thus, the hotels must exploit the guestroom amenities as an income stream rather than offering them as a complementary masterpiece. Moreover, there are certain guests who do not care about the type of amenities provided in the guestrooms and, thus, might not appreciate the provision of advanced amenities. For them, the efforts related to the upgradation of amenities will not bear any fruitful outcomes. Hotels managers should, therefore, try to select and curate amenities that are most useful and preferred by the guest in order to maximise their inclusive hotel experience and also reduce the significant cost incurred in providing them to the consumers. Conrad Hotels and Resorts introduced personal-choice amenities in 2012. In luxury five-star hotels in Delhi, therefore, guests are provided with a range of products to choose from with respect to the toiletries, mattresses, bath and bed linen, pillow menu, type of lighting, in-house technologies, etc., and based on the choices, rooms and supplies are provided which further makes the entire guest experience more personalised and exquisite. Hotel managers may also permit customers to select their own favoured amenity items and brands and diverge the room tariffs as per the guest choices. By doing this, hotels may perceive advanced consumer gratification and improved income.

Management of guestroom amenities can be considered as a type of software for five-star hotels. Altering the room amenities or their brands is comparatively simple compared to upgrading the physical facilities and setting of the hotel. Hotels can identify preferred brands and amenity items valued by their customers and place them in rooms in order to enhance perceived value and customer satisfaction. There are numerous other matters the hotel managers must also contemplate upon. Patrons staying at expensive high-end hotels may desire dissimilar items and products from the one staying at economical hotels. These findings of our study show that gender, type of vacation and customary preference of the type of hotel relate to the choice of guestroom amenities and guest's willingness to pay a price premium. A fascinating discovery of this chapter is that the female participants mostly remembered the amenities, which were personal upkeep products (such as Hermes perfume and Osim massagers); however, male respondents recalled technology-based products (for example, Bose headphones, electronic shavers and Samsung FHD television). This may also offer some advertising suggestions with regards to the display of recallable products in the printed or mass media advertisements so as to entice the guests.

References

Chen, C. F., & Rothschild, R. (2010). An application of hedonic pricing analysis to the case of hotel rooms in Taipei. *Tourism Economics*, *16*(3), 685–694.

Coenders, G., Espinet, J. M., & Saez, M. (2003). Predicting random level and seasonality of hotel prices: A latent growth curve approach. *Tourism Analysis*, *8*, 15–31.

Danziger, S., Israeli, A., & Bekerman, M. (2006). The relative contribution of strategic assets in determining customer perceptions of hotel room prices. *International Journal of Hospitality Management*, *25*(1), 129–145.

Dixit, S. K. (Ed.). (2017). *The Routledge handbook of consumer behaviour in hospitality and tourism*. London: Routledge.

Goldberg, S. M., Green, P. E., & Wind, Y. (1984). Conjoint analysis of price premiums for hotel amenities. *Journal of Business*, *57*(1), 111–132.

Gupta, V. (2019). Talent management dimensions and their relationship with retention of generation-Y employees in the hospitality industry. *International Journal of Contemporary Hospitality Management*, *31*(10), 4150–4169. doi:10.1108/IJCHM-10-2018-0859

Gupta, V., Khanna, K., & Gupta, R. K. (2018). A study on the street food dimensions and its effects on consumer attitude and behavioural intentions. *Tourism Review*, *73*(3), 374–388.

Gupta, V., Khanna, K., & Gupta, R. K. (2019). Preferential analysis of street food amongst the foreign tourists: A case of Delhi region. *International Journal of Tourism Cities*, *6*(3), 511–528. doi:10.1108/IJTC-07-2018-0054

Hagtvedt, H., & Vanessa, M. P. (2009). The broad embrace of luxury: The role of hedonic perceptions in the differential influence of a luxury (vs. value) brand concept on brand extension evaluations. *Journal of Consumer Psychology*, *19*, 608–618.

Heo, C. Y., & Hyun, S. S. (2015). Do luxury room amenities affect guests' willingness to pay? *International Journal of Hospitality Management*, *45*(2015), 161–168.

Hung, W., Shang, J., & Wang, F. (2010). Pricing determinants in the hotel industry: Quantile regressions analysis. *International Journal of Hospitality Management*, *29*(3), 378–384.

Israeli, A. (2002). Star rating and corporate affiliation: Their influence on pricing hotel rooms in Israel. *International Journal of Hospitality Management*, *21*(4), 405–424.

Israeli, A., Adler, N., Mehrez, A., & Sundali, J. (2001). Investigating the use of advertising for communicating hotels' strategic assets. *Journal of Hospitality and Leisure Marketing*, *7*(3), 23–38.

Israeli, A., & Reichel, A. (2001). The impact of location on the advertised room prices in an unstable environment: The case of Eilat, Israel. *Tourism Today*, *1*, 7–13.

Israeli, A., & Uriely, N. (2000). The impact of star ratings and corporate affiliation on hotel room prices. *International Journal of Tourism and Hospitality Research*, *2*(1), 27–36.

Johnson, E. J., & Russo, J. E. (1984). Product familiarity and learning new information. *Journal of Consumer Research*, *11*(1), 542–550.

Jones, T. J. A. (2005). *Professional management of housekeeping operations*. Hoboken, NJ: John Wiley.

Kaplan, A. M., & Haenlein, M. (2010). Users of the world, unite! The challenges and opportunities of Social Media. *Business Horizons*, *53*(1), 59–68.

Kim, W., Li, J., Han, J., & Kim, Y. (2015). The influence of recent hotel amenities and green practices on guests' price premium and re-visit intentions. *Tourism Economics*, *23*(3), 577–593.

Lennon, R., Rentfro, R., & Curran, J. (2012). Exploring relationships between demographic variables and social networking use. *Journal of Management and Marketing Research*, *2*(1), 1–16.

Miles, M. B., & Huberman, A. M. (1994). *Qualitative data analysis: An expanded sourcebook* (2nd ed.). London: SAGE Publications.

Monty, B., & Skidmore, M. (2003). Hedonic pricing and willingness to pay for bed and breakfast amenities in Southeast Wisconsin. *Journal of Travel Research*, *42*(2), 195–199.

Nueno, J. L., & Quelch, J. A. (1998). The mass marketing of luxury. *Business Horizon*, *4*(6), 61–68.

Park, C. W., & Lessing, V. P. (1981). Familiarity and its impact on consumer decision biases and heuristics. *Journal of Consumer Research*, *8*(2), 223–230.

Rao, A. R., & Monroe, K. B. (1988). The moderating effect of prior knowledge on cue utilisation in product evaluation. *Journal of Consumer Research*, *15*(1), 253–264.

Simoes, D., & Dibb, S. (2001). Rethinking the brand concept: New brand orientation. *Corporate Communications*, *6*(4), 217–224.

Vallen, G. K., & Vallen, J. J. (2005). *Check-in check-out: Managing hotel operation*. New Jersey, NJ: Pearson, Prentice-Hall.

Chapter 16

Michelin-starred Restaurants and Its Contribution to Luxury Gastronomy Tourism

Adela Balderas-Cejudo,[1] Diana Gavilan,[2] and S. Fernandez-Lores[3]

Abstract

The relationship between food and tourism has always been an increasingly important segment in the real worlds of hospitality, destination marketing and tourism development. However, only recently we have seen the acceptance of the growth in interest in gastronomic tourism as an academic field of study. Nowadays, there is a worldwide rise of concern for gastronomy and, specifically, for top gastronomy. The influence of gastronomy on tourism has provided new and exciting opportunities for major stakeholders in the tourism and travel industry.

This chapter focuses on the relation between Michelin-starred restaurants and luxury tourism in an attempt to understand the impact and influence of luxury restaurants in tourism. An evolutionary analysis of the literature and a Michelin-starred case study will be conducted with the aim of: (1) understanding Michelin star system and luxury gastronomy; (2) analyzing the potential link between top gastronomy and luxury gastronomy; and (3) listening to chefs' voices and perspectives on changes in consumer behaviour, perspectives and trends. Findings generated by this approach may help create innovative ways to address luxury tourism and hospitality.

[1]https://orcid.org/0000-0003-1703-068X
[2]https://orcid.org/0000-0002-5293-779X
[3]https://orcid.org/0000-0002-1164-0703

The Emerald Handbook of Luxury Management for Hospitality and Tourism, 321–334

doi:10.1108/978-1-83982-900-020211016

Keywords: Michelin star restaurants; gastronomy; luxury tourism; top chefs; food; sustainability

Introduction

'Gastronomy is not simply about gastronomic enjoyment or the practice or art of selecting, cooking, and relishing quality food. Rather, gastronomy is a form of symbolic communication within a community' (Dixit, 2019, p. 2). Gastronomy plays an increasingly important role in tourism. Wine and food have been an essential part of mainstream culture in the developing world since the 1960s; around the turn of the century, the relationship between tourism and gastronomy is gaining popularity. There has also been a focus on understanding the need for food producers and considering tourist needs when creating food-related experiences (Torres, 2003).

Wine and food consumption is an integral part of modern lifestyles, and it is often used to demonstrate the social status and the level of cultural capital (Hall & Mitchell, 2003). In recent decades, the relationship between gastronomy and tourism has grown rapidly. Gastronomy has evolved from a minor concern for destinations to one of the primary reasons for some visitors. Food tourism contributes significantly to the market destination's economy and interest in food tourism, and its growth has grown (Bessière, 1998; Dixit & Mawroh, 2019; Kivela & Crotts, 2006; Kim, Kim, & Goh, 2011). Food consumption is considered one of the significant factors in destination marketing growth because food has been shown to be an essential means of selling a destination's identity and culture (Quan & Wang, 2004).

According to Rand, Heath, and Alberts (2003), 'the local food holds much potential to enhance sustainability in tourism; contribute to the authenticity of the destination; strengthen the local economy; and provide for the environmentally friendly infrastructure' (p. 97). Furthermore, food as a cultural artefact and a maturing niche tourism market in its own right is also crucial to destination imaging (Hall & Sharples, 2004). Food is also an essential part of the tourist experience (Dixit, 2020; Hjalager & Richards, 2003). It can provide additional opportunities and occasions for visitors to have a more memorable and enjoyable vacation than they had expected. As a result, gastronomy is seen as a valuable source of marketable images and experiences for the tourism industry (Quan & Wang, 2004).

Gastronomy tourism has recently emerged as a significant growth sector, according to the World Tourism Organization's Second Global Report on Gastronomy Tourism (2017), not only because food and drink are central to the tourist experience but also because the concept of gastronomy tourism has developed and evolved to include a wide range of cultural practices.

> Gastronomy tourism represents an opportunity to constantly revitalize and diversify tourism, promote local economic development, involve many different professional sectors and

> bring new uses to the primary sector. Hence, gastronomy tourism contributes to promoting and branding destinations, maintaining and preserving local traditions and diversities, and harnessing and rewarding authenticity.
>
> (World Tourism Organization, 2017, p. 14)

Food Connoisseur

A phenomenon of food connoisseurs, 'foodies', has recently emerged 'The growing role of the consumer in the gastronomic experience production process is marked by the emergence of the "foodie", or food connoisseur, as a major influence in the development of food destinations and food tourism'. Foodies seem to be the new trend point to increasingly democratising tendencies regarding gastronomy. But, who are the foodies? Following Getz and Robinson (2014), Watson, Morgan, and Hemmington (2008) defined the terms as follows.

Harpers and Queen magazine coined the word 'foodie' and rose to popularity in Barr and Levy's (1984) paperback *The Official Foodie Handbook*, which described foodies as follows (p. 6):

> A Foodie is a person who is very very very interested in food. Foodies are the ones talking about food in any gathering – salivating over restaurants, recipes, radicchio. They don't think they are being trivial – Foodies consider food to be an art, on a level with painting or drama.

For Cairns, Johnston, and Bauman (2010), a foodie is a person 'with a passion for eating and learning about food' (p. 591).

The 'foodie' begins to use their travel experiences as a way to empathise their distinctive and unique knowledge of particular cuisines, foods or preparation methods. According to Balderas-Cejudo, Patterson, and Leeson (2019), the role of the media in shifting perspectives about tourism and gastronomy has been crucial through film, fashion, restaurant reviews, social media, television cookery shows (even for children) and the influence of individual personalities such as chefs, restaurateurs and travel authors (Hay, 2017).

Platforms, such as Instagram, for example, have taken on a significant role of how people communicate and also experience food. People share and influence food intake and actions in an increasingly connected age, ranging from 'Gen Z' to 'Millennials' to 'Baby Boomers'.

People nowadays are interested not only in eating but also in learning about food, its history, recipes and various cultural practices. Tourists are becoming more food conscious and savvy. They are eager to know what they consume and how real it is and how they can share it on social media, which has had a huge impact on food patterns and trends around the world.

Consistent with Barrère, Bonnard, and Chossat (2009), there is now a worldwide rise of concern for gastronomy and, specifically, for luxury and top gastronomy. Luxury gastronomy is distinct from food and traditional gastronomy, and it is associated with luxury consumption. Barrère et al. (2009) argue that top gastronomy defends the high importance of creativity and innovation, thereby contributing to the new conception of arts.

Top Gastronomy and Luxury Tourism Experience

Restaurants' position in society has evolved from simply providing food to now selling an experience (Johnson, Surlemont, Nicod, & Revaz, 2005). The restaurant industry is one of the most economies' oldest and most conventional industries, and it is quickly becoming a key component in attracting tourists and promoting tourism (López-Guzmán & Sánchez Cañizares, 2012).

Travellers think about going to a specific location to eat and learn about the regional and local cuisine (Daries-Ramon, Cristobal-Fransi, & Mariné-Roig, 2019). The rise in people's interest in different cuisines has led to increased luxury trips to different nations to gain knowledge of local food and culture and explore their cuisines. High-net worth individuals (HNWIs) highly seek local food experiences; however, they would have some difficulties defining themselves and finding something unique. Expert planners become genuinely relevant in that matter (ILTM, 2020).

Tourists seeking luxury experiences are not a homogeneous group, except that they are prepared and willing to pay a premium for exclusive experiences. This is a rapidly expanding market, fuelled primarily by HNWIs seeking greater authenticity in their travel destinations through local cuisine and tourism experiences. . However, middle-income earners, honeymoon travellers and foodies are also potential customer groups for luxury tourism and hospitality seeking unique, once-in-a-lifetime experiences.

Alternatives to satisfy the increasing demand to live unforgettable experiences are built around one of the four realms described by Pine and Guilmore (1999) in their seminal contribution 'The Economy of Experiences'. Since this chapter, experiences are sorted into four broad categories according to where they fall along the spectra of the two dimensions: the activeness of the client, customer or the leading actor of the experience – being an active or a passive participant – and their connection with the event – immersion or absorption. Accordingly, four food luxury experiences may be identified (Fig. 16.1):

- *Entertainment experiences* entail watching the activities and/or performances of others. The participant is not actively involved in the creation, but the mind is actively engaged during the event. This can happen in a wine blending demonstration, a museum and heritage visit or a concert in a vineyard.
- *Educational experiences* aim to increase the skills and enhance participants' knowledge through their active participation in, e.g., a wine testing, a culinary wine pairing event or a cooking class.

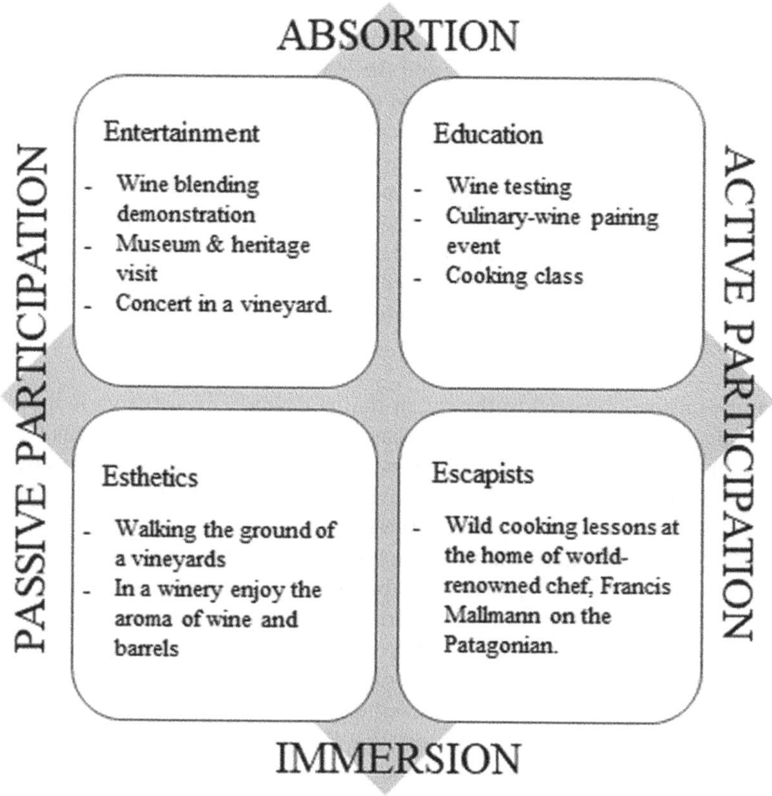

Fig. 16.1. The Four Realms of Experiences. *Source:* Adapted from Pine and Gilmore (1999).

- *Aesthetic experiences* capture participants' attention by the enriched sensory and sensual environment, the passive enjoyment of appreciating or 'just being in a setting' of the business, observing and absorbing all stimuli around such as consuming a 'winescape' or a rural road lined with vineyards.
- *Escapist experiences* require the most active participation in the events. This experience is a sort of co-creation that may result in a deep sense of fulfilment, a way of taking on a better self or even a new persona.

However, as Pine and Gilmore (1999) recognised, the most rewarding interactions include elements from all four realms. Usually, unique and memorable experiences are built around a mix of various or even all the four-realm described. The definition and notion of luxury are primarily based on personal experience. This implies that luxury is mainly an emotional experience and operates at an emotional level. Exploiting a single luxury item is less important than the general feeling of luxury and uniqueness. A sense of luxury is created and contributed by a combination of factors such as the local environment, people and their hospitality, the variety of culinary destinations and the overall hedonic experiences available.

Travel is often a sort of escapism, but as we enter a new age of technology and virtual reality, wholly immersive experiences become popular among young consumers. Nakamura.ke Mobile Kitchen became the world's first-ever glow-in-the-dark ramen shop, offering an unparalleled, immersive dining experience featuring high-end glow-in-the-dark food and drink, folklore and storytelling, created by award-winning chefs and designers according to its website (http://www.dashboard.us/glowinthedark-ramen-shop).

For its Los Angeles pop-up, Nakamura.ke offers an *omotenashi* VIP chef experience, limited to just six people per seating, including a performative element while the chefs prepare your food.

Haute cuisine is special to the extent that consumers expect from it some sort of 'magic of discovery' (Alcoba, Mostajo, Paras, Ebron, & Balderas-Cejudo, 2020). In order to become a mandatory call for both business and leisure luxury visitors, first – class chefs from all over the world must adjust their supply to luxury requirements (Barrère et al., 2009).

Since the rise of transformational luxury travel focused on self-improvement and personal enrichment, gastronomy becomes the goal of a learning journey. These travels provide the chance to return home with new knowledge and abilities aside from memories and stories to share. For instance, Plan South America – the expert planner of luxury travels with purpose – can arrange five days of private 'wild cooking' lessons at the home of world-renowned and South America's most beloved chef, Francis Mallmann, on the Patagonian island La Isla (https://www.plansouthamerica.com/).

Travelling and cooking lessons are becoming increasingly popular in the luxury segment. Black Tomato, a leading luxury travel company, provides innovative and inspiring experiences such as 'tasting notes'. This is a collection of glamorous, immersive journeys created and designed in collaboration with Michelin-starred chefs and food personalities around the world.

From the Peru's Sacred Valley with Virgilio Martinez to the Basque Country's secret culinary social club, Txokos with Chef Dani Lopez, 'tasting notes' on goes far beyond simply dining at Michelin-starred restaurants. (https://www.blacktomato.com/tasting-notes/).

The Michelin Star System

For travellers and gastronomes alike, the *Michelin Guide* has acted as their bible (Karpik, 2000). The Michelin star system potentially signals the institution as being part of the haute-cuisine sector (Surlemont & Johnson, 2005). These top-ranked restaurants can attract visitors and stimulate the economy and create a significant contribution regarding sustainability and contribute positively to many levels of the tourism value chain. According to Karpik (2000), 'a guidebook serves the economics of quality in markets where what is traded is quite singular, and where goods and services cannot be fully compared. It provides a meeting place for supply and demand and also redefines consumers and producers' (p. 369). Nowadays, the Michelin star guide is acknowledged internationally as the most serious (Harrington & Ottenbacher, 2010; Johnson et al., 2005). Furthermore, this conception of haute cuisine grows influential throughout the world (Bouty & Gomez, 2013).

Michelin was founded in 1891 by the Michelin brothers Edouard and Andre Michelin, from Clermont-Ferrand, as a tyre manufacturing company. The first edition of the *Red Guide* was published in the year 1900 (Gehrels, Kristanto, & Eringa, 2006). The *Red Guide* was introduced in the year 1900 for the first time (Gehrels et al., 2006).

It concentrated on interchangeable technical products and aided people in travelling by providing helpful car repair instructions. Telegraph stations and gas stations were among the items on the list (Johnson et al., 2005). It was turned into a tourist guide in 1909, with a focus on cultural content. Michelin established its three-star system based on the quality of the food in 1931, ranging from very good quality, excellent quality and fine and justly renowned cuisine (Harp, 2001). By 1934, the *Red Guide* had evolved into the modern-day gastronomic guide, concentrating on restaurants and cultural products (Karpik, 2000).

The *Michelin Guide* is focused on independence and anonymous inspections, according to Ottenbacher and Harrington (2007), and features a selection of the best hotels and restaurants in all comfort and price categories. Michelin stars are awarded to restaurants based on five criteria: product quality, taste mastery, cuisine personality, value for money and consistency, regardless of cuisine style. One star is considered 'a very good restaurant in its category', two stars reflect 'excellent cooking, worth a detour', while three stars display 'exceptional cuisine, worth a special journey' (Michelin, 2006). The guide features a selection of the best hotels and restaurants chosen by a rigorously selected and trained inspector team. The *Michelin Guide* (also known as the *Guide Rouge*) is the most prestigious rating system for fine cuisine and gastronomy in Europe (Johnson et al., 2005).

The culinary masters of the world regard the iconic *Red Guide* to fine dining to be 'the guide' when it comes to restaurant reviews. A Michelin star is the highest honour and award a chef can receive, and it has a significant effect on restaurant turnover and profitability (Surlemont & Johnson, 2005). The 'mystique' culture and clandestine essence of Michelin, which is an integral element of the *Red Guide* itself and helps to retain imagination in the section, is at the heart of this scheme (Alcoba et al., 2020). The *Michelin Guide* has a major impact on how people choose fine dining restaurants (Ottenbacher & Harrington, 2007).

The Basque Country has long been recognised for the excellence of its food. With the highest per capita concentration of Michelin star restaurants in the world, innovation and excellence also apply to haute cuisine in the Basque Country. Following the Basque Talent Observatory (https://www.bizkaiatalent.eus/en/pais-vasco-te-espera/senas-de-identidad/paraiso-estrellas-michelin/), the Basque Country has a total of 32 stars from 23 restaurants spread across the territory of Euskadi.

The reputation of Basque chefs owes much to the rise of the New Basque Cuisine, a renowned cuisine, combining traditional cooking with the sophistication of haute cuisine (Leizaola, 2006). The Basque Country's popularity is primarily due to the efforts of an early group of chefs – and those who followed in their footsteps – in bringing creativity and innovation to their work and disseminating Basque culture heritage through their cuisine (Aguirre García, Aldamiz-Echevarría González de Durana, & Aparicio de Castro, 2011).

Basque chefs have gained international recognition in recent decades, earning numerous international awards and Michelin stars, attracting a large number of gastronomy tourists and creating a robust culinary appeal not only for their profession but also for the region, serving as a valuable asset to the territory.

Case Study 16.1: Listening to the Chef's Voice: Eneko Atxa Azurmendi

The interview is defined as 'a conversation which proposes a particular purpose other than simply talk'. It is a precious technical tool in qualitative research for gathering data (Díaz Bravo, Torruco-García, Martínez-Hernández, & Varela-Ruiz, 2013) and deepening into the interviewee's voice to offer insights and explain individual behaviours and beliefs. Seven questions will lead us to Eneko Atxa's vision of gastronomy and the universe (Fig. 16.2).

Fig. 16.2. Chef Eneko Atxa.

Eneko Atxa was an award winner of five Michelin stars and chosen by Elite Traveler as one of the world's most influential chefs of the next decade. Eneko Atxa has three Michelin stars at his restaurant Azurmendi*** (Fig. 16.3). Azurmendi***, a unique bioclimatic building, is the most sustainable restaurant according to The World's 50 Best Restaurants. He won the Best Chef Award 2019 in Europe by Madrid Fusion, and Azurmendi*** secured the 14th position at the world ranking by The World's 50 Best Restaurants, and his cuisine is focused on the premise of home, culture and tradition, of the Basque producers and farmers and their work. A premise that is the base of Eneko's cuisine: flavour, sustainability and technique.

(1) *How do you define a three Michelin stars experience?*
Trying to talk about the three Michelin stars and the experiences that take place in every restaurant of these characteristics would be talking in a very generic way. I understand the three Michelin stars as a place that offers you something that is different, an experience that makes it unique, always around excellence, good food, comfort and many other things that make it different.

In our case, *we try to create a unique space that coexists in its own environment in the most sustainable way possible,* where one can find an experience that starts in a garden with a picnic in which we will put different snacks depending on the season, a visit to the kitchen with its corresponding snack made live in the kitchen while explaining how the house works. The customer keeps walking (this is a peculiarity that we started in 2012). The customer does not come to the restaurant, sits and leaves. On the contrary, the customer discovers the restaurant and the experience in an itinerant and dynamic way while discovering different proposals that are offered sometimes in the form of snacks or appetisers, etc., and arrives at the orangery where some of the local traditions are explained and converted into contemporary snacks.

Finally, the customer arrives at the dining room where, in permanent connection with the outside world and nature, he or she will be offered different gastronomic proposals for each season.

Likewise, all the details are essential. Thus, the service plays a fundamental role as well when it comes to an understanding, attending to and addressing the needs that the clients may have. These and many other qualities are fundamental for the proper functioning of a three-star Michelin, but in short, *it is about creating a unique experience with personality and identity and that always walks on the maximum excellence, almost obsessive.*

(2) *What is a customer like in a three Michelin stars restaurant?*
There is a very wide profile, and within the spectrum of clients that we receive today in *Azurmendi,* one can find large proportions of foreign clients who travel eager to discover new gastronomic proposals that can satisfy them. On the other hand, we also find the local customers who travel from a lesser or greater distance with the same objective: enjoying the experience.

(3) *What is the role of a three Michelin star restaurant in its area of influence? How does it affect the territory in which it is located?*
I believe that a *three Michelin star restaurant is a powerful active agent* that helps sectors such as the primary sector, such as tourism, transport, job creation, the enhancement of an environment, its inhabitants and other adjacent businesses, etc.

(4) *What role do Michelin restaurants play within luxury tourism? What relations exist between them and the other actors/integrators of luxury tourism?*
I believe that *Michelin restaurants are an important attraction for tourism.* Over the years, we understand that there is an increasingly important call or a call effect that comes from people who always travel with an eye on the fact that gastronomy is one of the factors that drive us to travel to certain countries, cities or places.

But those who travel (although gastronomy can be an important part of their trip) also look for other types of fundamental things in this aspect such as hotels that meet their expectations, accommodation, other types of alternative leisure that you can find in the place, etc. So I think that is something that we have understood practically all the agents that are directly influenced by tourism and we have created dialogue and we have known or at least we intend to give an answer to those market or business niches that can give us an added value with respect to our competing cities.

(5) *How do luxury and sustainability coexist? Is it a demand of new consumers?* It is not a demand; *sustainability is an absolutely necessary responsibility. It is an ingredient that seems intangible.* It is simply an act of responsibility that we should carry out to the best of our ability in any type of discipline: gastronomy or in tourism or in any other type of activity. We need to create a fairer world.

Fig. 16.3. Azurmendi***. *Source:* https://azurmendi.restaurant/en/sustainable-restaurant/.

(6) *How do you think luxury gastronomy will evolve? What trends are appearing?*
I do not believe in trends; *I have always believed in authenticity, honesty and personality.* I believe in something that you cannot find anywhere else; something that makes you authentic, makes you different, always bordering on excellence or walking on it but in a way that is far removed from global trends and deeply rooted in local distinctiveness.

(7) *What will the three Michelin star experience be like in 3 years, 5 years?* We are living in a moment of *absolute complexity* with a pandemic. I would not dare to draw a scenario, a possible future scenario for the next three or five years. What I do know is that haute cuisine will still be there. I do not know if it will wear the same suit it has worn in the last decade or if it will shed its skin or change its suit, but one way or another, haute cuisine will still exist.

Conclusion

Michelin star restaurants and chefs may be changemakers and active agents to a community. Following De Barnier, Falcy, and Valette-Florence (2012), Poelina and Nordensvard (2018) define luxury as having seven culturally specific common characteristic elements: exceptional quality; hedonism (beauty and pleasure); price (expensive); rarity (which is not scarcity); and exclusivity (which is not scarcity); selective distribution and associated personalised services; exclusive character (prestige, privilege) and creativity (art and avant-garde) (De Barnier et al., 2012).

Nowadays, destination actors recognise that they must go beyond Pine and Gilmore's (1999) prescription and engage the visitor as an active participant in developing and presenting gastronomic experiences. Tourists' ability to experience the 'authentic' and 'local' may also aid local residents in their attempts to preserve local culture and cuisine.

Just as sustainability appears multi-faceted, luxury has multiple facets as well (Kapferer & Michaut, 2015, p. 5). It is important to emphasise that luxury is often synonymous and associated with high quality, know-how, slow time, the preservation of hand-crafted traditions and the transmission of timeless goods and services from generation to generation: 'these connections will be in agreement with sustainability' (Kapferer & Michaut, 2015, p. 5).

Following the work on the role of luxury gastronomy in culinary tourism, an ethnographic study of Michelin-starred restaurants in France by Batat (2020), findings indicate that luxury restaurants engage in four major types of activities to improve destination attractiveness: (a) creating unique and exclusive luxury gastronomic experience offerings, (b) promoting terroir items and rural food tourism, (c) gastronomisation of rural destinations and (d) enhancing the destination's media capital.

The tourism industry's rising importance as an economic engine and its potential as a growth tool are undeniable (UNWTO, 2018). And Michelin-starred restaurants are not an exception; on the contrary, they are an essential asset in the gastronomic tourism of a territory. They may directly contribute to luxury tourism and can be change leaders and inspiration, bringing about different ways of doing things: respecting authenticity, promoting the culture, envisioning a better future, fostering sustainability and helping generate opportunities.

COVID-19 is challenging the world. Tourism and the restaurant industry are particularly vulnerable in this unprecedented situation. Nonetheless, the crisis sends vital messages about the tourism and hospitality industry's resilience, as well as about other ongoing crises that are potentially more devastating than COVID-19, such as climate change. The challenge is now to learn from this global tragedy to accelerate the transformation of sustainable tourism (Gössling, Scott, & Hall, 2020).

"Sustainability is a necessary responsibility" Eneko Atxa, 2020.

Acknowledgements

We would like to express our sincere gratitude to Chef Eneko Atxa.

References

Aguirre García, M. S., Aldamiz-Echevarría González de Durana, C., & Aparicio de Castro, M. G. (2011). Keys to success in an example of inter-competitor cooperation: The case of the "Big Seven" Basque chefs. Innobasque.

Alcoba, J., Mostajo, S., Paras, R., Ebron, R. A., & Balderas-Cejudo, A. (2020). On finding the voice of the customer in the digital traces of the Michelin-star gastronomy experience: Unveiling insights for service design. In *International conference on exploring services science* (pp. 21–34). Cham: Springer.

Balderas-Cejudo, A., Patterson, I., & Leeson, G. W. (2019). Senior foodies: A developing niche market in gastronomic tourism. *International Journal of Gastronomy and Food Science, 16*, 100152. doi:10.1016/j.ijgfs.2019.100152

Barrère, C., Bonnard, Q., & Chossat, V. (2009). Luxury gastronomy as an attractive activity for luxury tourism. In *Communication au colloque APIDT tourism, Lisboa 2009 et au colloque VDQS-Oenometrics.*

Barr, A., & Levy, P. (1984). *The official foodie handbook*. London: Ebury Press.

Batat, W. (2020). The role of luxury gastronomy in culinary tourism: An ethnographic study of Michelin-starred restaurants in France. *International Journal of Tourism Research*. doi:10.1002/jtr.2372

Bessière, J. (1998). Local development and heritage: Traditional food and cuisine as tourist attractions in rural areas. *Sociologia Ruralis, 38*(1), 21–34.

Bouty, I., & Gomez, M. L. (2013). Creativity in haute cuisine: Strategic knowledge and practice in gourmet kitchens. *Journal of Culinary Science & Technology, 11*(1), 80–95. doi:10.1080/15428052.2012.728979

Cairns, K., Johnston, J., & Bauman, S. (2010). Caring about food: Doing gender in the foodie kitchen. *Gender & Society, 24*(5), 591–615. doi:10.1177/0891243210383419

Daries-Ramon, N., Cristobal-Fransi, E., & Mariné-Roig, E. (2019). Deployment of restaurants websites' marketing features the case of Spanish Michelin-starred restaurants. *International Journal of Hospitality & Tourism Administration, 20*(3), 249–280. doi:10.1080/15256480.2017.1383961

De Barnier, V., Falcy, S., & Valette-Florence, P. (2012). Do consumers perceive three levels of luxury? A comparison of accessible, intermediate and inaccessible luxury brands. *Journal of Brand Management, 19*(7), 623–636. doi:10.1057/bm.2012.11

Díaz-Bravo, L., Torruco-García, U., Martínez-Hernández, M., & Varela-Ruiz, M. (2013). La entrevista, recurso flexible y dinámico. *Investigación en educación médica, 2*(7), 162–167.

Dixit, S. K. (Ed.). (2019). *The Routledge handbook of gastronomic tourism*. London: Routledge. doi:10.4324/9781315147628

Dixit, S. K. (2020). Marketing gastronomic tourism experiences. In S. K. Dixit (Ed.), *The Routledge handbook of tourism experience management and marketing* (pp. 322–335). London: Routledge.

Dixit, S. K., & Mawroh, H. (2019). Cultural manifestation of food in branding destination: A case of Khasi food of Meghalaya, India (Chapter 10). In E. Park, S. Kim, & I. Yeoman (Eds.), *Food tourism in Asia* (pp. 143–154). Singapore: Springer Nature Pte. Ltd.

Gehrels, S. A., Kristanto, S., & Eringa, K. (2006). Managing word-of-mouth communication in Michelin starred restaurants in The Netherlands. *Journal Manajemen Perhotelan*, *2*(2), 47–55.

Getz, D., & Robinson, R. N. (2014). Foodies and food events. *Scandinavian Journal of Hospitality and Tourism*, *14*(3), 315–330. doi:10.1080/15022250.2014.946227

Gössling, S., Scott, D., & Hall, C. M. (2020). Pandemics, tourism and global change: A rapid assessment of COVID-19. *Journal of Sustainable Tourism*, *29*(1), 1–20. doi: 10.1080/09669582.2020.1758708

Hall, C. M., & Sharples, L. (2004). The consumption of experiences or the experience of consumption? An introduction to the tourism of taste. In C. M. Hall, L. Sharples, R. Mitchell, N. Macionis, & B. Cambourne (Eds.), *Food tourism around the world* (pp. 13–36). London; New York, NY: Routledge.

Hall, M., & Mitchell, R. (2003). Tourism as a force for gastronomic globalization and localization. In *Tourism and gastronomy* (pp. 85–102). London: Routledge.

Harp, S. (2001). *Marketing Michelin: Advertising and cultural identity in twentieth-century France*. Baltimore, MD: The John Hopkins University Press.

Harrington, R. J., & Ottenbacher, M. C. (2010). Culinary tourism—A case study of the gastronomic capital. *Journal of Culinary Science & Technology*, *8*(1), 14–32. doi:10.1080/15428052.2010.490765

Hay, B. (2017). Gastronomy, tourism and the media. *Journal of Tourism Futures*, *3*(2), 184–185. doi:10.1108/JTF-09-2017-064

Hjalager, A.-M., & Richards, G. (2003). Still undigested: Research issues in tourism and gastronomy. In A.-M. Hjalager & G. Richards (Eds.), *Tourism and gastronomy* (pp. 238–248). London; New York, NY: Routledge.

Johnson, C., Surlemont, B., Nicod, P., & Revaz, F. (2005). Behind the stars: A concise typology of Michelin restaurants in Europe. *Cornell Hotel and Restaurant Administration Quarterly*, *46*(2), 170–187. doi:10.1177/0010880405275115

Kapferer, J.-N., & Michaut, A. (2015). Luxury and sustainability: A common future? The match depends on how consumers define luxury. *Luxury Research Journal*, *1*(1), 3–17.

Karpik, L. (2000). Le guide rouge Michelin. *Sociologie du Travail*, *42*(3), 369–389.

Kim, Y. H., Kim, M., & Goh, B. K. (2011). An examination of food tourist's behavior: Using the modified theory of reasoned action. *Tourism Management*, *32*(5), 1159–1165. doi:10.1016/j.tourman.2010.10.006

Kivela, J., & Crotts, J. C. (2006). Tourism and gastronomy: Gastronomy's influence on how tourists experience a destination. *Journal of Hospitality & Tourism Research*, *30*(3), 354–377. doi:10.1177/1096348006286797

Leizaola, A. (2006). Matching national stereotypes? Eating and drinking in the Basque borderland. *Anthropological Notebooks*, *12*(1), 79–94.

López-Guzmán, T., & Sánchez Cañizares, S. M. (2012). La gastronomía como motivaciónpara viajar. Un estudio sobre el turismo culinario en Córdoba. *PASOS: Revista de Turismo y Patrimonio Cultural*, *10*, 575–584.

Michelin. (2006). *Michelin guide Deutschland 2006*. Karlsruhe: Michelin Travel Publications.

Ottenbacher, M., & Harrington, R. J. (2007). The innovation development process of Michelin-starred chefs. *International Journal of Contemporary Hospitality Management*, *19*(6), 444–460. doi:10.1108/09596110710775110

Pine, B. J., & Gilmore, J. H. (1999). *The experience economy: Work is theatre, and every business a stage*. Boston, MA: Harvard Business Press.

Poelina, A., & Nordensvard, J. (2018). Sustainable luxury tourism, indigenous communities and governance. In M. A. Gardetti & S. S. Muthu (Eds.), *Sustainable luxury, entrepreneurship, and innovation (environmental footprints and eco-design of products and processes)*. Singapore: Springer.

Quan, S., & Wang, N. (2004). Towards a structural model of the tourist experience: An illustration from food experiences in tourism. *Tourism Management*, *25*(3), 297–305. doi:10.1016/S0261-5177(03)00130-4

Rand, G. E. D., Heath, E., & Alberts, N. (2003). The role of local and regional food in destination marketing: A South African situation analysis. *Journal of Travel & Tourism Marketing*, *14*(3–4), 97–112.

Surlemont, B., & Johnson, C. (2005). The role of guides in artistic industries: The special case of the "star system" in the haute-cuisine sector. *Managing Service Quality: International Journal*, *15*(6), 577–590.

Torres, R. (2003). Linkages between tourism and agriculture in Mexico. *Annals of Tourism Research*, *30*, 546–566.

Vásquez, C., & Chik, A. (2015). "I am not a foodie…": Culinary capital in online reviews of Michelin restaurants. *Food and Foodways*, *23*(4), 231–250.

Watson, P., Morgan, M., & Hemmington, N. (2008). Online communities and the sharing of extraordinary restaurant experiences. *Journal of Foodservice*, *19*, 289–302. doi:10.1111/j.1748-0159.2008.00110.x

World Tourism Organization. (2017). *Affiliate members report, volume sixteen – Second global report on gastronomy tourism*. UNWTO, Madrid.

World Tourism Organization and Organization of American States. (2018). *Tourism and the sustainable development goals – Good practices in the Americas*. UNWTO, Madrid. doi:10.18111/9789284419685

Part 4
Sustaining Luxury

Chapter 17

Managing Luxury Brand Creation, Communication and Sustainability: Evidence from the Four Seasons Hotels and Resorts Case

Anni Ding[1] *and Tiffany S. Legendre*[2]

Abstract

This chapter aims to introduce luxury brand management strategies applied in the hospitality industry and illustrate how luxury hospitality business operators can create, communicate and sustainably manage luxury brands. This chapter introduces the definitions and dimensions of luxury brands, followed by an exploration of the concept and importance of luxury brand sustainability. This chapter then explores how to maintain long-term luxury hospitality brand sustainability by applying a luxury brand framework. This chapter uses a case study featuring the Four Seasons Hotels and Resorts (FSHR) to understand how a luxury brand communicates and maintains its sustainability through its various dimensions through the process of brand creation, communication and management. The chapter ends by providing practical implications for existing luxury hospitality brand organisations in the arena of customer service.

Keywords: Luxury brand; management; sustainability; Four Seasons; hospitality; brand communication

[1]https://orcid.org/0000-0003-2923-2377
[2]https://orcid.org/0000-0002-7948-3281

The Emerald Handbook of Luxury Management for Hospitality and Tourism, 337–352

doi:10.1108/978-1-83982-900-020211017

Introduction

The luxury market has created notable market values across a wide range of categories (Cristini, Kauppinen-Räisänen, Barthod-Prothade, & Woodside, 2017; Gurzki & Woisetschläger, 2017). It has grown much faster than other customer goods categories in past decades (Fionda & Moore, 2009). In the hospitality industry, the luxury hotel sector is becoming one of the most crucial segments due to its rapid expansion in recent years (Liu, Wong, Tseng, Chang, & Phau, 2017). Despite the tremendous growth of the luxury market worldwide, research in the luxury field regarding the creation, development and management of luxury brands has been under-represented (Gurzki & Woisetschläger, 2017; Miller & Mills, 2012). Researchers have been calling for a need to investigate how to build a strong luxury hospitality brand (Liu et al., 2017).

For contemporary customers, the value of a luxury brand has shifted its paradigm from building a transactional relationship with benefits to creating a long-term relationship with a holistic experience (Atwal & Williams, 2009). Therefore, luxury hospitality brand managers need to understand how to bring brand management to the next level to build long-term relationships with customers and create brand sustainability (Schultz & Block, 2015). This chapter applies Dubois, Laurent, and Czellar's (2001) luxury brand framework to analyze an exemplary case to accomplish this goal. This framework suggests that a luxury brand can be created and managed sustainably from six dimensions: (1) excellent quality, (2) very high price, (3) scarcity and uniqueness, (4) aesthetics and polysensuality, (5) ancestral heritage and personal history and (6) superfluousness.

The Emergence of a Luxury Brand and the Need for Brand Sustainability

The word 'luxury' originates from the Latin word 'luxus', which means 'soft or extravagant living and (over-) indulgence' (Tynan, McKechnie, & Chhuon, 2010). Several researchers define luxury as anything beyond necessity (e.g. Goody, 2006; Sekora, 1977). However, the term 'luxury' is now commonly used by business operators to distinguish their brand and communicate their products and services to a particular tier of customers (Dubois, Czellar, & Laurent, 2005). There is a lack of agreement in the luxury brand literature regarding the definition of a luxury brand (Miller & Mills, 2012). Phau and Prendergast (2000) propose that luxury brands consist of evoked exclusivity, well-known brand identity, high brand awareness, perceived quality, sustained levels of sales and customer loyalty. Hagtvedt and Patrick (2009) define a luxury brand as a premium product provider that enhances pleasure as the central benefits and builds relationships with customers based on emotional connection. Husic and Cicic (2009) suggest that brand luxury is mainly influenced by brand image and quality. Kapferer and Bastien (2009) argue that luxury brands serve the purpose of fulfilling customers' pleasure and demonstrating their economic success. Fionda and Moore (2009) interpret luxury brands from nine perspectives: marketing communication, product integrity, design signature, premium price, exclusivity, heritage, environment and service, culture and clear brand identity. Keller (2009) proposes that luxury brands should have 10 characteristics: premium image, intangible brand associations and aspirational image, the alignment of the marketing program, brand elements, secondary associations, controlled distribution, premium pricing strategies, brand architecture, competition and legally protected trademarks. In summary, although the definition of luxury brands remain inconclusive, researchers agree that luxury brands are about offering an exclusive and unique experience that is different from regular brands (Cristini et al., 2017; Dixit & Abraham, 2020; Gurzki & Woisetschläger, 2017).

Based on the unique characteristics of luxury brands, managing the business is different and more challenging for luxury brands than for regular brands. This is because luxury brand management often meets with the long-term growth trade-offs that cannot be observed in non-luxury brand business (Cavender & Kincade, 2014). A luxury brand often faces the challenges of keeping the uniqueness and exclusivity as the brand's competitive advantages while, at the same time, being generalised and relevant to attract a bigger customer market in order to maintain the business growth over time (Keller, 2009). Therefore, the management of luxury brands should be different from non-luxury brands (Miller & Mills, 2012). Specifically, to achieve business success, luxury brand operators need to find the right strategies that maintain brand sustainability (Dixit, 2017; Fionda & Moore, 2009; Kapferer & Bastien, 2009). Brand sustainability refers to organisations' strategies to bring brand growth and loyalty to a new level and increase brand value over time (Schultz & Block, 2015). As Schultz and Block (2015) demonstrate, finding the right strategies to develop and maintain brands is essential because a brand can decline and fail without proper management. Luxury brands are no exception. For example, back in the 1990s, the Louis Vuitton Moët-Hennessy (LVMH) failed to apply correct long-term brand investment strategies and overpaid for brands during the acquisition spree, resulting in a tremendous financial loss; whereas in 2002, LVMH implemented a shift and applied effective brand strategies that focus on long-term brand growth, which successfully improved profitability and increased cash flow as results (Cavender & Kincade, 2014).

Maintaining brand sustainability is also critical to luxury hospitality brands. Previous literature shows that brand success is positively associated with financial performance in the luxury hotel sector (Kim & Kim, 2005). For luxury hospitality brands, brand sustainability is maintained by creating extraordinary service experience, increasing customer satisfaction and loyalty, and, thus, creating long-term relationships with customers (Wu & Liang, 2009). As Atwal and Williams (2009) suggest, instead of offering functional benefits and building a transactional relationship, luxury brands need to provide a holistic experience and maintain a long-term relationship with their customers to achieve business success. However, there is a lack of suggestive guidance for luxury hospitality brand operators to better understand and manage brands over time (Liu et al., 2017).

The current chapter adopts Dubois, Laurent, and Czellar (2001) luxury brand framework to demonstrate how luxury hospitality organisations can create, communicate and sustainably manage their brands. Compared to many other conceptual models, this luxury brand framework has the advantage of highlighting the components that contribute to a unique understanding of luxury hospitality brands. First, while many luxury brand frameworks are context-based and cannot be used to explain luxury hospitality brands, this framework consists of dimensions that are general enough (Ko, Costello, & Taylor, 2019). Second, this framework illustrates the dimensions of luxury brands from both tangible and intangible perspectives, which fits the features of luxury brands within the hospitality industry. Third, examples of luxury hotels were used in the model demonstration in the original study, suggesting that this framework can be appropriately applied to the luxury hospitality context. To summarise, Dubois et al.'s (2001) luxury brand framework give us a standard method for examining and understanding luxury hospitality brands.

Justification for the Case Selection

As mentioned above, in order to explain and illustrate the implication of brand strategies on luxury hospitality brands' creation, communication and maintenance, this chapter provides an analysis of a case study featuring our Four Seasons Hotels and Resorts (FSHR). There are several reasons this case was chosen. First, FSHR is known as one of the world's leading ultra-luxury hotel brands with more than 100 properties in more than 40 countries worldwide. Second, FSHR has earned the most Forbes Five-Star Ratings among all hotel brands and is recognised for offering an excellent service experience to its guests. Finally, the brand philosophy and performance of FSHR fits the luxury brand framework. This chapter aims to provide practical suggestions for industry practitioners through the exploration and review of the Four Seasons Hotel case.

The Communication and Sustainability Management of Luxury Brands – Evidence from the Four Seasons Hotels and Resorts Case

The luxury brand framework proposed by Dubois et al. (2001) is used for this case study analysis. This framework consists of six aspects (see Fig. 17.1). The case of FSHR is applied in each aspect to demonstrate how a luxury hospitality brand is created and communicated, and most importantly, how brand sustainability is maintained from different perspectives.

Excellent Quality

A successful luxury brand is expected to establish a premium image through high-quality performance that meets or even exceeds customer expectations to justify a luxury price (Keller, 2009). Moreover, customers expect the performance of products or services to be maintained at a high standard with a guarantee of reliability and durability over time (Dubois et al., 2001). In the hospitality

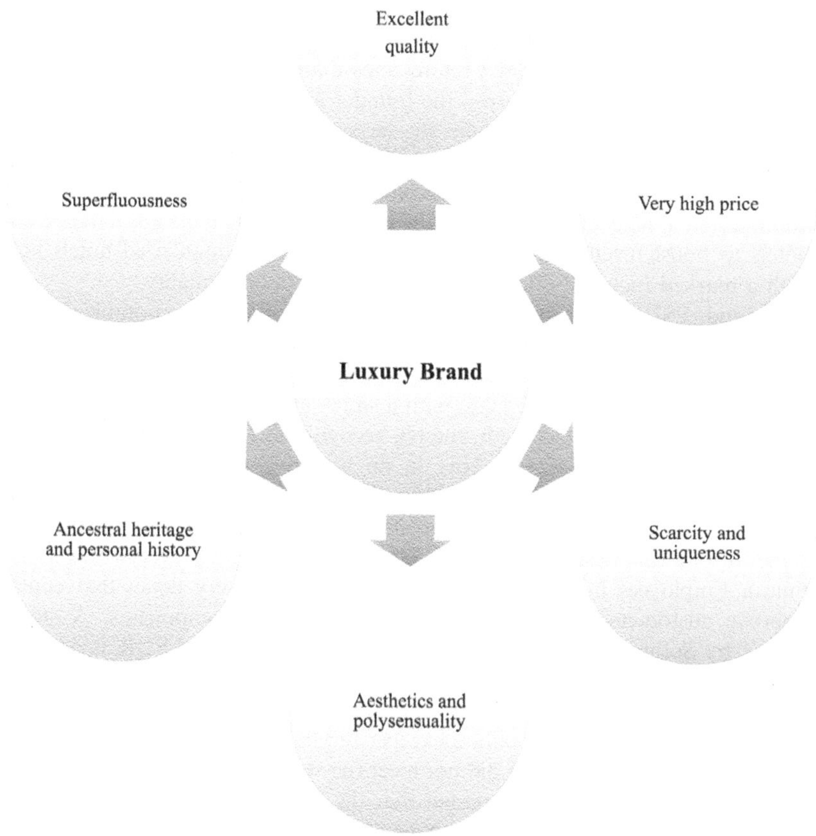

Fig. 17.1. Luxury Brand Framework Based on Dubois et al.'s (2001).

business, increasing customer satisfaction by delivering high-quality customer service is the key to maintaining long-term customer relationships. Although excellence in service is the primary objective for most business operators in the hospitality industry, the extent to which a brand can deliver quality service distinguishes it from its competitors. Therefore, the elaboration of luxury service calls for great attention to every detail (Dubois et al., 2001).

Very High Price

The premium price is believed to establish the luxury status of the brand (Fionda & Moore, 2009). Luxury brands are suggested to employ a premium pricing strategy to enhance quality and fewer discounts and markdowns (Keller, 2009).

Case Study 17.1: Excellent Service at FSHR

As a luxury brand that is widely known for its excellent service, FSHR considers service to be its priority. FSHR aims to use service quality as a competitive strength and creates impressions that can stay with customers for their lifetime. As Isadore Sharp – the founder and chairman of FSHR, mentions, the 'golden rule' that explains the success of FSHR is a simple idea that 'we should treat others the way we would want to be treated'. The application of FSHR's service philosophy is reflected in a variety of perspectives in operation. First, unlike other global chain hotel brands that include large-sized hotels to boost revenue, FSHR chooses to manage medium-sized hotels but with a marked focus on offering service with exceptional quality.

Second, FSHR has expertise in delivering premium and memorable service throughout every detail in the interaction with customers during their stay. At FSHR, the employees are trained and required to follow the seven SERVICE components: smile (greet the guest with a smile and speak clearly in a friendly manner), eye (eye contact with guests), recognition (acknowledge guests by using their name), voice (speaking in an attentive, courteous and natural manner with clear voice), informed (knowledgeable about hotel's services and products and able to take ownership of simple requests), clean (impeccable standard of cleanliness) and everyone (shows care for guests everywhere at all times). Employees are required to pay attention to every detail that could improve customers' staying experience based on the components. A story shared by Solomon (2013) about his friends experience at FSHR is a good example of excellent service quality illustrated by something as seemingly trivial as the click of a hotel door. One day during his stay at FSHR, when Solomon's friend closed his room door, a housekeeper serving the next room noticed that the closing sound of the door did not meet the standard FSHR gentle clicking sound, so she immediately requested that the engineering department fix the door for the guests. Such attention to detail among employees ensures FSHR's service consistency and creates a hassle-free experience for customers. As Homer Cruz, the assistant chef concierge, interprets, by 'quietly taking care of the details', their goal is to 'give guests precious time to relax and enjoy their stay' (Four Seasons Hotel Houston, n.d.).

Furthermore, as a unique strategy applied by FSHR, personal service functions as a source of superior quality. Employees are given the freedom and space to satisfy customers and meet their requests to the greatest extent possible. In one notable example shared by Ryan (2017), a guest at the hotel once requested to remove the door that joins two adjoining rooms. The concierge of FSHR made immediate arrangements to have the door taken out without question, regardless of how unreasonable this request may have sounded.

Third, FSHR introduces a service quality control system to ensure that the service meets the standards of FSHR in all aspects. To do so, management teams employ both external and internal auditors to constantly audit and evaluate service quality.

The reason is that customers commonly associate high price with high quality, the exclusivity of the product or service and a barrier to entry (Fionda & Moore, 2009). Since luxury brand customers traditionally have low price sensitivity, the strategy of offering discounts is not meaningful to the luxury brands' key segment of customers (Forgacs, 2003). Previous research even suggests that a luxury brand is applying a discount strategy to send an undesired signal to those target customers that ultimately decreases brand value (Keller, 2009). Moreover, a premium price is necessary to balance the high expenses incurred during the process of delivering a consistently high standard of service. Therefore, setting a high price is essential to the sustainability of luxury brands, and price promotion should not be a selling point during brand communication and in marketing campaigns (Forgacs, 2003).

Case Study 17.2: Premium Price at FSHR

FSHR sets a premium price for its services and offers few promotions. 'Not discounting rooms' is a policy of FSHR that is known in the luxury hospitality market (Turner, 2017). The FSHR 2002 Annual Report shows that, in 2001, while the entire hospitality industry experienced a downturn and many hotel companies chose to cut services and reduce prices, FSHR decided to maintain its services with no discounts to the room rates. As a result, FSHR improved its market share despite contrary predictions by industry experts. As John L Sharpe, the former COO of FSHR, once demonstrated, 'If we ever promote price, we are dead' (Forgacs, 2003). The average room rate at FSHR's top properties starts at $1,000 per night and above. For instance, the Georges V in Paris charges around $1,200 per night for a room, and the price of entry-level rooms of FSHR properties in New York starts at $850. In a list brought by Forbes regarding the most expensive hotel suites in 2012, the Four Seasons Hotel in New York ranked as number one with a room rate of $35,000 per night, which was $13,000 higher than the second most expensive luxury hotel listed on the ranking (Desai, 2012). However, FSHR delivers premium services that justify the high price. In its own unique way, FSHR transcends the quality-price matching philosophy, which creates additional profit for the business. The strategy that FSHR uses is to deliver beyond-standard service that fulfils customers' unique requirements with a premium fee attached to it. In the case mentioned above about a customer request to remove the doors of an adjoining room, the service operators of FSHR chose to meet customers' needs by charging a $50,000 additional fee. Such a strategy not only meets customers' needs and maintains customer relationships but also creates inventive opportunities for a rise in revenue.

Another strategy that FSHR applies to attract customers while maintaining a premium price is to offer special deals through carefully selected product distribution channels. Although FSHR rarely offers direct discount promotions, it provides special deals to attract its target customers by partnering with select travel agents. FSHR's unique partnership program, 'the Four Seasons Preferred Partner', is an exclusive and invitation-only program for luxury travel agencies (Four Seasons Preferred Partner, n.d.). Through this program, FSHR carefully selects travel agencies and offers its target customers exclusive deals, such as free room upgrades, breakfast, amenities and priority to access property facilities. In this way, FSHR maintains its relationship with target customers without discounting its products and decreasing the brand value.

Scarcity and Uniqueness

For luxury brand management, the scarcity and uniqueness elements are crucial to enhancing luxury brand identity, as it is the basis of long-term capitalisation and worldwide expansion (Phau & Prendergast, 2000). Scarcity and uniqueness should not be limited to the product itself but rather should inform the sophisticated experience that a luxury brand delivers to its customers through a variety of detailed perspectives, such as atmosphere, decoration, the way the product is presented to customers and the interaction with the service provider (Dubois et al., 2001).

Case Study 17.3: Scarce and Unique Experience at FSHR

Another practice that contributes to the exceptionalism of FSHR is that it presents its customers with an experience that is purposely integrated with the local culture, environment and customs of the surrounding community. Such a unique and immersive experience is delivered in detail through all dimensions, from the atmosphere of the hotel, from the products offered by the hotel, to the employees working in the hotel. As Isadore Sharp suggests, 'Four Seasons today is a brand that people equate with exceptional experiences and the lifelong memories that come with them' (Corporate Bios, n.d.).

Design of FSHR: The setting of culture starts with the interior and exterior design of FSHR (Tomchin, 2017). Each hotel of FSHR is unique to its location. As demonstrated by Tom Segesta, the general manager of Four Seasons in Houston, TX, 'each hotel is unique and different based on the location that it is in such that the customers can have a chance to immerse in the traditions and culture of the local community' (Tomchin, 2017). For example, the Four Seasons Resort Maui at Wailea displays a museum-quality contemporary art collection by a local artist that showcases Hawaiian culture. The Four Seasons Hotel Miami spent more than $3.3 million worth of commissions for the creation, acquisition and distribution of art pieces produced by local artists in its public areas. As a bridge to the local arts community, the Four Seasons Hotel Denver showcases an art gallery–like atmosphere to its customers and displays nearly 1,200 pieces of artwork mostly created by Colorado artists throughout its public and private spaces. As further explained by Kerri Sholly, the public relations director of Four Seasons Austin, TX, FSHR is one of the first companies to demonstrate its dedication to ensuring that the travelling public gets a genuine feel for their go-to destination (Tomchin, 2017).

Experiential tours offered by FSHR: FSHR designs its experiential products based on local culture and resources. Some examples of experiential tours offered by FSHR include a guided walking safari tour in the Serengeti region of Africa, a private yacht tour exploring the ocean surrounding the Maldives islands, a customised surfing safari in Kuda Huraa, Maldives, a horse-drawn carriage ride through some of the local markets and streets in Marrakech, Morocco, exclusive access to join the traditional *dabbawallas* lunch-box delivery trade in Mumbai, India, and a visit to an ancient local community in Colombia that still maintains its traditional lifestyle. Recently, FSHR has even launched a fully customised around-the-world trip named 'Word of Adventures 2021' that takes guests to eight different countries in 24 days. With the careful design of experiential products reflecting local cultural characteristics, FSHR ensures that each destination leaves customers with a unique and unforgettable experience, no matter which destination customers visit.

Employees at FSHR: FSHR names its employees' culture carriers, which means that the employees play a vital role in presenting the local culture to their customers. The employees at FSHR enjoy the opportunity to work all over the world. When arriving at a new location, the employees of FSHR are trained to adapt to the culture and customs of the host community to ensure consistency with the atmosphere of that location. For instance, before FSHR opened the Four Seasons Hotel George V in Paris, France, the management team, as well as the employees, had to be rigorously trained in the identification and understanding of cultural differences between France and America, especially with regard to the French people's philosophy and manner of doing business. Such efforts over time naturally lead to the expectation that the culture carriers will always carry out a memorable, respectful, culturally relevant and immersive experience that exceeds the expectations of the customers.

Aesthetics and Poly-sensuality

Some customers believe that the products of luxury brands should not just settle at meeting the requirements of quality, price and scarcity but should also provide an exceptional hedonic experience to customers (Dubois et al., 2001). As Kapferer (1997) suggests, luxury is an art that defines beauty, and luxury products could provide extra pleasure. Along with this statement, Dubois et al. (2001) further indicate that luxury products should be a source of sensual pleasure, which means they should be pleasant not only to look but also to hear, smell, taste and touch.

Case Study 17.4: Aesthetic Infusion at FSHR

Sensual aesthetic pleasure is a significant component of the overall FSHR experience, and it is always provided in service with the ultimate goal of creating a unique and memorable experience for customers.

Art of vision: In addition to creating a visually enjoyable experience by presenting a variety of local art throughout its public and private spaces, FSHR allocates spaces for art galleries that bring the pursuit of artistic beauty and excellence to a new and immediate level. For example, on the first floor in the Four Seasons Hotel Shenzhen, guests will find a full art gallery. FSHR also offers visually stimulating experiences that are designed to bring joy to customers. For instance, Four Seasons Hotel St. Louis provides an exclusive and customised firework show in the sky above the Mississippi River. Four Seasons Hotel Seattle offers a private outing to a glass artist's boathouse and studio.

Art of palate: FSHR expands the art experience to the exploration of the palate by offering exclusive culinary tours. Unlike a regular culinary experience, FSHR delivers the joys of taste with a combination of other senses to customers. For example, Four Seasons Hotel Chiang Mai in Thailand offers a cooking class guided by world-class Thai cuisine master chefs. Customers who take the class are able to pick fresh ingredients at the local market to incorporate into their dish through a hands-on cooking experience at the hotel kitchen. The Four Seasons Hotel Firenze in Italy offers a golden sunset dinner on the Ponte Vecchio, so customers can enjoy the pleasure from both the four-course gourmet meal designed by a Michelin-starred chef and the view of the sunset at Arno river. Grand-Hôtel du Cap-Ferrat in France and Four Seasons Costa Rica both offer to arrange a private dinner under a dome of stars. Four Seasons Hotel Riyadh in Saudi Arabia provides a dinner for two at Sky Bridge that enable customers to enjoy a panorama of the city below.

Art of music: Some locations of FSHR also provide exclusive opportunities for customers to enjoy the pleasure brought by music. For instance, customers at Four Seasons Hotel Casa Medina in Colombia have a chance to enrol in a private dancing class and learn salsa with professional Colombian salsa dancers.

Ancestral Heritage and Personal History

A luxury brand should be able to tell a story or even a legend to its customers (Dubois et al., 2001). History and stories describing a rich heritage are important to luxury brands because they create deep meaning and symbolic value to the brand (Keller, 2009). Moreover, the strategy of highlighting a brand's history and emphasising its long-standing fidelity to its place in history can elevate customers' perception of brand authenticity (Fionda & Moore, 2009).

Case Study 17.5: Brand Heritage and Distinctive Service at FSHR

Founded in 1960 and opening its first hotel in 1961, FSHR is a Canadian international luxury hospitality company headquartered in Canada that operates more than 100 hotels over the world. The story of FSHR can be interpreted as a tale of continuous dedication to the highest standards of service quality (Tuite, 2018). During the process of development worldwide, FSHR redefined the meaning of hospitality by pursuing and remaining committed to delivering a distinctive level of service. FSHR is the first hotel company in North America to introduce accommodation items such as bath amenities, robes and hair dryers that have now become standardised in the hotel industry worldwide. From 1985 on, FSHR has offered residential options with personalised service for customers who prefer the comforts of living at home. The prevailing strength of FSHR has been the assurance of the high quality of life. In 2012, FSHR expanded its service model and introduced private jet tours. To offer customers the best sleep experience, in 2014, FSHR introduced a customisable mattress for customers to choose from during their stay. Since 2015, FSHR has focused more on the innovation of technology and the adaptation of the most cutting-edge technologies to improve overall service quality. For example, FSHR introduced a mobile application to add to the convenience of customers during their stay and further enhanced the application by embedding an award-winning chat function that facilitates the communication experience between customers and hotel staff.

The secret of success along FSHR's history has been to define their business strategy in terms of 'redefining luxury as service' (Tuite, 2018). To help deliver its brand promise consistently over the last 60 years, FSHR established the 'three C's', 'connection, craftsmanship and characters', as its core values (Pougnet, n.d.). Such values help engender a memorable and emotional experience for both customers and employees alike through a sense-based, customer-focused approach. The practice of maintaining this kind of customer relationship has been shown to enhance the long-term loyalty of customers overall (Pougnet, n.d.). The key values of FSHR are summarised by Heather Jacobs, assistant manager of housekeeping: 'We believe our purpose is to create impressions that will stay with you for a lifetime. It comes from our belief that life is richer when we truly connect to the people and the world around us' (About Us, n.d.).

Superfluousness

The concept of luxury is also associated with superfluousness (Dubois et al., 2001). Some researchers believe that luxury contains the quality of superfluousness in that it transcends basic necessities (Sekora, 1977) and represents those things that are desirable but inessential (Goody, 2006). It is also believed that luxury products and services should derive their values from a more hedonic rather than functional perspective. One aspect of superfluousness is to create an extreme experience of convenience for customers to freely decide what they want to do without concern about the consequences of making a wrong decision (Dubois et al., 2001).

Case Study 17.6: FSHR's Superfluousness in Service

FSHR extends its luxury brand experience to superfluousness by expanding its product distribution channels horizontally and vertically. Horizontally, FSHR has created the Four Seasons Private Jet service to transport customers to themed tours with a stay at FSHR locations around the world. With the private jet service, customers' luxury experience does not start at the point of entering the hotel but rather start on the way to the hotel. FSHR has promoted a series of travel packages that take care of customers' transportation, accommodations and destinations with high FSHR standards, which creates a complete worry-free luxury travelling experience for the customers. Regarding its jet service, the official FSHR websites state, 'The only thing you'll have to think about is enjoying yourself and doing as much or little as you want. We'll take care of the rest' (The Pyramid, 2021). Vertical-wise, FSHR has developed a signature sleep collection that includes mattresses, bedding and other accessories such as towels, robes and books. FSHR spent more than 50 years partnering with experts to improve its mattresses and bedding quality to bring an unparalleled sleep experience to its customers. Moreover, the availability of sleep collection products allows customers to extend such a luxury experience to their own homes.

Towards a Future of Luxury Brand Sustainability

From the case study of FSHR explained by the luxury brand framework, this chapter illustrated how a luxury hospitality firm creates, communicates and maintains the brand over time. First, in the analysis of excellent quality, the study showed that the secret of FSHR's business success is offering exceptional quality of service. Specifically, the pursuits of service quality of FSHR are reflected in every detail during the process of service delivery. To FSHR, offering service is akin to craftsmanship, necessitating a constant passion for refinement and a keen eye for perfecting details. Although the importance of service quality is commonly emphasised in the hospitality industry, bringing the service quality to a distinguishing edge is what sets FSHR apart from other brands. Ensuring brand sustainability means improving the service quality to a level where every element can be consistently replicated to the highest possible standard and where every service provider is deeply committed to delivering this standard of services organically and with heart.

Second, in the analysis of premium pricing, it has been noted that while FSHR sets the premium price to maintain its luxury brand status, it also offers an extraordinary quality of service that merits customers' expectations of the price. Furthermore, FSHR applies the strategy of offering highly customised service options that can meet almost any customer desire at an appropriate charge. Through this strategy, customers' needs are fulfilled while FSHR gets the opportunity to create additional profits. This suggests that to maintain a sustainable luxury brand status, luxury brands should use fewer price promotion strategies that can decrease the brand value and make more of an effort to elevate their service quality to match the premium price. However, while referring to FSHR's strategy of service with an additional fee, the brands need to make sure that: (1) such service is highly personalised so that it remains beyond the service standard of a free charge and (2) the additional charge should match customers' price sensitivity and their perception of the service quality. Without a careful evaluation of service charge, customers might perceive an unfair price that fails to justify the service they received, and such perception can result in reduced customer satisfaction that is harmful to brand sustainability.

Third, in the analysis of scarcity and uniqueness, this study showcased how FSHR delivers a personalised and unique experience to its customers. Unlike other business operators, the uniqueness of the experience provided by FSHR is the immersion in the local culture and tradition at each destination. Achieving this goal requires great effort to understand the local culture, design the service based on the culture and allocate resources to create the most authentic experience for customers. Industry practitioners should carefully evaluate what types of experiences they can offer to their customers and how they can deliver such experiences with a consistently high quality of service. The ultimate goal of scarcity and uniqueness is to create and deliver an experience for customers that cannot be found elsewhere.

Fourth, in the analysis of aesthetics and poly-sensuality, the study showed how FSHR brings the customer experience to a new level representing the beauty of art and creates pleasure for all senses. To improve customer satisfaction, hospitality luxury business operators can think of how to create a service experience that merges different sensual elements by paying attention to the things customers see, smell, hear, taste and touch.

Fifth, in the analysis of ancestral heritage and personal history, the study suggested that, for a luxury brand, it is vital not only to have an appealing history and story to tell but also to show faith in the historical brand value and the perseverance of applying such value over time. The luxury hospitality organisation should be able to tell a brand story that brings out the meaning that is consistent with the history of the brand.

Lastly, in the analysis of superfluousness, this study introduced the extension of the FSHR service experience through different product distribution channels. Luxury is about bringing service to an 'unnecessary' but desirable level. To this end, FSHR has expanded its service model with the inclusion of private jet service and premium sleep collections. The implication here is that a brand can never be sustainable if it is complacent. Hospitality practitioners should always think further, pay extra attention to details and design short-term and long-term strategies to meet customers' needs to a transcendent level in an ongoing good faith effort to maintain greater luxury brand levels sustainability.

References

About Us. (n.d.). Retrieved from https://www.fourseasons.com/about_four_seasons/

Atwal, G., & Williams, A. (2009). Luxury brand marketing – The experience is everything! *Journal of Brand Management*, *16*(5–6), 338–346. doi:10.1057/bm.2008.48

Cavender, R. C., & Kincade, D. H. (2014). Management of a luxury brand: Dimensions and sub-variables from a case study of LVMH. *Journal of Fashion Marketing and Management*, *18*(2), 231–248. doi:10.1108/JFMM-03-2013-0041

Corporate Bios. (n.d.). Retrieved from https://www.fourseasons.com/about_four_seasons/corporate_bios/isadore_sharp

Cristini, H., Kauppinen-Räisänen, H., Barthod-Prothade, M., & Woodside, A. (2017). Toward a general theory of luxury: Advancing from workbench definitions and theoretical transformations. *Journal of Business Research*, *70*, 101–107. doi:10.1016/j.jbusres.2016.07.001

Desai, Y. (2012). The most expensive hotel suites so far this year. Retrieved from https://www.businessinsider.com/the-most-expensive-hotel-suites-for-2012-2012-2

Dixit, S. K. (Ed.). (2017). *The Routledge handbook of consumer behaviour in hospitality and tourism*. London: Routledge.

Dixit, S. K., & Abraham, A. (2020). Brand identity and positioning in selected India chain hotel companies. In M. A. Gardini, M. C. Ottenbacher, & M. Schuckert (Eds.), *The Routledge companion to international hospitality management* (pp. 260–270). London: Routledge.

Dubois, B., Czellar, S., & Laurent, G. (2005). Consumer segments based on attitudes toward luxury: Empirical evidence from twenty countries. *Marketing Letters*, *16*, 115–128. doi:10.1007/s11002-005-2172-0

Dubois, B., Laurent, G., & Czellar, S. (2001). *Customer rapport to luxury: Analyzing complex and ambivalent attitudes*. HEC Research Papers Series. Retrieved from https://ideas.repec.org/p/ebg/heccah/0736.html

Fionda, A. M., & Moore, C. M. (2009). The anatomy of the luxury fashion brand. *Journal of Brand Management*, *16*(5–6), 347–363. doi:10.1057/bm.2008.45

Forgacs, G. (2003). Brand asset equilibrium in hotel management. *International Journal of Contemporary Hospitality Management*, *15*(6), 340–342. doi:10.1108/09596110310488203

Four Seasons Hotel Houston. (n.d.). Retrieved from https://www.fourseasons.com/houston/service-and-amenities

Four Seasons Preferred Partner. (n.d.). Retrieved from https://www.kristennix-travel.com/four-seasons-preferred-partner

Goody, J. (2006). From misery to luxury. *Social Science Information*, *45*(3), 341–348. doi:10.1177/0539018406066526

Gurzki, H., & Woisetschläger, D. M. (2017). Mapping the luxury research landscape: A bibliometric citation analysis. *Journal of Business Research*, *77*, 147–166. doi: 10.1016/j.jbusres.2016.11.009

Hagtvedt, H., & Patrick, V. M. (2009). The broad embrace of luxury: Hedonic potential as a driver of brand extendibility. *Journal of Customer Psychology*, *19*(4), 608–618. doi:10.1016/j.jcps.2009.05.007

Husic, M., & Cicic, M. (2009). Luxury consumption factors. *Journal of Fashion Marketing and Management*, *13*(2), 231–245. doi:10.1108/13612020910957734

Kapferer, J. N. (1997). Managing luxury brands. *Journal of Brand Management*, *4*(4), 251–259. doi:10.1057/bm.1997.4

Kapferer, J. N., & Bastien, V. (2009). The specificity of luxury management: Turning marketing upside down. *Journal of Brand Management*, *16*(5–6), 311–322. doi: 10.1057/bm.2008.51

Keller, K. L. (2009). Managing the growth trade-off: Challenges and opportunities in luxury branding. *Journal of Brand Management*, *16*(5–6), 290–301. doi:10.1057/bm.2008.47

Kim, H. B., & Kim, W. G. (2005). The relationship between brand equity and firms' performance in luxury hotels and chain restaurants. *Tourism Management*, *26*(4), 549–560. doi:10.1016/j.tourman.2004.03.010

Ko, E., Costello, J. P., & Taylor, C. R. (2019). What is a luxury brand? A new definition and review of the literature. *Journal of Business Research*, *99*, 405–413. doi:10.1016/j.jbusres.2017.08.023

Liu, M. T., Wong, I. K. A., Tseng, T. H., Chang, A. W. Y., & Phau, I. (2017). Applying customer-based brand equity in luxury hotel branding. *Journal of Business Research*, *81*, 192–202. doi:10.1016/j.jbusres.2017.06.014

Miller, K. W., & Mills, M. K. (2012). Contributing clarity by examining brand luxury in the fashion market. *Journal of Business Research*, *65*(10), 1471–1479. doi: 10.1016/j.jbusres.2011.10.013

Phau, I., & Prendergast, G. (2000). Consuming luxury brands: The relevance of the 'Rarity Principle'. *Journal of Brand Management*, *8*(2), 122–138. doi:10.1057/palgrave.bm.2540013

Pougnet, S. (n.d.). Four Seasons: Connecting people in the hospitality industry. Retrieved from https://hospitalityinsights.ehl.edu/four-seasons-community-building

Ryan, J. (2017). The four seasons' approach to customer service. Retrieved from https://medium.com/the-ascent/the-four-seasons-approach-to-customer-service-a15d27450c2f

Schultz, D. E., & Block, M. P. (2015). Beyond brand loyalty: Brand sustainability. *Journal of Marketing Communications*, *21*(5), 340–355. doi:10.1080/13527266.2013.821227

Sekora, J. (1977). *Luxury: The concept in western thought, Eden to Smollett*. Baltimore, MD: Johns Hopkins University Press.

Solomon, M. (2013). Four Seasons' customer service: Consulting the systems behind the click of a hotel door. *Forbes*. Retrieved from https://www.forbes.com/sites/micahsolomon/2013/11/17/secret_shopping_four_seasons/#169dc5403f38

The Pyramid. (2021). Retrieved from https://m.fourseasons.com/content/fourseasons_mobile/en/privatejet/journeys/ancient-explorer-2021/the-pyramids/

Tomchin, R. (2017). Four Seasons: A global chain of excellence – Platform magazine. *Platform Magazine*. Retrieved from https://platformmagazine.org/2017/10/25/four-seasons-global-chain-excellence/

Tuite, S. (2018). The story of Four Seasons. Retrieved from https://press.fourseasons.com/news-releases/the-story-of-four-seasons/

Turner, S. (2017). How to book a stay at a Four Seasons hotel for less. *Forbes*. Retrieved from https://www.forbes.com/sites/sarahturner/2017/05/23/how-to-indulge-in-bargain-luxury-at-a-four-seasons-hotel/#15f158b07458

Tynan, C., McKechnie, S., & Chhuon, C. (2010). Co-creating value for luxury brands. *Journal of Business Research*, *63*(11), 1156–1163. doi:10.1016/j.jbusres.2009.10.012

Wu, C. H. J., & Liang, R. D. (2009). Effect of experiential value on customer satisfaction with service encounters in luxury-hotel restaurants. *International Journal of Hospitality Management*, *28*(4), 586–593. doi:10.1016/j.ijhm.2009.03.008

Chapter 18

Sustainable Luxury Tourism: Promises and Perils

Deep Jyoti Gurung,[1] Paridhi Brahma,[2] and Chandan Goswami[3]

Abstract

Recent decades have witnessed a rising concern regarding the prosperity of the environment and utilisation of resources. A sustainable approach is being promoted in all sectors. In the field of tourism, sustainable tourism is widely discussed among researchers and practitioners. On the other hand, luxury tourism is criticised for lavish resource utilisation to serve the few luxury tourists. There is a need to include sustainability in luxury tourism to benefit the environment, local communities, tourist destination and luxury tourists. However, sustainable luxury tourism is an emerging concept and needs more investigation. This chapter attempts to present the existing knowledge about sustainable luxury tourism by implementing a systematic literature review. Further, the opportunities and challenges associated with sustainable luxury tourism are being highlighted. This study has identified the factors that need to be considered to promote sustainable luxury tourism. Moreover, suggestions of the researchers are being presented to serve as guidelines. This study includes an example of the Diphlu river lodge, which has practised sustainable luxury tourism for many years. The viewpoint of luxury tourists are being understood by analysing the reviews of tourists from TripAdvisor using NVIVO-12 qualitative data analysis software. The combination of literature review and practical information provides insight into sustainable luxury tourism.

[1]https://orcid.org/0000-0003-4438-3210
[2]https://orcid.org/0000-0003-3539-927X
[3]https://orcid.org/0000-0002-3093-781X

The Emerald Handbook of Luxury Management for Hospitality and Tourism, 353–378

doi:10.1108/978-1-83982-900-020211018

Keywords: Sustainable luxury tourism; systematic literature review; tourist review; luxury tourism; Kaziranga National Park; Assam tourism

Introduction

Collins Dictionary has defined luxury as "very great comfort, especially among beautiful and expensive surrounding". The same dictionary also defines luxury as "something expensive that is not necessary but gives you pleasure". The definition projects luxury as something lavish that not all can afford. Luxury goods and services are generally very highly priced, and the quality of these is superior compared to normal goods and services. Such goods and services provide a status symbol to the users. If we go back to the days of Xuanzang, the main purpose of travel was the search for knowledge (Britannica, 2021). Leisure became one of the major purposes of travel giving birth to tourism in the seventeenth century (Walton, 2020). With the increase in the wealth of the people, luxury tourism is gaining popularity.

Sustainable consumption is defined as 'the consumption of goods and services that meet basic needs and quality of life without jeopardising the needs of future generations' (OECD, 2002). "Sustainable Development" was first used as an important term in the World Conservation Strategy, 1980 (IUCN UNEP & WWF, 1980). Sustainable development holds the potential to address the fundamental challenges for humanity at present and into the future (Hopwood, Mellor, & O'Brien, 2005). Applying the concept of sustainability to the tourism sector may be referred to as sustainable tourism. It is well accepted that tourism brings in a negative impact to the destination's eco-system, the culture of the locals and many other areas. As tourism brings in revenue, there is a tendency to keep on boosting tourism by governments of less developed areas. Short time benefits overshadow the long-term cost. Unplanned and uncontrolled growth of tourism may lead to many environmental issues. Increasing concern about sustainability has brought in the concept of carrying capacity in the tourism sector.

Luxury Tourism

Luxury tourism is defined by the consumption of high-priced and high-quality ever-evolving experiences (Page, 2012; Sundbo & Dixit, 2020). This implies that staying at high-end accommodation houses, eating at high-end restaurants, using private planes to travel is part of luxury tourism. The basic objective behind luxury tourism is to get such an experience that seemed unreal previously – a dream that has been converted to reality, the memory to be cherished later Luxury tourists are motivated to travel by the desire to experience luxury stays, far-off destinations, and excitement distinct from ordinary life apart from such places fulfilling the general needs for relaxation, visits and sightseeing, socialisation, economic and sports (Novotná & Kunc, 2019).

Luxury tourism can be defined as travelling away from the masses to experience uniqueness without considering the price to be paid for that. With the shift of travel motivations from leisure to experiences, new opportunities have emerged.

Luxury tourist value new offerings in the form of unique experiences through various programmes and services that cannot be experienced in one's home (Cerović, Pavia, & Floričić, 2019; Dixit, 2020). It is also observed that a rise in the standard of living has increased the demand for luxury experiences. It has triggered the qualitative development of services and facilities to the extent that it has surpassed the characteristics deemed mandatory as per the law. It becomes difficult to measure their quality using the traditional star rating system and therefore are valorised and positioned in the market as luxury accommodation (Cerović et al., 2019).

Sustainability and Sustainable Development

As per an estimate, some 300 definitions of 'sustainability' and 'sustainable development' are mostly limited to the domain of environment management and its associated disciplines (Johnston, Everard, Santillo, & Robert, 2007).

The mixed state of affairs of the increasing gap between the rich and the poor, and the simultaneously degrading biospheric systems summed up along with the catastrophic failures of the decision-making efforts brought in the idea of sustainability (Gibson, 2006).Sustainability is basically consuming with constraint (non-renewable resources and resources susceptible to degradation) so that future generations too can enjoy what we are enjoying now. It includes all aspects of life, starting from the environment to culture.

Sustainability connects the tripartite descriptions of society, environment and economy (Purvis, Mao, & Robinson, 2018). Development is human nature. Nevertheless, development puts pressure on the ecosystem. Hence the concept of sustainable development has come in. Sustainability is the desired goal of development and environmental management. It finds its place in numerous disciplines and a variety of contexts, such as sustainable yield of bioresources to a sustainable society in the form of a steady economy (Brown, Hanson, Liverman, & Merideth, 1987).

Sustainable Luxury Tourism

Luxury tourism is seen by a group of experts as destructive and wasteful activities and by another group as constructive, leading to conservation of eco-system and upliftment of indigenous people. The unique experience which luxury tourism offers to the tourists is termed by Smith as "superfluous items and wastefully lavish experience" (Raymen & Smith, 2019). On the other hand, Ryan and Stewart (2009), in their case on Al Maha, United Arab Emirates (UAE), found that a combined role of supplier vision of regeneration and restriction on visitor penetration into natural spaces promotes conservation.

It has also been seen that researchers who are in favour of luxury tourism generally talk about controlled tourism. Keeping the number of tourists restricted much below the maximum level may be the key to sustainable luxury tourism (D'hauteserre, 2005; Shah, 2011). As luxury tourism is high-priced tourism, keeping the number low and making it sustainable may be the ideal approach to developing luxury tourism.

However, there is a need to examine the dimensions of sustainable luxury tourism. This chapter conducts a systematic literature review to identify relevant research articles and identify the challenges and opportunities associated with sustainable luxury tourism. Further, an example from a luxury resort is provided, which practices sustainable luxury tourism. This example compares the findings of a systematic literature review and provides practical insight into the practices of sustainable luxury tourism.

Method of Systematic Literature Review

To understand the relationship between sustainability and luxury tourism, a systematic literature review was conducted by PRISMA (Liberati et al., 2009). It ensures that the process of the systematic literature review is transparently reported. Thus, this systematic literature review includes the following steps: research questions, search strategy, resources included, search process, search selection criteria and the number of studies included for synthesis.

Research Questions

This study attempts to answer the following research questions:

RQ1. How much exploration is done in the field of sustainable luxury tourism?

RQ2. Which factors can assist in the development of sustainable luxury tourism?

RQ3. What are the sustainability concerns associated with luxury tourism?

The answers to the above questions are expected to provide important information that can direct future research and also provide meaningful information to the stakeholders.

Search Strategy

To search for the relevant literatures that discuss the topic sustainable luxury tourism, the initial idea was to search using the exact phrase "sustainable luxury tourism". However, it was observed to be very rigid, and only limited results were found (e.g. SCOPUS-3, PROQUEST-1 and EBSCO-0). Therefore, the synonyms and associated words of the major terms were also identified, such as environment and ecology for sustainability. Moreover, the wildcard is used to include all the words that can be formed by different suffixes from the major terms such as tour, tourism; tourists can be derived from tour.

Boolean OR was used to include alternate synonyms and associated words. Boolean AND was used to connect the major terms. The following search string was formed and applied to search the online resources:

("ecolog*" OR "sustain*" OR "environ*") AND ("luxury tour*" OR "luxury travel" OR "luxury visit" OR "luxury trip" OR "luxury holiday")

Resources Considered for This Study

Web of Science has a collection of high-quality research of over 115 years (Clarivate, 2020), and SCOPUS is renowned for its wide coverage of interdisciplinary literature (Elsevier, 2020). Hence, Web of Science (WOS) and SCOPUS are considered as important resources for this study. Additional works of literature were added from EBCOHOST (Business Source Elite) and PROQUEST ABI/INFORM (refer Fig. 18.1).

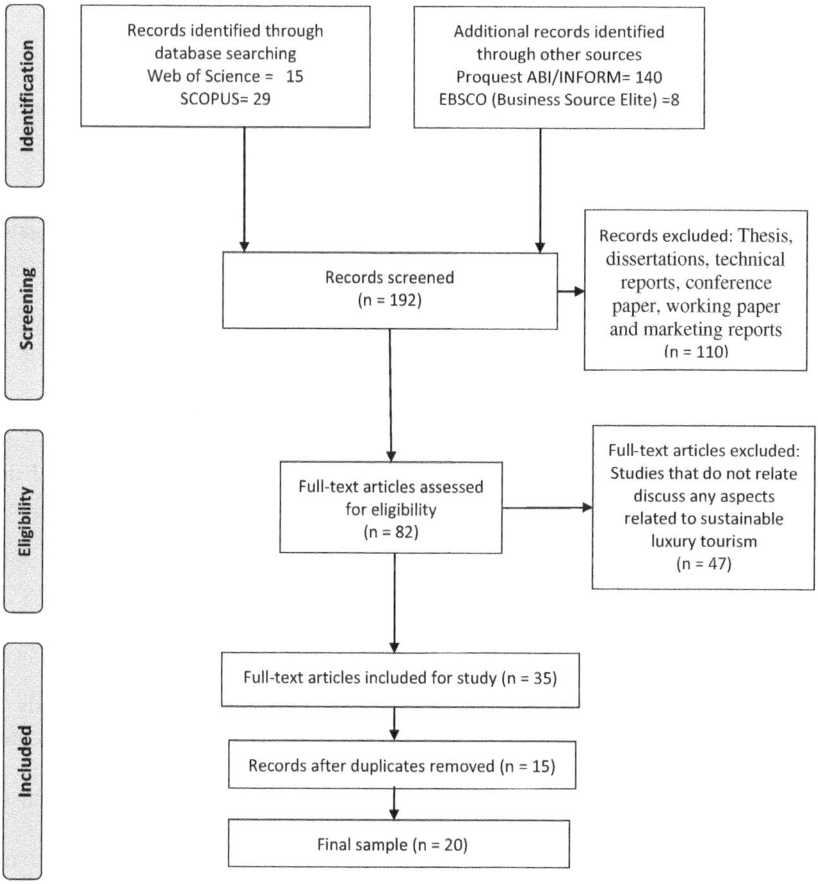

Fig. 18.1. Literature Search Process. *Source:* Adopted from Moher, Liberati, Tetzlaff, and Altman (2009).

Search Process

Three researchers have searched the resources mentioned above for valuable literature using the same search string (mentioned above) for consistency in the search results (i.e. identified literatures).

Study Selection Criteria and Procedures

The studies identified by the searches were evaluated based on the following inclusion and exclusion criteria.

Inclusion Criteria

(1) The studies that discuss sustainable, environmental or ecological concerns associated with the development of luxury tourism were considered to be included in this study (*RQ1*).
(2) The studies that discussed the prospects of sustainable luxury tourism (*RQ2*) were included for further examination. Further, the studies that investigated the problems associated with sustainable luxury tourism (*RQ3*) were considered for review.
(3) To be included for further review, the studies needed to be complete. In case there are different studies in the same geographical scope, each study will be treated as a separate study.
(4) Research articles on the theme 'sustainable luxury tourism' and related themes in any year and published in a peer-reviewed journal will be evaluated. Publications in a peer-reviewed journal ensures the findings and discussions of the articles are validated (Kelly, Sadeghieh, & Adeli, 2014).
(5) The articles need to be published in the English language.

Exclusion Criteria

The following criteria were followed to exclude a study:

(1) Studies in languages other than English were excluded from this study as none of the researchers are competent with any other major languages used in academia.
(2) Studies that solely concentrate on luxury tourism but do not highlight on sustainability, ecology or environment were excluded.
(3) Studies that focus on sustainability, ecology or environment but do not include luxury tourism, luxury stay, luxury holiday or luxury stay were excluded.
(4) Thesis, dissertations, technical reports, conference papers, working papers and marketing reports were excluded from this study.

Selecting Primary Sources

Initially the primary sources were selected by screening the abstract, keywords and titles by the three researchers. The studies which were found to be irrelevant by all the three researchers were excluded from the review. In case of any disagreement, the full text of the article was being reviewed by all the three researchers and after a detailed discussion, the final decision was taken to select or reject the articles for this study (refer Table 18.1).

The relevant documents were then further verified for duplicates, and final articles to be further reviewed were reduced to 20. The final sample of the articles was examined to find the answers to the above-mentioned questions.

RQ1. How much exploration is done in the field of sustainable luxury tourism?

The sources of articles reveal that studies related to sustainable luxury tourism are from various disciplines, but the number of articles also indicates that there is a need for more exploration on this theme. Moreover, this also indicates the lack of dedicated source of articles on sustainable luxury tourism (refer Table 18.2).

The relevant research articles were examined based on the trend of publication (on the basis of year), geographical scope, themes and locations (refer Figs. 18.2 and 18.3).

Publications of Research Articles

The collection of the literature suggests that most of the studies associated with sustainability and luxury tourism are conducted in Mexico ($n = 3$). Two studies have been found for each nation of Denmark, Morocco and the United Arab Emirates (UAE). One study was found for each of these countries – Monaco, Maldives, Malaysia, Switzerland, India, China, Costa Rica and the Kingdom of Saudi Arabia. One study included these three countries – St. Lucia, Barbados and Jamaica. Moreover, one study discusses about the possibility of luxury tourism in outer space that can benefit the preservation of earth resources.

Table 18.1. Sample of Articles.

Source of Literature	Documents Identified	Research Article	Not Relevant	Relevant
Web of Science	14	13	2	11
SCOPUS	29	14	3	11
Proquest ABI/INFORM	140	49	41	08
EBSCO (Business Source Elite)	8	6	1	5
TOTAL	187	82	47	35

Source: Researcher's own.

Table 18.2. Sources of Articles.

Sources	Number of Articles
Tourism Geographies	1
Human Organisation	1
Energy Policy	1
Journal of Sustainable Tourism	1
Journal of Tourism and Cultural Change	1
Asia Pacific Viewpoint	1
Policy Sciences	1
Journal of Coastal Conservation	1
Sustainability: Science, Practice, and Policy	1
International Journal of Culture, Tourism and Hospitality Research	1
Tourism Management Perspectives	1
Tourism Management	1
Journal of Coastal Research	1
The Journal of Economic Perspectives	1
International Journal of Contemporary Hospitality Management	1
Journal of Regional Research Investigaciones Regionales	1
Sustainability	1
Worldwide Hospitality and Tourism Themes	1
International Journal of Management Reviews	1
Frontiers in Marine Science	1
Coral Reefs	1

Source: Researcher's own.

RQ2. Which factors can assist in the development of sustainable luxury tourism?

Sustainable luxury tourism must include sustainable development of the environment and local community along with luxury tourism. The main objective of adopting sustainability in luxury tourism is to promote luxury tourism with reasonable consumption and to benefit the environment, culture and local community (Poelina & Nordensvard, 2017). Therefore, we examined the full text of selected articles to identify the promises that could benefit sustainable luxury tourism.

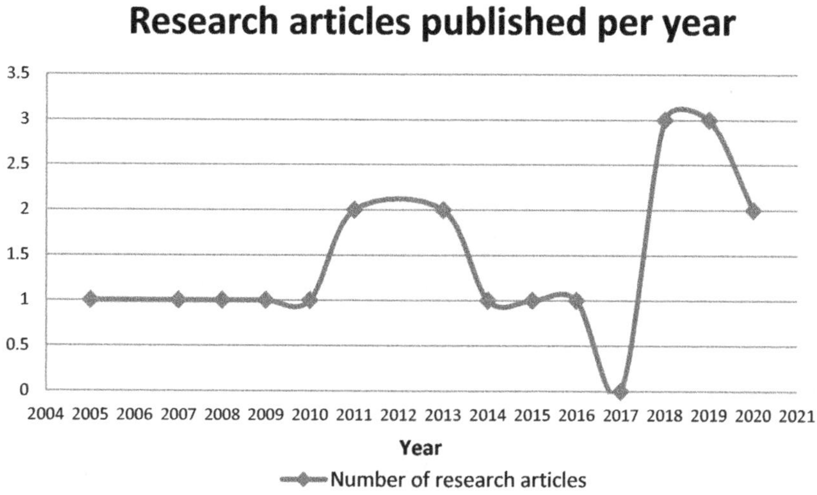

Fig. 18.2. Trend of Publication. *Source:* Researcher's own.

Key Players

According to the sample literature, the key players who can promote sustainable luxury tourism are socio-political leaders (D'hauteserre, 2005; Jensen & Andersen, 2013), customers (Athwal, Wells, Carrigan, & Henninger, 2019; Jensen & Andersen, 2013; Shah, 2011), academicians and conservationists (Riensche,

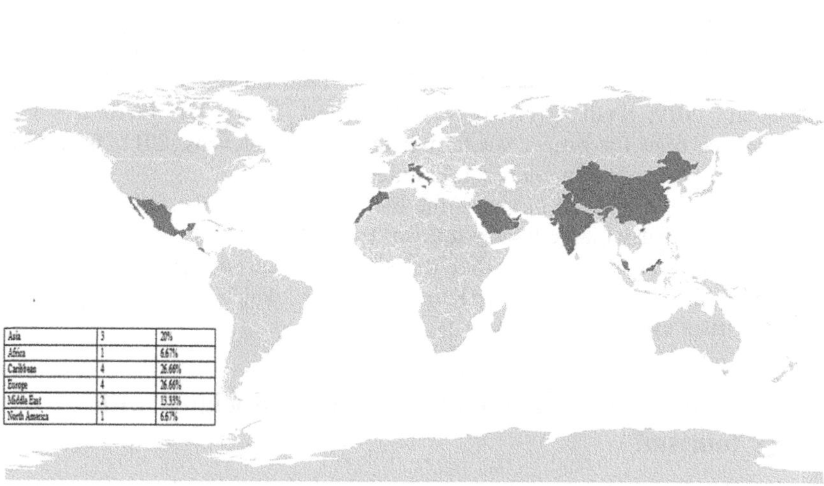

Fig. 18.3. The Countries Mentioned in the Articles (Marked in Blue). *Source:* Researcher's own.

Castillo, García-Frapolli, Moreno-Casasola, & Tello-Díaz, 2019), international communities (Omar, Othman, & Mohamed, 2014; Scheyvens, 2011) and managers and employees of hotels or resorts Shah (2011). The need for involvement of the local community and citizens in sustainable development are being highlighted in various studies (Buckley & Mossaz, 2016; D'hauteserre, 2005; Jensen & Andersen, 2013).

The opinions of researchers vary in terms of suitable ownership for sustainable luxury tourism. Scheyvens (2011) suggests public ownership, Shah (2011) suggests foreign and MNC ownership and D'hauteserre (2005) supports ownership by the locals. In fact, (Amo & Stefano, 2019) describes about the possibility of public–private partnership projects.

Resources

Mindful utilisation and preservation of resources is the backbone of sustainable luxury tourism (Omar et al., 2014). The resources discussed in the sample literature are renewable energy sources (Chalastani et al., 2020; Scheyvens, 2011), forest (D'hauteserre, 2005), wildlife (Buckley & Mossaz, 2016), marine life (Martínez-Castillo, Rodríguez-Troncoso, Santiago-Valentín, & Cupul-Magaña, 2020), biofuels (Jensen & Andersen, 2013), history and culture (González, 2018). Water is the most discussed resources, and the need for effective water management is suggested by many researchers (D'hauteserre, 2005; Ryan & Stewart, 2009; Tekken & Kropp, 2015).

Sustainable Practices for Sustainable Luxury Tourism

The need for strong environmental policies, action plans and strict legislations are being highlighted in most of the sample pieces of literature (Chalastani et al., 2020; D'hauteserre, 2005; González, 2018; Martínez-Castillo et al., 2020; Rienschе et al., 2019; Shah, 2011; Tekken & Kropp, 2015). The literatures have indicated the need for proper waste management (Scheyvens, 2011; Chalastani et al., 2020) and pollution control (D'hauteserre, 2005).

Analysis of the selected pieces of the literature revealed the following factors that can benefit sustainable luxury tourism (refer Table 18.3):

Thus, it can be understood that the success of sustainable luxury tourism depends on a well-planned and practical approach, which involves the collaborative effort of every stakeholder.

RQ3. What is the sustainability concerns associated with luxury tourism?

Tourism development in any destination has both positive and negative impact. The sample pieces of literature were evaluated to identify the concerns related to sustainable luxury tourism.

Table 18.3. Factors Identified in Previous Literature.

Researchers	Factors
D'hauteserre (2005)	• Socio-Political leadership • Sustainable practices • Pollution Control • Collaboration with local community • Inspection and certification • Water Management • Reforestation
Ryan and Stewart (2009)	• Water management • Conservation of environment
Scheyvens (2011)	• Carbon Neutral Strategy • International discussions on climate change • Investment in renewable energy • Waste management • Involvement of local community • Public ownership
Shah (2011)	• Strong environment policy (Corporate environment policy) • Prioritising financial, human resources and other resources for environmental activities by the hotel managers • Involving employees in idea generation for environmental activities • Targeting eco-conscious tourists • Concern of luxury tourists about environment • Size of the hotel, foreign and MNC ownership, • Luxury hotels with good financial performance are more conscious towards the environment • Luxury hotels with foreign and MNC ownership are more concerned about the environment
Tekken, Costa, and Kropp (2013)	• Controlled inflow of tourists
Jensen and Andersen (2013)	• Tax reduction on vehicles emitting less carbon dioxide (CO_2) • Involvement of politicians, citizens and consumers in seeking solutions for climate change • Promotion of bio-fuels
Omar et al. (2014)	• Involvement of international bodies like UNESCO in sustainable development • Improved water management policies

Table 18.3. *(Continued)*

Researchers	Factors
Tekken and Kropp (2015)	• Water tax for luxury tourism facilities • Commitment of hotels to manage water properly • Introduction of sustainable tourism certificates • Introduction of environmental audits • Public consultations to discuss adequacy of water • Survey of actual and future need for water
Buckley and Mossaz (2016)	• Wildlife conservation • Involvement of local community
Weinzierl (2018)	• Space tourism
González (2018)	• Promotion of culture and history of the destination • Strict control measure
Riensche et al. (2019)	• Protected Areas legislation • Assessment of impact on environment by academicians and conservationists • Implications of environment protection policies
Athwal et al. (2019)	• Focusing on consumers who wants to promote the contribution of their luxury purchases on societal welfare • The fear of negative publicity due to unethical practices encourages luxury organisation to address sustainability issues • Sustainable practices are becoming the measure of quality
Amo and Stefano (2019)	• Public–private partnership projects for development of sustainable tourism (including luxury tourism)
Chalastani et al. (2020)	• Renewable energy sources • Waste management system • Sustainable infrastructure • Action plans
Martínez-Castillo et al. (2020)	• Adequate management policies to protect marine life

Source: Researcher's own.

Impact of Luxury Tourism on Resources

Unplanned consumption of water resources is a major concern highlighted by many studies (González, 2018; Ryan & Stewart, 2009; Scheyvens, 2011; Tekken et al., 2013; Tekken & Kropp, 2015). Luxury tourism also increases demand for energy (Andersen, Christensen, Jensen, Kofoed, & Morthorst, 2008; Scheyvens, 2011).

Marine life is another valuable resource that has been facing a threat from tourism (Chalastani et al., 2020; González, 2018; Omar et al., 2014; Scheyvens, 2011). Loss of land resources due to the unplanned development of tourism facilities are highlighted by González (2018) and Muthusankar et al. (2018). Luxury tourism is also being blamed for neglecting the culture and history of the destination (Amo & Stefano, 2019; González, 2018; Stephenson & Ali-Knight, 2010). Moreover, exploitation of human resources in tourism has been reported in previous literatures (Gullette, 2007; Scheyvens, 2011; Stephenson & Ali-Knight, 2010).

Impact of Luxury Tourism on Natives

Sustainable luxury tourism must consider natives as they are being influenced in various ways by the luxury tourism development, such as loss of the previous occupation of the natives (Gullette, 2007; González, 2018; Omar et al., 2014; Riensche et al., 2019), unequal opportunities (Amo & Stefano, 2019; Riensche et al., 2019; Stephenson & Ali-Knight, 2010; Scheyvens, 2011) and unequal ownership opportunity (D'hauteserre, 2005). Moreover, Scheyvens (2011) clearly highlights that the locals only receive minimum benefits. Some major concerns highlighted by previous literatures are being presented below (refer Table 18.4).

The negative impact or concerns related to luxury tourism needs to be properly addressed to promote sustainable luxury tourism. Luxury tourists should be able to accept and popularise the combination of sustainability with tourism (Athwal et al., 2019). Pollution still remains a major concern to be addressed (González, 2018; Scheyvens, 2011) while promoting luxury tourism. Sustainable luxury tourism should find an alternative solution of promoting luxury tourism with minimum pollution.

Overview on the Findings of Systematic Literature Review

This study has attempted to integrate findings and suggestions related to the concept from past studies. Research articles from different disciplines have highlighted on the ill effects of tourism (especially luxury tourism) on the local communities, environment and important resources. On the other hand, past pieces of literature have also provided important suggestions about the required measures to develop sustainable luxury tourism.

Table 18.4. Concerns Discussed in Previous Literatures.

Researchers	Concerns
D'hauteserre (2005)	• Too ambitious plan to promote luxury tourism can burden the environment
Gullette (2007)	• Increase in informal employment • Previous agriculture and fishing have declined
Andersen et al. (2008)	• Increase in second homes can increase electricity consumption
Ryan and Stewart (2009)	• Consumes water resources
Stephenson and Ali-Knight (2010)	• Rapid urbanisation • Migration of labourers from different nations influences the demography of the urban population • Influence on heritage buildings • Unequal employment opportunity
Scheyvens (2011)	• Heavy demand on fresh water and energy • Lack of a proper waste management system • Pollution of the marine environment • Unequal distribution of employment (half foreigners) and ownership (elites and foreigners) opportunities • Exploitation of human labour force • Benefits to the local community is very minimum • Socio-political issues and weaknesses in the legal system
Tekken et al. (2013)	• Water demand is high in resorts for luxury tourists • Demand for additional tourism water can interfere with other usages, such as irrigated agriculture
Jensen and Andersen (2013)	• Uncertainties surrounding climate change • Less concern of consumers regarding CO_2 emissions • Costs and benefits of potential solutions is not done
Omar et al. (2014)	• Displacement destroyed the anchovy industry • Loss of livelihoods of the villagers • Choking of corals nearby
Tekken and Kropp (2015)	• Wastage of water resources

Table 18.4. *(Continued)*

Researchers	Concerns
Muthusankar et al. (2018)	• Increasing coastal erosion and extreme meteorological events
González (2018)	• Increase in the price of land • Expulsion of small local and regional commerce • Increased dust pollution due to overcrowd in tourist season • The environmental threat posed by increased cruises • Increase in the turbidity of the water • Increased tumour diseases • Risk of highly toxic benzopyrene • Vibrations, smoke, noise, electromagnetic pollution by non-stop radars • Damage and disturbing the perception of the city
Riensche et al. (2019)	• Inequality in earning for natives • Locals are being hindered from fishing • Only the wealthy, foreigners and politicians are reaping the benefits • The benefits of associations are limited to luxury accommodation facilities and need to involve others
Athwal et al. (2019)	• Many customers do not pay much concern on the sustainable practices while making luxury purchases. • Luxury tourists often perceive sustainability as compromise with luxury
Amo and Stefano (2019)	• Inequality and unemployment to the natives • Uncontrolled development projects • Limited access to beach and potable water for the locals • Damage of archaeological sites • Environmental concerns • Labour law violations
Chalastani et al. (2020)	• Development of tourism facilities has a negative impact on marine life
Martínez-Castillo et al. (2020)	• Use of bleach in golf courses at luxury tourist places impacts coral life

Source: Researcher's own.

Diphlu River Lodge: An Example of Sustainable Luxury Tourism in Practice

Diphlu River Lodge (DRL) is a luxury accommodation facility near Kaziranga National Park in the Indian state of Assam. Kaziranga National Park is a UNESCO-World Heritage Site (Natural), known for its conservation success story and the highest concentration of the Indian rhinoceros population. The National Park is home to many Schedule-I and Schedule-II species that receive the highest degree of protection under the Wildlife (Protection) Act, 1972, enacted by the Parliament of India. The National Park is one of the most popular tourist attractions in the state.

DRL is sandwiched between National Highway No. 37 and the Diphlu river, a small perennial tributary of the river Brahmaputra which flows along the southern boundary of the National Park. DRL is a reincarnated version of Kamrup Lodge, a pioneering ecotourism lodge which was established by Mr. Jagdish Phookan during the 1970s. The lodge ceased operations after running for seven years and then was revived into its present *avatar* in the year 2008 by his son Mr. Ashish Phookan and his daughter-in-law Mrs Jahnavi Phookan. The lodge is marketed as a responsible tourism endeavour. It was adjudged as one of the 'Best Wildlife Resort in India' by *Lonely Planet* magazine, India (The Hindu, 2013), and has hosted high-profile guests like the Duke and Duchess of Cambridge, HRH Prince William and Catherine Middleton, during their visit to India in the year 2016.

Considering its proximity to the National Park and its surrounding, some of the sustainable practices adopted by the lodge include the following.

Small-scale business – DRL is spread across 10 acres of land with a total carrying capacity of just 24 guests at a time. Guests can opt for two all-inclusive packages. One package applies when the National Park remains open during the winter from November to April which includes activities like jungle jeep safari, dolphin watching, village tours, etc. and the other when the National Park remains closed during the Monsoon from May to October each year. The lodge tries to discourage walk-in guests and, for this reason, has maintained a low profile with zero signboards along the highway.

Traditional architecture – The lodge has 12 cottages and a dining area built on stilts around a paddy field/mustard field with a pond in the middle (refer Fig. 18.4). The structures on stilt are inspired by the traditional houses of the Mising indigenous community living in the area who build their homes like that to safeguard their life and property during the annual floods. The lodge premises get flooded during the rainy season. In the year 2020 itself, it has flooded thrice. The structures remain very similar to the traditional designs: it's a semi-concrete structure with its base, floors and pillars made of concrete while the rest of the structure having a thatched roof is made from locally sourced building materials such as wood and bamboo. This gives an authentic look to the cottages and the entire lodge blends very well with its rural surroundings. The fusion between modernity and tradition can also be seen inside these structures. All the rooms (individual cottages) have en-suite open-air bathrooms, cane furniture, air-conditioning, large glass pane sliding doors that open up to the individual

Fig. 18.4. Inside One of the Traditional Cottages. *Source:* Authors.

balcony, garden furniture, lavish and varied amenities, decorative bamboo pillars, walls adorning traditional artefacts, boudoir pillows having motifs of different indigenous communities of Assam, etc. Machan, the main sitting area of the lodge having the dining area and the bar, is also made on similar lines with an authentic touch to it. All these structures are connected with raised/bamboo pathways.

Natural boundary – The lodge is built on elevated land beside the Diphlu river. The river forms a natural boundary between the lodge and the National Park. There is no additional man-made boundary around the lodge premises. This keeps the animal corridors free from any artificial barrier. This helps in providing safe refuge and passage to the National Park wildlife during annual floods when they are forced to migrate outside the Protected Area to the nearby Karbi Anglong hills that lie on the southern side of the National Park.

Low light and low noise policy – The lodge has adopted a low light and low noise policy between 9 p.m. and 10 a.m. This practice definitely helps in reducing anthropogenic induced disturbance for wildlife, as they are usually sighted very close to the lodge during those hours. However, due to the lack of any man-made barrier around the premises, wildlife occasionally ventures inside the lodge premises at night. To avert any incidence of Man–Animal Conflict (MAC), especially when the guest remains unaware regarding such events, guards are kept to patrol the premises at night. The lodge provides a torch to each guest, and staff accompany them if they come out of their cottage at night.

Water management – The lodge has its own grey water recycling system in place for optimum productivity from the pumped groundwater. Wastewater from the bathroom, kitchen sinks and laundry are covered under this programme. The recycled water is drained to the central pond and used for irrigation and for washing guest vehicles. The lodge also requests its guests to use the laundry service instead of washing their own clothes in their individual bathrooms to maintain the aesthetic appeal of the lodge, preventing any misuse of water resource.

Waste management – Wide range of waste is generated by guests as well as the management. This includes biodegradable, non-biodegradable and chemical wastes. The lodge has its sewage treatment plant, and the biosolids derived are used as fertilisers in the in house farm. Wet waste generated is either used as fertilisers or given to local pig farms. To reduce the use of non-biodegradable waste, the lodge tries to minimise single-use plastics wherever possible. DRL uses and encourages guests to use sterilised glass bottles and flasks that can be refilled at the water dispenser stations. All waste generated in the lodge is collected in oxo-biodegradable bags. Non-biodegradable wastes are collected by ragpickers who know the best use of such wastes.

Local community involvement – Assam Bengal Navigation (ABN) is the parent body of the lodge. Since April 2017, 5% of the amount charged for each room booked goes to the corpus fund of the ABN foundation. The foundation aims for excellence in service and for uplifting the rural communities of the areas they operate in. This fund is also used to collaborate with NGOs that work for animal welfare. The lodge employs 34 employees, out of which 32 belong to Assam, most of them from neighbouring areas; one from Tripura, another state in the Northeast region and one from Jharkhand, a state in central India. New recruits have to go through intensive in-house hospitality and English-speaking training programmes before taking up their roles; they are given key positions of the lodge, i.e. manager, cooks and housekeeping staff.

The lodge also has its own group of naturalists who belong from the nearby villages and thus have in-depth knowledge about their subject and place. The lodge has tried to make the best use of its workforce by adding local dishes in its meal plans; naturalists who know the local language take tourists for village tours, tea plantation visits, coffee plantation visits and rubber plantation visits. Cultural performances such as the Bihu dance and Santali dance on occasions are also performed by the local dance troupes from nearby villages. The use of local art and culture benefits local entrepreneurs and artisans and promotes cottage industries like cane and bamboo craft. The textiles used for lodge decors and utilities are all made and supplied by in-house fair-trade weaving units run by local women. The lodge also sells artefacts and souvenirs made by locals in Machan. All such practices help in the preservation of local culture and traditional practices. Indirect incomes are also earned by locals like the vehicle drivers and owners who are hired for guest transfers and local vegetable and meat sellers from whom the lodge purchases majority of its supplies.

The impact of adopting these practices on guests was measured by doing a thematic analysis using NVivo 12 qualitative data analysis software. This exercise involved the collection of 187 TripAdvisor reviews (Only English, as on 8 August 2020) and manually coding them under-identified themes. A semantic approach was taken for coding the data (refer Tables 18.5 and 18.6).

Table 18.5. Demographic Information of Travellers.

Traveler Type (Families/Couples/Solo/Business/Friends/ Missing data)	59/65/9/6/32/ 16
Place of Origin (Domestic/Inbound/Missing data)	65/101/21

Table 18.6. Frequency Table.

Themes	Positive Codes	Counts	Negative Codes	Counts
Environment-friendly practices	Lodge *Location: Next to river Diphlu, close to all the National Park range gates/peaceful, relaxing and tranquil atmosphere/wildlife sighting on the bank of river Diphlu/privacy of guests/overlooking farmland/domestic elephant bathing experience in Diphlu river*	143	Lodge *Location: Noise due to the nearby highway traffic/20km from the central range/unsafe to walk in lodge premises at night/finding the location due to the absence of signboard/ quiet atmosphere/ noisy atmosphere due to arrival of river cruise guest for meals*	14
	Low light and low noise policy	5	Low light and low noise policy	0
Promotion of local culture and tradition	Architectural fusion *Rooms: Spacious/ comfortable/high on amenities/clean and spotless/traditional design/en-suite bathroom/decors/build quality/insect-free/ quality furniture/ modern facility/ veranda/sufficient amount of bathwater* *Machan: Good bar/ open sitting area overlooking the river/ decors/library*	200	Architectural fusion *Rooms: Insects in the room/small geyser/ maintenance and upgradation of bathroom/needs updates with facilities and amenities/lacks Wi-Fi* *Premises: Prone to mosquitos/bungalows overlooking the river are fairly small/build quality/power cuts/can be better looked after/ suggestion to grow*	23

Table 18.6. *(Continued)*

Themes	Positive Codes	Counts	Negative Codes	Counts
			more trees/lack of modern amenities like swimming pool/ luxurious appeal	
	Other Activities *Village walk/tea plantation visits/orchid farm visit*	5	Other Activities *Guest found the activities limited*	1
	Food *Delicious/local cuisines/food variety/ fresh/buffet service*	139	Food *Taste/limited food variety/buffet service-preference to other types of service over buffet service, food doesn't remain hot, food poisoning/quality and freshness/value for money/dining setting*	21
Local community involvement	Cultural dance troupe performance	15	Cultural dance troupe performance	0
	Hospitality by local Staff *Warm/welcoming/ friendly/caring/ courteous/attentive/ polite/kind/efficient/ cheerful/enthusiastic/ helpful/curious/non-obstructive/punctual/ personalised service/ knowledgeable naturalist and guide/ good transfer vehicle driver/good safari driver, spotter/well organised/manager – good humor, mindful, knowledgeable, attentive, fantastic*	265	Hospitality by local staff *Communication/value for money/ unprofessional, poorly trained staff (old review)/late opening of restaurant for breakfast/some staff were not helpful, friendly/manager – no clue regarding the price of items on sale in the souvenir shop/ not sufficient naturalist/ maintenance of premises/lack of activity in the lodge/*	26

Table 18.6. *(Continued)*

Themes	Positive Codes	Counts	Negative Codes	Counts
	conversationist/ friendly owner		*booking through website/too many staff/poor condition of some safari vehicles/ room allotment*	
Other sustainable practices	*Luxurious yet environment friendly/ exclusive/built with locally available sources/wastewater recycling/association with responsible tourism organisations/ appreciative of eco-tourism practice/the economic benefit to local communities/ fairly priced in-house souvenir shop selling local textiles/a common tipping box for all staff*	18	*Fridge in individual rooms/new oxo-biodegradable plastic bags in dustbin everyday/no limit on the amount of water use/elephant safari*	3

Source: Researcher's own.

Thematic analysis has revealed that most of the luxury tourists have given positive feedbacks towards the adoption of sustainable practices in the lodge. The analysis also reveals that such responsible practices don't go unnoticed; however, the guests also become more critical with their observations once organisations start putting the responsible tourism tag in their business. Consecutive year-wise reviews have shown that accommodation units can transform their weakness into strengths by taking subjective customer feedback seriously to improve their services. This strategy was adopted by DRL and most of the negative feedback may not be applicable at present as the reviews were collected from 2009 till 2020.

Elephant safari was referred in 13 positive reviews by the guest. However, the DRL administration does not encourage elephant rides. International guests are mostly against it, but few domestic guests and others insist, due to which they are forced to aid them in securing an elephant safari against their will as the National Park offers it as a part of the safaris on offer (refer Table 18.7).

Table 18.7. SWOT Analysis.

Strengths	**Weakness**
• Location of the Lodge, Nearness to all the Ranges of the National Park and Close Encounters with Wildlife from the Safety of the Room • Privacy for the Lodge guest Helping it Cater to High Profile guests • Excellent Staff • The Assemblage of luxury with local tradition • Responsible tourism Practices	• High Maintenance Cost of the Lodge • Non- Smoking Rooms due to the Highly Flammable Building Materials • Noise due to Nearness to the National Highway
Opportunities	• **Threats**
• The region faces frequent power disruption, solar energy can be harvested in the lodge premises making it even more sustainable. • Introduction of electric powered vehicles for guest transfers from airports or for jungle safaris	• Competition from new luxury eco-tourism lodges that might come up in the near future • Continuation of elephant safari as an activity could create a negative image towards the property. • The National Park authorities may create a hostile situation for the lodge location during successive additions subject to a rise in wildlife population

Source: Researcher's own.

Discussion and Conclusion

The number of literature pieces included in this study indicates that there are very limited regions where luxury tourism has been analysed from a sustainability viewpoint. Moreover, authors have found a lack of dedicated research articles on sustainable luxury tourism. This also indicates the need to explore the dimensions of sustainable luxury tourism by researchers and practitioners.

The findings of this study reveal the dangers associated with developing luxury tourism without considering its long-lasting effect on the local community, resources and other stakeholders. Sustainable luxury tourism needs to consider these ill effects and find remedies to mitigate the same. The findings of this study also provide a pathway to promote luxury tourism while preserving and edifying the people and resources associated with the destination. The solutions provided by the literatures are practical in nature and needs to be implemented. However, the suggestions from the literatures are specific to a destination; the practitioners and researchers need to evaluate their destination of interest before implementing the suggestions.

This study also presents a case study of the Diphlu river lodge with its challenges and scope of sustainable luxury tourism. Proper utilisation of resources and local community participation has strengthened the image of the Diphlu river lodge as a sustainable luxury tourism practitioner. On the other hand, luxury tourists are willing to participate in sustainable practices. Not all luxury tourists regard sustainability as a compromise with luxury; rather proper information makes them an active participant in sustainable practices. Moreover, the role of hotel management in promoting sustainable practices can be observed. The case study supports the findings of a systematic literature review and reveals the role of different key players for sustainable luxury tourism. The sustainable practices are the unique selling proposition of the Diphlu river lodge. The tourist's reviews, rewards and recognitions received by Diphlu lodge and the positive reviews of tourists are the evidence of a promising future in sustainable luxury tourism.

There is a need to promote sustainable practices as exclusive and unique experiences to attract more luxury tourists. Thus, there is a need to project sustainable practices as a selling point instead of a substitute of luxury.

Future Research

To explore the benefits and challenges associated with sustainable luxury tourism, more research studies are required. Future studies should investigate and identify the segments of luxury tourists based on their interest to participate in sustainable practices. The studies also need to identify marketing techniques to promote sustainability as a selling point and replace its image as a compromise with luxury. Sustainable luxury tourism is a multidisciplinary field and will always need collaboration from different fields. Research from different destinations on sustainable luxury tourism practices and challenges would help in the development of knowledge.

Limitations of the Study

Although attempts have been made to reduce error in this study, the limitations include lack of resource on the theme 'sustainable luxury tourism'. The availability of resources could have helped in comparing the results of this study. Moreover, the results of systematic literature review need to be carefully evaluated because they are confined to the destinations of the studies and lack universality. The uniqueness of each destination needs to be considered before applying the findings of this study.

On the other hand, the findings of thematic analysis also need to be evaluated with caution. Most of the guest reviews in TripAdvisor came from tourists arriving from countries (24 in total) that are performing well in the Environment Performance Index-2020 (Yale University, 2020), the range remained between 1 and 180 rank, with 55 and 34 inbound tourists from countries falling under the range group of 1–20 and 21–40, respectively (Yale University, 2020). Apart from India, which ranks at 168 in the EPI, 2020, only 12 reviews came from countries between 41 and 180 range (Yale University, 2020). Hence, the result may not reflect attitudes of luxury tourists towards sustainable practices from countries that are not performing well in the EPI.

Acknowledgements

All the observations were made during the corresponding author's visit to the lodge. Content analysis of lodge website and brochures were done recently; the findings were substantiated with the additional facts provided by the owner of the lodge Mrs Jahnavi Phookan and Ms Antara Phookan, Director, Marketing and Communications, without whose support and contribution this case study would not be a success.

References

Amo, M. D. H., & Stefano, M. C. D. (2019). Public–private partnership as an innovative approach for sustainable tourism in Guanacaste, Costa Rica. *Worldwide Hospitality and Tourism Themes*, *11*(2), 130–139. doi:10.1108/whatt-11-2018-0078

Andersen, F. M., Christensen, M. S., Jensen, O. M., Kofoed, N.-U., & Morthorst, P. E. (2008). Second-home electricity consumption. *Energy Policy*, *36*(1), 280–289. doi:10.1016/j.enpol.2007.09.013

Athwal, N., Wells, V. K., Carrigan, M., & Henninger, C. E. (2019). Sustainable luxury marketing: A synthesis and research agenda. *International Journal of Management Reviews*, *21*(4), 405–426. doi:10.1111/ijmr.12195

Britannica, T. Editors of Encyclopaedia. (2021, January 1). Xuanzang. Encyclopedia Britannica. Retrieved from https://www.britannica.com/biography/Xuanzang

Brown, B., Hanson, M., Liverman, D., & Merideth, R. (1987). Global sustainability: Toward definition. *Environmental Management*, *11*(6), 713–719. doi:10.1007/bf01867238

Buckley, R., & Mossaz, A. (2016). Decision making by specialist luxury travel agents. *Tourism Management*, *55*, 133–138. doi:10.1016/j.tourman.2016.02.002

Cerović, M., Pavia, N., & Floričić, T. (2019). The importance of innovation in designing a luxury tourism offering: The perspective of tourism destination stakeholders. In 5th International Scientific Conference ToSEE-Tourism in Southern and Eastern Europe 2019 "Creating Innovative Tourism Experiences: The Way to Extend the Tourist Season", Faculty of Tourism and Hospitality Management, University of Rijeka, Opatija, Croatia, 16–18 May 2019, (pp. 177–189).

Chalastani, V. I., Manetos, P., Al-Suwailem, A. M., Hale, J. A., Vijayan, A. P., Pagano, J., … Duarte, C. M. (2020). Reconciling tourism development and conservation outcomes through marine spatial planning for a Saudi giga-project in the Red Sea (the Red Sea project, vision 2030). *Frontiers in Marine Science*, *7*, 1–15. doi:10.3389/fmars.2020.00168

Clarivate. (2020). Trusted publisher-independent citation database–Web of science group. Retrieved from https://clarivate.com/webofsciencegroup/solutions/web-of-science/

D'hauteserre, A.-M. (2005). Tourism, development and sustainability in Monaco: Comparing discourses and practices. *Tourism Geographies*, *7*(3), 290–312. doi:10.1080/14616680500164716

Dixit, S. K. (2020). *The Routledge handbook of tourism experience management and marketing*. London: Routledge.

Elsevier. (2020). About Scopus–Abstract and citation database. Elsevier. Retrieved from https://www.elsevier.com/solutions/scopus?dgcid=RN_AGCM_Sourced_300005030

Gibson, R. (2006). Sustainability assessment: Basic components of a practical approach. *Impact Assessment and Project Appraisal*, *24*(3), 170–182. doi:10.3152/147154606781765147

González, A. T. (2018). Venice: The problem of overtourism and the impact of cruises. *Journal of Regional Research Investigaciones Regionales*, *42*, 35–51.

Gullette, G. (2007). Development economics, developing migration: Targeted economic development initiatives as drivers in international migration. *Human Organization*, *66*(4), 366–379. doi:10.17730/humo.66.4.05h1427k901386ph

Hopwood, B., Mellor, M., & O'Brien, G. (2005). Sustainable development: Mapping different approaches. *Sustainable Development*, *13*(1), 38–52. doi:10.1002/sd.244

IUCN-UNEP-WWF. (1980). *World conservation strategy: Living resource conservation for sustainable development* [Ebook] (2nd ed.). Retrieved from https://portals.iucn.org/library/efiles/documents/wcs-004.pdf

Jensen, M., & Andersen, A. H. (2013). Biofuels: A contested response to climate change. *Sustainability: Science, Practice and Policy*, *9*(1), 42–56. doi:10.1080/15487733.2013.11908106

Johnston, P., Everard, M., Santillo, D., & Robert, K. (2007). Reclaiming the definition of sustainability. *Environmental Science and Pollution Research International*, *14*(1). Retrieved from https://www.researchgate.net/publication/6455179_Reclaiming_the_Definition_of_Sustainability

Kelly, J., Sadeghieh, T., & Adeli, K. (2014). Peer review in scientific publications: Benefits, critiques, and a survival guide. *The Journal of the International Federation of Clinical Chemistry and Laboratory Medicine*, *25*(3), 227–243.

Liberati, A., Altman, D. G., Tetzlaff, J., Mulrow, C., Gøtzsche, P. C., Ioannidis, J. P. A., … Moher, D. (2009). The PRISMA statement for reporting systematic reviews and meta-analyses of studies that evaluate health care interventions: Explanation and elaboration. *PLoS Medicine*, *6*(7). doi:10.1371/journal.pmed.1000100

Martínez-Castillo, V., Rodríguez-Troncoso, A. P., Santiago-Valentín, J. D., & Cupul-Magaña, A. L. (2020). The influence of urban pressures on coral physiology on marginal coral reefs of the Mexican Pacific. *Coral Reefs*, *39*(3), 625–637. doi:10.1007/s00338-020-01957-z

Moher, D., Liberati, A., Tetzlaff, J., & Altman, D. G. (2009). Preferred reporting items for systematic reviews and meta-analyses: The PRISMA statement. *PLoS Medicine*, *6*(7), e1000097. doi:10.1371/journal.pmed.1000097

Muthusankar, G., Proisy, C., Balasubramanian, D., Bautès, N., Bhalla, R. S., Mathevet, R., & Vasudevan, S. (2018). When socio-economic plans exacerbate vulnerability to physical coastal processes on the south east coast of India. *Journal of Coastal Research*, *85*, 1446–1450. doi:10.2112/si85-290.1

Novotná, M., & Kunc, J. (2019). Experience matters! Luxury consumption patterns and motivation of the Czech affluent society. *Deturope*, *11*(2), 121–142. Retrieved from https://www.deturope.eu/img/upload/content121-142.pdf

OECD. (2002). Towards sustainable household consumption? Trends and policies in OECD countries. Retrieved from https://www.oecd.org/env/consumption-innovation/1938984.pdf

Omar, S. I., Othman, A. G., & Mohamed, B. (2014). The tourism life cycle: An overview of Langkawi Island, Malaysia. *International Journal of Culture, Tourism and Hospitality Research*, *8*(3), 272–289. doi:10.1108/ijcthr-09-2013-0069

Page, S. (2012). *Tourism management: Managing for change* (3rd ed.). Amsterdam: TaylorandFrancis.

Poelina, A., & Nordensvard, J. (2017). Sustainable luxury tourism, indigenous communities and governance. In M. Gardetti & S. Muthu (Eds.), *Sustainable luxury, entrepreneurship, and innovation. Environmental footprints and eco-design of Products and processes*(1st ed., pp. 147–166). Singapore: Springer. doi:10.1007/978-981-10-6716-7_8

Purvis, B., Mao, Y., & Robinson, D. (2018). Three pillars of sustainability: In search of conceptual origins. *Sustainability Science*, *14*(3), 681–695. doi:10.1007/s11625-018-0627-5

Raymen, T., & Smith, O. (2019). *Deviant leisure-criminological perspectives on leisure and harm*. S.l.: PalgraveMacmillan.

Riensche, M., Castillo, A., García-Frapolli, E., Moreno-Casasola, P., & Tello-Díaz, C. (2019). Private over public interests in regional tourism governance: A case study in Costalegre, Mexico. *Sustainability*, *11*(6), 1760. doi:10.3390/su11061760

Ryan, C., & Stewart, M. (2009). Eco-tourism and luxury–The case of Al Maha, Dubai. *Journal of Sustainable Tourism*, *17*(3), 287–301. doi:10.1080/09669580802366587

Scheyvens, R. (2011). The challenge of sustainable tourism development in the Maldives: Understanding the social and political dimensions of sustainability. *Asia Pacific Viewpoint*, *52*(2), 148–164. doi:10.1111/j.1467-8373.2011.01447.x

Shah, K. U. (2011). Strategic organisational drivers of corporate environmental responsibility in the Caribbean hotel industry. *Policy Sciences*, *44*(4), 321–344. doi:10.1007/s11077-011-9130-x

Stephenson, M. L., & Ali-Knight, J. (2010). Dubai's tourism industry and its societal impact: Social implications and sustainable challenges. *Journal of Tourism and Cultural Change*, *8*(4), 278–292. doi:10.1080/14766825.2010.521248

Sundbo, J., & Dixit, S. K. (2020). Conceptualisations in tourism experience. In S. K. Dixit (Ed.), *The Routledge Handbook of Tourism Experience Management and Marketing* (pp. 15–26. London: Routledge.

Tekken, V., Costa, L., & Kropp, J. P. (2013). Increasing pressure, declining water and climate change in north-eastern Morocco. *Journal of Coastal Conservation*, *17*(3), 379–388. doi:10.1007/s11852-013-0234-7

Tekken, V., & Kropp, J. P. (2015). Sustainable water management–Perspectives for tourism development in north-eastern Morocco. *Tourism Management Perspectives*, *16*, 325–334. doi:10.1016/j.tmp.2015.09.001

The Hindu. (2013). Assam's jungle lodge Diphlu in Lonely Planet's awards list. Retrieved from https://www.thehindu.com/news/national/other-states/assams-jungle-lodge-diphlu-in-lonely-planets-awards-list/article4686314.ece

Walton, J. K. (2020, November 10). *Tourism*. Encyclopedia Britannica. Retrieved from https://www.britannica.com/topic/tourism

Weinzierl, M. (2018). Space, the final economic frontier. *Journal of Economic Perspectives*, *32*(2), 173–192. doi:10.1257/jep.32.2.173

Yale University. (2020). *Environment performance Index*. Yale University. Retrieved from https://epi.yale.edu/epi-results/2020/component/epi

Chapter 19

Conceptualising Social Sustainability in Sustainable Luxury Tourism in Martuwarra, Northern Western Australia

Johan Nordensvärd[1] *and Anne Poelina*[2]

Abstract

Sustainable luxury has often been seen to offer both environmental sustainability and the possibility for innovative entrepreneurial development of natural and cultural heritage. The possibility and challenges of sustainable luxury tourism for Indigenous groups have been discussed by Poelina and Nordensvärd (2018) at some length by including a cultural governance perspective that brings culture and nature together. They stressed how protecting our shared human heritage and human culture can be aligned with a new wave of sustainable luxury tourism. To achieve this, we need to create links to both management and protection of landscapes and ecosystems as vital parts of heritage protection and social development. This chapter explores how and why we need to integrate social sustainability into sustainable luxury tourism, where we can foresee potential pitfalls and conceptualise nature-based and Indigenous tourism to empower local Indigenous communities and provide them with sustainable employment, economic development and community services. The sustainable tourism model provides brokerage necessary to strengthen their capacity for innovation, entrepreneurship and transformational change. This transformational change requires tourist visitors and non-Indigenous tourism operators to be open to a new experience with Indigenous guides and tourism operators to see, share and learn how to feel 'Country' (Poelina, 2016; Poelina & Nordensvärd, 2018). We will use Martuwarra (Fitzroy River) and its communities in Kimberley (Western Australia) as a case study to develop a sociocultural sustainable

[1]https://orcid.org/0000-0001-8508-4212
[2]https://orcid.org/0000-0001-6461-7681

The Emerald Handbook of Luxury Management for Hospitality and Tourism, 379–402

doi:10.1108/978-1-83982-900-020211019

luxury tourism framework that includes governance, legal and management and social policy perspective.

Keywords: Sustainability; sustainable luxury tourism; Indigenous communities; Indigenous governance; Indigenous livelihoods and innovation

Introduction

Sustainable luxury has often been seen to offer both environmental sustainability and the possibility of innovative entrepreneurial development in areas of natural and cultural heritage. The possibility and challenges of sustainable luxury tourism for Indigenous groups have been discussed by Poelina and Nordensvärd (2018) at some length by including a cultural governance perspective in bridging the divide between culture and nature, explaining how and why management and protection of landscapes and ecosystems are integral to human heritage, culture and a new wave of sustainable luxury tourism (Poelina, 2020). There can be a way forward to take of visited places by combining ethics of care and sustainable luxury tourism practices (Mair & Laing, 2013; Miller, Rathouse, Scarles, Holmes, & Tribe, 2010; Walker & Moscardo, 2016) while providing Indigenous employment. This could be achieved by focussing on small and diverse, community-based sustainable luxury 'conservation, science and cultural tourism' (RiverOfLife, Poelina, Alexandra, & Samnakay, 2020b).

Indigenous nations in the Kimberley region of Western Australia believe the multiple forms of culture and conservation, and science can create new sustainable economies (RiverOfLife et al., 2020b). WAITOC, the lead Indigenous tourism representative body for Indigenous tourism in Western Australia, continues to respond to a range of challenges and pitfalls. Sustainable luxury tourism promotes fair revenue for Indigenous groups in remote and marginalised areas (Poelina & Nordensvärd, 2018). Some of the most severe obstacles range from land tenure, difficulties in raising finance, lack of market profile and market skills and the overall design of tourist itineraries (Altman, 1993; Altman & Finlayson, 2003; Dyer, Aberdeen, & Schuler, 2003; Zeppel, 2001). Schmiechen and Boyle argue that Indigenous tourism 'remains an extremely fragile and tenuous sector of the tourism industry' (2007, p. 60). Higgins-Desbiolles, Trevonow, and Sparrow (2014) have highlighted the importance of including and incorporating an understanding of Indigenous culture and Indigenous community aspirations. Importantly, there is a need as part of the ethical business of experienced operators and investors to make a commitment to partner and or invest in Indigenous business, innovation, mentoring and entrepreneurship (Gunya, 2007).

Indigenous tourism faces the challenge of a gap between the belief in what Indigenous tourism can attract in revenue and the actual demand for these services. There is a strong belief that Indigenous tourism has a growing demand (Department of Industry, Tourism & Resources, 2005; Tourism Research Australia, 2010). At the same time, the market for Indigenous tourism is relying on international visitors, which means that visitors will go to some lengths both in terms of cost and transport to reach remote Indigenous communities. We, therefore, need to calculate the demand for the services vis-à-vis the costs when it comes to remoteness from suppliers and markets (Young, 1988). There are other issues such as access to start-up and developmental capital (Finlayson & Madden, 1995; Fiszbein, 1997), and government schemes have been difficult for Indigenous people to access (Buultjens, Waller, Graham, & Carson, 2002).

We can see that these start-up challenges become even more difficult when many potential customers are international rather than national. 'Conventional wisdom is that Indigenous tourism is much more popular among international tourists, especially from Northern European countries than among domestic tourists. Moreover, Indigenous ecotourism enterprises are to some degree protected from competition due to their relative remoteness' (Coria & Calfucura, 2012, p. 51). Ryan and Huyton have found in a study of visitors to Katherine that only about a third of tourists indicated a high level of interest in Indigenous tourism products. They argue that the target group that has a higher interest in visiting Indigenous sites tend to be female, younger, with better education and not seldomly from North America or Northern Europe (2002). This argument highlights the fact that sustainable luxury tourism might not just be desirable but also necessary.

Bennett et al. (2016), Bunten (2010), Colton and Whitney-Squire (2010), Fuller, Buultjens, and Cummings (2005), and Turner, Berkes, and Turner (2012) highlight that tourism itself can become the way to link up economic, social, cultural and environmental objectives while building capacity and promoting Indigenous community development. Still, the success of these endeavours will rely on peoples' willingness and ability to take risks. Bendell (2012) raises concern regarding the values embraced by some emerging companies within the luxury tourism industry. It is time for entrepreneurs to consider fundamental transformations for their activities to meet ethical and sustainable standards (Egri & Herman, 2000). A critical factor for sustaining a luxury tourism activity in a remote location is Indigenous earth-centred values. Active dialogue, partnerships and cooperative agreements between the tourism industry and Indigenous entrepreneurs are required to secure investment for implementing and evaluation of Indigenous Tourism Nation-Building Projects.

Creating Indigenous tourism that is educative, experiential, sustainable and lucrative with high ethical, social and environmental standards is challenging. It goes beyond normal understandings and experiences of tourism by including an ethic of care for the Country (Mair & Laing, 2013; Miller et al., 2010; Sundbo & Dixit, 2020, pp. 15–26; Walker & Moscardo, 2016). This chapter aims to explore how and why we need to integrate sociocultural sustainable, transformative cultural and environmental learning experience to add value to sustainable luxury tourism. In this way, we can conceptualise how to overcome the pitfalls of nature-based and Indigenous tourism in a reciprocal way that will empower local Indigenous communities and provide them with employment, sustainable economic development and community development and improved tourism visitor experience. These testimonies provide evidence for grounding models for sociocultural transformational change. We consider the opportunity to engage with the lifeways of the Martuwarra, the Fitzroy River, in the vast Kimberley, northern

Western Australia, regarding developing a sustainable luxury tourism framework that includes governance, legal, economic and management and sociocultural policy perspective.

The chapter is presented in four parts: the first part discusses the challenges and opportunities Indigenous communities experience. Tourism provides a learning experience for the visiting tourist. Furthermore, tourism enhances the opportunity for sustainable livelihoods, economic and community development. We highlight the benefits of cross-cultural knowledge and the unique experience of engaging with Indigenous people and learning to see, feel and hear nature by empathising with an ethics of care. The second part focusses on a social development framework for Indigenous sustainable luxury tourism.

In the third and fourth part, we develop principles for underpinning sustainable luxury Indigenous tourism and the types of investment required to support the Kimberley to reach its full potential, particularly in regard to how sustainability principles compare to the actual reality the ground in the Kimberley.

Background

Indigenous population globally continue to suffer from discrimination, marginalisation, extreme poverty and conflict. It is pertinent to understand the history of colonial development in remote northern regions such as the Kimberley. Before 1838, Indigenous people in the region lived in Eutopia; their world was plentiful and peaceful and regulated by First Law, the Law of the Land. These original Australians co-existed in a world of circular economy, trade, ceremony and balance with nature, managing their estates from the beginning of time through to modernity, 2020. 'Viewed solely through a euro-centric, development lens, the Martuwarra, and the broader Kimberley region, became a development venture to the coloniser intent on using its resources for economic profit' (RiverOfLife, McDuffie, & Poelina, 2020a, p. 83). Today after 150 years of extensive development and profiteering by others, Martuwarra Indigenous people live with extreme disadvantage, poverty and multiple forms of violence (RiverOfLife et al., 2020a).

Indigenous Australian people continue to experience extreme social, economic, juridical and political disadvantages in Australian Society (Aboriginal & Torres Strait Islander Affairs, 2003; Ivory, 2003; SCRGSP, 2016, 2020). This is illustrated by higher unemployment vis-a-vis non-Indigenous groups. Furthermore, the average income and rates of business ownership are significantly lower than corresponding non-Indigenous groups (Aboriginal & Torres Strait Islander Affairs, 2003; SCRGSP, 2016, 2020). There are also particular issues concerning how Indigenous communities access quality social services and how this is limited by barriers such as discrimination. A large part of the Indigenous population resides in areas that are often remote with less chance of employment. More and more Australian government agencies see tourism as a possible route to promote Indigenous entrepreneurship and employment (SCRGSP, 2016, 2020). A major priority has been to increase tourism investment in regional and remote areas through a diverse array of Federal and State government programmes.

Evidence from the Western Australian Indigenous Tour Operators Corporation (WAITOC) identifies socially sustainable luxury tourism as an emerging mechanism for creating life-sustaining activities for empowering remote Indigenous groups. Kimberley members have been operating in remote habitats such as the Martuwarra with its unique spiritual, cultural and environmental assets (Poelina & Nordensvärd, 2018). Sustainable luxury can be understood as a vehicle for increasing respect for the environment and social development and as a portal into the culture, art and innovation and a legacy of local craftsmanship (Gardetti, 2011).

The Martuwarra (Fitzroy River) is a unique remote river system that spans 733 km long, and during the wet season, the river is second to Amazon River when it comes to volume of water (Watson et al., 2011). The Martuwarra Fitzroy River valley tract starts high up in the mountainous hill country. It flows past the Great Sandy Desert and 'empties into King Sound to form the largest tide-dominated delta in the world. The total valley from Fitzroy Crossing to King Sound is globally unique in its biotic and abiotic diversity, remoteness and arid setting' (Semeniuk & Brocx, 2011, pp. 151–160). The river originates in the Wunaamin Miliwundi Ranges (formally known as King Leopold Range) and comprises a vast catchment area of almost 100,000 km. This encompasses more than 20% of the Kimberley region. What makes the region unique is its mega-diversity from a biogeographic perspective. Through this landscape, the river is the centrepiece which is up to 1 km wide in parts and is flanked by a 300 km long floodplain spanning up to 15 km wide. The river empties its water into King Sound south of Derby (Pepper & Keogh, 2014).

The river itself runs through some of the most impressive and pristine landscapes of northern Western Australia, such as sandstone plateaux and 350 million-year-old Devonian limestone of former barrier reefs, which has now become deep and dramatic gorges. This river landscape offers significant fauna in and around the river, such as 18 endemic fish species (the endangered freshwater sawfish and freshwater crocodiles, sharks, rays, turtles, mussels, waterbirds, falcons, bats and quolls) (Vogwill, 2015, pp. 6–11). Important freshwater mangroves play important roles as sources of food and traditional medicine (Watson et al., 2011). In 2011, the Australian Government announced the National Heritage Listing of 19 million hectares of the West Kimberley, including the entire Martuwarra, for its outstanding cultural significance to the nation.

The Martuwarra

The Martuwarra Fitzroy River Council (Martuwarra Council) has produced a Conservation and Management Plan for the National Heritage Listed Fitzroy River Catchment Estate. The exceptional natural and cultural values within the 96,000 square kilometres of River Catchment demonstrate a qualitative dimension to the region, state, nation and globally. The entire Fitzroy River Catchment was added to the National Heritage Listing in 2011 by the Australian Government. The Fitzroy River is also listed as an Indigenous Heritage Site under the Western Australian Indigenous Heritage Act 1972 (RiverOfLife et al., 2020b).

We defined luxury in regard to sustainable luxury tourism as focussing on the uniqueness of access to a spiritual and environmental experience created through Indigenous engagement with the environment (Poelina & Nordensvärd, 2018). The Kimberley region is our primary focus because it is considered to be one of the world's last great wilderness areas, acknowledged as significant and unique by National Heritage listing in 2011.

Sustainable entrepreneurial management can bring together culture and nature for responsible protection of landscapes and ecosystems. These integral elements of human heritage and culture ground a new wave of sustainable luxury tourism. The value of Indigenous governance and the rights to their land are central to building a case for sustainable luxury and promoting multi-layered ethical care for the places people visit and becoming part of protecting both the land and their cultures. This could happen if we turn conceptual, social sustainability into sustainable luxury tourism to empower local Indigenous communities. As the Martuwarra, Fitzroy River is globally unique. There are real opportunities to work with Indigenous people across six Indigenous nations to invest and create models such as geoparks, biospheres and marine parks and extend the opportunity for larger Indigenous Protected Areas. This type of partnership requires building trust, co-design and investment to engender a culture of change towards sustainable economic development and social change.

Investor partnerships with Indigenous people are strengthened when a diverse range of investments are applied. In addition to funding, responsible investment requires ethical partnerships with Indigenous leaders and provides business mentoring. Establishing ethical partnerships will be critical for reimagining diverse cultural models for science, conservation and new economies. There are real opportunities to create an ethical and profit-making business in this type of luxury tourism, especially as the last remnants of cultural and biodiversity remain in the guardianship of Indigenous peoples across the globe (RiverOfLife et al., 2020b).

Conceptual Framework

The discourse of sustainability has become an increasingly important lens through which to understand development projects in general as it combines cultural, environmental, social and economic factors. The sustainable development and environmental management discourse merge humans within the natural environment. Kate Kearins and colleagues argue that sustainability is 'the concept of a system that has at its heart ecological sustainability and the longevity of biophysical systems that support human life' (Kearins, Collins, & Tregidga, 2010, p. 519). Sustainability was popularised by the Brundtland Report in 1987, which became synonymous with 'development that meets the needs of the present without compromising the ability of future generations to meet their own needs'

(UN WCED, 1987, p. 43). There was an interest in integrating the social, economic and environmental dimensions of development. These dimensions were seen as the core pillars of sustainable development. Even though sustainability has become one of the most important environmental concepts in development, the social pillar was often seen as less of a priority than both the environmental and economic pillars. For a long time, the social pillar was also seen as being vague and marginal (Casula Vifell & Soneryd, 2012; Dempsey, Bramley, Power, & Brown, 2011).

Some argue that the social pillar is the most elusive aspect of sustainable development (Thin, 2002). The original Brundtland Report did highlight the social pillar relatively strongly and argued that it was important that 'the essential needs of the world's poor' should be given an 'overriding priority' (UN WCED, 1987, p. 43). At the same time, it was seen that the state of technology and social organisation directly impacted how the environment could meet the needs of the poorest (UN WCED, 1987, pp. 43/37). Many scholars highlight the importance of developing a greater link between the social and environmental pillars (Dobson, 2003; Gough et al., 2008; Littig & Grießler, 2005). Murphy (2012) has undertaken an extensive review of the literature related to sustainable development and has suggested four pre-eminent policy concepts for social sustainability such as fairness (equity), sustainability awareness, participation and social cohesion in relation to food access and welfare. There have been attempts to link social sustainability to the concept of environmental justice. This makes sense as environmental justice has a more developed social justice perspective compared to the original social pillar in sustainable development. Julian Agyeman and Bob Evans (2004) argue that 'just sustainability' needs a clear linkage between sustainable development and environmental justice to prevent the social pillar from becoming one-sided. David Harvey (1996) points out that addressing environmental injustices needs to be a first priority on the sustainability agenda.

Environmental justice differs from other environmental discourses by defining the environment as the set of linked places 'where we live, work and play' (Turner & Wu, 2002, p. 4). The environmental justice framework originated in a US context that focussed around issues of race and ethnicity and how these were intertwined with the distributions of environmental bads, such as pollution and technological risk (Bullard, 1999). Distributive justice became a focal point in framing how communities of colour have been exposed to environmental hazards, such as toxic waste or other environmental issues. Categories such as gender, race and class and the impact of environmental bads and access to environmental goods such as quality of life, natural resources and a clean environment became central to the discussion of distributive justice and the environment (Boström, 2012, p. 5). The focus on distribution has often been expanded upon with other dimensions, such as participation, recognition and capabilities. Daniel Faber (2005) and David Schlosberg (2007) have highlighted the importance of process and production in environmental justice.

There is an overarching perception that most pollution and degradation is caused by the more affluent and powerful, and that the environmental consequences hit the poor disproportionately. It is therefore important that the decision-making processes in environmental policy and specific projects are transparent, just and participative. Distributive justice is concerned with whether environmental resources are allocated in a fair and equitable way among diverse groups of citizens and stakeholders (Maiese, 2003). Distributive justice has often focussed on how marginalised groups have been exposed to environmental hazards, such as toxic substances or other environmental issues. The environmental justice movement started in California to struggle against the effects of industrial contamination on 'poorer communities of colour' (Lohmann, 2009, p. 1075). Procedural justice is less about equal distribution and more about the direct empowerment and participation of different stakeholders in environmental processes (Boström, 2012, p. 5).

Schlosberg argues that we also need to add a capability dimension to environmental justice; this would 'enrich conceptions of environmental and climate justice by bringing recognition to the functioning of these systems, in addition to those who live within and depend on them' (Schlosberg, 2013, p. 44). This approach has created an attempt to include environmental concerns in an environmental justice movement that has often been perceived to be anthropocentric (Shrader-Frechette, 2002). We need to rethink injustice not only in relation to human beings and communities but also to all of those non-humans that depend on the system's integrity for their own functioning (Schlosberg, 2013, p. 44). That means that interrupting, corrupting or defiling the functioning of ecological support systems is a larger scale of injustice. This larger approach to environmental injustice has meant that Indigenous people have used this framework to an increased degree to explain their struggles to claim their ancestral lands and protect these lands. Native Americans are now seen as an essential 'part of the US environmental justice movement, and Indigenous peoples have used environmental justice language and organising themes in other countries and regions as well' (Schlosberg & Carruthers, 2010, p. 12). We think that environmental justice creates a link between the struggle of Indigenous people and social sustainability.

In this chapter, we argue that we need to develop a strong framework for understanding just social sustainability to develop Indigenous luxury tourism that empowers and strengthens Indigenous communities. Boström created a model based around social sustainability/environmental justice (2012) that has been adapted by Urban, Nordensvärd, Siciliano, and Li (2015) and Nordensvard, Urban, and Mang (2015) to analyse the social implications of hydropower dam projects and their social impact in developing countries.

Substantive justice issues, on the other hand, encompass basic needs such as sufficient food, adequate housing, income and extended needs, employment/ opportunities, basic services and social services and fair distribution of environmental 'bads' and 'goods'/equality of rights.

Procedural justice differs in its focus on the processes of policy rather than the policy itself or the outcome. We can see such a focus move towards how communities are involved in the overall policy process. This often involves aspects such as accountable governance and management of the policy, planning and standard-setting process. Is the implementation done according to the processes and actors held to account for the promises and pledges? This often includes monitoring of the overall process (Boström, 2012; Nordensvard et al., 2015; Urban et al., 2015).

We argue that both substantive justice and procedural justice are important dimensions to understand the situation of Indigenous people vis-a-vis sustainable luxury tourism. Still, we argue that we need to expand our understanding of environmental justice to understand some underlying challenges and pitfalls of Indigenous community tourism.

Schlosberg and Carruthers (2010) have pointed out clearly that Indigenous request for environmental justice is both about distributional equity and also functioning of Indigenous communities. Therefore, it is important to see the connection between Indigenous people and their ancestral lands and prime the traditions and practices among Indigenous communities. One will here prime both the functioning and capabilities of both the community and the environment (Schlosberg & Carruthers, 2010). Cooke, Nordensvärd, Bin Saat, Siciliano, and Urban (2017) have developed two dimensions based on Indigenous communities' capabilities based around sustaining traditional livelihoods and environments on one side and strengthening the bond between the Indigenous groups and their land on the other side.

We will add these factors to our just social sustainability analysis in analysing whether, how and to what extent sustainable luxury tourism is able to protect the self-reliance of Indigenous groups and their environment as well as the relationship to their ancestral lands. The first factor goes beyond just redistributing resources but discusses how Indigenous groups can sustain traditional livelihoods and relationships. The second factor discusses how socially sustainable luxury tourism can strengthen the bond between the Indigenous groups and their land. These lead to assessments of how and to what extent social sustainable luxury tourism 1 are able to protect and strengthen Indigenous groups related to their ancestral lands and social sustainable luxury tourism 2 can enable the self-reliance of Indigenous groups and their environment to sustain traditional livelihoods.

Challenges, Pitfalls and Opportunities for Indigenous Sustainable Luxury Tourism

There are many challenges in creating a truly sustainable Indigenous tourism where some are down to the nature of luxury tourism in relation to the position of Indigenous people in Australia. Indigenous Australians experience higher unemployment than their non-Indigenous peers, and the average income and rate of business ownership are significantly lower, which explains why tourism is seen as a development tool for creating 'much-needed opportunities for employment, social stability and preservation of culture and traditions' (Commonwealth of Australia, 2003, p. 41). Tourism has become a viable option for Indigenous people to establish themselves in the economy (Fuller, Antella, Cummings, Scales, & Simon, 2001). Cultural tourism could be of importance if it could integrate the actual land and Indigenous cultural assets. There is a fine line between protecting cultural integrity and develop commercial activities in a balanced way

(Altman & Finlayson, 2003). Therefore, it is important to create tourism centred around collaboration and sharing tourism visitors' experiences with diverse Indigenous peoples, culture and landscapes.

To avoid the exploitation of ecosystems and Indigenous communities, it is important to have a strong normative foundation for sustainable luxury tourism. Culture conservation economy is an important framework that primes strong community leadership and a sustainable approach to maintain and upkeep healthy country and culture in Indigenous communities (Hill et al., 2008). We will argue in this chapter that tourism must go further to support that native title claim should not just involve the claim of the land but also their cultures and the services that they are built on. Hill et al. (2008) highlight that a culture conservation economy needs to recognise the culture and rights of Indigenous people, (re-)builds and supports meaningful livelihoods and sustainable communities and fosters environmental conservation and restoration (Hill et al., 2008).

Still, the challenges are plentiful and often rather contradictory. Jacobsen sees that Indigenous entrepreneurs in remote regions tend to have very good levels of business acumen and success (2017), but this does not always translate to more benefits and wealth for their Aboriginal communities. There is also the argument that state and federal government support Aboriginal owned tourist enterprises tend not to be sustainable as they often end once government grant funding ends (Birdsall-Jones, Wood, & Jones, 2007; Foley, 2004). There has been a rise in research to find the lack of success among Indigenous tourism entrepreneurs (Buultjens & Gale, 2013; Buultjens & White, 2008; Ruhanen, McLennan, & Whitford, 2016). Often this is due to low start-up finance, capital (Buultjens & White, 2008; Ruhanen et al., 2016) on one side and, perceived by some, lack of business skills, including marketing and difficulty to fit into the overall tourist marketw

There is an overall challenge to create a commercially viable product that competes with other tourism products, which seems to be sometimes contrary to the ethos of a community-owned culture developed and used by traditional owners over thousands and thousands of years.

There have been doubts around how this process is actually translated into a quality product (Bunten, 2010). Some argue that the results are too often not a reflection of the rich cultural heritage from the oldest existing culture but actually often homogenous products that do not reflect the cultural differences of tourism operators (Foley, 2004) and in low levels of demand for Indigenous tourism products (Ruhanen, Whitford, & McLennan, 2015). Some scholars lift up racism as a challenge towards particular Aboriginal enterprises (Collins, Morrison, Basu, & Krivokapic-Skoko, 2017; Ruhanen et al., 2016). Often these businesses are, in the end, impeded by a lack of financial capital, which is a sign of colonialism that impacts abilities to build a sustainable tourism market in remote Australia (Schmallegger & Carson, 2010). Of course, other challenges are down to a lack of education on one side and then the challenges of being in remote areas with extreme weather, bad infrastructure, distance from policymakers and distance from markets (Akbar & Hallak, 2019).

Protecting and Strengthening Indigenous Groups Relationship to Their Ancestral Lands

Coria and Calfucura (2012), Bunten (2010), Weaver (2009, and Colton and Whitney-Squire (2010) have argued 'that control or security of tenure and recognition of legal rights to ancestral or traditional lands and waters is a key issue influencing the success of Indigenous tourism enterprises' (Fletcher et al., 2016, p. 1106). Coria and Calfucura (2012) claim that to reach the full potential of social and economic benefits of Indigenous tourist projects, it is important that Indigenous communities assert control over land and resources. Langton, Palmer, and Ma Rhea (2014) argue that such 'guaranteed land security and the ability of Indigenous and local peoples to exercise their own governance structures are central' to maintaining traditional knowledge systems upon which Indigenous livelihoods depend (24). The debate has been very much focussed on physical and geological land, and little of our understanding of tenure and legal rights have focussed on the actual culture, cultural governance and spirituality.

There is a discussion on the extent and speed of how the Australian Government have invested directly into Indigenous tourism operations, which can be detrimental to branding and marketing as unique and complementary to sustainable luxury tourism, which aims to give unique experiences, and the focus has been on 'creating homogenous experiences via large organisations that target mass international tourist markets promoting Indigenous tourism (involving varying levels of Indigenous engagement) and result in limited opportunities for the development of small, diversified local tourism enterprises and employment opportunities' (Akbar & Hallak, 2019, p. 5). A recent study by Ruhanen et al. (2015) reveals that visitor demand is very low, as neither domestic nor international tourists prioritise Indigenous tourism experiences in their travel plans. Low prioritisation of Indigenous tourism experiences in visitors' travel planning has important implications for Indigenous tourism business models, particularly with regard to the value proposition and constellation (Ruhanen et al., 2015, p. 81; Yunus, Moingeon, & Lehmann-Ortega, 2010).

This is, of course, very much contradictory as there is an overall large interest in Australian Indigenous images and experiences that is very much exploited by the Australian state in general and Tourism Australia in specific. The Indigenous heritage of Australia is very much intertwined in how Australia is sold to the world, and often these glitzy campaign and adverts rely on high budgets that remote Indigenous enterprises lack. 'Indigenousity has become an important component in the marketing of Australia as a tourist destination' (Healy, 2008, p. 177). What we see here is that remote Indigenous enterprises struggle with low interest while at the same time government tourism authorities are using Indigenous symbols and culture to lure people to Australia itself. Cultural tourism has been growing in the tourism industry (cf. Tonnaer, 2008), and culture has become important in how destinations portray themselves as different.

What we can see is that Indigenous culture is becoming part of branding Australia as a different tourism destination. Branding 'is the orchestrated attempt to communicate [a] brand-linked relevance and meaning to key audiences, and nation branding might be seen as an attempt to link relevance and meaning to a nation-state' (Pomering & White, 2011, p. 167). Australia was the first country in the world to think of their country as a brand, which according to Graeme Turner is problematic as it 'turns a nation into a commodity, but it is not a commodity it's a social community' (Turner in Thornton, Redwood, Punchard, & Silvestrin, 2002). Blundell (1994) explains that nations such as Australia, the United States and Canada have used the cultural artefacts of original inhabitants to differentiate themselves from other countries, and such objects become 'national symbols' (p. 252).

In cultural tourism, Indigenous people's commodification, traditions and cultures are especially prominent (Adams, 2006; Bunten, 2008). In other words, there is an 'Indigenousization' that is prominent in the semiotic landscape of Australia, which 'gives the impression that Indigenous culture is a quintessential representation or icon of local and national life with the corollary, given its suggested primordiality, that it has always been thus' (Franklin, 2010, p. 196). Indigenous culture is clearly part of the Australian tourism brand to lure people to the country and we can also see that Indigenous art and motifs are used extensively to market both country and corporations as distinctive in a global market. The participation of this branding either benefiting Indigenous businesses or open to participation in the branding process seems to be neglected. According to Heyd (2003), 'cultural appropriation came into the spotlight in countries such as Canada, the United States and Australia' in the 1990s when non-Indigenous people copied, mimicked or stole Indigenous and minority 'ideas, images and art styles' (p. 37).

Many challenges and opportunities are facing Indigenous sustainable luxury tourism. Fletcher et al. (2016), for instance, claim that three significant challenges could undermine Indigenous tourism endeavours, such as control or security of tenure and recognition of legal rights to ancestral or traditional lands on one side, the overall governance and management of the land on the other and finally the policy context. We do not only have the issues of security of tenure, legal rights to land but we have also the question whether Indigenous people can own and profit from their culture. We can here a struggle to own the images of culture and also the actual land itself. For Indigenous tourism to become an essential mode of development, Indigenous will have to achieve a native title claim to all Indigenous images and symbols used in Tourism Australia or any other governmental or state tourist campaigns. We will expand on this argument below that protecting ancestral lands and Indigenous ways of life will mean taking control of Indigenous imagery used in tourism and other commercial enterprises.

To protect and strengthen Indigenous groups' relationship to ancestral lands, there has to be a reclaim of culture, cultural governance spirituality from Tourism Australia and all similar endeavours. Repatriation of Australian Indigenous cultural artefacts has been high on the agenda. 'Despite the absence of dedicated legislation, repatriation of special categories of Australian Indigenous cultural material – in particular human remains and secret/sacred objects – has explicit support from all levels of government' (Pickering, 2015, p. 428). If there is support from repatriating land and artefacts, there is also a dimension of repatriating cultural imagery and spaces to strengthen the connection between Indigenous groups and ancestral lands in more ways. For example, if Indigenous culture needs to reach higher visibility, it is through ownership of that visibility and recreate that visible bond towards ancestral land and the cultural link to the land visible. It will need to repatriate the Indigenous imagery from State and Corporate Australia, not just images of Australian Indigenous culture but also the images of landscapes, ecosystems and nature that the Indigenous cultural governance has nurtured, shaped and cared for. Indigenous ontological and epistemological explanations of land understand landscapes as storied and living (Tuck, McKenzie, & McCoy, 2014); and it adds to Indigenous critical, creative pedagogies (for instance, Santana & Akhurst, 2019) for transformative environmental education applications.

For instance, general usage of the term 'environment' or 'nature' assumes a place 'out there', separate from people. There is no Noongar or Nyikina (Indigenous) words for the notions conveyed by either of these terms. Rather, from a Noongar and Nyikina language perspective, the place is already a nature culture (as described by Whitehouse, Watkin Lui, Sellwood, Barrett, & Chigeza, 2014), which carries an understanding of landscape inclusive of systems of human interdependence. There could be an argument that Indigenous groups need to be able to reclaim some of the environmental and social goods of Australian tourism on one side and a right to the benefits of Tourism Australia using images and landscapes of Australia to lure tourists and to create Australia as a brand on the other side. For Indigenous tourism in remote areas to take off, Indigenous people will not only take control or security of tenure and recognition of legal rights to ancestral or traditional lands and waters but also to images, symbols and landscapes on one side and participation and sharing of the resources generated by these images, symbols and landscapes on the other side.

Self-reliance of Indigenous Groups and Their Environment Sustain Traditional Livelihoods

Much of the dominant discourse around Indigenous tourism in Australia has been based around marketing, product development and economic benefits to justify why Indigenous tourism development should be supported (Whitford & Ruhanen, 2010). Few have paid attention to the Quality of Life Indigenous communities experience or their role as environmental stewards maintaining the world's biodiverse ecologies (Whitford & Ruhanen, 2010).

> Transformative Sustainability Education (TSE)… recognises meanings and stories in landscape and celebrating Indigenous ways of knowing, being and doing. Indigenous-led environmental education links with one of the longest continuous environmental education systems in the world. Researchers contend that because of its ongoing history, environmental education carries a cultural obligation where Indigeneity and colonial forces are active. In Australia, every landscape is Indigenous and storied, all Australians have an inherent right to learn that joy in place, along with the responsibility to care for it.
>
> (Wooltorton, Collard, Horwitz, Poelina, & Palmer, 2020, p. 1)

If we look at the different factors presented earlier, such as both tourism as a way to combat social exclusion, strengthen the capabilities of Indigenous remote communities and protect the environment and ecosystems, whilst incorporating Sustainable Transformative Education (TSE) as a double educative.

There are signs that Indigenous business owners are a small but important part of Indigenous livelihoods, where 6% of the combined urban, rural and remote Indigenous workforce is comprised of business owners. We do not need to forget that in regional and remote areas there are estimated to be 12,000 Indigenous business owners and businesses that deliver important services (Salt, 2018). 'Both urban and remote Indigenous and Torres Strait Islander tourism enterprises are typically small to medium in size, with varying operational models' (Akbar & Hallak, 2019, p. 3). We often see small-scale business coupled with a more holistic view of man and nature, and there is a tendency to see all things connected, which affects business practices (Garnett, Austin, Shepherd, & Zander, 2016). This is particularly the case for Indigenous corporations that has natural experiences as their main product. Often these business ventures start from a world view where man has 'a reciprocal relationship with the natural world and the intangible elements within the world they inhabit' (Suchet-Pearson, Wright, Lloyd, & Burarrwanga, 2013). Smith, Scherrer, and Dowling (2009) recognise that this leads to engaging tourists with engagement with intangible elements, including local spirits or rock art sites. Often, some Indigenous tourism businesses have an underlying purpose of facilitating the repatriation of Indigenous community members to their traditional land, which is also shared by Indigenous social entrepreneurship and social businesses (Yunus et al., 2010). The tourism industry is there to support traditional livelihoods in ancestral lands.

We will use the case study of Kimberley, Western Australia and specifically the diverse Indigenous nations within the Martuwarra Fitzroy River catchment. The Martuwarra Fitzroy River Council (Martuwarra Council) is a newly formed alliance between six Indigenous nations who have united to champion alternative visions for sustainable futures. The Council represents Indigenous people who have sustainably managed land and water resources (Country) for tens of thousands of years, based on a deep association with their River Country (RiverOfLife et al., 2020b). According to First Nations' scholar Kyle Whyte,

> Indigenous peoples, collectives are not anthropocentric. That is, they do not exclude animals, plants, and ecosystems as members with the responsibilities of active agents bound up in moral relationships of reciprocal responsibilities with humans and other non-humans.
>
> (2018, p. 155)

The Martuwarra Council's Directors are community leaders and senior elders together with emerging leaders from across the River's Indigenous nations who are united with 'One Mind and One Voice'. In July 2020, Council endorsed a strategic approach to exploring alternative futures for the region that capitalises on Earth-centred approaches to 'development' that respect nature's contributions to non-human and human well-being. The Council's alternative paradigm for sustainable futures activities gives voice to realising this vision. These nations refer to the river as their sacred ancestral living serpent being whom they have a collective responsibility to love and protect. First Law acknowledges the serpent being has a right to life and, as a sacred River, the right to live and flow (RiverOfLife et al., 2020b).

This way of thinking about livelihoods and relationships includes the river as a significant 'totem', the most important ingredient for sustaining the greater good for non-human, human and ecological well-being. Deborah Bird Rose's construction of totemism as an 'ecological management system' for distributing rights and responsibilities links humans with non-humans in specific ways. Whyte (2018, p. 159) affirms that the term ecologies 'are systematic arrangements of human, non-human beings…entities…and landscapes that are conceptualised and operate purposefully to facilitate as society's capacity to survive and flourish in a particular landscape and watershed'.

The Convention on Biological Diversity's (CBD) '2050 Vision' aims for a world living in harmony with nature by the year 2050 (CBD, COP Decision x/2). Yet, biodiversity is threatened globally, increasing to an unprecedented extent (Diaz et al., 2019; IPBES, 2019). Australia is one of the most diverse places on the planet; it is also rated as having one of the world's worst extinction rates (Woinarski, Burbidge, & Harrison, 2014). In this context, biodiversity conservation has been identified as one of society's most important planetary challenges (Steffen et al., 2015).

In the 100th Sustainability Podcast interview, Poelina shared with listeners the pressure her and other leaders in the Martuwarra Fitzroy River Catchment face regarding protecting some of the world's last remaining biodiverse hotspots. In a podcast 'Valuing and Respecting Sustainability Principles from the Beginning of Time through to Modernity' (Offland & Poelina, 2020), the message across the planet is an urgency to shift away from the destruction of the planet to alternative economies which are sustainable and result in sharing of the wealth creation, so all of us can reach our full potential and prosper as human beings. Poelina affirms Martuwarra Fitzroy River Indigenous peoples live within world views, values and ethics where it is all about relationships where everything is connected and where all species have a right to justice.

We, therefore, have to recognise that the Martuwarra and its distinct cultural and environmental heritage could be lost as threats of development could transform parts of the river into intensive farming areas. This could lead to both extensive land clearing and water extraction. There is also the potential of coal mining, gas fracking and industrial irrigated agriculture in the Martuwarra Catchment Area (Zonderland, 2017). We can see here that sustainable Indigenous tourism could supply the communities with both an ethical and commercial support of Indigenous native title rights and direct support of Indigenous communities in managing the environment in their sustainable stewardship (RiverOfLife et al., 2020b). This is critical as the livelihoods and well-being of both Indigenous and non-Indigenous people are reliant on a healthy and living river and ecological system. We highlight the opportunity to incorporate Indigenous wisdom into planning and development and for investment into new sustainable economies.

The Martuwarra Council is focussed on strengthening the capacity of our emerging leaders, some not so young, who are keen and ready to step up to leadership roles. They are beginning an entrepreneurial pathway as Storytellers & Community Researchers through the Arts, creating powerful and beautiful stories. These Storytellers promote hope, resilience and transformation through sharing their learnings and positive spirit. Together the young are equally valued with the elders. The Council is growing a new dream for equity, human and climate justice in the balance between human and non-human beings – the birds, fish, kangaroos, sawfish and crocodiles and the trees, grasses and snakes. Within Indigenous collectives such as the Council, sustainable luxury tourism will evolve planning processes to include diverse complementary activities. The Councils strategic plan has identified 15 Positions Statements to reflect their vision for sustaining their traditional livelihoods and relationships (RiverOfLife et al., 2020b).

One of the essential aspects of protecting the traditional way of life through tourism is to protect these from not only alternative development mode such as mining and industrial agriculture but also to protect Indigenous tourism as an activity to be practised and advertised by Indigenous members. Kimberley has some of the lowest extinction rates in Australia, and there is a need to link up the livelihoods of Indigenous communities to the care of the land. In Norway and Sweden, some traditional livelihoods such as reindeer herding have become exclusive to Indigenous Sami communities. There are around 6,500 Sami involved in traditional reindeer herding. 'The Sami do not have to compete with the majority population in reindeer herding as the right to reindeer herding is reserved only for Sami in both Norway and Sweden' (Koivurova et al., 2015, p. 13). There are the arguments to protect Indigenous tourism as a livelihood through legal protection, which would need to be based around cultural governance and devolve some of the decision-making from the government, state and tourism boards to local development of a more diverse, sustainable and social tourism business that is more geared towards the high end. Indigenous communities and enterprises must be protected legally to cooperate with different stakeholders to develop sustainable and profitable luxury products that can support communities. There is an argument to ensure that Indigenous tourism is a livelihood that engages Indigenous communities when it comes to participation and also to enjoy the goods of this possible livelihood.

Conclusion

This chapter has defined a discussion around how to make sustainable luxury tourism socially sustainable from an Indigenous perspective. We have argued that we need to look not just on participation and the substantive dimension of social sustainability but also at how sustainable luxury can protect and strengthen Indigenous groups' relationship to their ancestral lands, and self-reliance of

Indigenous groups and their environment sustain traditional livelihoods. The chapter brings up pitfalls that are faced by Indigenous tourism such as being underfunded; tourist products do not always reflect the diverse, rich cultural history, not branded very well towards foreign or national tourist; some highlight a lack of interest in Indigenous tourist product overall. This is often countered that Tourism Australia and tourist relevant corporations brand themselves and Australia as closely linked to Indigenous culture, art, history and managed nature. There is, therefore, a mismatch between how the country is branding Indigenous culture as a tourist attraction and the issue to actually brand and develop diversified and exclusive sustainable luxury tourism in remote areas. We argue that to achieve this, there is a need for Indigenous groups to have larger participation and control over national and corporate Indigenous branding and to use this branding to strengthen both Indigenous groups relationship to ancestral land and also to help create better branding for actually Indigenous products. There is also a need to develop stronger cultural protection of Indigenous sustainable luxury tourism to be an exclusive Indigenous community activity in the same way as Sami reindeer husbandry is exclusive to Sami people in Norway and Sweden.

References

Aboriginal and Torres Strait Islander Affairs. (2003). *Guide to provide advice for Indigenous tourism business—Media release*. Canberra, ACT: Minister for Immigration and Multicultural and Indigenous Affairs. Retrieved from http://www.atsia.gov.au/media/ruddock_media03/r03064.htm

Adams, K. (2006). *Art as politics: Re-crafting identities, tourism, and power in Tana Toraja, Indonesia*. Honolulu: University of Hawai'i Press.

Agyeman, J., & Evans, B. (2004). 'Just sustainability': The emerging discourse of environmental justice in Britain? *Geographical Journal*, *170*(2), 155–164. doi: 10.1111/j.0016-7398.2004.00117.x

Akbar, S., & Hallak, R. (2019). Identifying business practices promoting sustainability in aboriginal tourism enterprises in remote Australia. *Sustainability*, *11*(17), 4589. doi:10.3390/su11174589

Altman, J. (1993). *Indigenous Australians in the National Tourism Strategy: Impact, sustainability and policy issues*. Canberra, ACT: Centre for Aboriginal Economic Policy Research, ANU.

Altman, J., & Finlayson, J. (2003). Aborigines, tourism and sustainable development. *Journal of Tourism Studies*, *14*(1), 78–91. doi:10.3316/IELAPA.200305727

Bendell, J. (2012). *Elegant disruption: How luxury and society can change each-other for good*. Brisbane, QLD: Griffith University.

Bennett, E. M., Solan, M., Biggs, R., MacPhearson, T., Norstrom, A., Olsson, P., … Xu, J. (2016). Bright spots: Seeds of a good anthropocene. *Frontiers in Ecology and the Environment*, *14*(8), 441–448. doi:10.1002/fee.1309

Birdsall-Jones, C., Wood, D., & Jones, R. (2007). Great expectations: Indigenous land-based tourism in regional Western Australia. In *Striving for sustainability: Case* (pp. 187–210). Lismore, NSW: Cross University Press.

Blundell, V. (1994). “Take home Canada”: Representations of aboriginal peoples as tourist souvenirs. In S. H. Riggins (Ed.), *The socialness of things: Essays on the socio-semiotics of objects* (pp. 251–284). Berlin and New York, NY: Mouton de Gruyter.

Boström, M. (2012). A missing pillar? Challenges in theorising and practising social sustainability: Introduction to the special issue. *Sustainability: Science, Practice and Policy*, *8*(12), 3–14. doi:10.1080/15487733.2012.11908080

Bullard, R. D. (1999). Dismantling environmental racism in the USA. *Local Environment*, *4*(1), 5–19. doi:10.5250/amerindiquar.34.3.285

Bunten, A. C. (2008). Sharing culture or selling out? Developing the commodified persona in the heritage industry. *American Ethnologist*, *35*(3), 380–395. doi: 10.1111/j.1548-1425.2008.00041

Bunten, A. C. (2010). More like ourselves: Indigenous capitalism through tourism. *American Indian Quarterly*, *34*(3), 285–311.

Buultjens, J., & Gale, D. (2013). Facilitating the development of Australian indigenous tourism enterprises: The business ready program for indigenous tourism. *Tourism Management Perspective*, *5*, 41–50. doi:10.1016/j.tmp.2012.09.007

Buultjens, J., Waller, I., Graham, S., & Carson, D. B. (2002). Public sector initiatives for Aboriginal small business development in tourism: Occasional paper number 6. Centre for Regional Tourism Research, Southern Cross University. Retrieved from https://researchportal.scu.edu.au/discovery/fulldisplay/alma991012821269102368/61SCU_INST: ResearchRepository

Buultjens, J., & White, N. (2008). Indigenous tourism: The possibilities into the future. In Proceedings of the Proceedings Desert Knowledge Symposium, Alice Springs, 3–6 November.

Casula Vifell, Å., & Soneryd, L. (2012). Organising matters: How the ‘social dimension’ gets lost in sustainability projects. *Sustainable Development*, *20*(1), 18–27. doi: 10.1002/sd.461

Collins, J., Morrison, M., Basu, P. K., & Krivokapic-Skoko, B. (2017). Indigenous culture and entrepreneurship in small businesses in Australia. *Small Enterprise Research*, *24*, 36–48. doi:10.1080/13215906.2017.1289855

Colton, J. W., & Whitney-Squire, K. (2010). Exploring the relationship between aboriginal tourism and community development. *Leisure/Loisir*, *34*(3), 261–278. doi:10.1080/14927713.2010.521321

Commonwealth of Australia. (2003). *Tourism white paper: A medium to long-term strategy for tourism*. Canberra, ACT: Department of Industry, Science and Resources.

Cooke, F., Nordensvärd, J., Bin Saat, G., Siciliano, G., & Urban, F. (2017). The limits of social protection: The case of hydropower dams and indigenous peoples' land. *Asia and the Pacific Policy Studies*, *4*(3), 437–450. doi:10.1002/app5.187

Coria, J., & Calfucura, E. (2012). Ecotourism and the development of indigenous communities: The good, the bad, and the ugly. *Ecological Economics*, *73*, 47–55. doi:10.1016/j.ecolecon.2011.10.024

Dempsey, N., Bramley, G., Power, S., & Brown, C. (2011). The social dimension of sustainable development: Defining urban social sustainability. *Sustainable Development*, *19*(5), 289–300. doi:10.1002/sd.417

Department of Industry, Tourism and Resources. (2005). *Request for tender: Research project—Business ready program for Indigenous tourism*. Canberra, ACT: Department of Industry, Tourism & Resources.

Diaz, S., Settele, J., Brondízio, E. S., Ngo, H. T., Agard, J., Arneth, A., … Zayas, C. N. (2019). Pervasive human-driven decline of life on earth points to the need for transformative change. *Science, 366*(6471), 1327. doi:10.1126/science.aax3100

Dobson, A. (2003). *Citizenship and the environment*. New York, NY: Oxford University Press.

Dyer, P., Aberdeen, L., & Schuler, S. (2003). Tourism impacts on an Australian indigenous community: A Djabugay case study. *Tourism Management, 24*(1), 83–95. doi:10.1016/S0261-5177(02)00049-3

Egri, C. P., & Herman, S. (2000). Leadership in North American environmental sector: Values leadership styles and contexts of environmental leaders and their organisations. *Academy of Management Journal, 43*(4), 523–553. doi:10.2307/1556356

Faber, D. R. (2005). Building a transnational environmental justice movement: Obstacles and opportunities in the age of globalisation. In J. Bandy & J. Smith (Eds.), *Coalitions across borders: Transnational protest and the neoliberal order* (pp. 43–68). Lanham, MD: Rowman & Littlefield.

Finlayson, J., & Madden, R. (1995). Regional tourism case studies: Indigenous participation in tourism in Victoria. In Bureau of Tourism Research (Ed.), *Tourism research and education in Australia. Proceedings from the tourism and education conferences* (pp. 269–275). Canberra, QLD: BTR.

Fiszbein, A. (1997). The emergence of local capacity: Lessons from Colombia. *World Development, 25*(7), 1029–1043. doi:10.1016/S0305-750X(97)00020-X

Fletcher, C., Pforr, C., & Brueckner, M. (2016). Factors influencing Indigenous engagement in tourism development: An international perspective. *Journal of Sustainable Tourism, 24*, 1100–1120. doi:10.1080/09669582.2016.1173045

Foley, D. (2004). *Understanding indigenous entrepreneurship: A case study analysis.* Brisbane, QLD: University of Queensland.

Franklin, A. (2010). Aboriginalia: Souvenir wares and the "aboriginalization" of Australian identity. *Tourist Studies, 10*(3), 195–208. doi:10.1177/1468797611407751

Fuller, D., Antella, J., Cummings, E., Scales, J., & Simon, T. (2001). *Ngukurr field report: Stage two*. Research Report No. 5. Adelaide, SA: Flinders University.

Fuller, D., Buultjens, J., & Cummings, E. (2005). Ecotourism and indigenous microenterprise formation in northern Australia opportunities and constraints. *Tourism Management, 26*(6), 891–904. doi:10.1016/j.tourman.2004.04.006

Gardetti, M. A. (2011). *Sustainable luxury in Latin America, conference dictated at the seminar sustainable luxury & design*. Madrid: Instituto de Empresa (Business School).

Garnett, S. T., Austin, B. J., Shepherd, P., & Zander, K. K. (2016). Culture-based enterprise opportunities for indigenous people in the Northern Territory, Australia. In K. Iankova & A. Hassan (Eds.), *Indigenous people and economic development: An international perspective* (pp. 111–132). London: Routledge.

Gough, I., Meadowcroft, J., Dryzek, J., Gerhards, J., Lengfeld, H., Marandya, A., & Ortiz, R. (2008). JESP symposium: Climate change and social policy. *Journal of European Social Policy, 18*(4), 325–344. doi:10.1177/0958928708094890

Gunya Discussion Paper: Indigenous Economic Development Scheme – A Solution for Indigenous communities. (2007). Gunya Australia. Retrieved from https://aiatsis.gov.au/sites/default/files/docs/research-and-guides/native-title-research/GunyaDiscussionPaper.pdf

Harvey, D. (1996). *Justice, nature and the geography of difference*. Oxford: Blackwell.

Healy, C. (2008). *Forgetting aborigines*. Sydney: UNSW Press.

Heyd, T. (2003). Rock art aesthetics and cultural appropriation. *Journal of Aesthetics and Art Criticism*, *61*(1), 37–46. doi:10.1111/1540-6245.00090

Higgins-Dasbiolles, F., Trevonow, G., & Sparrow, S. (2014). The Coorong Wildemess Lodge: A case study of planning failures in Indigenous tourism. *Tourism Management*, *44*, 46–57. doi:10.1016/j.tourman.2014.02.003

Hill, R., Harding, E. K., Edwards, D. J., Hill, D. O., Martin, A., & McIntyre-Tamwoy, S. (2008). *A cultural and conservation economy for Northern Australia*. Canberra, ACT: Land & Water.

IPBES. (2019). *Global assessment report on biodiversity and ecosystem services*. Bonn: IPBES Secretariat.

Ivory, B. (2003). Poverty and enterprise. In S. C. Carr & T. S. Sloane (Eds.), *Poverty and psychology: Emergent critical practice*. New York, NY: Kluwer Academic/ Plenum Publishers.

Jacobsen, D. (2017). Tourism enterprises beyond the margins: The relational practices of Aboriginal and Torres Strait Islander SMEs in remote Australia. *Tourism Planning & Development*, *14*, 31–49. doi:10.1080/21568316.2016.1152290

Kearins, K., Collins, E., & Tregidga, H. (2010). Beyond corporate environmental management to a consideration of nature in visionary small enterprise. *Business & Society*, *49*(3), 512–547. doi:10.1177/0007650310368988

Koivurova, T., Masloboev, V., Hossain, K., Nygaard, V., Petrétei, A., & Vinogradova, S. (2015). Legal protection of Sami traditional livelihoods from the adverse impacts of mining: A comparison of the level of protection enjoyed by Sami in their four home states. *Arctic Review*, *6*(1), 11–51. doi:10.17585/arctic.v6.76

Langton, M. L., Palmer, L., & Ma Rhea, Z. (2014). Community-oriented protected areas for Indigenous peoples and local communities: Indigenous protected areas in Australia. In S. Stevens (Ed.), *Indigenous peoples, national parks, and protected areas. A new paradigm linking conservation, culture, and rights* (pp. 84–107). The University of Arizona Press.

Littig, B., & Grießler, E. (2005). Social sustainability: A catchword between political pragmatism and social theory. *International Journal of Sustainable Development*, *8*(1–2), 65–79. doi:10.1504/IJSD.2005.007375

Lohmann, L. (2009). Climate as investment. *Development and Change*, *40*(6), 1063–1083. doi:10.1111/j.1467-7660.2009.01612.x

Maiese, M. (2003). "Distributive justice", beyond intractability. In G. Burgess & H. Burgess (Eds.), *Conflict research consortium*. Boulder, CO: University of Colorado. Retrieved from www.beyondintractability.org/essay/distributive_justice/

Mair, J., & Laing, J. H. (2013). Encouraging pro-environmental behaviour: The role of sustainability-focused events. *Journal of Sustainable Tourism*, *21*(8), 1113–1128. doi:10.1080/09669582.2012.756494

Miller, G., Rathouse, K., Scarles, C., Holmes, K., & Tribe, J. (2010). Public understanding of sustainable tourism. *Annals of Tourism Research*, *37*(3), 627–645. doi:10.1016/j.annals.2009.12.002

Murphy, K. (2012). The social pillar of sustainable development: A literature review and framework for policy analysis. *Sustainability: Science, Practice and Policy*, *8*(1), 15–29. doi:10.1080/15487733.2012.11908081

Nordensvard, J., Urban, F., & Mang, G. (2015). Social innovation and Chinese overseas hydropower dams: The nexus of national social policy and corporate social responsibility. *Sustainable Development*, *23*, 245–256. doi:10.1002/sd.1591

Offland, B., & Poelina, A. (2020, July 17). Episode 2, Wake Up the Snake: Waking Up the Consciousness of the People, Natures Insight. Retrieved from https://youtu.be/-swe6fKnYHk. Accessed on January 19, 2021.

Pepper, M., & Keogh, S. (2014). Biogeography of the Kimberley, Western Australia: A review of landscape evolution and biotic response in an ancient refugium. *Journal of Biogeography*, *41*(8), 1443–1455. doi:10.1111/jbi.12324

Pickering, M. (2015). 'The big picture': The repatriation of Australian Indigenous sacred objects. *Museum Management and Curatorship*, *30*(5), 427–443.

Poelina, A. (2016). A lucky country and a fair go. In I. Tadros (Ed.), *This is my country* (pp. 18–23). New York, NY: Foto Evidence Press. Retrieved from http://fotoevidence.com/book/27/hard-copy

Poelina, A. (2020). A coalition of hope! A regional governance approach to Indigenous Australian cultural wellbeing. In A. Campbell, M. Duffy, & B. Edmondson (Eds.), *Located Research: Regional places, transitions and challenges* (pp. 153–180). Singapore: Springer Singapore. doi:10.1007/978-981-32-9694-7_10

Poelina, A., & Nordensvärd, J. (2018). Sustainable luxury tourism, indigenous communities and governance. In M. Gardetti & S. Muthu (Eds.), *Sustainable luxury, entrepreneurship, and innovation. Environmental footprints and eco-design of products and processes*. Singapore: Springer.

Pomering, A., & White, L. (2011). The portrayal of Indigenous identity in Australian tourism brand advertising: Engendering an image of extraordinary reality or staged authenticity? *Place Branding and Public Diplomacy*, *7*(3), 165. doi:10.1057/pb.2011.19

RiverOfLife, M., McDuffie, M., & Poelina, A. (2020a). *Martuwarra country: A historical perspective (1838–present)*. Martuwarra Fitzroy River Council; Nulungu Research Institute, The University of Notre Dame Australia. doi:10.32613/nrp/2020.5

RiverOfLife, M., Poelina, A., Alexandra, J., & Samnakay, N. (2020b). *A conservation and management plan for the National Heritage listed Fitzroy River Catchment Estate (No. 1)*. Martuwarra Fitzroy River Council, Nulungu Research Institute, The University of Notre Dame Australia. doi:10.32613/nrp/2020.4

Ruhanen, L., McLennan, C., & Whitford, M. (2016). Developing a sustainable indigenous tourism sector: Reconciling socio-economic objectives with market driven approaches. In *Indigenous people and economic development: An international perspective* (pp. 205–222). London: Routledge.

Ruhanen, L., Whitford, M., & McLennan, C. (2015). Indigenous tourism in Australia: Time for a reality check. *Tourism Management*, *48*, 73–83. doi:10.1016/j.tourman.2014.10.017

Ryan, C., & Huyton, J. (2002). Tourists and aboriginal people. *Annals of Tourism Research*, *29*(3), 631–647. doi:10.1016/S0160-7383(01)00073-1

Salt, B. (2018). *Accelerating indigenous entrepreneurship*. KPMG. Retrieved from https://assets.kpmg/content/dam/kpmg/au/pdf/2018/census-insights-accelerating-indigenous-entrepreneurship.pdf. Accessed on March 9, 2018.

Santana, C. R., & Akhurst, G. (2019). Critical creative pedagogies: A decolonial and Indigenous approach using visual arts and creative writing. *Australian Journal of Indigenous Education*, *50*(1), 1–8.

Schlosberg, D. (2007). *Defining environmental justice*. Oxford: Oxford University Press.

Schlosberg, D. (2013). Theorising environmental justice: The expanding sphere of a discourse. *Environmental Politics*, *22*(1), 37–55. doi:10.1080/09644016.2013.755387

Schlosberg, D., & Carruthers, D. (2010). Indigenous struggles, environmental justice, and community capabilities. *Global Environmental Politics*, *10*(4), 12–35. Retrieved from https://www.muse.jhu.edu/article/404381

Schmallegger, D., & Carson, D. (2010). Is tourism just another staple? A new perspective on tourism in remote regions. *Current Issues in Tourism*, *13*(3), 201–221. doi:10.1080/13683500903359152

Schmiechen, J., & Boyle, A. (2007). Aboriginal tourism research in Australia. In R. Butler & T. Hinch (Eds.), *Tourism and indigenous peoples: Issues and implications* (pp. 58–72). Burlington, ON: Butterworth-Heinemann.

SCRGSP (Steering Committee for the Review of Government Service Provision). (2016). *Overcoming indigenous disadvantage: Key indicators 2016*. Canberra, ACT: Productivity Commission.

SCRGSP (Steering Committee for the Review of Government Service Provision). (2020). *Overcoming Indigenous disadvantage: Key indicators 2020*. Canberra, ACT: Productivity Commission.

Semeniuk, V., & Brocx, M. (2011). King sound and the tide-dominated delta of the Fitzroy river: Their geoheritage values. *Journal of the Royal Society of Western Australia*, *94*, 151–160. doi:10.1016/S0272-7714(02)00214-7

Shrader-Frechette, K. (2002). *Environmental justice: Creating equality, reclaiming democracy*. New York, NY: Oxford University Press.

Smith, A. J., Scherrer, P., & Dowling, R. (2009). Impacts on aboriginal spirituality and culture from tourism in the Kimberley coastal waterways of the Kimberley region, North West Australia. *Journal of Ecotourism*, *8*(2), 82–98. doi:10.1080/14724040802696007

Steffen, W., Richardson, K., Rockström, J., Cornell, S. E., Fetzer, I., Bennett, E. M, … Sörlin, S. (2015). Sustainability. Planetary boundaries: Guiding human development on a changing planet. *Science*, *347*(6223), 1259855, 736–747. doi: 10.1126/science.1259855

Suchet-Pearson, S., Wright, S., Lloyd, K., & Burarrwanga, L. (2013). Caring as country: Towards an ontology of co-becoming in natural resource management. *Asia Pacific Viewpoint*, *54*(2), 185–197. doi:10.1111/apv.12018

Sundbo, J., & Dixit, S. K. (2020). Conceptualisations in tourism experience. In S. K. Dixit (Ed.), *The Routledge handbook of tourism experience management and marketing*. London: Routledge.

Thin, N. (2002). *Social progress and sustainable development*. London: ITDG Publishing.

Thornton, S., Redwood, J., Punchard, E., & Silvestrin, L. (2002). Film Australia, prospero productions, films for the humanities & sciences. Selling Australia. The brand. Films for the Humanities & Sciences.

Tonnaer, A. (2008). *Tourism dreaming: A study of the encounter culture in indigenous cultural tourism in Australia*. Århus: University of Aarhus. Retrieved from http://hdl.handle.net/2066/71875

Tourism Research Australia. (2010). *Indigenous tourism in Australia: Profiling the domestic market*. Canberra, ACT: Tourism research Australia. Retrieved from http://www.ret.gov.au/tourism/

Tuck, E., McKenzie, M., & McCoy, K. (2014). Editorial: Land education: Indigenous, post-colonial, and decolonising perspectives on place and environmental education research. *Environmental Education Research*, *20*(1), 1–23.

Turner, K. L., Berkes, F., & Turner, N. J. (2012). Indigenous perspectives on ecotourism development: A British Columbia case study. *Journal of Enterprising Communities: People and Places in the Global Economy*, *6*(3), 213–229. doi:10.1108/17506201211258397

Turner, R. L., & Wu, D. P. (2002). *Environmental justice and environmental racism: An annotated bibliography and general overview, focusing on U.S. literature, 1996–2002*. Department of Environmental Science, Policy and Management, Universityof California, Berkeley.

UN WCED (World Commission on Environment and Development). (1987). *Our common future*. Oxford: Oxford University Press.

Urban, F., Nordensvärd, J., Siciliano, G., & Li, B. (2015). Chinese overseas hydropower dams and social sustainability: The Bui Dam in Ghana and the Kamchay Dam in Cambodia. *Asia and the Pacific Policy Studies*, *2*(3), 573–589. doi:10.1002/app5.103

Vogwill, R. (2015). *Water resources of the Mardoowarra (Fitzroy River) catchment*. Perth, WA: Wilderness Society.

Walker, K., & Moscardo, G. (2016). Moving beyond sense of place to care of place: The role of indigenous values and interpretation in promoting transformative change in tourists' place images and personal values. *Journal of Sustainable Tourism*, *24*(89), 1243–1261. doi:10.1080/09669582.2016.1177064

Watson, J., Watson, A., Poelina, A., Poelina, N., Watson, W., Camilleri, J., & Vernes, T. (2011). Nyikina and Mangala Mardoowarra Wila Booroo: Natural and cultural heritage plan. Nyikina-Mangala Aboriginal Corporation and World Wildlife Fund for Nature Australia.

Weaver, D. (2009). Indigenous tourism stages and their implications for sustainability. *Journal of Sustainable Tourism*, *18*(1), 43–60. doi:10.1080/09669580903072001

Whitehouse, H., Watkin Lui, F., Sellwood, J., Barrett, M. J., & Chigeza, P. (2014). Sea country: Navigating Indigenous and colonial ontologies in Australian environmental education. *Environmental Education Research*, *20*(1), 56–69.

Whitford, M., & Ruhanen, L. (2010). Australian indigenous tourism policy: Practical and sustainable policies. *Journal of Sustainable Tourism*, *18*(4), 475–496. doi:10.1080/09669581003602325

Whitford, M., & Ruhanen, L. (2014). Indigenous tourism businesses: An exploratory study of business owners perceptions of drivers and inhibitors. *Tourism Recreation Research*, *39*(2), 149–168. doi:10.1080/02508281.2014.11081764

Whyte, K. P. (2018). Reflections on the purpose of indigenous environmental education. Retrieved from http://www.academia.edu/download/57760824/2018_Whyte_-_Reflections_on_Indigenous_Environmental_Education.pdf

Woinarski, J. C. Z., Burbidge, A. A., & Harrison, P. L. (2014). *The Action Plan for Australian Mammals 2012*. Melbourne: CSIRO Publishing.

Wooltorton, S., Collard, L., Horwitz, P., Poelina, A., & Palmer, D. (2020). Sharing a place-based Indigenous methodology and learnings. *Environmental Education Research*, *26*(7), 917–934. doi:10.1080/13504622.2020.1773407

Young, E. (1988). Aboriginal economic enterprises: Problems and prospects. In D. Wade-Marshall & P. Loveday (Eds.), *Contemporary issues in development: North Australia: Progress and prospects* (Vol. 1, pp. 182–200). Darwin, NT: NARU Darwin.

Yunus, M., Moingeon, B., & Lehmann-Ortega, L. (2010). Building social businessmodels: Lessons from the Grameen experience. *Long Range Planning*, *43*(2/3), 308–325. doi:10.1016/j.lrp.2009.12.005

Zeppel, H. (2001). Aboriginal cultures and Indigenous tourism. In N. Douglas, N. Douglas, & R. Derrett (Eds.), *Special interest tourism* (pp. 232–259). Milton, QLD: Wiley.

Zonderland, S. (2017, November 7). Protecting the kimberley with the wilderness society of Western Australia. Retrieved from https://www.shewentwild.com/blogposts/the-kimberley-fracking. Accessed on September 25, 2019.

Chapter 20

Sustainable High Yield Luxury Tourism Experiences

Jacqualine Spence,[1] Pierre Benckendorff,[2] and P. Monica Chien[3]

Abstract

Luxury tourism is an emerging area of research and deserves consideration for its implications for tourism and hospitality management and policy development. This chapter reviews the phenomena of luxury tourism in the academic literature and in particular, its links to the concepts of high yield, sustainability and tourist experiences. The global hiatus caused by the COVID-19 pandemic has presented an opportunity for tourism policy-makers and industry stakeholders to reconsider how luxury tourism can contribute to optimal economic, social and environmental outcomes with desired yield and sustainability aspirations. A renewed understanding of luxury tourism in terms of its production and consumption processes, as well as the associated value, emotion and narrative, is therefore of critical importance. The value of this chapter lies in synthesising a number of strands of inquiry across disparate bodies of literature to identify a research agenda. Areas that are proposed for further research include the conceptualisation of luxury tourism, the evolving nature of luxury experiences; value co-creation across all stages of luxury tourism; and the nexus between luxury tourism and destination image. Managerial implications of luxury tourism are also discussed, including the necessary conditions for cultivating luxury tourism; the need to measure the social and environmental impact of luxury tourism; and the important relationships between luxury tourism, innovation and market leadership.

Keywords: Luxury; tourism impact; sustainability; high yield tourism; value co-creation; exclusivity

---

[1]https://orcid.org/0000-0001-7403-3866
[2]https://orcid.org/0000-0002-2794-8534
[3]https://orcid.org/0000-0003-1666-0479

The Emerald Handbook of Luxury Management for Hospitality and Tourism, 403–423

doi:10.1108/978-1-83982-900-020211020

Introduction

With annual global tourism export earnings estimated at USD 1.7 trillion in 2019, most governments have formulated policies designed to build this industry (WTO, 2019). At the same time, concerns about the downsides of overtourism are increasing. As nations wish to make the most of their investments in tourism at the industry sector and individual business levels, it is essential to develop a greater understanding of how different tourism market segments can optimise economic, social and environmental contributions.

In recent years, many governments and destinations have shifted their tourism marketing strategy from simply encouraging the number of tourists towards enhancing 'quality' tourism growth. The aim is to achieve greater revenues and profitability while avoiding potential negatives that can be associated with tourism, such as congestion, intergroup conflicts and environmental degradation (Chien & Ritchie, 2018; Dwyer et al., 2006; OECD, 2020). In financial terms, developing 'high yield' tourism is often articulated as moving away from price-conscious mass tourism towards encouraging categories of tourism where income contributed to the host economy from each tourist is relatively higher (Dwyer et al., 2006).

One approach to achieving high yield from tourism is to focus on the expensive, luxury end of the market. But if providing luxury tourism services and experiences is so high yield, why do all governments and service providers not prioritise the luxury market in their tourism strategy? One reason is that other forms of tourism can provide volume and help to achieve other objectives, such as the diversification of risk, industry reliance and employment opportunities. Another reason might be that a luxury strategy is very difficult, risky and costly to execute and sustain (Kapferer & Bastien, 2009).

A luxury tourism strategy tends to require an inordinate amount of ongoing investment by providers into assets, competencies and capabilities, such as expensive real estate locations, higher levels of staffing and training, upscale amenities, brand building and other kinds of organisational capabilities that are essential to the provision of premium service levels, and which contribute to a tourist's overall luxury experience. However, these challenges contribute to the exclusivity and profitability potential of the luxury tourism market opportunity. Since successful luxury positioning is complex and costly to imitate, it can be a significant source of competitive advantage and economic benefit to providers (Crouch & Ritchie, 1999).

With the added value to be gained through a better understanding of the luxury tourism market, it is surprising that more academic research has not focused on it. This chapter first examines the relationship between luxury tourism and yield concepts, followed by a review of how luxury services and experiences have been covered in tourism and hospitality literature will be presented. An agenda for future research will be presented, and, finally, implications for the management of luxury tourism experiences will be outlined.

Background

Across cultures, many scholars have framed tourism as a luxury since its earliest occurrence – an experiential opportunity that is desired and highly valued yet not available to many people (Gunter & Netto, 2015; Smeral, 2003). The earliest forms of tourism have been described as travel experiences undertaken by privileged individuals. Amongst the earliest civilisations known – ancient China (Yan, 2018), Egypt (Casson, 1974), Greece (Romero, 2013) and Rome (Lomine, 2005; Friedländer, 1908) – tourism was undertaken for purposes such as event participation, health, pleasure, sightseeing and visiting others.

In the eighteenth century, 'The Grand Tour' referred to tourism across the continent of Europe, predominantly at first, by young aristocratic males (Towner, 1985). However, particularly after 1780, a considerable number of the Grand Tour travellers were from other social groups, such as the professional middle classes, clergy and retired members of the armed forces. Towner (1985) describes reasons for undertaking the Grand Tour as related to business, career, culture, economy, education, health and science. At that time, there was no tourism industry as such. Along their journey, tourists accessed what they needed together with citizens. As the number of Grand Tourists increased, some locals along well-worn routes began to regularly provide goods and services to meet the recurring needs of these travellers, such as accommodation or money exchange services. Few supplier records have survived from that time. In contrast, many of the Grand Tourists chronicled their travels, leaving behind detailed accounts of, and reflections upon, their experiences. Hence, most of what is known about this nascent tourism industry is from the perspective of literate male tourists from relatively privileged backgrounds (Towner, 1985).

Since that time, the popularity and availability of tourism has extended to many socio-economic groups across the world, although not for everyone. The freedom and means to undertake leisure travel has always set some individuals apart from others, and exclusivity remains a characteristic of some forms of tourism. These days, across many market segments, tourism is regarded as luxurious when undertaken to make it indulgent, high in quality and exclusive in some way that is either monetary, social or hedonic (Iloranta, 2019; Wirtz, Holmqvist, & Fritze, 2020). This is the tourist's experience of luxury.

Luxury Tourism and Yield Concepts

High Yield Tourism

Tourism can be understood from many perspectives. One of these is economic. Among economic perspectives are the direct commercial impacts of tourism, tourism-related employment and livelihood diversification, tourism spending contribution to gross domestic product (GDP), regional competitiveness and sustainability (i.e., achieving a balance of environmental, social and economic objectives) (Dwyer et al., 2006). Studies of the relationship between tourism development and economic growth indicate that it is not always strong, predictable, reliable and positive (Lee & Chang, 2008). In the COVID-19 pandemic era, Holzner's (2011) warnings about the potential pitfalls of any economy's over-reliance on tourism (the 'Dutch Disease Effect') seem prescient. Nonetheless, many governments have continued to seek economic growth through tourism development, given the potential economic gains.

One macro-economic clientele focus is the so-called high yield tourist segments. Dwyer et al. (2006) define high yield tourists as those representing high per capita spending measured as either expenditure per visitor per day, Gross Operating Surplus (GOS) per visit per day, GDP per visitor per day, net benefits per visitor per day or employment per $1 million.

Income is a proxy measure for identifying high yield target markets (Lim, 1997). The largest growth in luxury tourism is expected in Asia due to increasing numbers in the region of high-net-worth individuals (HNWIs) with financial assets in excess of USD one million and ultra HNWIs with financial assets in excess of $30 million (Capgemini, 2019). Many HNWIs are luxury tourism purchasers who, from a supplier's perspective, represent high yield. At the other end of the scale are price-conscious budget consumers.

As travel becomes increasingly possible once again post COVID-19, it is expected that forms of high yield luxury tourism will be the most pandemic resistant (Statista, 2020). ILTM (2020) survey results indicate that the pandemic has not dampened the desire to travel amongst HNWIs. Even during the pandemic, individuals who have been able to afford and make special arrangements, to access private jets and luxury accommodations, have continued to travel, even choosing to self-quarantine in remote luxury homes or entire islands that can be rented or purchased (Jasper, 2020; Lastoe & Vora, 2020; Smyth, 2020).

Looking forward, in some luxury markets such as destination weddings, industry experts report that the theme for post-pandemic travel is 'bigger and better' due to pent up desires and more time to plan and set money aside for a splurge (Spranklen, 2020). If these trends eventuate, there will be even more enthusiastic demand for lavish, extended destination weddings where guests stay in a dream location for several days. As isolation during the pandemic has caused people to romanticise spending more time with family and friends, a wedding celebration is being perceived by some as a perfect opportunity to do so in luxury (Petter, 2020). Further, as more countries open their borders to international visitors, it has also been reported that wealthy individuals are already planning how they can spend time in countries that are perceived to have handled the pandemic relatively well through the application of strict protocols, such as New Zealand (Sharma, 2020); and they are prepared to pay for the privilege (Karl, Chien, & Ong, 2020). These travellers have the capacity to pay for additional services related to health and safety if it enables their travel (Globetrender, 2020).

It is often assumed that high yield tourism pertains to status-driven, affluent tourists who choose to travel in luxury that is blatantly expensive (Makkar & Yap, 2018). However, it cannot be assumed that willingness to pay amongst all high yield tourist segments is always driven by materialism, demands for opulence and conspicuous consumption (Eckhardt, Belk, & Wilson, 2015), or that high-end facilities and amenities will be sufficient to attract them (Iloranta, 2019). Commenting on trends in luxury tourism, Koch (2011, p. 179) observes that, for many high yield tourists, 'luxury is not meant for the "exhibition" of rather superficial status symbols, but instead focuses on self-realisation'.

It is essential for providers who target high yield tourists to understand the nature of experiences that are apropos to relevant market segments. It is the willingness to pay more for attributes and experiences they value at any point in time, which makes tourists high yield by definition.

Sustainable Yield Tourism

As in most other industries, tourism and hospitality providers are under pressure from all kinds of stakeholders to consider the environmental, sustainability and governance (ESG) implications of their strategies and operations (Su & Chen, 2020). The challenge for those promoting tourism (governments, destination marketing organisations [DMOs] and other stakeholders) as an economic, societal and environmentally sustainable 'good' is to achieve an optimisation of positive net outcomes over time. As such, a more useful quantification of the yield – 'sustainable yield' – is proposed by Dwyer et al. (2006). Tourism market segments with a high sustainable yield are those where profitability can be balanced with ESG implications. From a governmental viewpoint, there are often trade-offs to be made, for example, between economic growth and ESG-related policies. Within a sustainable yield framework, the overall attractiveness and merits of different types of tourism can be evaluated in terms of contributions made to the host destination in relation to economic, social and environmental goals (Buhalis, 2000; Koren-Lawrence, Collins-Kreiner, & Israeli, 2020).

As an industry, tourism has suffered from an image of wasteful, excessive consumption (Gössling, 2002). On occasion, luxury tourism purchased by high yield tourists has been at the extreme end of that image (Smith, 2019). This is due to some highly visible forms of high yield tourism being expensive in terms of ESG costs. For example, after the direct experience of holding the 2008 Formula One (F1) Grand Prix in Singapore, residents largely supported hosting the event again but called for the introduction of additional measures during future races to help effectively manage negative externalities, such as congestion and socio-cultural disruptions (Cheng & Jarvis, 2010). When a location such as Singapore considers hosting an event like the Formula 1 racing championship, expected advantages include the attraction of high yield tourists who fill local luxury hotels, restaurants and retailers before, during and after the race. Just as importantly, this event can contribute to international marketing of the destination (Henderson, Foo, Lim, & Yip, 2010) and innovation in relevant sectors, such as manufacturing, technology and tourism (Kim et al., 2017).

Authors such as Dwyer (2016) and Pollock (2012) concede that the notion of the 'ideal tourist' or 'ideal kind of tourism' is elusive. If the preferred activities of high yield tourists degrade the environment or reduce access to limited resources for other social groups, they can be costly visitors to hosts. Conversely, some destinations strategically target tourists with a high willingness to pay because the revenue surplus can be applied to address sustainable development objectives (Grilli, Tyllianakis, Luisetti, Ferrini, & Turner, 2021). Some authors identify tourism offerings where a judicious focus on high yield tourism dynamics helps multiple stakeholders to optimise positive returns on tourism investments (Wamboye, Nyaronga, & Sergi, 2020). Examples include tourists whose high willingness to pay is driven by strong personal interest, such as conservation (Knowles, 2019), culture (Throsby, 2009), family events (Correia, Kozak, & Del Chiappa, 2020; Polovitz Nickerson, Jorgenson, & Boley, 2016), learning and personal growth (Ndou, Mele, & Del Vecchio, 2019) and philanthropy (Stainton, 2016). Some high yield tourists pursue adventure, immersion in nature, tranquillity or non-material enrichment in ways that are minimally invasive for the host destination (Sukmawati, Ernawati, & Nadra, 2018).

Being environmentally friendly does not always require sacrifice on quality and comfort (Moeller, Dolnicar, & Leisch, 2011). Some tourists elect to pay premium prices for trips that entail the most sumptuous offerings possible in environments that necessitate cramped accommodation, precarious situations and reduced food and beverage options – for example, as part of a small expedition group on an other-worldly private yacht cruise around the southern coast of Papua New Guinea or weeks of glamping with an esteemed expert guide crossing the Rub' Al Khali (Empty Quarter) desert of Oman. Research demonstrates that high yield tourists often forgo grand and luxuriant accommodation and amenities for rare or highly valued experiences (Iloranta, 2019). Luxury is in the eye of the beholder and is situationally contingent.

Given the significant market demand for luxury tourism offerings and the economic value of it to suppliers, a greater focus on achieving a systematic understanding of luxury in recent tourism research is warranted.

Luxury Tourism in Existing Literature

Conceptualising Luxury in Tourism

For more than a decade, researchers have found that traditional identifiers of luxury tourism – such as expensive accommodation or modes of transport – do not necessarily hold (Yeoman, 2008). Concepts such as 'unconventional luxury' (Thomsen, Holmqvist, Wallpach, & Hemetsberger, 2020) and 'new luxury' (Grönroos & Leppänen, 2009; Rytilahti, 2008) have been proposed to represent broader notions of what luxury means (Eckhardt et al., 2015). Sometimes luxury is associated with everyday pleasures or where common goods become so scarce that they are perceived as a luxury (Thomsen, Holmqvist, Wallpach, Hemetsberger, & Belk, 2020). For example, for most people, privacy would not be regarded as a luxury. Yet in these times of social media and a 24/7 news cycle, for some tourists seeking luxury, such as celebrities, privacy may be a key value. Reasons for this may also include concerns for safety and security (Murphy, 2020; Pallini, 2020), social concealment of wealth (Atkinson, 2015) or desire to escape for rest and relaxation out of the public eye (Iloranta, 2019).

Much discussion about luxury in tourism draws upon work in other disciplines, such as anthropology (Ross, 2014), marketing (Atwal & Williams, 2009), psychology (Dubois, Jung, & Ordabayeva, 2020) and sociology (Kapferer, 2015). Since the earliest literature on the subject, luxury is presented as highly dependent upon psycho-social factors (De Barnier, Falcy, & Valette-Florence, 2012). While some authors emphasise the social construction and social signalling value of luxury consumption (Correira et al., 2020), others highlight its personal hedonic (Holbrook & Hirschman, 1982) and self-realisation value (Kapferer, 2015). In each case, the focus has shifted from striving to define the concept of luxury in absolute terms towards framing luxury as occurring along a continuum with other levels of luxury (Vigneron & Johnson, 2004), ordinariness (De Barnier et al., 2012) or necessity (Heine, 2012). The questions outstanding for researchers include: What are the perceived dimensions of luxury that occur along continua, and in what combination, and of what relative importance are they? (Wirtz et al., 2020). These questions also apply to luxury in the context of tourism and hospitality.

Demand Perspectives on Luxury Tourism

Luxury tourism and hospitality studies have been predominantly demand oriented, with the primary sampling unit being the tourist (Iloranta, 2019). Unfortunately, this has not yet led to a systematic framework for understanding luxury tourism as a phenomenon or luxury tourism consumption. Most demand-side research aims to understand consumer attitudes, values, behaviours, motivations, intentions or relative satisfaction, and relationships between these.

While attempts at overall conceptual definitions of luxury tourism are scarce, many researchers attempt to isolate and define important factors in luxury tourism consumption (Correia et al., 2020). In attempts to characterise luxury in tourism, authors use a wide range of terms, including abundance, escape, exclusive, extraordinary, impeccable, indulgent, pleasure, prestige, private, quality, status, thrill, transformational and unique (Harkison, Hemmington, & Hyde, 2018; Yeoman, 2008). One of the definitional challenges is that many of these terms could also apply to tourism that would not be categorised as a luxury that is neither rare, opulent or exclusive. This raises questions about the nature of luxury tourism as a category that requires more examination.

Most demand-side researchers report on the relative significance of factors that are characteristic of the particular luxury experience under examination, reflecting the quantitative methodological approaches that dominate these studies (Chu, Tang, & Luo, 2016). Many of these researchers propose models for the measurement of luxury brand prestige, service quality or relationships between consumer perceptions of attributes and outcomes, such as satisfaction, willingness to pay or intentions to return (Han & Hyun, 2018; Hwang & Han, 2014). Even when the same kind of phenomenon (for example, luxury hotel stays) is the research context across multiple studies, there is little commonality in focus. Findings presented in these studies represent how polymorphous is the conceptualisation of luxury generally (Thirumaran & Raghav, 2017). The variable and fragmented nature of the research results limits any ability for the findings to be generalisable.

Nonetheless, many of the studies provide thought-provoking contributions to understanding select aspects of luxury tourism and hospitality, such as the relevance to luxury tourism of social value, emotion, narrative, luxury services and luxury experiences, experiential tourism and value co-creation. A brief comment will now be provided on each of these examples.

Social Value

Since luxury services and experiences are intangible and transitory in nature, concepts such as conspicuous consumption might be considered less salient. However, some studies of luxury tourism services and experiences indicate that the social value of consuming them can be important. Social status, face consciousness, other consumers' perceptions (OCPs) and perceived symbols of success are identified as drivers of willingness to pay for luxury services in the context of luxury domestic tourism (Correia & Kozak, 2012), restaurant dining (Kiatkawsin & Han, 2019) and cruises (Hwang & Han, 2014; Hyun & Han, 2015). It seems that the value of bragging rights spans both luxury goods and services.

Emotion

Luxury is about emotion and memorability (Kapferer, 2015). In the context of luxury tourism, it is the tourist's emotional predispositions, reactions, decisions and memories that endure and prevail (Otto & Ritchie, 1996). Emotional aspects of luxury tourism have been found to connect with consumer behaviours such as seeking pleasure and comfort, pursuing extraordinary experiences, enjoying exclusive-brand membership (tribalism) or rewarding oneself with an indulgence; and personal hedonic and eudemonic connections with people, places, products and processes (Atwal & Williams, 2017; Lee & Kim, 2019). For example, in some studies, satisfaction with luxury accommodation and dining services is reported as stronger when consumers also report high levels of emotion related to absorption, awe, engagement, involvement, learning, discovery, thrill and nostalgia (Arnould & Price, 1993; Han & Hyun, 2018; Hwang & Hyun, 2013; Kiatkawsin & Han, 2019).

Research in luxury tourism consumption has also found that emotions which might be regarded as negative – such as guilt, shame and envy – can also function in ways that are constructive, or at least benign. For example, Mkono and Hughes (2020) highlight that feelings of guilt and personal responsibility in relation to adverse impacts of tourism (ecoguilt and ecoshame) may lead consumers to make more environmentally friendly tourism investments. Studies on consumer social networking also find that envy is a motivator, heightening anticipation and enjoyment of luxury travel experiences, both vicariously and in practice (Belk, 2011; Liu, Wu, & Li, 2018).

Narrative

Many researchers have studied how consumers' perceptions of luxury tourism offerings are influenced by the different narrative structures and approaches used by suppliers in their marketing and service delivery (Megehee & Spake, 2012; Ryu, Lehto, Gordon, & Fu, 2018). Stories are important for all marketing and all kinds of tourism but, in the luxury tourism market, heightened significance is given to narrative transportation – which refers to the supplier's ability to facilitate the consumer's immersion in a whole other world (Ryu, Lehto, Gordon, & Fu, 2019). Customer revisits, intentions to return and word of mouth referrals are important success metrics for suppliers (Han & Hyun, 2018). However, ultimately, it is the tourists' own recollections and storytelling about the experience that will drive these outcomes (Arnould & Price, 1993; Manthiou, Kang, & Hyun, 2017). Whether or how, and to what extent, this is unique to luxury tourism requires further research.

Luxury Services and Luxury Experiences

The essential nature of tourism is the experience (Prentice, Witt, & Hamer, 1998), and tourism is fundamentally a service-intensive industry (Otto & Ritchie, 1996). Across luxury consumption studies generally, there has been a shift in emphasis towards the role of experience (Holmqvist, Ruiz, & Peñaloza, 2020) and consideration of luxury services as distinct from products (Wirtz et al., 2020). Wirtz et al. (2020) offer a definition of luxury services that are based upon 'extraordinary hedonic experiences that are exclusive' (Wirtz et al., 2020, p. 20). 'Exclusive' is the key term here. Otherwise, the definition by Wirtz et al. (2020) is broadly applicable to a wide range of services and experiences. The definition illustrates the semantic conflation of terms such as 'service' and 'experience' throughout relevant literature. Even though this definition of luxury services borders on being tautological, Wirtz et al. (2020) make an important contribution by sharpening the focus on luxury services as an important, distinct concept. In tourism, luxury services are more a supplier-centric concept while luxury experiences are more a consumer-centric concept.

Experiential Tourism

Some researchers distinguish 'experiential tourism' as a distinct phenomenon (Chan, To, & Chu, 2016; Gilovich & Gallo, 2020; Huffman, 2019; King & Zhang, 2020; Sundbo & Dixit, 2020). While there is a lack of consensus around the definition, scholars and industry writers use the term 'experiential tourism' to describe a relatively high level of active involvement by participants in touristic activities that are personally meaningful and therefore more highly valued by individuals (Fu, Tanyatanaboon, & Lehto, 2015; Packer & Gill, 2017).

Some kinds of experiential travel are expensive so there are many instances where luxury and experiential travel are intertwined. Some kinds of experiential tourism will also be luxury tourism. However, they are not always one and the same. Some experiential tourism, such as backpacking, would not be described as luxurious. Other examples may be debatable as to the extent of the luxuriousness, such as a private yacht charter that is indulgent and unique, with a high level of service, yet where sleeping quarters are cramped and food options constrained.

Value Co-creation

Many tourism and hospitality studies support the view that the tourist is a co-producer, and not just the consumer, of touristic experiences generally (Buhalis & Sinarta, 2019; Cabiddu, Tsz-Wai, & Piccoli, 2013; Chathoth, Ungson, Harrington, & Chan, 2016; Prebensen & Xie, 2017). This also applies to luxury in tourism and hospitality, particularly since the consumer is the ultimate arbiter of luxury. Some luxury tourism studies also focus on value created in customer-to-customer (C2C) interactions (Arnould & Price, 1993). How value co-creation functions in luxury tourism provides abundant opportunities for further research.

Supply Perspectives on Luxury Tourism

Being in the business of luxury tourism sounds like a glamorous enterprise, but it is not for the faint-hearted provider. Luxury tourism is high priced because it is expensive to develop and deliver. High-quality components and high levels of service add cost to offerings that will, by design, be consumed by relatively smaller numbers of tourists. In short, the costs of provision are higher while the overall target market is smaller. Significant investment into luxury infrastructure and offerings may not succeed in attracting the target consumers and, even at the luxury end of the travel market, most consumers have a sense of value for money (Hwang & Lyu, 2018; Yeoman & McMahon-Beattie, 2005). The economics of luxury tourism provision are fraught with risk. That luxury perceptions are dependent upon the subjective tastes and whims of tourists adds to the design and fulfilment challenges for suppliers.

While supply-side literature on luxury tourism is more scarce, it does include coverage of non-Western examples of luxury tourism, although largely in relation to specific contexts – cultural destinations (Armstrong, Mok, Go, & Chan, 1997), the building of infrastructure to attract tourists (Scheyvens, 2011) and prestigious non-Western accommodation and dining brands (Bharwani, 2015; Chathoth, Mak, Sim, Jauhari, & Manaktola, 2011).

Most luxury tourism studies from the supply perspective pertain to one of four different units of analysis, either: (1) governments and government agencies, including policy formation (Dwyer et al., 2006); (2) destinations, their development and management (Thirumaran & Raghav, 2017); (3) facilities and amenities, including service culture (Harkison, 2017; Sukmawati et al., 2018) or (4) tourism service providers – either online (Rayman-Bacchus & Molina, 2001) or offline (Chiu, 2009).

Because many kinds of high yield tourists tend to be well-informed, often frequent, travellers, they are reputed to be demanding customers, constantly on the lookout for the novel, unusual and authentic (Dixit, 2020; Dwyer, Edwards, Mistilis, Adler, & Scott, 2009). Therefore, at each level of the industry, luxury tourism service providers must continually seek innovative ways to differentiate themselves to gain the attention of customers and build emotional attachments in order to win their business, repeat purchases and referrals (Hjalager, 2015; Peters & Pikkemaat, 2006; Razumova, Ibáñez, & Palmer, 2015; Reitsamer & Brunner-Sperdin, 2015).

Positive consumption experiences often depend upon a supplier's ability to harness customers' active participation in the co-creation process (Prahalad & Ramaswamy, 2004). Effective luxury tourism providers act as facilitators, developing potential value that customers will actualise (Holmqvist, Visconti, Grönroos, Guais, & Kessous, 2020). In many ways, the provider is in the business of facilitating an experience of escapism for the luxury tourism consumer. This is a critical capability – or set of capabilities – for providers in this intensely competitive and global sector. These capabilities are being extended by greater levels of personalisation for a traveller through all phases of a trip – before, during and after – using 'smart tourism' approaches, such as technology applications that enable providers to have access to real-time information and feedback loops (Buhalis & Amaranggana, 2015). However, even when smartness concepts are deployed, luxury tourism experience provision requires levels of finesse and flexibility – for both provider and consumer – that are not always easy to execute in an organisational setting. As Crompton states,

> (t)he real challenge for the tourism industry is to create the right psychological environment, not just worry about technical things.
> (Crompton, 1994, in Otto & Ritchie, 1996, p. 165)

Research Agenda

Based upon the market opportunities that exist for luxury tourism experiences as well as the paucity of existing literature on the subject, there is an abundance of research questions to be addressed. Areas for further research include the following.

Conceptualisation and Key Distinctions

Luxury tourism has various meanings. Further research is required into the definition and meaning of luxury tourism to consumers, suppliers and other stakeholders. Further research studies into non-Western supplier and consumer perspectives, and non-Western luxury offerings, are required to fully appreciate and understand the meaning of luxury tourism.

Scholars have distinguished 'old luxury' from 'new luxury'. Can 'old luxury' tourism be differentiated from 'new luxury' tourism? Might some individuals be a consumer of both 'old luxury' tourism in some ways and 'new luxury' tourism in others?

All tourism can be framed as an experience. Luxury tourism differs from ordinary tourism along a continuum. Luxury tourism has been described as a psycho-social phenomenon that is indulgent, high in quality and hedonic value and exclusive. Further work is required to identify the characteristics of luxury tourism and how it is distinguished from ordinary tourism.

High yield tourism and sustainable yield tourism have been defined in terms of economic impact and sustainability impact. Further work is required to understand the trade-offs between high yield and sustainable yield tourism.

Certain kinds of luxury cannot be bought – being able to spend an extra week with someone special or spending quiet moments in a pleasurable, tranquil place. Such luxuries may be components of tourism, yet each can be accessed without a supplier or monetary exchange. Can a luxury tourism experience be disentangled from its commercial transaction value? What are the implications for luxury tourism suppliers?

Studies indicate the importance of other participants to the luxury tourism experience. More research is required to understand the incremental value related to customer-to-customer (C2C) exchanges in luxury tourism experiences.

Luxury tourism services are a subset of luxury tourism experiences. The distinction between luxury services as part of luxury tourism experiences requires further clarification. The co-creation by tourists and providers of social value, emotion and narrative warrants further research, particularly in relation to how each of these influences perceptions of luxury in tourism experiences.

Unit of Analysis

In studies of luxury tourism and hospitality, the unit of analysis is usually the consumer, the supplier, the intermediary (such as a travel agent) or the product or service (hotel, cruise, restaurant, event or tour). More research providing a holistic view of the end-to-end luxury tourism experience would provide insights into the luxury tourist perspective, including the important consumption stages of planning and pre-trip and post-trip sharing with others.

Demand-oriented studies are predominant in luxury tourism and hospitality literature. More research is required into the supplier perspective given the challenges involved in addressing this market and, in particular, high yield tourism and sustainable yield tourism.

Destination Image

Consumer perspectives on luxury tourism is a major area of interest within the field of destination image and warrant further research.

Managerial Implications

Though this chapter is conceptual and exploratory, four major managerial implications emerge from the discussion and analysis. First, as an important influence on tourism outcomes generally, governments can set and control the environment within which luxury tourism operates, including safety fundamentals, the extent and quality of supporting infrastructure and the facilitation of reciprocal touristic rights (where appropriate) between countries. The necessary conditions for facilitating luxury tourism activities, and drivers of growth in the sector, need to be better understood for further cultivation and development.

Second, the economic benefits of expanding the sustainable luxury tourism market need to be identified with respect to location-specific revenue retention and multiplier effects, employment rates (numbers employed, location thereof and skill development) and drivers of innovation. This will help to provide a better basis for assessing the impact value of different kinds of tourism options and for making targeted policy and investment decisions in relation to them.

Third, luxury tourism consumers tend to seek novel and unique experiences, which is a major driver of innovation in the sector. The issue of innovation as a contributing factor to luxury tourism experiences is largely unexplored. The principal manifestation of tourism innovation will be the need for robust processes to continuously generate and manage new or fundamentally reconceptualised product/service offerings. Luxury tourism is about pushing boundaries to provide an extraordinary experience. Axiomatically, it has a demonstration effect in a market, potentially introducing new ideas, new ways of doing things and new ways of conceptualising value.

Fourth, the role of limited or unique access to assets for creating luxury tourist experiences represents a distinctive competency and a core competitive advantage. In addition, it entails issues of governance, ethics, equity and sustainability that are captured in ESG aspirations. The extent to which unique experiences are able to remain unique (be the sole possible provider) and be sustainable may depend upon 'exclusive rights' provided by government concessions, contractual arrangements (including access or use limits, or quotas), innovative processes or specialised human resources training. The extent to which such arrangements form the basis for luxury tourism experiences warrants further exploration.

Conclusion

The academic study of luxury tourism is nascent. The clear implication of this is that luxury tourism offers many future research opportunities and that the various strands of current enquiry represent significant opportunities for the further clarification of concepts and their definitional understanding. As luxury tourism is better understood, policymakers at all levels will be able to more clearly define tourism-related objectives in terms of development, sustainability and promotion, the contributions to various stakeholder groups to be sought and the sustainability of luxury tourism assets, environment and culture.

As understandings of luxury tourism develop and coalesce, research opportunities will present themselves in both qualitative and quantitative forms. If the trajectory of other emergent management disciplines is a guide, the movement will likely be towards more large-scale quantitative and longitudinal studies, which will provide input into macroeconomic policymaking. Simultaneously, it can be expected that qualitative studies will become more targeted and refined, especially in the areas of understanding consumer behaviour and experiences sought and the ways in which these are matched by luxury tourism providers.

Substantial opportunity exists to develop the concept of luxury tourism as an increasingly important phenomenon and to explore the dynamics amongst luxury tourism consumers and providers. Competition for the custom of high yield tourists is and will be globally intense as the world emerges from pandemic-related restrictions. Accordingly, luxury tourism and hospitality suppliers have much to gain from an enhanced understanding of the needs and wants of those market segments relevant to high yield and sustainable yield, luxury tourism. Further research is needed to help luxury tourism and hospitality sectors to develop across global, national, regional destination and supplier perspectives.

References

Armstrong, R. W., Mok, C., Go, F. M., & Chan, A. (1997). The importance of cross-cultural expectations in the measurement of service quality perceptions in the hotel industry. *International Journal of Hospitality Management*, *16*(2), 181–190. doi: 10.1016/S0278-4319(97)00004-2

Arnould, E., & Price, L. (1993). River magic: Extraordinary experience and the extended service encounter. *Journal of Consumer Research*, *20*, 24–45. doi:10.1086/209331

Atkinson, R. (2015). Limited exposure: Social concealment, mobility and engagement with public space by the super-rich in London. *Environment and Planning A: Economy and Space*, *48*(7), 1302–1317. doi:10.1177/0308518X15598323

Atwal, G., & Williams, A. (2009). Luxury brand marketing–The experience is everything! *Journal of Brand Management*, *16*(5–6), 338–346. doi:10.1007/978-3-319-51127-6_3

Atwal, G., & Williams, A. (2017). Luxury brand marketing – the experience is everything! In J.-N. Kapferer, J. Kernstock, T. Brexendorf, & S. Powell (Eds.), *Advances in luxury brand management*. (pp. 43–57). London: Palgrave Macmillan. doi:10.1007/978-3-319-51127-6

Belk, R. (2011). Benign envy. *AMS Review*, *1*, 117–134. doi:10.1007/s13162-011-0018-x

Bharwani, S. (2015). Hospitality innovations in the emerging experience economy: A case study of the Oberoi resorts. In *Hospitality, travel, and tourism: Concepts, methodologies, tools, and applications* (pp. 672–694). Pennsylvania, PA: IGI Global. doi:10.4018/978-1-4666-6543-9.ch040

Buhalis, D. (2000). Marketing the competitive destination of the future. *Tourism Management*, *21*(1), 97–116. doi:10.1016/S0261-5177(99)00095-3

Buhalis, D., & Amaranggana, A. (2015). Smart tourism destinations enhancing tourism experience through personalisation of services. In I. Tussyadiah & A. Inversini (Eds.), *Information and communication technologies in tourism* (pp. 377–389). Cham: Springer. doi:10.1007/978-3-319-14343-9_28

Buhalis, D., & Sinarta, Y. (2019). Real-time co-creation and nowness service: Lessons from tourism and hospitality. *Journal of Travel and Tourism Marketing*, *36*(5), 563–582. doi:10.1080/10548408.2019.1592059

Cabiddu, F., Tsz-Wai, L., & Piccoli, G. (2013). Managing value co-creation in the tourism industry. *Annals of Tourism Research*, *42*, 86–107. doi:10.1016/j.annals.2013.01.001

Capgemini. (2019). *World Wealth Report 2019*. Retrieved from https://worldwealthreport.com. Accessed on September 20, 2020.

Casson, L. (1974). *Travel in the ancient world*. Toronto, ON: Hakkert.

Chan, W., To, C., & Chu, W. (2016). Desire for experiential travel, avoidance of rituality and social esteem: An empirical study of consumer response to tourism innovation. *Journal of Innovation and Knowledge*, *1*(1), 24–35. doi:10.1016/j.jik.2016.02.001

Chathoth, P. K., Mak, B., Sim, J., Jauhari, V., & Manaktola, K. (2011). Assessing dimensions of organisational trust across cultures: A comparative analysis of U.S. and Indian full service hotels. *International Journal of Hospitality Management*, *30*(2), 233–242. doi:10.1016/j.ijhm.2010.09.004

Chathoth, P. K., Ungson, G. R., Harrington, R. J., & Chan, E. S. W. (2016). Co-creation and higher order customer engagement in hospitality and tourism services: A critical review. *International Journal of Contemporary Hospitality Management*, *28*(2), 222–245. doi:10.1108/IJCHM-10-2014-0526

Cheng, E., & Jarvis, N. (2010). Residents' perception of the social-cultural impacts of the 2008 Formula 1 Singtel Singapore Grand Prix. *Event Management*, *14*(2), 91–106. doi:10.3727/152599510X12766070300849

Chien, P. M., & Ritchie, B. (2018). Understanding intergroup conflicts in tourism. *Annals of Tourism Research*, *72*, 177–179. doi:10.1016/j.annals.2018.03.004

Chiu, C.-K. (2009). Understanding relationship quality and online purchase intention in e-tourism: A qualitative application. *Quality and Quantity*, *43*(4), 669–675. doi:10.1007/s11135-007-9147-6

Chu, Y., Tang, L., & Luo, Y. (2016). Two decades of research on luxury hotels: A review and research agenda. *Journal of Quality Assurance in Hospitality & Tourism*, *17*, 1–12. doi:10.1080/1528008X.2015.1047076

Correia, A., & Kozak, M. (2012). Exploring prestige and status on domestic destinations: The case of the Algarve. *Annals of Tourism Research*, *39*(4), 1951–1967. doi:10.1016/j.annals.2012.06.005

Correia, A., Kozak, M., & Del Chiappa, G. (2020). Examining the meaning of luxury in tourism: A mixed-method approach. *Current Issues in Tourism*, *23*(8), 952–970. doi:10.1080/13683500.2019.1574290

Crouch, G., & Ritchie, J. (1999). Tourism, competitiveness and societal prosperity. *Journal of Business Research*, *44*, 137–152. doi:10.1016/S0148-2963(97)00196-3

De Barnier, V., Falcy, S., & Valette-Florence, P. (2012). Do consumers perceive three levels of luxury? A comparison of accessible, intermediate and inaccessible luxury brands. *Journal of Brand Management*, *19*(7), 623–636. doi:10.1057/bm.2012.11

Dixit, S. K. (Ed.). (2020). *The Routledge handbook of tourism experience management and marketing*. London: Routledge.

Dubois, D., Jung, S., & Ordabayeva, N. (2020). The psychology of luxury consumption. *Current Opinion in Psychology*, *39*, 82–87. doi:10.1016/j.copsyc.2020.07.011

Dwyer, L. (2016). Who is the ideal tourist? *Acta Turistica*, *28*(2), 151–183. doi:10.1016/j.tourman.2008.04.002

Dwyer, L., Edwards, D., Mistilis, N., Adler, C., & Scott, N. (2009). Destination and enterprise management for a tourism future. *Tourism Management*, *30*, 63–74. doi:10.1016/j.tourman.2008.04.002

Dwyer, L., Forsyth, P., Fredline, L., Jago, L., Deery, M., & Lundie, S. (2006). *Concepts of tourism yield and their measurement*. Gold Coast, QLD: CRC for Sustainable Tourism. doi:9781920965181

Eckhardt, G. M., Belk, R. W., & Wilson, J. A. J. (2015). The rise of inconspicuous consumption. *Journal of Marketing Management*, *31*(7–8), 807–826. doi:10.1080/0267257X.2014.989890

Friedländer, L. (1908). *Roman life and manners under the early empire*. London: Routledge. doi:10.4324/9781315123011. Retrieved from https://archive.org/details/cu31924026955256/page/n15/mode/2up. Accessed on August 20, 2020.

Fu, X., Tanyatanaboon, M., & Lehto, X. Y. (2015). Conceputualizing transformative guest experience at retreat centers. *International Journal of Hospitality Management*, *49*, 83–92. doi:10.1016/j.ijhm.2015.06.004

Gilovich, T., & Gallo, I. (2020). Consumers' pursuit of material and experiential purchases: A review. *Consumer Psychology Review*, *3*(1), 20–33. doi:10.1002/arcp.1053

Globetrender (2020). Eight experts discuss the trends shaping luxury travel of tomorrow. *Globetrender*, 11 June. Retrieved from https://globetrender.com/2020/06/11/expert-insider-luxury-travel-trends/. Accessed on September 24, 2020.

Gössling, S. (2002). Global environmental consequences of tourism. *Global Environmental Change*, *12*(4), 283–302. doi:10.1016/S0959-3780(02)00044-4

Grilli, G., Tyllianakis, E., Luisetti, T., Ferrini, S., & Turner, R. K. (2021). Prospective tourist preferences for sustainable tourism development in Small Island Developing States. *Tourism Management*, *82*, 104178. doi:10.1016/j.tourman.2020.104178

Grönroos, C., & Leppänen, H. (2009). *The hybrid consumer: Exploring the drivers of a new consumer behaviour type*. Hanken School of Economics Working Papers, No. 543. Hanken School of Economics, Helsinki. Retrieved from http://hdl.handle.net/10227/465

Gunter, U., & Netto, A. P. (2015). International travel to and from Brazil–Overseas tourism as a luxury good and a status symbol. *Tourism Economics*, *22*(5), 1151–1160. doi:10.5367/te.2015.0473

Han, H., & Hyun, S. S. (2018). Role of motivations for luxury cruise traveling, satisfaction, and involvement in building traveler loyalty. *International Journal of Hospitality Management*, *70*, 75–84. doi:10.1016/j.ijhm.2017.10.024

Harkison, T. (2017). The importance of the human touch in the luxury accommodation sector. *Research in Hospitality Management*, *7*(1), 59–64. doi:10.1080/22243534

Harkison, T., Hemmington, N., & Hyde, K. F. (2018). Luxury accommodation–Significantly different or just more expensive? *Journal of Revenue and Pricing Management*, *17*, 231–243. doi:10.1057/s41272-017-0085-1

Heine, K. (2012). The concept of luxury brands. Retrieved from www.conceptofluxurybrands.com. Accessed on August 20, 2020.

Henderson, J. C., Foo, K., Lim, H., & Yip, S. (2010). Sports events and tourism: The Singapore Formula One Grand Prix. *International Journal of Event and Festival Management*, *1*(1), 60–73. doi:10.1108/17852951011029306

Hjalager, A. M. (2015). 100 innovations that transformed tourism. *Journal of Travel Research*, *54*(1), 3–21. doi:10.1177/0047287513516390

Holbrook, M., & Hirschman, E. (1982). The experiential aspects of consumption: Consumer fantasies, feelings and fun. *The Journal of Consumer Research, 9*(2), 132–140. doi:10.1086/208906

Holmqvist, J., Ruiz, C. D., & Peñaloza, L. (2020). Moments of luxury: Hedonic escapism as a luxury experience. *Journal of Business Research, 116*, 503–513. doi: 10.1016/j.jbusres.2019.10.015

Holmqvist, J., Visconti, L. M., Grönroos, C., Guais, B., & Kessous, A. (2020). Understanding the value process: Value creation in a luxury service context. *Journal of Business Research, 120*, 114–126. doi:10.1016/j.jbusres.2020.07.002

Holzner, M. (2011). Tourism and economic development: The beach disease? *Tourism Management, 32*(4), 922–933. doi:10.1016/j.tourman.2010.08.007

Huffman, M. (2019). New study finds most consumers prefer experiences over things. *Consumer Affairs*. Retrieved from https://www.consumeraffairs.com/news/new-study-finds-most-consumers-prefer-experiences-over-things-101519.html. Accessed on June 14, 2020.

Hwang, J., & Han, H. (2014). Examining strategies for maximising and utilising brand prestige in the luxury cruise industry. *Tourism Management, 40*, 244–259. doi: 10.1016/j.tourman.2013.06.007

Hwang, J., & Hyun, S. S. (2013). The impact of nostalgia triggers on emotional responses and revisit intentions in luxury restaurants: The moderating role of hiatus. *International Journal of Hospitality Management, 33*, 250–262. doi:10.1016/j.ijhm.2012.09.001

Hwang, J., & Lyu, S. O. (2018). Understanding first-class passengers' luxury value perceptions in the US airline industry. *Tourism Management Perspectives, 28*, 29–40. doi:10.1016/j.tmp.2018.07.001

Hyun, S. S., & Han, H. (2015). Luxury cruise travelers: Other customer perceptions. *Journal of Travel Research, 50*(1), 107–121. doi:10.1177/0047287513513165

Iloranta, R. (2019). Luxury tourism service provision–Lessons from industry. *Tourism Management Perspectives, 32*, 100568. doi:10.1016/j.tmp.2019.100568

ILTM. (2020). ILTM data reveals key global insights. *View from ILTM*, 3 July. Retrieved from https://view.iltm.com/2020/07/03/iltm-data-reveals-key-global-insights/. Accessed on October 20, 2020.

Jasper, C. (2020). Rich Americans are staring to travel again–In private jets. *Time*, 15 May. Retrieved from https://time.com/5837291/rich-americans-travel-private-jets-coronavirus/. Accessed on September 20, 2020.

Kapferer, J.-N. (2015). *Kapferer on luxury: How luxury brands can grow yet Remain Rare*. London: Kogan Page.

Kapferer, J. N., & Bastien, V. (2009). *The luxury strategy: Break the rules of marketing to build luxury brands*. London: Kogan Page. Retrieved from https://EconPapers.repec.org/RePEc:hal:journl:hal-00786813

Karl, M., Chien, P. M., & Ong, F. (2020). Impulse buying behaviour in tourism: A new perspective. *Annals of Tourism Research*, 103136. doi:10.1016/j.annals.2020.103136

Kiatkawsin, K., & Han, H. (2019). What drives customers' willingness to pay price premiums for luxury gastronomic experiences at Michelin-starred restaurants? *International Journal of Hospitality Management, 82*, 209–219. doi:10.1016/j.ijhm.2019.04.024

Kim, M. K., Kim, S.-K., Park, J.-A., Carroll, M., Yu, J.-G., & Na, K. (2017). Measuring the economic impacts of major sports events: The case of Formula One Grand Prix (F1). *Asia Pacific Journal of Tourism Research, 22*(1), 64–73. doi: 10.1080/10941665.2016.1176061

King, B., & Zhang, H. (2020). Experiential tourism and hospitality learning: Principles and practice. In P. Benckendorff & A. Zehrer (Eds.), *Handbook of teaching and learning in tourism*. Cheltenham: Edward Elgar.

Knowles, N. L. B. (2019). Targeting sustainable outcomes with adventure tourism: A political ecology approach. *Annals of Tourism Research, 79*, 102809. doi: 10.1016/j.annals.2019.102809

Koch, K.-D. (2011). Luxury tourism–Does this market segment still work? In R. Conrady & M. Buck (Eds.), *Trends and issues in global tourism 2011. Trends and issues in global tourism*. Berlin: Springer. doi:10.1007/978-3-642-17767-5_19

Koren-Lawrence, N., Collins-Kreiner, N., & Israeli, Y. (2020). The future of the past: Sustainable management of archaeological tourist sites–The case study of Israel. *Tourism Management Perspectives, 35*, 100700. doi:10.1016/j.tmp.2020.100700

Lastoe, S., & Vora, S. (2020). The wealthy forge ahead with (slightly altered) travel plans in spite of "stay at home" directives. *CNN Travel*, 28 March. https://www.cnn.com/travel/article/wealthy-people-remote-hotels-coronavirus/index.html. Accessed on August 15, 2020.

Lee, C. C., & Chang, C. P. (2008). Tourism development and economic growth: A closer look at panels. *Tourism Management, 29*(1), 180–192. doi:10.1016/j.tourman.2007.02.013

Lee, Y., & Kim, I. (2019). A value co-creation model in brand tribes: The effect of luxury cruise consumers' power perception. *Service Business, 13*, 129–152. doi: 10.1007/s11628-018-0373-x

Lim, C. (1997). Review of international tourism demand models. *Annals of Tourism Research, 24*(4), 835–849. doi:10.1016/S0160-7383(97)00049-2

Liu, H., Wu, L., & Li, X. (2018). Social media envy: How experience sharing on social networking sites drives millennials' aspirational tourism consumption. *Journal of Travel Research, 58*(3), 355–369. doi:10.1177/0047287518761615

Lomine, L. (2005). Tourism in Augustan Society. In J. K. Walton (Ed.), *Histories of tourism: Representation, identity and conflict* (pp. 71–87). Clevedon: Channel View Publications.

Makkar, M., & Yap, S.-F. (2018). Emotional experiences behind the pursuit of inconspicuous luxury. *Journal of Retailing and Consumer Services, 44*, 222–234. doi:10.1016/j.jretconser.2018.07.001

Manthiou, A., Kang, J., & Hyun, S. S. (2017). An integration of cognitive appraisal theory and script theory in the luxury cruise sector: The bridging role of recollection and storytelling. *Journal of Travel and Tourism Marketing, 34*(8), 1071–1088. doi:10.1080/10548408.2016.1277575

Megehee, C. M., & Spake, D. F. (2012). Consumer enactments of archetypes using luxury brands. *Journal of Business Research, 65*(10), 1434–1442. doi:10.1016/j.jbusres.2011.10.009

Mkono, M., & Hughes, K. (2020). Eco-guilt and eco-shame in tourism consumption contexts: Understanding the triggers and responses. *Journal of Sustainable Tourism, 28*(8), 1223–1244. doi:10.1080/09669582.2020.1730388

Moeller, T., Dolnicar, S., & Leisch, F. (2011). The sustainability–Profitability trade-off in tourism: Can it be overcome? *Journal of Sustainable Tourism*, *19*(2), 155–169. doi:10.1080/09669582.2010.518762

Murphy, J. (2020). Upscale motels are the perfect prescriptions for a pandemic vacation. *Bloomberg*, 29 July. Retrieved from https://www.bloomberg.com/news/articles/2020-07-29/upscale-motels-offer-safety-and-luxury-during-covid-19-travel. Accessed on September 20, 2020.

Ndou, V., Mele, G., & Del Vecchio, P. (2019). Entrepreneurship education in tourism: An investigation among European Universities. *Journal of Hospitality, Leisure, Sport and Tourism Education*, *25*, 100175. doi:10.1016/j.jhlste.2018.10.003

OECD. (2020). *OECD tourism trends and policies 2020*. Paris: OECD Publishing. doi:10.1787/6b47b985-en

Otto, J. E., & Ritchie, J. R. B. (1996). The service experience in tourism. *Tourism Management*, *17*(3), 165–174. doi:10.1016/0261-5177(96)00003-9

Packer, J., & Gill, C. (2017). Meaningful vacation experiences. In S. Filep, J. Laing, & M. Csikszentmihalyi (Eds.), *Positive tourism*. (pp. 19–34). Oxon: Routledge.

Pallini, X. (2020). 9 ways flying on a private jet is safer than flying on a commercial airliner in the age of coronavirus. *Business Insider*, 20 June. Retrieved from https://www.businessinsider.com/why-private-jet-travel-is-safer-than-flying-commercial-2020-5. Accessed on October 10, 2020.

Peters, M., & Pikkemaat, B. (2006). Innovation in tourism. *Journal of Quality Assurance in Hospitality and Tourism*, *6*(3–4), 1–6. doi:10.1300/J162v06n03_01

Petter, O. (2020). Wedding trends for 2021: What will the industry look like after coronavirus? *Independent*, 10 May. Retrieved from https://www.independent.co.uk/life-style/love-sex/wedding-trends-2021-coronavirus-marriage-a9503326.html. Accessed on October 3, 2020.

Pollock, A. (2012). Conscious travel: Signposts towards a new model for tourism. In *Contribution to the 2nd UNWTO Ethics and Tourism Congress Conscious Tourism for a New Era*, Quito, Ecuador, September 12, 2012. Retrieved from https://skift.com/wp-content/uploads/2012/09/presentacion-anna-meira-pollock.pdf. Accessed on October 31, 2020.

Polovitz Nickerson, N., Jorgenson, J., & Boley, B. B. (2016). Are sustainable tourists a higher spending market? *Tourism Management*, *54*, 170–177. doi:10.1016/j.tourman.2015.11.009

Prahalad, C. K., & Ramaswamy, V. (2004). Co-creation experiences: The next practice in value creation. *Journal of Interactive Marketing*, *18*, 5–14. doi:10.1002/dir.20015

Prebensen, N., & Xie, J. (2017). Efficacy of co-creation and mastering on perceived value and satisfaction in tourists' consumption. *Tourism Management*, *60*, 166–176. doi:10.1016/j.tourman.2016.12.001

Prentice, R. C., Witt, S. F., & Hamer, C. (1998). Tourism as experience: The case of heritage parks. *Annals of Tourism Research*, *25*(1), 1–24. doi:10.1016/S0160-7383(98)00084-X

Rayman-Bacchus, L., & Molina, A. (2001). Internet-based tourism services: Business issues and trends. *Futures*, *33*(7), 589–605. doi:10.1016/S0016-3287(01)00003-9

Razumova, M., Ibáñez, J. L., & Palmer, J. R. M. (2015). Drivers of environmental innovation in Majorcan hotels. *Journal of Sustainable Tourism*, *23*(10), 1529–1549. doi:10.1080/09669582.2015.1062016

Reitsamer, B. F., & Brunner-Sperdin, A. (2015). Tourist destination perception and well-being: What makes a destination attractive? *Journal of Vacation Marketing*, *23*(1), 55–72. doi:10.1177/1356766715615914

Romero, F. G. (2013). Sports tourism in ancient Greece. *Journal of Tourism History*, *5*(2), 146–160. doi:10.1080/1755182X.2013.828784

Ross, G. (2014). Meaning making in religious and spiritual travel. In S. Filep & P. Pearce (Eds.), *Tourist experience and fulfilment: Insights from positive psychology. Routledge advances in tourism 31* (pp. 91–109). Oxon: Routledge.

Rytilahti, P. (2008). Social value of old and new luxury. In T. Keinonen (Ed.), *Design connections: Knowledge, value and involvement through design* (pp. 46–53). Helsinki: University of Art and Design.

Ryu, K., Lehto, X. Y., Gordon, S. E., & Fu, X. X. (2018). Compelling brand storytelling for luxury hotels. *International Journal of Hospitality Management*, *74*, 22–29. doi:10.1016/j.ijhm.2018.02.002

Ryu, K., Lehto, X. Y., Gordon, S. E., & Fu, X. X. (2019). Effect of a brand story structure on narrative transportation and perceived brand image of luxury hotels. *Tourism Management*, *71*, 348–363. doi:10.1016/j.tourman.2018.10.021

Scheyvens, R. (2011). The challenge of sustainable tourism development in the Maldives: Understanding the social and political dimensions of sustainability. *Asia Pacific Viewpoint*, *52*, 148–164. doi:10.1111/j.1467-8373.2011.01447.x

Sharma, S. (2020). Post-pandemic travel: The rich are looking to buy access to COVID-19 safe zones. *Hindustan Times*, 21 July. Retrieved from https://www.hindustantimes.com/travel/post-pandemic-travel-the-rich-are-looking-to-buy-access-to-covid-19-safe-zones/story-WuXemitLhswEUSLU0UjaTL.html. Accessed on October 10, 2020.

Smeral, E. (2003). A structural view of tourism growth. *Tourism Economics*, *9*(1), 77–93. doi:10.5367/000000003101298286

Smith, O. (2019). Luxury, tourism and harm: A deviant leisure perspective. In T. Raymen & O. Smith (Eds.), *Deviant leisure: Criminologist perspectives on leisure and harm* (pp. 305–323). London: Palgrave Macmillan. doi:10.1007/978-3-030-17736-2

Smyth, J. (2020). Wealthy buyers snap up "safe haven" private islands to flee pandemic. *Financial Times*, 23 July. Retrieved from https://www.ft.com/content/40e54149-7483-41b5-85cf-36ff200aafc6. Accessed on September 10, 2020.

Spranklen, A. (2020). From huge dresses to old Hollywood-style pool parties, these are the biggest blowout wedding trends for 2021. *The Tatler*, 13 August. Retrieved from https://www.tatler.com/article/wedding-trends-2021. Accessed on October 10, 2020.

Stainton, H. (2016). A segmented volunteer tourism industry. *Annals of Tourism Research*, *61*, 256–258. doi:10.1016/j.annals.2016.09.011

Statista. (2020). *Luxury travel and tourism amid the pandemic*. Hamburg: Statista.

Su, C.-H., & Chen, C.-D. (2020). Does sustainability index matter to the hospitality industry? *Tourism Management*, *81*, 104–158. doi:10.1016/j.tourman.2020.104158

Sukmawati, N., Ernawati, N., & Nadra, N. (2018). Luxury tourism: A perspective of facilities and amenities. *International Journal of Applied Sciences in Tourism and Events*, *2*(1), 32–37. doi:10.31940/ijaste.v2i1.900

Sundbo, J., & Dixit, S. K. (2020). Conceptualisations in tourism experience. In S. K. Dixit (Ed.), *The Routledge handbook of tourism experience management and marketing* (pp. 15–26). London: Routledge.

Thirumaran, K., & Raghav, M. (2017). Luxury tourism, developing destinations: Research review and trajectories. *Asian Journal of Tourism Research*, *2*(2), 137–158. doi:10.12982/AJTR.2017.0013

Thomsen, T., Holmqvist, J., Wallpach, S., Hemetsberger, A., & Belk, R. (2020). Conceptualising unconventional luxury. *Journal of Business Research*, *116*, 441–445. doi:10.1016/j.jbusres.2020.01.058

Throsby, D. (2009). Tourism, heritage and cultural sustainability: Three golden rules. In L. F. Girard & P. Nijkamp (Eds.), *Cultural tourism and sustainable local development* (pp. 13–30). Farnham: Ashgate Publishing.

Towner, J. (1985). The grand tour: A key phase in the history of tourism. *Annals of Tourism Research*, *12*(3), 297–333. doi:10.1016/0160-7383(85)90002-7

Vigneron, F., & Johnson, L. (2004). Measuring perceptions of brand luxury. *Journal of Brand Management*, *11*, 484–506. doi:10.1057/palgrave.bm.2540194

Wamboye, E., Nyaronga, P., & Sergi, B. (2020). What are the determinants of international tourism in Tanzania. *World Development Perspectives*, *17*, 100175. doi:10.1016/j.wdp.2020.100175

Wirtz, J., Holmqvist, J., & Fritze, M. P. (2020). Luxury services. *Journal of Service Management*, *31*(4), 665–691. doi:10.1108/JOSM-11-2019-0342

WTO. (2019). *International tourism highlight, 2019 edition*. Madrid: United Nations World Tourism Organisation. doi:10.18111/9789284421152

Yan, L. (2018). Origins of nature tourism in imperial China. *Journal of Tourism Futures*, *4*(3), 265–274. doi:10.1108/JTF-04-2018-0016

Yeoman, I. (2008). *Tomorrow's tourist*. Oxford: Elsevier.

Yeoman, I., & McMahon-Beattie, U. (2005). Luxury markets and premium pricing. *Journal of Revenue and Pricing Management*, *4*(4), 319–328. doi:10.1057/palgrave.rpm.5170155

Chapter 21

50 Shades of the Luxury Hospitality Industry

Katarzyna Minor[1] *and Andy Heyes*[2]

Abstract

The world of luxury hospitality is predominately perceived in a positive format; however, what this chapter discusses is the negative side to the luxury hospitality. The chapter critically examines the hotels' role in breaches of human rights and their possible involvement in illegal practices such as human trafficking and modern slavery. It provides an overview of the problem, the key terms and the stages of hotel involvement in this procedure. It examines the underlying reasons for the status quo, including discretion and secrecy accompanying luxury service. It highlights the vulnerability of the luxury sector relating to repercussions in terms of reputation and loss of trust. The chapter further examines the concept of duty of care and the breaches of it in relation to hotel staff. Particularly, it examines the industry-wide problem of alcohol and drug abuse among employees and the possible reasons behind it, followed by possible best practice solutions.

Keywords: Human rights; human trafficking; modern slavery; exploitation; duty of care; drug and alcohol abuse

Introduction

Luxury is often associated with disposable money and the purchasing of goods and services that we actually do not need in order to survive (Swarbrooke, 2018). The literature on luxury is often seen to be positive, with stereotypical definitions of the concept relating to high quality, decadence and somewhat excessiveness in

---

[1]https://orcid.org/0000-0002-9047-7587
[2]https://orcid.org/0000-0002-5843-849X

The Emerald Handbook of Luxury Management for Hospitality and Tourism, 425–442

doi:10.1108/978-1-83982-900-020211021

terms of comfort (Bellaiche, Mei-Pochtler, & Hanisch, 2010; Frank, 1999; Hoffmann & Coste-Marniere, 2012; Thomas, 2007). Luxury is said to be desired even by those who cannot afford it. As pointed out by Berry (1994), the essence of desirability mixed with want and excessive need is mixed with notions of obtaining something above and beyond the everyday normality of life (Hoffmann & Coste-Marniere, 2012), thus making the process of consuming and experiencing luxury a 'must' for some people, especially in cultures which are rich economically and socially (Berry, 1994). However, this desirability could be argued to be excessive and, in some cases (as this chapter will discuss), illegal. The impact of human desire is more substantial than a lot of people believe. The essence of money and the luxuries of life which it provides are to be discussed. Luxury hospitality has been providing services to a large array of people, from the rich and famous to the couple celebrating their 50th wedding anniversary; the perception of luxury is associated with the quality of the tangibles delivered with an essence of the intangibles such as empathy, reassurance and reliability (Parasuraman, Zeithaml, & Berry, 1985).

Along with luxury comes the essence of excessive need (Berry, 1994) – far passing those of the human need to survive (Maslow, 1954). While luxury has and is still prevalent when it comes to the excessive needs of individuals (Berry, 1994), the inclusion of money, in particular, is still an issue when it comes to the excessiveness of the luxury consumer. The typical British expression of 'Money Talks' can be said to be prevalent in this/these circumstances, whereby the power of someone's wallet far suppresses the expectations of human morality and ethics. Unfortunately for some, it could be argued, the power of money and the exclusive nature which it brings (Heyes & Lashley, 2017) far outweigh their own individual morals and make them believe they can afford/achieve everything. High price and cost alone can indicate the selfish needs of individuals – that is to say, that price can create exclusivity, separating those who can from those who cannot (Chandon, Gilles, & Pierre, 2016; McKinsey and Co., 1990). On the one hand, while high price can be said to emphasise exclusivity, on the other hand, it could also be considered that it offers other benefits to individuals, including that of discretion and, most importantly, for those in high disposition, secrecy. The exclusive nature of luxury can be argued to be created because of extravagance in wealth and the ability to spend excessively (Heyes & Lashley, 2017), bringing into context the needs of the 'super rich'. While other people across the world are struggling with survival, others are spending extravagantly to showcase their superiority and/or social and economic 'power' – therefore consuming conspicuously (Veblen, 1899). It is this power which some could argue will lead people into illegal practices and thus making them consume exponentially items/services which others would deem immoral and illegal. The world of hospitality, unfortunately, is being and will continue to be used to accommodate such activities, which will be discussed later in this chapter.

As pointed out by Lashley (2015), the world of hospitality can, is and has been commercialised to the fullest extent; the exchange of money indeed makes the true hospitality no more. Furthermore, it is the exchange of money which is of the most importance in the context of luxury hospitality. The consumption of luxury goods can be seen as a mark of distinction between classes and incomes. With some hotel suites costing in excess of US$20,000 (Heyes & Lashley, 2017), it is the availability of the money that is making luxury hospitality possible. Yet, it is the money itself that is often hidden when bills, payments and tips are discreetly hidden from view (Hajdakova, 2013). While conspicuous consumption has been prevalent in society for decades (Veblen, 1899), the so-called need for discretion, privacy and secrecy can be said to be a new form of conspicuous – that of inconspicuous consumption (Eckhaldt, Belk, & Wilson, 2015). Discretion, privacy and secrecy, some could argue, can also be seen through the forms of tipping. These tips, however, can be argued to be in contradiction with the morality of luxury – that is to say, there is a need for individuals to understand the fundamentals of tipping and whether these so-called tips are in contradiction to the morals of human life.

Hospitality, in a pure form of the word, is heavily connected to social morality (Derrida, 2000), yet, the availability and easy access of money makes it a commercial exchange (Lashley, 2015), putting into consideration morality of such, which is often criticised. This chapter deals with the often-omitted notion of the darker side of luxury hospitality, where 'external', often illegal, services are provided to the guests by the hotels to limit the risk (i.e., via concierge) and/or, particularly important within the context of luxury hospitality, have a strong desire to willingly respond to customer demands without moral boundaries (COMBAT Project, 2016). It looks at three 'behind the door' areas or sub-themes where the dark side of hospitality emerges: its connection to human trafficking, human rights violations and rampant drug and alcohol abuse.

Concept of Discretion

Central to this chapter's discussions are concepts of discretion and entitlement that luxury hospitality operates within. Luxury hospitality sells intangible services of a premium atmosphere underpinned by a high degree of discretion. Discretion can be defined as 'the ability to behave without causing embarrassment or attracting too much attention, especially by keeping information secret' (Cambridge Dictionary, 2020). Thus, the luxury hotels require their staff to be skilled not only in 'providing everything' to the guest but doing so by employing the art of privacy and secrecy. Furthermore, Sherman (2007) points out that luxury hospitality brings together subordination of workers with corresponding guests' entitlement to attention, labour and quite often concealment of excessive material property. When considered from the viewpoint of the darker side of luxury hospitality, this entitlement is extended to the staff's own moral convictions. In other words, the staff may need to adopt what Partico (2009) referred to as 'moral styles', where morality is situational rather than derived from one's cultural background. The activity that would be considered immoral and unacceptable in the employee's private life becomes acceptable in the workplace; thus, the shifting morality is considered 'necessary skills' in the provision of a discreet luxury hospitality product.

Human Trafficking

What Is Human Trafficking?

By design, human trafficking is considered to be a hidden crime, one that the hospitality industry inevitably is highly suited to support, by offering anonymity, privacy and policy of open doors to anyone willing to pay for the services, and often giving silent permission or engage in enabling illicit behaviours for those willing to pay high enough price.

European Union (European Directive, 2011/36/EU, Article 2, online) describes Human Trafficking as follows:

> The recruitment, transportation, transfer, harbouring or reception of persons, including the exchange or transfer of control over those persons, by means of the threat or use of force or other forms of coercion, of abduction, of fraud, of deception, of the abuse of power or of a position of vulnerability or of the giving or receiving of payments or benefits to achieve the consent of a person having control over another person, for the purpose of exploitation.

Human trafficking is often confused with human smuggling and with voluntary prostitution. Human smuggling an act of smuggling migrants across the border. Prostitution does not equal human trafficking. First, prostitution is legal in many countries. Secondly, whether prostitution is deemed legal or illegal, it may be performed voluntarily the sex worker. There are two main reasons why a person may be trafficked: for sexual exploitation (59%) and/or for forced labour (34%); however, other forms of trafficking are also recognised, such as forced marriage, organ harvesting and forced criminality to name a few. Trafficking is a notoriously difficult crime to detect; thus, obtaining data with regards to the extent of this procedure is difficult. United Nations Office for Drugs and Crime (UNODC) (2018) notes that women and girls are disproportionately affected by human trafficking, accounting for more than 70% of the total detected victims. Women and girls tend to be trafficked for sexual exploitation (96%), whilst men and boys for forced labour (75%). Human trafficking is a geographically diverse crime, affecting different parts of the globe in different ways, as illustrated in Fig. 21.1.

The following section will discuss sexual exploitation in the context of luxury hotels; forced labour is discussed within the Human Rights section of this chapter.

Sex Trafficking

No part of the hospitality industry is immune to the exposure to human trafficking, with Businesses Ending Slavery and Trafficking (BEST) (n.d.) reporting that, overall, 63% of sex trafficking incidents involve hotels, ranging from economy to luxury hotels. The definition of human trafficking identifies three stages of the act: the recruitment stage, the movement stage and the exploitation stage. Luxury hospitality is particularly vulnerable during the first and last stage. Within the former, the traffickers use the premises to buy and sell the victims. There are a number of ways in which the person can become a victim of traffickers: threat or

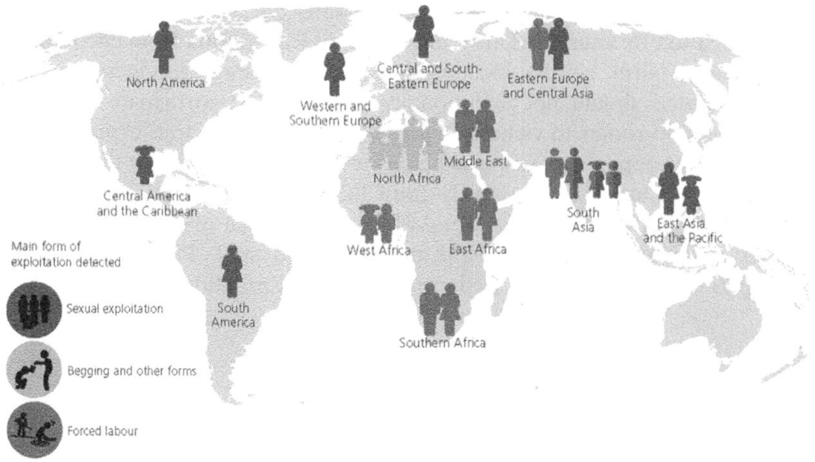

Fig. 21.1. Main Forms of Exploitation and Profiles of Detected Victims, by Subregions. *Source:* UNODC, Global Report on Trafficking in Persons, 2018, p. 11.

use of force, coercion, abduction, fraud, deception, abuse of power or vulnerability. The hospitality outlets are often used as a part of a deception process to trick vulnerable persons into being trafficked; often, the person believes to be in a relationship with a trafficker, a method known as 'lover boy' (Europol, 2016). The latter, the exploitation stage, happens within the confines of the hotel room. Whilst the hotel rooms seem to give an aura of anonymity to the users of prostitution, the high-end hotels often give silent permission for this act to happen due to the previously discussed 'moral styles' and the expectations of luxury through discretion. Furthermore, as Sherman (2007) points out, there is a tendency to treat luxury guests as 'privileged subjects' who are entitled to any extravagance.

As noted, prostitution is not illegal in many countries, and particularly in those countries, it is extremely difficult to recognise a person who is forced into this procedure. This is compounded by the fact that only 14% of hotels report having specific policies on dealing with the risk of sexual exploitation of their premises (Minderoo, 2019). This is despite heightened efforts by various organisations worldwide to provide guidance and training manuals to businesses. More often than not, hotels actively take part in the process of supply, often misguided by the understanding that luxury means giving everything to the high-paying customer, even if it is illegal or morally wrong. Indeed, brands of companies such as Hilton and Intercontinental were among 12 hotel chains named in a court case at the basis of which were accusations of ignoring signs of sexual exploitation and direct profiteering for this activity (Hodal, 2019). Legally, the hotels often find themselves in a grey area, where the hotel is not legally culpable unless it is proven it intentionally facilitated trafficking. This may often prove to be difficult to do since the luxury end of hospitality relies on the previously discussed discreet service.

Morally, however, van der Graaf (2018) points out a hotel has a four-level ethical obligation to: cause no harm, prevent or at least refrain from encouraging others to cause harm, combat existing harm and encourage goodness. In other words, the hotels should not intentionally facilitate and participate in human trafficking or rent rooms to victims or perpetrators of human trafficking. A hotel should, however, report any suspected cases of human trafficking to the police in a timely manner and should actively train their staff in recognising signals of human trafficking. However, the moral obligation is often overshadowed by the monetary considerations of both the managers and staff. Please refer to the following *Human Rights Issues* section for further information.

Sex Tourism and Exploitation of Minors

Child sexual exploitation is geographically fragmented; globally, each region faces different challenges, with some areas more vulnerable to the practice than others. This is particularly true in areas where tourism is developing, and the tourist hotspots are developed near pockets of poverty; however, it is important to note that not all victims of child sexual exploitation come from disadvantaged backgrounds. Similarly, whilst most of the victims are girls, in some areas, exploitations of boys is more prevalent. This is a particularly hard crime to detect, and indeed some areas, such as the Middle East and North Africa, tend not to record the crimes. Even if the crimes are recorded, as they are in Europe, they tend to be under-reported, making it hard to estimate the true degree of child sexual exploitation (Hawke & Raphael, 2016).

Hawke and Raphael (2016) report that sexual exploitation of minors is on the rise and is becoming more invisible. They suggest that sexual child exploitation is shifting from the streets to indoor locations, often including hotels, and is increasingly harder to detect as it is facilitated by the Internet and social media. One of the main issues as to why this is a very hard crime to detect is to do with the way the hotels operate their procedures. Hotels are easy to access as very often they do not require registration of children if they are accompanied by an adult or indeed check-in procedures are more frequently performed online, eliminating any need for the customers to meet the staff. Indeed, in Southeast Asia (the Philippines and Thailand have traditionally been the go-to places for child sex offenders), it has been reported that some of the hoteliers actively advertise child sex services as part of the travel package (Bergman, 2013). Conversely, some of the global chains, such as ACCOR, have learnt from the experiences of their properties in Southeast Asia and have deployed global strategies to counteract child sexual exploitation (ACCOR, 2016).

Contributing Factors

The luxury hotel sector is not immune to the pan-hospitality problems related to staff turnover and training. The sector is characterised by high turnover and a transient workforce, where the skills that one employee had are not immediately replaced. This is particularly problematic within the hotels which rely on 'on-the-job' training, where there is a heightened attention to job-specific, technical tasks, then other supplementary training designed to safeguard reputation. Specialised training, such as ones relating to tackling human trafficking, is often provided at a specific time of the year; thus, it may not form a part of the introductory training as it requires resources and knowledge, which often are outsourced.

Furthermore, in the event of staff shortages, the employees are often supplemented by agency or zero-hour contract staff, who tend to have less organisational commitment and/or may not be trained to detect human trafficking nor how to handle suspected cases. This means that even if a hotel has a robust policy to deal with human trafficking, it cannot be effectively executed (Hughes, 2015).

Another contributing factor is economic consideration for hotels managers/owners, where the hotels sometimes offer external 'services' to guests, which may be commissionable, bringing in direct profit to the hotels. This direct involvement of the management creates an environment of a silent permission, in which the employees are less likely to come forward to report a crime being committed. This, in turn, is creating a work culture which is unaffected by illegality and where it is a status quo of operations.

The staff as well can share profits in the proceeds or be given tips by the customers in exchange for arranged, illegal services. The economic consideration in relation to staff is particularly important in light of habitual low pay within the industry (see below Human Rights Issues section), where the staff may be included to partake in the illegal activities to supplement their wages.

Dangers

The most obvious danger for the hotel is not adhering to the letter of the law. A number of countries have legislation designed to counteract human trafficking, such as Modern Slavery Act 2015 in the United Kingdom and Directive 2011/36/EU in European Union. Not understanding or lack of awareness about the law is not a mitigation for not adhering to it. Indeed, there have been cases reported of hotels in the United Kingdom who have been sectioned under this act for repeatedly allowing exploitation of minors to happen on their premises.

There are severe reputational dangers related to the hotels engaging with human trafficking. Since the hotels have a significant moral and ethical responsibility to their customers and are placed in the position of trust, this trust can be broken if the hotel knowingly engages in illegal activities, particularly connected with human trafficking and sexual exploitation. Trust and value of the brand are often built over the years but can be lost in a relatively short amount of time if the hotel is proven to knowingly allow exploitation to happen on the premises. Proactive management of the issues and transparency of the policies, contrary to the perception, is not an admission of guilt but shows the strength of the brand and increases the trust of the customers. Furthermore, the proactive approach of the companies strengthens the hotel's image of support for the current societal movement against exploitation.

Instances of human trafficking and lack of training designed to tackle it on the premises may result in staff de-motivation and increased turnover (BEST, n.d.). Indeed, Mills, Tortez, and Blanton (2019) point out that if the staff are trained, they have the most opportunities to curb trafficking on the premises, which will result in heightened employee morale and sense of empowerment.

Modern Slavery and Human Rights

What Is Modern Slavery?

Slavery is illegal in every country in the world, yet, it is still happening; it is considered to be a specific form of human trafficking, and these days it is referred to as modern slavery. It can take many forms such as forced labour (i.e., where people are forced to work against their will under a threat of punishment), bonded labour (i.e., forced labour as a repayment for a loan or advancement of a loan), child labour (i.e., any forced work performed by a child to include bonded labour, prostitution, pornography and illicit activities and fighting), descent-based labour (i.e., occurs when people are forced to work or are treated as someone property because of their ethnicity or background) and forced early marriage (i.e., mostly affecting girls who are forced to marry and live a life of servitude, often accompanied by physical violence) (UNA UK, 2016).

Minderoo foundation (2019) reported that hotels in general are failing to address the risks of modern slavery in their direct operations and supply chains. In fact, they reported out of 71 hotels studied, only 6 had policies banning recruitment fees, which protect migrants from debt bondage, 76% of hotels did not disclose supply chain information and none had actionable plans to tackle the issue. Indeed, this issue is not isolated to any particular type of hotel, yet it is the luxury hotels which may fail to address human rights issues due to the misplaced belief that addressing and employing appropriate code of conduct equals admitting a guilt that these practices exist on the premises. This can affect brand reputation and will lead to damaging the business (COMBAT Project, 2016).

Contributing Factors

It is estimated that around 18,000 are forcibly working in hotels, restaurants and bars in Europe alone (COMBAT Project, 2016). According to Human Trafficking Foundation (2018), the hotel industry is particularly vulnerable to instances of modern slavery due to the employment practices of a vast number of temporary and/or agency workers, who may not be aware of their own or others' rights, yet are performing key operational roles (COMBAT Project, 2016). This is compounded by the use of culturally diverse labour pools where language barriers often prevent reporting and/or an understanding of their rights (COMBAT Project, 2016).

In 2015, an investigation uncovered that housekeeping departments are particularly vulnerable to exploitative working conditions (Minderoo, 2019) due to the high percentage of agency, migrant and female employment. The nature of housekeeping work makes it highly suited for this crime to remain hidden, as the work generally is performed out of sight of the customers and in isolation from other hotel workers. Since housekeeping is not a revenue-generating department and is considered a cost, the labour is often subcontracted where the hotels have little or no control over the employment practices. Particularly in light of the post-COVID-19 pandemic of 2020, where the hotels will likely need to increase their housekeeping effort whilst maintaining very tight cost control, there is a heightened danger for abuses of labour to happen in this department.

Unfortunately, a lot of the hotels have poor human resources practices and often do not perform checks or through vetting on subcontractors supplying the staff. This means that the signs of forced labour, such as checking if the persons receive adequate pay for their work or if indeed receive their pay, are not performed. This is particularly crucial as often the victims work to 'repay' a so-called debt to their enslavers, even if they had no say in agreeing to this debt.

Whilst the hotels often lack training designed to spot modern slavery, some members of staff may also be hesitant to come forward if the hotel has not implemented policies protecting whistleblowers, or they may fear retribution from others who are complicit.

Human Rights Issues

Human rights are fundamental principles and standards that aim to secure dignity, freedom and equality for all people (Universal Declaration of Human Rights, 1948). They underpin everything we do, from the way staff are treated and their working conditions to how guests are treated, in the supply chain and how goods and services for the hotel are produced in the communities where a hotel is based and in the way a hotel business is run. Some of the key issues related to hospitality and hotel work are related to the right to work, right to decent working conditions and the right to participate.

Within the context of luxury hospitality, considering the high price of services, questions have started emerging relating to the ethical, moral, economic and social power consumers can have over those whose lives depend on jobs serving the rich and famous. The imbalance of power can be highlighted through an example of daily room rates when often they can far suppress the staff's weekly or even annual salary. Indeed, hospitality, in general, is characterised by low-wage, long-hour culture. This, as pointed out by Lashley (2016), is coupled with low union membership, hence, limited options for collective resistance existing. This makes for a group of workers who are forced into jobs in which they have limited powers and choices. This status quo forces a lot of people who are working in the hospitality industry into living conditions which are often below an acceptable standard of living; Lashley refers to people who are finding themselves in these situations as 'non-slaves'. At the same time, the hospitality industry has struggled with high levels of operational staff turnover, partially associated with the low wages and working conditions. This is compounded by some of the management treating turnover as a monetary, not operational problem (Madanoglu, Moreo, &

Leong, 2004). This makes turnover an acceptable reality for many operators rather than a systemic problem of ill-treatment and broken rights of their employees.

Further consideration needs to be given to the gender inequality within the industry. Ryle (2017) defines gender inequality as a way in which the meanings assigned to sex and gender as social categories create disparities in resources such as income, power and status. Hospitality work is gendered; for example, in the United Kingdom, there are 1.05 million women working in the hospitality industry, of which 49% work in operational roles and only 12% work in managerial roles (People 1st, 2017). Whilst equality in terms of employment is growing in other sectors, there has been a sharp decrease in the number of female hospitality managers between 2009 and 2015 (People 1st, 2016). This also relates to the issue of equal pay. Despite laws of equal pay existing in the United Kingdom since the 1970s and the Equality Law 2010 ensuring equal pay, there is still gender-based operational and managerial staff pay difference within the industry (People 1st, 2017).

Contributing Factors

People 1st (2017) identifies that 65% of all part-time roles in the hospitality and tourism sector are filled by women, rising to 82% in contract food and service management and 72% in hotels. This means that the industry disproportionately relies on women part-time workers. Whilst the industry is claiming that this way of employment provides cost-effective workforce which can be employed in line with peaks of the trade, part-time employment creates a dual approach to employees, with 'core' employees likely to progress within the industry and reaping the benefits of increased wages and training, while 'peripheral', part-time staff constantly remain de-skilled and detached from organisational commitment. This means, due to higher female representation in part-time positions, that women are disadvantaged in terms of career progressions and training opportunities. It is noteworthy that whilst some employees choose to work on a casual or part-time basis due to work-life balance and other commitments, these employees still want to be involved in decision-making and day-to-day running of the businesses (Sobaih, Coleman, Ritchie, & Jones, 2011).

Furthermore, hotels tend to be more reactive than proactive in the way they address the labour shortages, particularly in the short term, and this opens the industry to a transient workforce which fuels already high levels of turnover, skills gaps and contributes to a low-wage culture (Wong, 2004). 'Zero-hour' contracts, i.e., a contract with no guaranteed hours or obligation by the employer to provide work, have risen, with a 174% increase in the use of this type of contract within the industry between 2005 and 2012 (ACAS, n.d.; Spanjers, 2013), and more recently there is an increased use of the gig economy within the industry (GIG, 2018). This is likely to fuel negative perceptions of the industry and contribute to unfair HR practices.

The career trajectory within the hospitality industry of men and women is similar until the age of 22; the number of women in the sector then starts to fall slightly before there is a steep decline at the age of 33 (People 1st, 2017). Therefore, the problem is not attracting women to the industry per se but how to make the industry attractive enough for women to come back from maternity leave. One of the biggest contributing factors is the unsociable hours that hospitality work commands; these often clash with family obligations. Whilst in a lot of countries most men and women believe that women and men should have equal opportunities and access to work, in reality, child and home responsibilities still remain largely in the female domain.

Substance Abuse and Duty of Care

What Is Substance Abuse?

In the United States, the rates of drug and alcohol abuse by full-time hotel employees are reported to be as high as 19.1% and 11.8%, respectively (Bush & Lipari, 2015); the sector struggles to address the underlying issues which lead the employees to substance misuse. Whilst in most industries, intoxication at work provides reasons for gross dismissal, in the hospitality industry, substance abuse seems to be a commonplace. With the likes of high-profile chef such as Anthony Bourdain (2013) admitting that the hospitality industry is drenched in drugs and alcohol, the industry seems to give a silent permission and turns a somewhat 'blind eye' to these types of behaviours and permissions for the abuse to take place, both from an employee and a customer perspective. From an employee/employer perspective, the effects of alcohol and substance abuse on employers have been well-documented throughout, including that of higher absenteeism and costs as a result (Aasland & Riise, 1990). A more recent study by Tressider and Martin (2018) classified substance abuse as deviant behaviour, alongside of theft, sex and service sabotage. Documentaries such as Gordon Ramsay on Cocaine (ITV, 2017) identified substantial cocaine use within both staff and customer toilets, with the drug in some cases being used in less than one hour of inspection.

Alcohol is the mostly wieldy used drug in the world; the abuse of alcohol has for decades been known to be in existence within the hospitality industry. Traditionally, it was thought that since alcohol is readily available in the hospitality for cooking and sale, it explains the fact behind heightened alcohol consumption and abuse within the industry. However, Pizam (2010) pointed out that such claims are based mostly on anecdotal information. Indeed, there are a number of studies which point out that alcohol abuse amongst hospitality workers is far higher when compared with other occupations and that hospitality workers have the highest levels of alcohol consumption associated with the risk of harm (e.g., Pidd et al., 2006). In fact, as Pizam (2010) points out, in the United Kingdom, innkeepers and bartenders have the highest death rates recorded caused by liver cirrhosis, which is associated with alcoholism, amongst all occupational groups. Furthermore, Isralowitz, Reznik, and Belhassen (2012) highlighted in their study that within a period of 30 days, 43% of hospitality workers engaged in binge drinking.

Most people responsibly interact with alcohol, and indeed alcohol consumption is an enjoyable social pastime for many. However, alcohol misuse happens when a person drinks in a way that is harmful or becomes dependent upon alcohol (NHS, 2018). Binge drinking, i.e., drinking a large amount of alcohol in a short space of time or drinking to get drunk, is considered a harmful way of consuming alcohol. It is associated with a heightened risk of personal injury, violence, unsafe sex, loss of possessions and alcohol poisoning. Alcoholism means a person loses control over their drinking and has an excessive desire to drink. Thus, alcohol abuse happens when the drinking continues despite health, social, interpersonal or legal difficulties.

All drugs affect the way that people perceive the world as they produce psychophysiological changes (Kelly & Perrone, 2014). Some people use drugs to enhance consciousness and others to alter their moods, from relaxation to inducing excitement. Kelly and Perrone (2014) point out that as a substance, no drugs by themselves are good or bad; their perception depends upon the legality and social context of use. However, there is little doubt that the use of illicit drugs (such as marijuana, heroin, crack cocaine, methamphetamines, psilocybin mushrooms, LSD, ecstasy and others) can lead to addiction; this can lead to harmful effects to the human body and mind and can lead to serious diseases, physical and mental disabilities and death (Pizam, 2012).

The accommodation industry was recorded to have 16.9% of employees using drugs (Pizam, 2012). Pidd, Roche, and Buisman-Pijlman (2011) reported that hospitality workers are two to three times more likely than employees of the industries to attend work whilst under the influence of drugs or take drugs during their workday. This places the work within the industry as 'high risk' when dealing with drugs.

Contributing Factors

The hospitality industry is characterised by long working hours, high stress and intense emotional and physical labour. Pidd, Roche, and Fisher (2015) note that chefs tend to be particularly affected as they experience higher than most occupational stress, compounded by commonplace bullying and harassment. Front of house employees often experience emotional strain as a direct result of intense emotional labour. Particularly within the luxury end of hospitality where there is a heightened need to deal with need for status of customers, be discreet and shift own morals (see the concept of discretion section above) and often the employees are abused by the customers (Sherman, 2007). Kotera, Adhikari, and van Gordon (2018) points out that these occupational factors contribute to poor mental health in the industry, yet, note that hospitality workers tend not to seek professional mental health help. Instead, as noted by Height, Gajjar, and Okumus (2019), the combination of unhappiness at work and home causes some to abuse drugs and alcohol as a coping mechanism.

Conversely, Corsun and Young (1998) link alcohol consumption in hospitality to organisational culture. Tressider and Martin (2018) refer to hospitality as having a hedonistic work culture, where 'work hard, play hard' attitudes exit alongside of general acceptance of drug taking.

Pizam (2012) points out that one of the contributing factors is lack of policies in place for drug and alcohol detection. To deal with substance abuses in the workplace educational programmes, robust policies on drug and alcohol coupled with random testing were by far the most effective practices to reduce substance abuse in the workplace (Pizam, 2012). Some of those policies could be as suggested by Thomas, Kitterlin, and Moll (2014) pre-employment drugs trials, which have proven to be effective in tackling the problems which the industry is currently facing.

Engagement with specialised, external organisations is needed as a part of Employee Assistance Programmes; in fact, it was pointed out as early as 1989 by Lazarus and Adler that these types of programmes are effective in tackling substance abuse at workplace. It needs to be recognised that the often hotels on their own lack skills necessary to effectively tackle substance abuse problems. Particularly smaller organisations lack funds and skills to tackle the problem on their own. Larger organisation or organisations with larger partial behind them, to include luxury hotel, often have policies and external engagement in place. Particularly important is that the external organisation understands the specific problems of the industry and the environment the employees work in; thus, generic drug and alcohol organisation may not be able to help when faced with industry-specific problems. Charities such as Hospitality Action in the United Kingdom have been created in order to help and support for those suffering from substance abuse.

What Is Duty of Care?

Duty of care can be defined as a moral or legal requirement to ensure the well-being of others. There is a substantial duty of care placed upon the employers in relation to employees. Thus, ignoring the problems related to drug and alcohol abuse can be understood as breaches of duty of care. Particularly if considered from younger workers' perspective, working environment culture is inflectional upon the shaping of young person's drinking habits (Pidd et al., 2006). This is particularly important when considered employment patterns and possible implications; i.e., for example, across the EU 20% of hospitality workforce is made up from people under 25 (Hortec, 2018).

Duty of care can be evident in the way the employers decide to tackle substance abuse in the workplace. Since these problems are often hidden, emphasis on sanctions will only make them employees hide their struggles more. Thus, part of the duty of care should be a right balance of policies, testing and training and helps from external specialist organisations.

Signs of Drug and Alcohol Use at Work

Hospitality Action noted the following three signs of substance abuse at work: first, reduced work performance, where the employees make more mistakes, lack judgement and are less able to handle stress. This can also include difficulty concentrating on work, confusion, taking longer to do jobs and/or periods of high/low productivity. Work performance also would include the sign of poor timekeeping, increased absenteeism and avoiding supervisors.

Second are physical signs. These will include smelling of alcohol and loss of appetite. Particular attention should be paid to change in appearance, such as unkempt look, characterised by lack of hygiene/sweating, shaking and frequent flu-like symptoms.

Third area to consider is personality changes. These include sudden mood changes, irritability and aggression. Also of consideration can be increased overreaction to criticism and frictions with colleagues.

Action Plan

The issues presented in this chapter can be tackled in a number of ways. Here are four best practice advice.

Firstly, there is a need for hotels to admit that problems may exist, thus allowing for appropriate actions to take place. There needs to be an understanding that by implementing policies and training, the organisation not only safeguards the customers and employees but also their own reputation and builds an image of a responsible hotel with regards to a range of social issues.

Secondly, a key factor is a substantial need for training frontline employees to, firstly, identify such practices which may be taken place within the hospitality environment and, secondly, to deal with such ill practices, if they were to arise within the working environment. It could be argued, therefore, that HR departments are needing to provide an exponential amount of training resources to facilitate with such measures. Such training could be based on internal training schemes or also devised by partnering with external organisation that specialise in certain measures.

Therefore, the third action point is to proactively engage with external specialist organisations. For example, the European Council Hospitality Restaurant and Institutional Education have devised a training mechanism which help to tackle that of sexual exploitation (EuroCHRIE, 2020). BEST has a range of free resources on which hotels can base their own human trafficking detection training. Hospitality Action (2020) works with hospitality organisation to tackle substance abuse. Such external organisations are set in motion to try to tackle such issues; however, there is a need for the hospitality industry to tackle such issues by partnering with such academic institutions/partnerships to deal with the contemporary issues which the industry is currently facing.

Last but not the least is creating a set of robust policies, both external and internal facing. These should guide the staff on correct procedures, points of contact, persons responsible, point to training material and training opportunities, as well as legislation designed to protect them and the customers.

However, what needs to be highlighted is that the best practice implementation is not without its problems. These usually lay in the economic costs of such a specialised training programme. There are some calls that such training should be subsidised by the government or be reimbursed/subsidised by other areas of the industry, i.e., professional management associations (e.g., Institute of Hospitality). Based on the luxury world and the stereotypical high prices/revenue, which is associated with such, there are arguments that such companies should, therefore, look to invest in such training programmes by themselves. Nonetheless, the authors argue there is a need to address such issues on a corporate global scale, adopting such training programmes within/amongst a multi-national level.

Conclusion

What this chapter has looked at uncovering is a wide range of issues exploring the darker side of luxury hospitality. While luxury has and continues to be seen in a positive light due to the hedonistic values it brings to the customers, issues discussed in this chapter are considered to be controversial, and against moral obligations, the industry has to both customers and employees.

Of critical importance is the consideration whether the criminality led income boost is more important than the general heath of reputation. Additionally, what the industry needs to consider is that running a business at the costs of the employees' health is really good for business. It is therefore hoped that the issues discussed in this chapter have and will continue to bring forward further levels of discussion along with research which can greatly contribute to tackling the outlined issues.

References

Aasland, O. G., & Riise, G. (1990). Alkohol og arbeid. *Tidskrift for Norsk Loegeforening*, *13*(110), 1697–1699.

ACAS. (n.d.). Zero-hours contracts. Retrieved from https://www.acas.org.uk/zero-hours-contracts

ACCOR. (2016). *Watch programme, combating sexual exploitation of children in travel and tourism: A historic commitment from ACCOR hotel group*. ECPAT International.

Bellaiche, J. M., Mei-Pochtler, A., & Hanisch, D. (2010). *The new world of luxury: Caught between growing momentum and lasting change*. Paris: The Boston Consulting Group.

Bergman, J. (2013). Can Burma avoid the curse of sex tourism? *Time Magazine*, 12 April. Retrieved from http://world.time.com/2013/04/12/can-burma-avoid-the-curse-of-sex-tourism/

Berry, C. (1994). *The idea of luxury: A conceptual and historical investigation*. Cambridge: Cambridge University Press. doi:10.1017/CBO9780511558368

Bourdain, A. (2013). *Kitchen confidential*. London: Bloomsbury Publishing.

Bush, D., & Lipari, N. (2015). Substance use and substance use disorder by industry. Short Report, The Substance Abuse and Mental Health Services Administration. Retrieved from https://www.samhsa.gov/data/sites/default/files/report_1959/Short-Report-1959.html.

Businesses Ending Slavery and Trafficking (BEST). (n.d.). Retrieved from https://www.bestalliance.org/

Cambridge Dictionary. (2020). *Discretion.* Retrieved from https://dictionary.cambridge.org/dictionary/english/discretion

Chandon, J.-L., Gilles, L., & Pierre, V. F. (2016). Pursuing the concept of luxury: Introduction to the JBR special issue on "luxury marketing from tradition to innovation". *Journal of Business Research, 69*(1), 299–303. doi:10.1016/j.jbusres.2015.08.001

COMBAT Project. (2016). Combat: Trafficking in human beings in the hotel industry. Oxford: Oxford Brookes University.

Corsun, D., & Young, C. (1998). An occupational hazard: Alcohol consumption among hospitality managers. *Marriage & Family Review, 28*(1–2), 187–211. doi:10.1300/J002v28n01_11

Derrida, J. (2000). Re-reading: Of hospitality: Anne Dufourmantelle Invites Jacques Derrida to respond (R. Bowlby, Trans.). Stanford, CA: Stanford University Press.

Eckaldt, G., Belk, R., & Wilson, A. J. (2015). The rise of inconspicuous consumption. *Journal of Marketing Management, 31*(7–8), 807–826. doi:10.1080/0267257X.2014.989890

EuroCHRIE. (2020). Combat human trafficking. archive.eurochrie.org. Retrieved from https://archive.eurochrie.org/education/combat-human-trafficking-toolkit/

European Union. (2011). Directive 2011/36/EU. Retrieved from https://eur-lex.europa.eu/legal-content/EN/ALL/?uri=CELEX:32011L0036

Europol. (2016). *Situation report: Trafficking in human beings in the EU.* Retrieved from https://www.europol.europa.eu/publications-documents/trafficking-in-human-beings-in-eu.

Frank, R. (1999). *Luxury fever: Why money never fails to satisfy in an era of excess.* New York, NY: The Free Press.

GIG. (2018). *The gig economy & hospitality: A future to prepare for.* GIG App Ltd. Retrieved from https://docs.wixstatic.com/ugd/6c51ad_61d31e1cc9034c19b057e3ceacd0289b.pdf

Hajdakova, I. (2013). Discreet economy: Luxury hospitality in the context of post-socialist transformation of Czech Society. *Laboratorium, 5*(1), 73–97.

Hawke, A., & Raphael, A. (2016). The global study report on sexual exploitation of children in travel and tourism, ECPAT international jointly with defence for children-ECPAT Netherlands, Bangkok, Thailand. Retrieved from https://www.ecpat.org/wp-content/uploads/2019/06/Offenders-on-the-move-Global-Study-on-the-Sexual-Exploitation-of-Children-in-Travel-and-Tourism.pdf

Heyes, A., & Lashley, C. (2017). Price, luxury and exclusivity: Exploring London's luxury hotels. *Research in Hospitality Management, 7*(1), 17–24. doi:10.1080/22243534.2017.1355470

Hight, K., Gajjar, T., & Okumus, F. (2019). Managers from "hell" in the hospitality industry: How do hospitality employees profile bad managers? *International Journal of Hospitality Management, 77*, 97–107. doi:10.1016/j.ijhm.2018.06.018

Hodal, K. (2019). Major global hotel brands accused of profiting from sex trafficking. *The Guardian.* Retrieved from https://www.theguardian.com/global-development/2019/dec/11/major-global-hotel-brands-accused-of-profiting-from-sex-trafficking

Hoffmann, J., & Coste-Marniere, I. (2012). *Luxury strategy in action.* Hampshire: Palgrave Macmillan. doi:10.1057/978020361546

Hortec Hospitality Europe. (2018). Facts and figures: The hospitality industry's contributions to European economy society. Retrieved from https://www.hotrec.eu/facts-figures-2/

Hospitality Action. (2020). Retrieved from https://www.hospitalityaction.org.uk/

Hughes, F. (2015). Human trafficking: Why it's time for the hotel industry to act? In A. L. Torres & M. . Á. Gardetti (Eds.), *Sustainability in hospitality: How innovative hotels are transforming the industry* (pp. 249–259). Saltaire: Taylor & Francis Group.

Human Trafficking Foundation. (2018). Tackling modern slavery in the hotel sector. Retrieved from https://www.humantraffickingfoundation.org/news/2018/10/1/tackling-modern-slavery-in-the-hotel-sector

Isralowitz, R., Reznik, A., & Belhassen, Y. (2012). Binge drinking among Israeli hospitality workers: Gender and country of origin differences. *Drugs: Education, Prevention and Policy*, *19*(2), 181–183. doi:10.3109/09687637.2011.649498

ITV. (2017). Gordon Ramsay on Cocaine: Gordon Ramsay investigates the cocaine trade for a new documentary. Retrieved from https://www.itv.com/news/2017-10-13/gordon-ramsey-investigates-the-cocaine-trade-for-new-documentary/

Kelly, B., & Perrone, D. (2014). Alcohol and other drugs. In J. Trevino (Ed.), *Investigating social problems* (pp. 308–336). Thousand Oaks, CA: Sage.

Kotera, Y., Adhikari, P., & van Gordon, W. (2018). Motivation types and mental health of UK hospitality workers. *International Journal of Mental Health and Addiction*, *16*, 751–763. doi:10.1007/s11469-018-9874-z

Lashley, C. (2015). Hospitality and hospitableness. *Research in Hospitality Management*, *5*(1), 1–7. doi:10.1080/22243534.2015.11828322

Lashley, C. (2016). Liberating wage slaves: Towards sustainable employment practices. In C. Lashley (Ed.), *The Routledge handbook of hospitality studies* (pp. 386–400). London: Routledge.

Lazarus, B., & Adler, H. (1989). A study of workplace substance abuse programs in the hotel and motel industry. *Journal of Hospitality and Travel Research*, *13*(3), 457–462. doi:10.1177/109634808901300347

Madanoglu, M., Moreo, P., & Leong, J. (2004). Reasons for employee turnover among room attendants. *Journal of Human Resources in Hospitality and Tourism*, *2*(1), 17–38. doi:10.1300/J171v02n01_02

Maslow, A. H. (1954). *Motivation and personality*. New York, NY: Harper.

McKinsey and Co. (1990). *The luxury industry: An asset for France*. Pairs: McKinsey.

Mills, M., Tortez, L., & Blanton, R. (2019). Moving beyond employees: Anti-trafficking training as facilitating social change. *Industrial and Organisational Psychology*, *12*, 34–38. doi:10.1017/iop.2019.4

Minderoo Foundation. (2019). Hotel sector failing to protect workers ahead of Australian MSA anniversary. Retrieved from https://www.minderoo.org/walk-free/news/hotel-sector-failing-to-protect-workers-ahead-of-australian-msa-anniversary/

NHS. (2018). Alcohol misuse. Retrieved from https://www.nhs.uk/conditions/alcohol-misuse/

Parasuraman, A., Zeithaml, V., & Berry, L. L. (1985). A conceptual model of service quality and its implications for future research. *Journal of Marketing*, *49*(4), 41–50. doi:10.2307/1251430

Partico, J. (2009). Spinning the market: The moral alchemy of everyday talk in postsocialist Russia. *Critique of Anthropology*, *29*(2), 205–224.

People 1st. (2016). *Insight report: Women working in the hospitality and tourism sector*. London: People 1st.

People 1st. (2017). *The performance & talent management revolution: Driving productivity in hospitality & tourism*. London: People 1st.

Pidd, K., Berry, J., Harrison, J., Roche, A., Driscoll, T., & Newson, R. (2006). *Alcohol and work: Patterns of use, workplace culture and safety*. Canberra, ACT: The Australian Institute of Health and Welfare.

Pidd, K., Roche, A., & Buisman-Pijlman, F. (2011). Intoxicated workers: Findings from a national Australian survey. *Addiction, 106*, 1623–1633. doi:10.1111/j.1360-0443.2011.03462.x

Pidd, K., Roche, A., & Fischer, J. (2015). A recipe of good mental health: A pilot randomized controlled trial of psychological wellbeing and substance use intervention targeting young chefs. *Drugs: Education, Prevention & Policy, 22*(4), 352–361. doi:10.3109/09687637.2015.1016400

Pizam, A. (2010). Alcoholism amongst hospitality workers. *International Journal of Hospitality Management, 29*, 547–548. doi:10.1016/j.ijhm.2010.06.004

Pizam, A. (2012). Illicit drug use among hospitality employees. *International Journal of Hospitality Management, 31*, 631–632.

Project Combat. (2016). *Combat: Trafficking in human beings in the hotel industry*. Oxford: Oxford Brookes University.

Ryle, R. (2017). Gender. In J. Treviono (Ed.), *Investigating social problems* (2nd ed.). Thousand Oaks, CA: Sage.

Sherman, R. (2007). *Class act: Service inequality in luxury hotels*. Berkley, CA: University of California Press.

Sobaih, A. E., Coleman, P., Ritchie, C., & Jones, E. (2011). Part-time restaurant employee perceptions of management practices: An empirical investigation. *The Service Industries Journal, 31*(11), 1749–1768.

Spanjers, S. (2013). Legal implications of zero hours contracts. *HR Magazine Online*. Retrieved from https://www.hrmagazine.co.uk/article-details/legal-implications-of-zero-hours-contracts

Swarbrooke, J. (2018). *The meaning of luxury in tourism, hospitality and events*. Oxford: Goodfellow Publishers.

Thomas, D. (2007). *Deluxe: How luxury lost its luster*. London: Penguin Books Ltd.

Thomas, L., Kitterlin, M., & Moll, L. (2014). Pre-employment drug testing for hospitality sales positions: Who's buying in? *Journal of Human Resources in Hospitality and Tourism, 13*(3), 297–322.

Tressider, R., & Martin, E. (2018). Deviant behaviour in the hospitality industry: A problem of space and time. *Hospitality & Society, 8*(1), 3–22.

United Nations Association – UK (UNA UK). (2016). Anti-slavery day: Factsheet. Retrieved from https://www.una.org.uk/anti-slavery-day-factsheet

United Nations Office for Drugs and Crime. (2018). *Global report in trafficking in persons*. New York, NY: United Nations.

Universal Declaration of Human Rights. (1948). United Nations. Retrieved from https://www.un.org/en/universal-declaration-human-rights/

Van der Graaf, E. (2018). Human trafficking in the hotel industry in the Netherlands: The fine line between service, responsibility and liability. *Research in Hospitality Management, 8*(1), 11–16.

Veblen, T. (1899). *The theory of the leisure class*. New York, NY: Macmillan.

Wong, K. K. F. (2004). Industry specific and general environmental factors impacting on hotel employment. *Asia Pacific Journal of Tourism Research, 9*(1), 19–28.

Part 5
Luxury in a Post-Pandemic World

Chapter 22

Reinventing Luxury Travel Imaginaries: Early Responses of Travel Influencers to the COVID-19 Pandemic

Christian S. Ritter[1]

Abstract

This chapter assesses how luxury travel imaginaries were modified in the aftermath of the outbreak of the coronavirus pandemic. Drawing on long-term fieldwork among travel influencers, the chapter presents their response strategies to the COVID-19 pandemic. Based on an analysis of evidence from participant observation in tourist sites, network visualisations, in-depth interviews and platform profiles, I trace the transformations luxury travel imaginaries have undergone since the beginning of 2020. Before this global crisis, travel influencers became new puissant players in the highly globalised tourism industry as they regularly received assignments from tourism boards and hotels. Although brand sponsorship was considered a substantial source of revenue for travel influencers, their collaborations in travel destinations and the monetisation of travel content on YouTube were further assets to secure a livelihood. The coronavirus outbreak, however, turned their life-worlds upside down. This ethnographic investigation identified three main responses of travel influencers to the current long-term crisis of tourism: (1) diversification of content creation and orientation towards other influencer genres, (2) support for local tourism organisations and online promotion of staycations and (3), finally, travel to tourist sites for circulating online content on safe travel standards. Digital platforms became a major arena where the future of tourism has been re-negotiated in the wake of the COVID-19 outbreak. The in-depth investigation suggests that travel influencers were in a position to create new powerful representations of luxury as safe travel

[1] https://orcid.org/0000-0001-7394-2646

The Emerald Handbook of Luxury Management for Hospitality and Tourism, 445–461

doi:10.1108/978-1-83982-900-020211022

since they acquired the skills to establish stable storyworlds for their travel experiences, which attracted the attention of large platform audiences.

Keywords: Digital platforms; professionalisation; tourism imaginaries; transmedia storytelling; travel influencers; safe zones

Introduction

The Czech travel vlogger Adam strolls through a restored medieval city gate into Tallinn's Old Town. He continues walking through the pedestrian zone and soon reaches café Rukis, where Juula, the Social Media Manager of the local Tourist Information Centre, awaits him. She has arranged a meeting to discuss the creation of video content on tourist sites in the city. Adam is a successful travel influencer with high follower figures on digital platforms who has created dozens of travel vlogs providing information and practical advice on popular travel destinations. Enjoying coffee and cake in the rustic atmosphere of the café, Juula and Adam negotiate the terms of the video production. Juula suggests showcasing St. Olaf's Church, a prominent historical feature of the Old Town, in a 12-minute video. They finally agree that the vlog shall be uploaded onto the Estonian tourist board's website. Leaving the café, Adam is eager to get the creative process started and meets his cameraman Jan at a nearby luxury hotel to discuss the project. Minutes later, the two are carrying video equipment to the church and inspecting the surrounding area.

In recent years, a growing body of research has been dedicated to the rise of influencers in numerous industries around the globe. The culture of influencers was initially inspired by the notion of personal influence (Katz & Lazarsfeld, 2009). Early studies of influencers mostly address the stardom of the highly successful subgroup. While influencer marketing has evolved into a multi-billion-dollar industry, the work of influencers has been widely associated with their self-branding practices and the micro-celebrity status (e.g., Khamis, Lawrence Ang, & Welling, 2016). Influencers manage to establish enduring, parasocial relationships with their followers and consequently generate sustained interest from corporations and associated brands (Yuan & Lou, 2020). The collaborations between destination marketers and travel influencers increasingly involve intermediaries that seek to formalise the agreements between both parties (Stoldt, Wellman, Ekdale, & Tully, 2019). As influencers strive to establish substantial followings on digital platforms, their self-presentation is enacted within the constraints of the sociotechnical systems of the Internet (Carter, 2016). The global rise of influencers, however, requires further investigation at the middle and lower levels of their socioeconomic pyramid. Although the Internet has been identified as a central driving force of tourism imaginaries (e.g., Lo & McKercher, 2016), the role of travel influencers in reinforcing such imaginaries has not been thoroughly studied in terms of transmedia storytelling. Combining a transmedia approach with perspectives from the anthropology of professionals, the present chapter describes how travel influencers responded to the predicaments of the COVID-19 crisis and reconfigured luxury travel imaginaries. The purpose of this chapter is twofold. Firstly, I seek to understand further the role of digital platforms in the transformations of tourism in the age of the coronavirus pandemic. Secondly, I trace disaster response strategies by drawing on three cases depicting how they adjusted their platform practices during the pandemic. In recent years, travel influencers, mainly comprising travel bloggers and travel vloggers, established themselves as puissant actors in global tourism. Based on evidence collected in the early phase of the coronavirus crisis, this transdisciplinary investigation raises a number of empirical questions: How do hobbyist creators of travel content advance to become professional travel influencers? In what ways have travel influencers responded to the COVID-19 pandemic? And, finally, how do answers to these questions provide insights into the modifications of luxury travel imaginaries since the coronavirus outbreak?

Committed to a mixed-method approach (e.g., Born & Haworth, 2017), I integrate ethnographic fieldwork with computational social network analysis. Between summer 2018 and spring 2020, I accompanied numerous travel influencers to their travel destinations in Southeast Asia and Europe. Throughout the investigation, but in particular during the period of lockdown, I studied travel influencers remotely (Walton, 2018). In doing so, I arranged video chats, exchanged direct messages and followed their platform profiles. From the early days of the investigation, scrolling, liking and posting on Instagram, Twitter and YouTube complemented the observational tactics in physical settings, enabling me to alternate between a physical presence in field sites and remote ethnographic research (Postill, 2017). Tourism is, by definition, rooted in mobility, and travel influencers are constantly moving targets. After meeting eight travel influencers during my initial fieldwork in Estonian tourist sites, I decided to accompany them on their travels to explore their daily routines more closely. By conducting multi-sited fieldwork, I learned about their encounters with local tourism professionals, how they produced travel content and gained insight into their skilled practices for strategically disseminating travel stories on digital platforms and websites. Participant observation was conducted in numerous tourist sites. I explored the skilled practices of several travel influencers, shadowing them to various travel destinations, including Angkor Wat, Bangkok, Berlin, Hanoi, Helsinki, Kiev, Kuala Lumpur, Riga, Singapore, Sofia, Stockholm, Tallinn, Tartu and Vilnius. This investigation mainly draws on observational data on tourist sites, 30 in-depth interviews and network visualisations. Pseudonyms are used for all individuals described in this chapter to protect their personal identities. I documented my experiences of local places and platforms through observation protocols, photographs and screen recordings. The qualitative data sets, including observation records, interview transcripts and audio-visual materials, were analysed in accordance with the principles and procedures of grounded theory (Corbin & Strauss, 2008).

To incorporate the methods of the medium into this investigation (Born & Haworth, 2017; Rogers, 2013), the qualitative data sets were complemented by digital data sets. Considering the web as an insightful data source, I collected social network data from digital platforms, such as Twitter and Instagram. Social network data were scraped from Instagram and retrieved from the public Twitter application programming interface and with the Excel add-in NodeXL Pro. Using the layout algorithm ForceAtlas2, the network graphs were generated in the network visualisation software Gephi (Jacomy, Venturini, Heymann, & Bastian, 2014). Based on long-term fieldwork in tourist sites and on digital platforms, I argue that travel influencers were in a position to create new powerful representations of luxury as safe travel since they acquired the skills to establish stable storyworlds for their travel experiences, which attracted the attention of large platform audiences. While the first section of this chapter describes the professionalisation of travel influencers in the past few years, the second part assesses their responses to the coronavirus pandemic restrictions. Finally, I discuss the modifications luxury travel imaginary have undergone since the beginning of the global health crisis.

The Twitter accounts VisitTallinn and VisitEstonia launched an outreach campaign on the occasion of the 2019 *Tallinn Old Town Days*. Numerous local businesses, journalists and international tourism organisations retweeted their messages. The network visualisation in Fig. 22.1 illustrates the online status of the users 'VisitEstonia', 'VisitTallinn' and the travel influencers Adam and Jan (RealHonestGuide) within this temporary network. The data set shows that a tweet sent by VisitEstonia was retweeted five times. A message by VisitTallinn was retweeted four times, and the travel influencers were mentioned in three tweets. By continuously engaging in coordinated platform practices, Estonian tourism professionals, such as Juula, raise the visibility of the travel destination Tallinn on digital platforms.

The influencer became a central actor of the social web, aiming to gather large followings and promote lifestyles, products, organisations and indeed political campaigns. Rejecting an office-based work life in the 'cubicle farm', multitudes of travel influencers work as freelancers and are 'on the road' throughout most of the year. In contrast to previous groups of nomadic travellers, such as 'drifters', 'backpackers' and 'flashpackers', travel influencers do not travel to spend money, rather to generate it. For numerous travel influencers, the most common way of earning money is through product sponsorship, whereby they become a brand ambassador. For instance, posting photographs on Instagram can be complemented by affiliate links, which redirect viewers to online shops. Given the high aesthetic quality of photographs on Instagram, travel influencers often post imagery of luxury products in exquisite locations to attract likes and comments. They can furthermore generate a substantial stream of revenue by monetising audio-visual content on YouTube. A dream common to most of the researched travel influencers is to share highly influential videos on YouTube; however, the platform sets restrictive rules on how users can monetise their uploaded content. Only registered users are eligible to apply for membership in the YouTube Partner Program, provided they have attained 1,000 subscribers and achieved 4,000 public watch hours over 12 months (YTC, 2020). Travel vloggers who meet these criteria can receive a substantial stream of revenue from their uploaded vlogs. Such revenue, however, depends on the measured popularity of their videos. The 'costs per 1,000 impressions' are calculated for each uploaded video. Members of the

Case Study 22.1: From Travel Passion to Tourism Profession

On 30 May 2019, a short video portraying the tower of Tallinn's St. Olaf's Church appeared on Twitter. The audiences watch Adam as he climbs the 258 steps to the top of the tower and revels in the magnificent view over the city on a sunny summer day. Comparing the historic church tower to a present-day skyscraper, he extols the church as a must-see tourist attraction of Tallinn. An abridged 1-minute version of the vlog was posted by Juula, who manages the official Twitter account for the travel destination Estonia. The same tweet also announces the start of *Tallinn Old Town Days*, a cultural festival showcasing dancing, theatre and musical performances. Adam retweets the message. Festivals are highly significant for the Estonian tourist board, which aspires to increase the visibility of the country's tourist places on digital platforms. Indeed, tourists around the world are increasingly inspired by travel content posted on the Internet. To contextualise the early responses of the studied travel influencers to the coronavirus pandemic, I begin by investigating their career paths and strategies for securing long-term streams of revenue. In doing so, I examine how hobbyist creators of travel content expand their skills to become professional travel influencers.

Drawing on the anthropology of professionals (Hull, 2020; Kiik, 2019; Mosse, 2011), the first case study describes the professional lives of the travel vloggers Adam and Jan. Their collaboration with the Estonian tourist board epitomises a common stream of revenue that travel influencers can generate to earn a livelihood. By attracting large numbers of followers on digital platforms, they increase their chances of being invited to travel destinations by tourist boards, tour guides or hotels. Such deals may include agreements on content sharing via digital platform handles of travel influencers in exchange for an agreed fee or free transportation and accommodation. The majority of travel influencers consider such arrangements as a part of the 'gig economy', enabling them to secure other work coinciding with their schedules and travel itineraries. Collaborations between travel influencers and tourism boards, tour guide companies or hotels can also involve the direct selling of travel content, which is used on the websites or platform profiles of tourism organisations.

By announcing the cultural festival *Tallinn Old Town Days* on Twitter, Juula sought to amplify the travel imaginaries about the travel destination Tallinn. Extensive outreach campaigns on digital platforms can make tourists aware of local events. A social network analysis of the 2019 *Tallinn Old Town Days* shows how local collaborations between destination marketers and travel influencers were enhanced on the platform Twitter. The data set for Fig. 22.1 was assembled by the search term 'Tallinn Old Town Days'. It contains 29 nodes, which represent accounts of Twitter users, and 62 edges, which refer to various platform practices. The relatively small sample size makes it possible to manually check the quality of the relevant data, which increases the reliability of the network analysis (Lomborg & Bechmann, 2014). The edges relate to four types of platform practices: tweeting, retweeting, mentioning and following. The size of the nodes reflects the popularity of a Twitter account within this particular network.

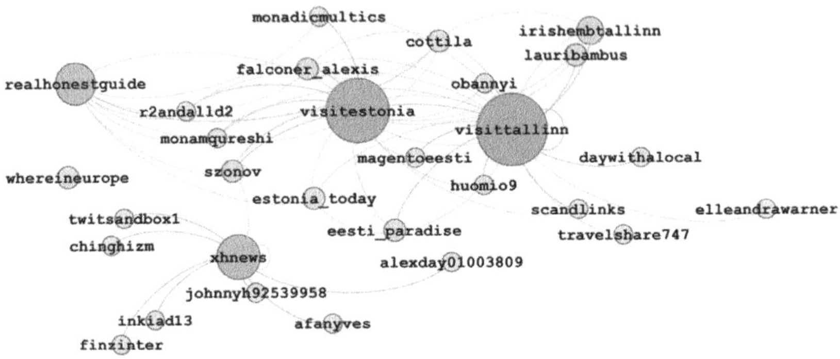

Fig. 22.1. Network Graph Displaying Twitter Users during the Local Event 'Old Town Days Tallinn'.

YouTube Partner Program are granted access to YouTube Studio, which consists of a set of analytics tools in the backend of their user account. Another metric that travel influencers regularly monitor is the 'impressions click-through rate', indicating the percentage of impressions that are turned into views of a given video. Thumbnails and titles of YouTube videos are displayed when users search for a clip on the platform. These two features of a video can also appear in the viewer's feed or be added to a list of suggested videos. If YouTube users click on the thumbnail or title of a video clip, the influencer's popularity and revenue are increased. For example, 1,000 views of a video clip were given a specific monetary value.

Travel influencers often interconnect a variety of media channels, such as websites, video platforms and social media platforms. In the last couple of years, their forms of transmedia storytelling predominately included the photo-sharing platform Instagram and the video platform YouTube. All of the researched travel influencers pursued transmedia strategies for disseminating their travel stories. Transmedia approaches have burgeoned in the social sciences and humanities since the early 2000s. One of the first traces of transmedia storytelling can be found in a film project developed in 1999. The low-budget independent film *The Blair Witch Project* built a large fan base on the info web. Initially, transmedia storytelling referred to a then-new aesthetic that emerged in response to the convergence of media and the art of world-making (Jenkins, 2006). A fictional story might be introduced in a film, for example, and expanded through novels, comics, games, amusement parks or perhaps television series. However, each media form remains self-contained. Hence, it is not necessary to watch the movie to enjoy the video game and vice versa. Nonetheless, all media forms provide entry points to the whole storyworld. Viewers can follow the main story across differing delivery channels. Transmedia storytelling can be seen as a socio-cultural process inherent in the dispersal of integral elements of a fictional story across multiple channels for the purpose of creating a unified entertainment experience (Jenkins, 2006). Transmedia storytelling is grounded in an overarching storyworld. For example, the storyworld of the drama 24 was comprised of films, TV series, comics and novels, amongst others (Scolari, 2009). Travel influencers do not create fictional accounts but regularly emphasise the authenticity of their travel content. They create a complex yet coherent storyworld, in which audiences can engage across different platforms. The main plot line of most travel influencers is composed of a series of linear, sequential events. A storyworld can include individual personas, settings, events, subplots and background stories. Travel influencers often create narratives about travel destinations, their travels to such destinations and their encounters with other influencers.

The researched travel influencers could make a living as location independent, self-employed solo entrepreneurs by delivering attractive travel content to large audiences. This in-depth study of travel influencers revealed three main revenue sources. For many travel influencers, a substantial stream of revenue derived from collaborations with destination marketers. Other viable options were the monetisation of video content on YouTube and product sponsorship. Many hobbyist travel content creators primarily managed to establish themselves as travel influencers by mastering the skilled practices of creating stable online personas that come into being in transmedia storyworlds. Combining pictorial, audiovisual and textual travel content across multiple platforms, they construct coherent storyworlds mirroring their personal travel experiences. The career paths and transmedia storyworlds of travel influencers provide the socio-professional context for the dramatic transformations luxury travel imaginaries have recently undergone. In the following section, I shall describe the reactions of travel influencers to the outbreak of the coronavirus.

The Future of Luxury Travel after COVID-19

Many of the researched travel influencers had made considerable headway en route to a successful career when the outbreak of COVID-19 triggered the first lockdowns across several countries at the beginning of 2020. Attaining the figure of 10,000 followers is perceived as a crucial landmark in the process of becoming a microinfluencer. This status is widely considered the sine qua non for starting a successful career as a travel influencer. Against the background of these professional career ambitions held by many travel influencers, I assess how they responded to the COVID-19 outbreak. The beginning of the coronavirus pandemic and subsequently the widespread ban on international travel forced travel influencers into a state of stasis. On 11 March 2020, the World Health Organization officially declared the outbreak of the coronavirus a pandemic. By the end of March 2020, the majority of the world's airlines' fleets were grounded, and cruise ships were immobilised, mostly stranded in harbours. International tourism ground to a virtual halt. Consequentially, the livelihoods of tourism professionals around the world were put at high risk. In the immediate aftermath of the last global economic crisis, which began in 2008, the number of tourists dropped by approximately 40 million worldwide (WTO, 2020). Although the severe consequences of the continuously evolving global health crisis cannot yet be fully appreciated, it is feasible that pandemic may take an even larger toll on the global tourism industry than the 2008 financial crisis. Outbreaks of infectious diseases, such as typhus, smallpox, cholera and influenza, have a long history of disrupting travel due to the risk of contagion. In the past, the ramifications of pandemics often reified the stigmatisation of marginalised social groups, othering, gendered concepts of responsibility and racism. Since the beginning of the COVID-19 pandemic, the rhetoric for limiting the spread of the disease often entailed discriminatory classifications. Given the high level of interconnectedness in today's globalised world, a planetary health approach stabilising the balance between all living creatures is required to overcome the current global recession (e.g., McCloskey, Dar, Zumla, & Heymann, 2014).

The outbreak of the COVID-19 pandemic has widely been perceived as a drastic disaster within the global tourism industry, and considerable layoffs have taken place in all major segments, including accommodation, guided touring and gastronomy. The myriad of travel influencers who were stranded in their travel destinations largely adhered to the advice of their government and returned to their country of citizenship. Between mid-March and May 2020, many of the researched travel influencers were confined to their homes, and initially, their output on digital platforms increased sharply. However, posting relatively fresh travel videos and photographs, which were produced in early 2020, was deemed inappropriate due to the rising death toll around the world. In the early weeks of the lockdown, many travel influencers had all of their 2020 trips, already planned and sponsored, cancelled. The situation of travel influencers was somehow paradoxical. On the one hand, they were grounded, and travel appeared to be impossible in the foreseeable future. However, on the other hand, the screen time of their followers on digital platforms reached an all-time high.

The experience of a pandemic varies considerably from person to person. While for many there is no immediate threat to their own lives, anxieties and worries about their professional prospects may begin to reach into the present. Future and present form a continuum, which makes it necessary to study them in concert (Bryant & Knight, 2019). The ways possible futures are imagined may affect the present and vice versa. People regularly imagine how life will look like on the other side when the pandemic has come to an end. Building on a philosophical tradition tracing back to antiquity, the study of orientations can illuminate the digital cultures of travel influencers who have to cope with increased uncertainty about their future career plans. The six major orientations towards the future include anticipation, expectation, speculation, potentiality, hope and destiny, representing different ways in which the future may affect our present (Bryant & Knight, 2019). Public debates about the current global health crisis have been dominated by future orientations. By sharing stories about the consequences of COVID-19 on tourism on digital platforms, travel influencers already inhabited the future of their lives. In the rapidly shifting world of contemporary tourism, one of their central aspirations has been to promote good standards for safe travel. A plethora of travel influencers claimed that they would be among the first to consistently travel during the coronavirus pandemic. In recent blog posts and vlogs, they sought to instil the feeling of safety in their audiences. For example, a few of the researched influencers travelled by plane during the course of the pandemic and posted photographs that promoted air travel as a safe mode of transport.

Between April and May 2020, I conducted a series of in-depth online interviews with travel influencers. While most interlocutors initially focussed on their adjusted platform practices, the future of their careers emerged as a second central theme. After the initial shock, the vast majority of the research participants reported on an abundantly positive effect of the lockdown measures. With their readers and viewers forced to stay at home, their numbers of likes, subscriptions and comments skyrocketed on various platforms. All travel influencers were delighted while checking the analytics of the web traffic they could attract. Despite the joys of sharing travel throwback content, it soon became clear that the world was plunging into a deep recession that only a few living people had experienced. Building on their following and online status on digital platforms, various travel influencers shifted gear and engaged with new business partners. The turmoil following the outbreak of the coronavirus pandemic ironically encapsulated the dependency of travel influencers on platform metrics. Having lost their bread and butter topic, some travel influencers partnered up with online mental health advisers, food companies or fashion brands. A further subgroup produced new podcast series or switched to other digital platforms, such as TikTok and Twitch. While sponsored posts represented about 35% of the influencer posts in mid-February 2020, sponsored content dropped to about 4% of their overall posts in mid-April 2020 (Whateley, 2020). The initial response of travel influencers was mainly a diversification of their content creation on two different levels. Some rapidly adjusted their influencer genre, while others wandered to different platforms.

In the Irish tourism scene, Twitter has become one of the most popular platforms for sharing information on tourist sites and events in recent years. When the Irish government began to officially endorse local tourism in June 2020, many tourism organisations took the opportunity to announce tourist attractions and promote nature tourism. Such efforts were widely considered to be a way of increasing visitor numbers to tourist sites while maintaining high safety standards. The network graph displays how communities were connected within the hashtag public. In addition to tourists, various professional groups, including photographers, travel bloggers, journalists and tourism organisations, sought to shape the communication within the hashtag public. For instance, the account of TV weather presenter Barra Best can be interpreted as a hub in this particular network. This account could mediate between various subnetworks and gain the attention of various professional groups.

Case Study 22.2: Stasis and Staycation

In June 2020, the lockdown measures were steadily eased in some parts of the world, and many of the researched travel influencers could leave their homes for outdoor activities other than buying groceries. The second case study about the South Korean travel influencer Jang-mi illustrates how perceptions of luxury travel were transformed during the pandemic. Longing for new travel experiences and fresh travel content, many travel influencers decided to travel locally. Hiking and camping were increasingly considered safe and suitable alternatives to the more glamorous world of hotels, cruises and shopping malls. Imaginings of luxury began to slowly change. Experiences of rural areas and nature enjoyed unprecedented popularity, promising to sharply limit the contact with other tourists. Safe travel that minimised the risks of catching and spreading the coronavirus became a highly valued service. In early July, Jang-mi wrote a blog entry covering a camping trip during the night of summer solstice at Vartry Reservoir, Ireland. This piece introduced a new subplot to her storyworld. At that time, she had been stranded in Ireland for several months. Hiking off the beaten track was the only possible way of satisfying her need to travel. Unexpectedly, camping became a tangible form of luxury, which attracted great attention on digital platforms. While camping and hiking were traditionally perceived as low-cost activities, the meaning attached to these travel practices suddenly changed. Given the dangers and difficulties of reaching remote tourist sites, the calmness attached to hiking and camping paradoxically rendered them luxury experiences in such turbulent times. Jang-mi expressed the unprecedented immobility as 'staying put and camping locally'. She further explained that she enjoyed the wildlife more consciously and used the stasis to re-explore some of her cultural traditions, such as Korean food and medicine. Indeed, governments and tourist organisations encouraged people to opt for 'staycations' rather than travelling abroad for a holiday during 2020. In June and July, Jang-mi slowed down her platform practices. Only occasionally would she post photographs of red deer or greenfinches on Instagram and Twitter, adding the newly popular hashtag staycation. The performance of extravaganza on digital platforms was, step by step, replaced by stoic moments of mindfulness.

The hashtag staycation became a trending topic on Twitter in many countries between June and August 2020. For this hashtag public, a data set was assembled by the query '#staycation'. All data were collected on 8 August 2020. Based on the rules for the public Twitter API, data from the previous seven days were included in the data set, which is comprised of 14,585 nodes and 21,331 edges. The graph below (Fig. 22.2) shows the giant component of the directed network. The edges of the graph represent various platform practices, including tweeting, retweeting and mentioning. The breadth of the edges indicates the frequency of platform practices between two nodes, while the direction of platform practices between nodes is symbolised by curved edges. The size of the nodes reflects the number of all platform practices for a Twitter account. Accordingly, the large nodes stand for the most influential accounts within this hashtag public.

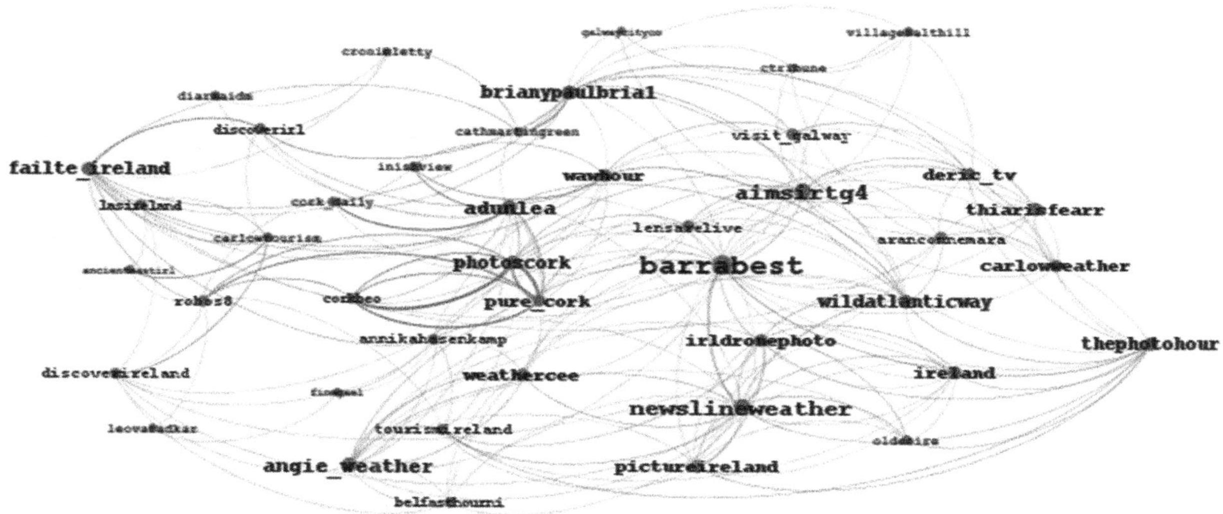

Fig. 22.2. Connected Communities within the Hashtag Public Staycation on Twitter.

Based on evidence from the platform profiles and in-depth interviews with travel influencers, I primarily identified three types of response to the coronavirus pandemic. Firstly, travel influencers sought to maintain their status by diversifying their content creation. Secondly, another response strategy involved embracing local tourism to deliver travel stories about luxurious staycations and nature tourism to platform audiences. Finally, many travel influencers aspired to promote practices of safe travel. Many of the researched travel influencers voiced hope that they could play a crucial role in rebuilding the travel industry while acting as role models for hygiene standards and social distancing rules. Tourism boards, governments and hotel chains showed sustained interest in partnering with travel influencers and envisioned them as new ambassadors of safe travel. Such ambitions gave rise to the creation of new luxury tourism imaginaries, which is assessed in greater detail in the next section.

Imagining Safe Zones

A major anthropological approach to tourism revolves around the imaginaries that tourism service providers and tourists co-construct (e.g., Dixit, 2020; Gravari-Barbas & Graburn, 2012; Gravari-Barbas & Nelson, 2016; Mostafanezhad & Norum, 2019; Sundbo & Dixit, 2020). The term tourism imaginary refers to a socially transmitted representational assemblage which is used as a meaning-making and world-shaping device and interacts with personal imaginings (e.g., Salazar & Graburn, 2014). Luxury travel imaginaries are a specific segment of the manifold, overlapping tourism imaginaries that circulate around the world through travel magazines, TV shows, newspaper articles, blog posts and digital platforms. Drawing on a travel story about the reopening of tourist sites in the United Arab Emirates, I assess the transformations of luxury travel imaginaries since the coronavirus outbreak. The study of tourism imaginaries reveals worldviews, discourses, stereotypes and fantasies. Tourism imaginaries can also take on the form of socio-cultural frameworks for encounters with others. The main components of such representational assemblages include the imaginings and expectations of individual travellers, globally disseminated images of local places, the collective identity of the visited communities and, finally, the beliefs that travellers develop about locals, and vice versa (Leite, 2014). The Lacanian theory of registers, which is based on a fundamental distinction between the real, the symbolic and the imaginary, is an underlying basis of the overwhelming majority of empirical studies about tourism imaginaries. This conceptual triad provides insight into the formation of the sense of self, otherness and the world (e.g., Lacan, 1936). Based on the ability of humans to think about themselves as objects, socio-cultural identities are formed through a twofold process of externalisation and internalisation. The proliferation of digital platforms in everyday life has altered the formation of socio-cultural identities. Through the sharing of blogs, photographs and vlogs, travel influencers perform self-presentations. Digital platforms provide a secure backstage area for disseminating travel stories (Lo & McKercher, 2016). Luxury travel imaginaries are a collection of images and representations, which travellers mainly create for digital platforms. Due to the genre-specific claim of authenticity, travel vloggers and bloggers primarily generate storyworlds that centre around their own experiences and selves.

The story of luxury travel imaginaries prior to the outbreak of the coronavirus could not be told without acknowledging the veritable explosion of travel photographs on Instagram. Celebrated by some and despised by others, Instagram users uploaded vast collections of high-quality photographs portraying top-notch tourist attractions. This platform serves as a circulation engine and provides a crucial infrastructure of movement for luxury travel imaginaries. Digital platforms accelerate the dislocation of images from travel destinations. Travel influencers often depict themselves while indulging in luxury goods in extraordinary settings. In the last five years or so, Instagram has become a tool for both promoting luxury travel and circulating stereotypes about travel destinations. By posting photographs or short videos on Instagram, travel influencers reshape luxury travel imaginaries. Their enduring platform practices reconfigure the relations between the different components of such representational assemblages, including photo and video cameras, tourist attractions, travel magazines, restaurants, tourists, tour guides and TV shows. Luxury travel imaginaries traditionally oscillated between indulgence in exquisite food and mesmerising places offering outstanding views. Luxury items in extravagant settings became a powerful genre on Instagram. In recent years, the visual culture of Instagram amplified luxury travel imaginaries. High-quality photographs displayed travel influencers in exclusive hotels or on cruise ships, enjoying high-end cuisine and views over epic landscapes. Images of extravagant travel influencers circulated via millions of devices, attracting millions of comments and likes. Textual narratives, photographs and videos mostly represented tourist places. Such imaginaries of places also play a crucial role in the current transformation of luxury travel.

Sometime between March and May 2020, the Instagram posts portraying influencers with bottles of champagne lounging beside hotel pools or on rooftop terraces began to fade away. However, the platform practices of travel influencers did not decrease. They directed their attention towards new forms of luxury emerging during the pandemic. In doing so, they transformed visual representations of luxury travel and, ultimately, the meaning of luxury. Depictions of extravaganza on digital platforms began to slowly be replaced by images of safe travel. The COVID-19 crisis of tourism, which may have just begun, provides crucial insights into the adjusted self-presentations of travel influencers and the newly emerging representations of luxury travel. Reaching out to their numerous contacts in the travel industry, many travel influencers engaged in online discussion sessions staged on Instagram and other platforms. Such sessions were dedicated to reflections on how trust in the safety of travelling could be rebuilt. In June 2020, many travel influencers publicised blog posts that reviewed hotels and holiday homes in terms of safety standards. They knew that the stakes were high and feared losing their online status, for which they had worked hard over the previous years. Hotels and destination marketers quickly began to collaborate with travel influencers to promote their safety standards. Stories about touch-free

Case Study 22.3: Broadcasting Safe Travel

The third case study illuminates the modifications that luxury travel imaginaries underwent in the aftermath of the COVID-19 outbreak. In mid-July 2020, the Palestinian-Israeli travel couple Omar and Amira travelled to Dubai, United Arab Emirates. The travel restrictions had only been lifted a few days earlier, but in a short video shared on Instagram, they walked their viewers through the different stages of their journey. Throughout the video, they emphasised the different measures that were in place to make the trip as safe as possible. They began by inspecting the conditions of air traffic. Only 80% of the seats were occupied by passengers, allowing for social distancing during the flight. The cabin crew wore personal protective equipment, while all passengers were obliged to use the supplied face coverings and gloves. Upon arrival, the travel couple took a free COVID-19 test. As their test results were negative, they could enter the hotel. The choice of their leisure activities also took social distancing rules into account as they mainly engaged in kayaking and cycling. In the final sequence of the video, Omar praised Dubai for welcoming tourists back and expressed his appreciation of the safety measures that were in place. The travel destination Dubai established safe touristic zones, in which intensive testing and strict social distancing rules significantly limited the spread of the coronavirus. The safe zone was widely portrayed as an idyllic micro-paradise for luxury experiences. Omar's and Amira's video amplified the newly emerging imaginary of the safe place, where holiday makers can escape their day-to-day troubles. The safe travel place has become a new embodiment of luxury. The video about the safe zone epitomises a new trend among travel influencers. Safety narratives are increasingly shaping representations of luxury travel and personal imaginings of luxury among platform users. A few days after the video was published on Instagram, Omar and Amira wrote a blog post. This piece contained background stories about their trip and served as a further entry point to their storyworld.

Imaginaries of extravagant indulgence in exquisite places have slowly been replaced by the imaginary of the safe place. Extravagant consumption of luxury goods was substituted by a desire for safety. A myriad of travel influencers underlined the safety of their first pandemic trips in their online travel stories. Safe zones can, however, become porous. Many newly established travel bubbles could not fully repel the coronavirus. For instance, the crews and passengers of a number of cruise ships fell victim to virus outbreaks after the industry had restarted their touristic journeys in June 2020. An analysis of the researched travel influencers indicates a change from imaginings of extravagant lifestyles to imaginings of safety. Luxury travel imaginaries have shifted from an orientation towards the performative consumption of luxury goods to an orientation towards infrastructures of travel safety. Accordingly, the self-presentations of travel influencers increasingly involved stories about safe travel places. They prioritised the circulation of information on safety rules and healthy travel routines, hoping to maintain their online status and their identities as influencers.

check-ins, regular temperature checks, social distancing signs, low-contact breakfast deliveries and sealed hotel room doors mushroomed on the Internet.

Conclusion

Prior to the coronavirus outbreak, travel influencers established themselves as new players in the global tourism industry. Although brand sponsorship has predominantly been considered the most substantial source of revenue for travel influencers, this in-depth investigation revealed alternative ways of securing a livelihood, including collaborations with local tourism professionals in travel destinations and the monetisation of travel content on YouTube. The researched travel influencers mainly responded in three ways to the current, long-term crisis in tourism. Firstly, one subgroup shifted their priorities as influencers and diversified their content creation. Secondly, others embraced local tourism and promoted staycations on digital platforms. Finally, numerous travel influencers envisioned themselves as ambassadors of safe travel. The recent platform practices of travel influencers have reshaped luxury travel imaginaries. The digital performance of extravaganza was widely replaced by images of safe zones. Over the last five years, the platform Instagram has enjoyed growing popularity within the global tourism industry, providing a new infrastructure for telling visual travel stories. The various components of luxury travel imaginaries have been reshuffled. Travel magazines, tour guides and travel TV series have increasingly been integrated into a new media ecology of global tourism, and the power relations between traditional and digital media have been reversed. The pandemic seems to rather accelerate than mitigate this process. Digital platforms became a major arena where the future of tourism has been re-negotiated in the wake of the COVID-19 outbreak. Based on an ethnographic investigation into travel influencers in the era of the global pandemic, I argue that travel influencers were in a position to create new powerful representations of luxury as safe travel since they acquired the skills to establish stable storyworlds for their travel experiences, which attracted the attention of large platform audiences.

The culture of influencers involves both gendered practices and gender stereotypes. Gender perspectives can shed further light on how gender is performed among travel influencers, unveiling gendered access to professional networks as well as empowerment and constraints of remote labour. Furthermore, tourism research faces new methodological challenges. While tourism is, in general, at risk of dwindling for a long time, the deteriorating effects of the COVID-19 outbreak on research practices complicate in-depth investigations into tourist sites. Given the present uncertainty for travel-related research, tourism scholars are likely to continuously integrate digital methods and remote co-presence into their research agendas. Tourism practitioners should align their managerial strategies with the ongoing platformization trend, of which the rise of travel influencers is but one example. The ethnographic investigation into responses to the coronavirus disaster points towards a further major change, of course. Concerns of travel safety for both transportation and accommodation will not disappear in the foreseeable future, and tourism practitioners shall establish appropriate guidelines for safe travel that can protect the public health in tourist places as well as local business interests. The coronavirus pandemic has already transformed work life, travel and digital cultures in unprecedented ways. An assessment of the early responses of travel influencers and other tourism professionals to the pandemic indicates a great versatility in the travel industry and an even greater demand for sustainable leisure travel. While the COVID-19 crisis is undoubtedly the greatest challenge tourist organisations have faced in decades, the hope of contributing to the re-birth of a world where sustainable tourism can flourish again has not been extinguished.

References

Born, G., & Haworth, C. (2017). Mixing it: Digital ethnography and online research methods—A tale of two global digital music genres. In L. Hjorth, H. Horst, A. Galloway, & G. Bell (Eds.), *The Routledge companion to digital ethnography* (pp. 96–112). New York, NY: Routledge.

Bryant, R., & Knight, D. (2019). *The anthropology of the future*. Cambridge: Cambridge University Press.

Carter, D. (2016). Hustle and brand: The sociotechnical shaping of influence. *Social Media + Society*, 1–12. doi:10.1177/2056305116666305

Corbin, J., & Strauss, A. (2008). *Basics of qualitative research: Techniques and procedures for developing grounded theory*. Thousand Oaks, CA: Sage.

Dixit, S. K. (2020). *The Routledge handbook of tourism experience management and marketing*. London: Routledge.

Gravari-Barbas, M., & Graburn, N. (2012). Tourist imaginaries. *Via, 1*. doi:10.4000/viatourism.1180

Gravari-Barbas, M., & Nelson, N. (2016). *Tourism imaginaries at the disciplinary crossroads: Place, practice, media*. New York, NY: Routledge.

Hull, E. (2020). Professionals. In F. Stein, S. Lazar, M. Candea, H. Diemberger, J. Robbins, A. Sanchez, & R. Stasch (Eds.), *The Cambridge encyclopedia of anthropology*. New York, NY: Cambridge University Press. doi:10.29164/20pros

Jacomy, M., Venturini, T., Heymann, S., & Bastian, M. (2014). ForceAtlas2, a continuous graph layout algorithm for handy network visualization designed for the Gephi software. *PLoS One*, *9*(6), e98679. doi:10.1371/journal.pone.0098679

Jenkins, R. (2006). *Convergence culture: Where old and new media collide*. New York, NY: NYU Press.

Katz, E., & Lazarsfeld, P. (2009). *Personal influence: The part played by people in the flow of mass communications*. New Brunswick, NJ: Transaction Publishers.

Khamis, S., Lawrence Ang, L., & Welling, R. (2016). Selfbranding, 'micro-celebrity' and the rise of social media influencers. *Celebrity Studies*, *8*(2), 191–208. doi:10.1080/19392397.2016.1218292

Kiik, L. (2019). Conservationland: Toward the anthropology of professionals in global nature conservation. *Critique of Anthropology*, *39*(4), 391–419. doi:10.1177/0308275X18821177

Lacan, J. (1936). Au-delà du "principe de réalité." In J. Lacan (Ed.), *Écrits* (pp. 73–92). Paris: Seuil. 1966.

Leite, N. (2014). Afterword: Locating imaginaries in the anthropology of tourism. In N. Salazar & N. Graburn (Eds.), *Tourism imaginaries: Anthropological approaches* (pp. 260–279). London: Berg.

Lomborg, S., & Bechmann, A. (2014). Using APIs for data collection on social media. *The Information Society*, *30*(4), 256–265. doi:10.1080/01972243.2014.915276

Lo, S., & McKercher, B. (2016). Beyond imaginary of place: Performing, imagining, and deceiving self through online tourist photography. In M. Gravari-Barbas & H. Graburn (Eds.), *Tourism imaginaries at the disciplinary crossroads: Place, practice, media* (pp. 231–260). Burlington, VT: Ashgate.

McCloskey, B., Dar, O., Zumla, A., & Heymann, D. (2014). Emerging infectious diseases and pandemic potential: Status quo and reducing risk of global spread. *The Lancet Infectious Diseases*, *14*(10), 1001–1010. doi:10.1016/S1473-3099(14)70846-1

Mosse, D. (2011). *Adventures in aidland: The anthropology of professionals in international development*. London: Berg.

Mostafanezhad, M., & Norum, R. (2019). The anthropocenic imaginary: Political ecologies of tourism in a geological epoch. *Journal of Sustainable Tourism*, *27*(4), 421–443. doi:10.1080/09669582.2018.1544252

Postill, J. (2017). Remote ethnography: Studying culture from afar. In L. Hjorth, H. Horst, A. Galloway, & G. Bell (Eds.), *The Routledge companion to digital ethnography* (pp. 61–69). New York: Routledge.

Salazar, N., & Graburn, N. (Eds.). (2014). *Tourism imaginaries: Anthropological approaches*. London: Berg.

Rogers, R. (2013). *Digital methods*. Cambridge, MA: MIT Press.

Scolari, C. (2009). Transmedia storytelling: Implicit consumers, narrative worlds, and branding in contemporary media production. *International Journal of Communication*, *3*, 586–606. Retrieved from http://ijoc.org/index.php/ijoc/article/view/477/336

Stoldt, R., Wellman, M., Ekdale, B., & Tully, M. (2019). Professionalising and profiting: The rise of intermediaries in the social media influencer industry. *Social Media + Society*, *5*(1), 1–11. doi:10.1177/2056305119832587

Sundbo, J., & Dixit, S. K. (2020). Conceptualisations in tourism experience. In S. K. Dixit (Ed.), *The Routledge handbook of tourism experience management and marketing* (pp. 15–26). London: Routledge.

Walton, S. (2018). Remote ethnography, digital co-presence: Exploring visual anthropological methods for research on the web. In C. Costa & J. Condie (Eds.), *Doing research in and on the digital: Research methods across fields of inquiry* (pp. 116–133). Oxford: Routledge.

Whateley, D. (2020) Data reveals how large the decline in paid Instagram influencer posts has been in recent weeks, as the ad market sputters. 1 May, (Lauchmetrics).

WTO. (2020). *World Tourism Barometer*. January 2020. Vol 18(1), 1–5. Madrid: UNWTO.

YouTube Creators. (2020). Intro to making money on YouTube. Retrieved from https://www.youtube.com/watch?time_continue=184andv=bIngfKyJyUwandfeature=emb_logo. Accessed on December 7, 2019.

Yuan, S., & Lou, C. (2020). How social media influencers foster relationships with followers: The roles of source credibility and fairness in parasocial relationship and product interest. *Journal of Interactive Advertising*, *20*(2), 133–147. doi:10.1080/15252019.2020.1769514

Chapter 23

Luxury Hospitality and Tourism during the Pandemic

Kuan-Huei Lee[1]

Abstract

The growth of luxury tourism has been brought to a grinding halt due to the COVID-19 pandemic affecting most parts of the world since early 2020. Although the time for recovery of hospitality and tourism industries is still unclear, experience from the SARS outbreak in 2003 showed that the bounce back from consumers could be fast. The group of most affluent consumers, mostly known as HNWIs (high net worth individuals), will resume their original consumption behaviour much sooner than the rest of the market; these affluent consumers are the main target market of luxury hospitality and tourism industries.

This chapter presents different types of luxury tourism, luxury lifestyle, luxury tourists' decision-making and luxury hospitality products. The luxury food and beverage business in Singapore is presented to illustrate the commercial environment during the pandemic.

Keywords: Luxury tourism; luxury hospitality; Michelin star; pandemic; food and beverage; Singapore

Introduction

In the past few decades, the segment of luxury hospitality and tourism consumers has increased rapidly, especially in luxury loading and dining. This segment is identified as high net worth individuals (HNWIs) who have sufficient disposable income to spend on tangible and intangible products to meet their wealth and social status. The luxury tourism segment is opposite to the mass tourism segment, where products and services are tailored to a niche group selling at a

[1]https://orcid.org/0000-0003-4722-109X

The Emerald Handbook of Luxury Management for Hospitality and Tourism, 463–477

doi:10.1108/978-1-83982-900-020211023

premium price compared to those targeting the mass tourism market. This chapter introduces luxury tourism, luxury lifestyle, the decision-making process of luxury tourists and, finally, luxury hospitality products. The case of Singapore's luxury food and beverage business during the pandemic is presented.

Luxury Tourism

Luxury represents everything that is unique and exclusive. The group of HNWIs is the main consumer in the luxury tourism segment. The market of luxury tourism is estimated to be worth US$2.05 trillion, covering not only hotels and restaurants but also arts and cultural fields around the world.

Luxury is usually associated with disposable income and with tangible items, the money that a person can dispose of after buying all the necessary goods for living. Luxury is also associated with uniqueness, exclusivity, prestige, status, indulgence, rarity and status. When the word is related to tourism and hospitality, it is the disposable income and the disposable time that one can spend in travelling or engaging in a specific experience that is not related to essential physiological need but to a much higher order of need. Recent association to luxury includes experiences, authenticity, sustainability and state of mind (Liu & Li, 2020).

The concept of luxury changes according to the person's lifestyle and mindset. More consumers are concerned about the environment, social responsibility and green consumption (Armoni, Ernawati, & Jendra, 2018). Nevertheless, the concept of luxury in tourism is far from mass tourism but closer to niche tourism, where tourists are aiming to explore uncommon experiences, visit exotic places or taste exclusive food.

There are different segmentations against tourists that exercise luxury tourism. In a study by Ni Made Rai, Ni Made, and Nyoman Mastiani (2018), tourists were classified into the Super Elite and the Classical Elite. While the Super Elite group require upper-class facilities and tangible things on travels, the Classic Elite group put more emphasis on value which it is more difficult to measure monetarily, for example, a secret hidden off the beaten path.

Swarbrook (2018) proposed another approach to segment the luxury market into four approaches: (1) Attitudes towards risk; (2) Attitudes to luxury consumption; (3) Disposable time available and (4) Personal motivators.

Attitudes towards risk are associated with political instability, terrorism, natural disaster, crime rate or disease that might happen with the consumption of a particular luxury product or service. *Attitudes to luxury consumption* are the preference for formal or informal settings; lodging in places with professional servers; formal or informal service; avoidance of sharing experience with others or not; exploring and interacting with people from the local community or avoiding it. *Personal motivators* are values, beliefs, ethical standards, status and eagerness for personal development that shape a person. *Disposable time available* refers to the time spent solely for leisure than the time needed to spend to maintain contact with the office or other official activities during leisure time.

In this approach, luxury tourists seek the sensation of feeling safe in a destination, how much spending they are willing to commit for possession of items, their personal values and desires toward different offerings and the available time to have the complete experience in the destination. The seeking of sensation is closely related to the lifestyle that the tourist is living or aspiring to have.

Luxury Lifestyle

Lifestyle is the pattern that an individual is living in a society. The lifestyle is related to the mode of living of a person, and the surrounding micro- and macro-environments influence this mode of living. According to the French sociologist Bourdieu (1990), lifestyle is linked to taste based on the notion of habitus. Behaviours are influenced by *habitus*, which is the accumulation of experience and exposure since an individual was born. The individual's choice is a product of habitus, which is the prior social and historical experiences that the person was exposed to. The luxury lifestyle, in this case, is influenced by the person's high net worth and decision-making.

A luxury lifestyle is associated with the freedom of behaviour and choices that one can make. The degree of freedom and the number of choices are again associated with the person's high net worth. With the freedom and the power of choices, a person can decide that kind of lifestyle to follow. The luxury lifestyle is also associated with Maslow's hierarchy of needs that explains the different priorities that one might need to fulfil before achieving the top set need, that is, the need for self-actualisation.

Psychographics studies are usually referred to as studies focusing on people's lifestyle choices. Psychographics include personality traits, interests, opinions, perceptions, values, needs and activities. It is widely used in market segmentation to systematically categorise consumers into different groups according to a psychographic variable (Bray, 2008).

There are several quantitative instruments developed based on psychographic variables. VALS2, RISC, 4C and LOV are some of the instruments developed to find out activities, interest and opinions of consumers by classifying them into a value map or attitude map. For instance, Wahlers and Etzel (1985) used AIO to study the relationship between preferences of holiday activities and individual stimulus' necessities. Shih (1986) employed VALS as a tool to assess tourist's trip experience to Pennsylvania.

Decision-making Process for Luxury Tourists

Every consumer needs to make the decision for different products or services. Different motivators and determinants influence the decision. These factors can be internal, from the consumer's own point of view, or external, influenced by the environment that the consumer is living in and interacting with (Decrop, 2010).

Since decision-making is a process, the consumer reaches the final choice after a series of consideration, evaluation and study of constraints.

In this process, a luxury tourist might ponder all the different alternatives in the market. For example, considering where to go for the next vacation, the tourist lists all possible destinations and evaluates all these possible destinations after considering those found in the evoked set. The evaluation will be determined by the tourist's internal and external motivators such as personality, preference and lifestyle. Destinations in the evoked set are assessed instinctively by the luxury tourist, the surrogate set contains destinations not prioritised by the tourist but saved as alternatives, and in the exclusion set are destinations absolutely excluded by the tourist. The evoked set would serve as the basis for the tourist to start pondering, and the final choice would arise from the available set that contains all destinations feasible after considering different constraints. In the Fits-Like-a-Glove (FLAG) choice model of decision-making, the individual makes a decision according to the past experience and the environment that he or she is experiencing (Lee, Scott, & Packer, 2015).

Both Practice theory and Micro-tipping Point (MTP) theory propose that an individual's decision-making is related to the conscious or unconscious mental condition of the individual at the moment of making the decision. Other theories, such as rational and constructive theories, suggest individuals ponder different considerations before making a decision.

Knowing the values, lifestyle and decision-making process of luxury tourists, the provider of these products and services will need to design the offerings to match expectations and beyond from the demand side.

Four factors influence the decision-making of a luxury tourist. The tourist is influenced by internal variables, external variables, situational and the marketing mix in the design of the luxury product (Swarbrooke, 2018).

Design of the Luxury Product

The supply side offers luxury products in the market formed by destination marketing organisers (DMOs), hotels, travel agent, airlines, cruise and restaurants. There are several features that the providers need to consider in order to offer the right luxury product to attract the correct segment in the market. The composition of the luxury products are pretty complex and comprises both tangible and intangible elements. In the case of a restaurant, tangible features are the dining room décor, tables, chairs and menu. Intangible features are elements such as the ambience and service of the waiter. Tangible features display directly and symbolise elements of luxury to consumers. The material and quality of each tangible item is reflecting the standard that luxury tourists are looking for. With only tangible items matching the standard is not enough unless intangible features are also meeting the standard required by this group of consumers.

The target market is referring to how to identify this special group of consumers that are willing to spend on luxury products. Market segmentation techniques can be used by segmenting based on geographic, demographic or psychographic variables. Geographic segmentation allocates consumers according to the location where they live, work or hand out. Demographic segmentation allocates consumers according to each individual profile such as gender, age, ethnic group and income level. Psychographic segmentation allocates consumers according to their interest and preference, such as the sport they like and the food they consume. Luxury tourists generally have a higher high net worth compared to the average and have higher expectations in meeting their demands.

Price is usually the least of the concerns for this special group of consumers if they find products that can meet their need. This group is less sensitive to the price, which has lower price elasticity compared to the mass market; the change in price leads to a smaller change in the demand quantity. Different pricing strategies include cost-based pricing, demand-based pricing, experience curve pricing, psychological pricing, skimming pricing and competition-oriented pricing. When setting up the pricing strategy, the most important aspect to consider is the ability to pay for the target market segment.

Reputation can only build over time, maintaining a particular characteristic that is strongly linked with the company's name. Reputation is essential in the decision-making process of luxury consumers who will rely on past experience and reviews from different sources. With the rapid development of technology, it is not difficult to find published opinions about a certain product from the Internet. Especially with this niche group of luxury consumers, word of mouth (WoM) from people they trust is important and will influence their final choice of purchasing.

The channel selected to approach the group of luxury consumers is key to convey the right message to them. Promotion and communication deal with the selection of the right channel and the design of the message. It is important to study what form of news this group of more affluent consumers read and where they frequent to send the message using the right tool at the right time. Do this group of HNWIs use Twitter, Instagram, Facebook or read an e-Newspaper? Which magazines do they read? If there is a promotion about luxury hotel discount activity, which is the right channel to choose?

Values are things that one believes are important and dictate one's judgement and behaviour in life. Luxury consumers have individual and social luxury values (Loureiro & de Araújo, 2014). Individual values are materialism, hedonism and self-identity. Social values are related to individual issues, desires and want that manifest through social position and status (Ajitha & Sivakumar, 2017). The image that the individual wants other people to see him or her.

Different Types of Hotels

Hotel groups categorise their hotels into different brands according to the target market. According to the STR Chain Scale, which group hotels primarily according to actual average room rates. The Chain Scale segments are: Luxury, Upper Upscale, Upscale, Upper Midscale, Upper Midscale, Midscale, Economy and Independent. According to the hotel industry in Singapore, there are five categories: Luxury, Upper Upscale, Upscale, Midscale and Budget. The highest category for these scales is the Luxury segment.

Luxury Hotels' daily room rate is the top 15% among all hotels as these hotels offer an exclusive experience, first-class bedrooms and quality amenities. Nevertheless, the definition of luxury has changed over the years with the increase of well-informed travellers seeking value and peculiarly curated service offerings. Travel advisors pointed out that hotels are required to fulfil certain criteria to be considered a luxury hotel. These criteria include the quality of room and bathroom, plenty of amenities and facilities such as fitness centre and pool, as well as round-the-clock room service.

Another way of classification of hotels is using a star rating system from one-star to the highest of five-star classification. Five-star hotels are considered luxury hotels and it is based on pricing, location, type of visitors, service offering and facilities. Different countries might adopt some variations in the classification of stars. Table 23.1 illustrates the European Hotelstar's Union classification of hotels.

Luxury Restaurants

Luxury in gastronomy is associated with fine dining and an exclusive food experience that is unusual in the common marketplace. Luxury restaurants are frequently visited by HNWIs that offers full table service, dedicated atmosphere and high-quality menus. Many of these restaurants are run by well-known celebrity chefs. One expression of luxury is the Michelin Guide for restaurants.

Like hotels use stars to rank different categories, the restaurant business, as per the famous Michelin Guide, also awards a star to restaurants according to the different levels of services and offerings from one to three stars. The Michelin Guide for restaurants was introduced in the 1900s by car tire manufacturer's

Table 23.1. European Hotelstar's Union Classification.

Star	Category
★	Tourist
★S	Superior Tourist
★★	Standard
★★S	Superior Standard
★★★	Comfort
★★★S	Superior Comfort
★★★★	First Class
★★★★S	First Class Superior
★★★★★	Luxury
★★★★★S	Superior Luxury

Source: Adapted from Hotelstars (2020).

company Michelin Tyres in France, and the intention was to promote more usage of cars with a guide targeting motorists covering maps, a list of mechanics, petrol stations, hotels and restaurants. The categorisation of the star was introduced in 1926 with only one single star recommendation for fine dining establishments. Later in 1931, two and three stars were introduced. Stars are awarded by under-covered inspectors that consumed at these restaurants without revealing the time of the visit and their identities.

Once an inspector is appointed by Michelin, he or she will spend up nine to 18 months pairing up with a senior inspector to learn the process. Michelin chooses these inspectors among thousands of resumés that the company receives yearly. The group of inspectors are composed of different nationalities who will need to follow the following criteria when reviewing a restaurant: (1) quality of the ingredients used; (2) mastery of flavour and cooking techniques; (3) the personality of the chef reflected in the cuisine; (4) value for money and (5) consistency between visits (Michelin Guide, 2020).

To be awarded one Michelin star is considered a very good restaurant in its category, the two-star restaurant is excellent in cooking, worth a detour, and three stars is an exceptional cuisine, worth a special journey. Michelin Guide covers about 30 countries/cities across the Western and Eastern parts of the world. Starting from 2014, gastropubs was introduced in a separate list. In 2016, two street food stalls in Singapore were awarded for the first time with one star. A special interest group travels around the world to taste different restaurants awarded with Michelin stars that are called star hunters.

In 1955, Michelin Guide introduced the Michelin Bib Gourmand category of recognition. This category awards the least expensive restaurants with good quality at a value price. To be considered in this category, the meal with two courses, one glass of wine and dessert cannot exceed US$40 per person.

Michelin Plate or L'Assiette is the youngest category of recognition introduced in the 2018 edition. Restaurants that cannot meet the three-Star or Bib Gourmand recognition are usually much less expensive, which in this category include many 'street food' eateries which were recognised for their quality of food.

A longitudinal study of 35 Michelin one- to three-star restaurants in France by Batat (2020) showed that chefs are important change makers; there are four activities in these luxury restaurants to enhance visitor's attractiveness to the venue: (1) the design of unique gastronomic experience; (2) the promotion of terroir products and rural food tourism; (3) the gastronomisation of rural destinations and (4) the augmentation of the media capital of the destination.

The Slow Food Movement

The Slow Food movement might be associated with luxury food consumption if 'time' is considered a precious element to enjoy the eating experience. Researchers found Slow Food members are usually white-collar, middle to high social class, namely High Net worth, since the cost of participating in Slow Food activities is considered high (Gaytán, 2004). The speed of modern life is faster and faster, with advancing technologies changing our daily lives. The philosophy of Slow Food is embracing a lifestyle that is more aware of the surrounding environment with factors of soil, climate and topography in mind.

The founder of the movement Carlo Petrini emphasised that Slow Food is seeking to repair damages done to the environment, to protect and sustain the surroundings, animals, culture and people (Slowfood.com, 2020). The movement encourages consumption of local produce, establishes linkage between the land and the food, creates food memory and teaches people to learn how to appreciate local food. Slow Food's central philosophy is 'good, clean and fair'. 'Good food is defined as being tasty and diverse and produced in such a way as to maximize its flavour and connections to a geographic and cultural region. Clean food is sustainable, helping to preserve rather than destroy the environment, while fair food is produced in socially sustainable ways that emphasize social justice and fair wages' (Schneider, 2008, p. 390).

Carlo Petrini is a strong believer that a possible solution to the current global food crisis is to prioritise subsistence agriculture instead of export, and by following the central philosophy of 'good, clean and fair' in daily food consumption (Petrini, 2007). By following the above, Petrini is convinced that local economics would grow, get stronger and favour small-scale local production and food security. In addition, it shows respect for the environment and local traditions (Petrini, 2007).

Case Study 23.1: The Influence of Covid-19 in Luxury Food and Beverage Industry in Singapore

The luxury industry sector was badly affected by the coronavirus pandemic that spread across most parts of developed countries in early 2020. With most countries implementing lockdown measures of entry to borders as well as the elimination of all dining in activities in restaurants and eateries. In Singapore, different degree measures were implemented from February 2020, with the 'circuit breaker' implemented from 21 April to 1 June, where only essential services were opening. In the food and beverage businesses, only takeaway was allowed for all eateries, no dine-in services. Many restaurants decided to close completely during this period of time. The measure of dining in was lifted after 19 June, allowing a small group of five people and safety measures were implemented across all restaurants in business.

There are about 12,000 restaurants in Singapore, contributing 0.8% of GDP (Statistics, 2019). The food and beverage market was worth over S$8.8 million before Covid-19. The compound annual growth rate (CAGR) was estimated to be 2.1% until 2023. Due to the pandemic starting in early 2020, 93% of restaurants in Singapore suffered a decline in revenue as much as 80%, and many were closing permanently after the pandemic (Statista, 2020).

Many luxury restaurants, including Michelin, awarded dine-in restaurants that have chosen to expand their takeaway services by using food delivery services for the first time, such as GrabFood, Food Panda or Deliveroo. For GrabFood delivery, an additional icon was created within the application called Michelin-recommended listing restaurants and eateries awarded or recommended by the Michelin Guide Singapore 2019 ranging from Michelin Plate, Michelin Bib Gourmand, one and two Michelin Stars (Table 23.2).

Table 23.2. List of Restaurants Recommended by Michelin Guide Using a Delivery Platform.

Restaurant/Eatery	Type of Cuisine	Michelin Recommended as
Buki Nero	Italian	Michelin Plate
The 1950's Coffee, Chinatown	Café	Michelin Plate
Boon Tong Kee Kway Chap	Asian/Local	Michelin Plate
Gaig Restaurant	Spanish	Michelin Plate
Golden Peony	Asian/Local	
The Song of India	Indian/Vegetarian	1 Michelin Star
FOC Restaurant	Chicken, Vegetarian	Michelin Plate
Beni Singapore	French/Japanese	1 Michelin Star
Restaurant Zen	Western/French	2 Michelin Star
Alma by Juan Amador	Western	1 Michelin Star
To-Ricos Guo Shi	Chinese/Local	Michelin Bib Gourmand
Esquina	Spanish	Michelin Plate
R&B Express	BBQ/Asian/ Chicken	Michelin Plate
Maguro-Donya	Japanese/Local	Michelin Plate
Famous Kitchen	Asian/Local	Michelin Plate
Cheek Bistro	Werstern	1 Michelin Star
Gunther's	French	Michelin Plate
Sik Bao Sin-Desmond's Creation	Chínese	Michelin Bib Gourmand
Summer Pavilion, The Ritz Carlton	Fine Dinning	1 Michelin Star
Fleur De Sel Le Restaurant	French/Western	Michelin Plate
Yen Yakiniku	Japanese	Michelin Plate
Imperial Treasure Super Peking Duck	Chinese	Michelin Plate

Table 23.2. *(Continued)*

Restaurant/Eatery	**Type of Cuisine**	**Michelin Recommended as**
Hua Kee Hougang Famous Wan Ton Mee	Chinese/Noodles	Michelin Plate
Sin Ho Sai	Local/Seafood	Michelin Plate
L'Angelus	French	Michelin Plate
Yan Ting – The St Regis Singapore	Chinese	Michelin Plate
Lagnaa Barefoot Dining	Indian	Michelin Bib Gourmand
Min Jiang – Goodwood Park Hotel	Chinese	Michelin Plate
Nouri	Western/Pasta	1 Michelin Star
A Noodle Story	Chinese/Local	Michelin Bib Gourmand
Zaffron Kitchen	Indian/Asian	Michelin Bib Gourmand
Madame Fan	Chinese	Michelin Plate
The Blue Ginger Restaurant	Asian/Peranakan	Michelin Bib Gourmand
Imperial Treasure Fine Shanghai Cuisine	Chinese/Asian	Michelin Plate
Indocafe	Asian/Peranakan	Michelin Bib Gourmand
Summer Palace – Regent Singapore	Asian	1 Michelin Star
Nana Curry	Curry/Chinese	Michelin Bib Gourmand
Garibaldi Italian Restaurant & Bar	Italian/Pasta	1 Michelin Star
Min Jiang at Dempsey	Chinese/Seafood	Michelin Plate
Hua Ting Restaurant	Chinese/Asian/ Seafood	Michelin Plate
Hjh Maimunah	Local/Halal/ Chicken	Michelin Bib Gourmand
Chen's Mapo Tofu	Chinese/Asian	Michelin Bib Gourmand

Table 23.2. *(Continued)*

Restaurant/Eatery	Type of Cuisine	Michelin Recommended as
Man Man	Japanese/Seafood/ Asian	Michelin Bib Gourmand
Geylang Famous Beef Kway Teow	Chinese/Noodle/ Local	Michelin Plate
Kok Sen Restaurant	Chinese/Asian	Michelin Bib Gourmand
Muthu's Curry	Indian/Chicken	Michelin Bib Gourmand
Bismillah Biryani	Indian/Chicken	Michelin Bib Gourmand
Song Fa Bak Kut Teh	Chinese/Local	Michelin Bib Gourmand
Akira Back	Japanese/Fusion	Michelin Plate
Anglo Indian Café & Bar	Indian	Michelin Bib Gourmand
Putien	Asian/Noodles	1 Michelin Star
Man Fu Yuan – Hotel Intercontinental	Asian/Chinese/ Local	Michelin Plate
91 Fried Kway Teow Mee	Chinese/Noodles	Michelin Plate
Balestier Road Hoover Rojak	Local	Michelin Bib Gourmand
Rolina Traditional Hainanese Curry Puffs	Local	Michelin Plate
Sheng Seng Fried Prawn Noodles	Asian/Local/ Noodles	Michelin Plate
Koh Brother Pig's Organ Soup	Local/Chinese	Michelin Bib Gourmand
Depot Road Zhen Shan Mei Claypot Laksa	Local	Michelin Bib Gourmand
Hong Peng La Mian Xiao Long Bao	Chinese/Noodles	Michelin Plate
Sin Kee Chicken Rice	Local/Chinese	Michelin Bib Gourmand

Source: GrabFood (2020).

Safety measures were implemented after circuit breaker, temperature taking, safe distancing, tracing app, regular sanitation and mask-wearing were applied in every food and beverage establishment since 19 June 2020 when restaurants were allowed to open for dine-in services. A luxury restaurant such as one Michelin Star restaurant Burnt Ends was fully booked for weeks right after the lifting was announced.

Buffet restaurants in luxury hotels have changed to à la carte ordering, and the change also applied to the international steamboat chain Haidilao which started operation opening seven outlets fully booked on weekends. The free condiment area for customers to choose freely now can only be prepared by restaurant staff. Food can only be shared on the table using communal servicing utensils provided (Lalley, 2020).

Customers are required to check into the Government's SafeEntry system using QR-code to declare their entry to the different business outlets and read contactless menus on their mobile devices to lower possible contamination through hard copy menus. Every restaurant entrance has posters and signs reminding customers to check in; all tables in the restaurant tables are situated at least one metre from each other with a maximum of five chairs per group (Fig. 23.1).

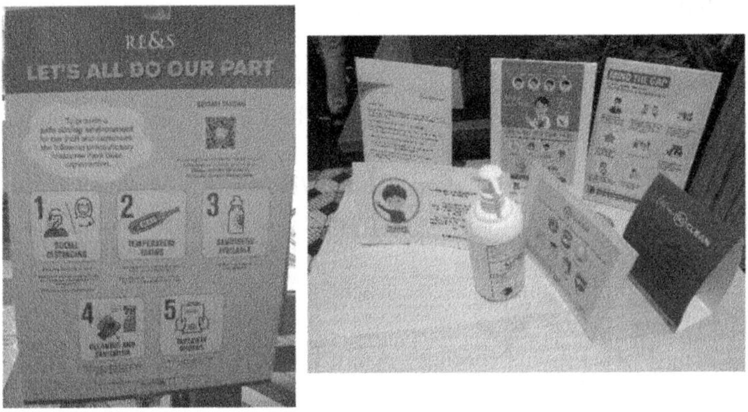

Fig. 23.1. New Steps Implemented After the Covid-19 Pandemic Adopted by the F&B Industry in Singapore. *Source:* The Author, 2020.

The iconic hotel Marina Bay Sands (MBS) re-opened most of the businesses, including all restaurants operated by MBS, on 17 July 2020 after three months of closure. Restaurants operated under MBS are Black Tap, Bread Street Kitchen, CUT, Mott 32, Yardbird, KOMA and db Bistro and Oster Bar, but require that customers make a reservation first. MBS is using a developed entry system that allows capturing of live updates of the entire population across different properties. The system is in sync with Singapore Government's Safe Entry system for contact tracing purpose.

The process during this period of time to enter any dine-in restaurant is as follows:

(1) Safe distancing of at least one metre between tables and seat configurations of up to five customers per table
(2) Incorporation of SafeEntry QR-Code
(3) Health screening upon entry with temperature checks and sanitisation of hands
(4) Mandatory wearing of masks always with the exception of eating and drinking
(5) Digital menu via a personalised QR Code at each table
(6) Salt and pepper pots, as well as any promotional flyers, are removed from each table

Conclusion

Luxury tourists are individuals with high net worth who can afford to spend on niche products and services and pay a premium for them. Many academics have classified this group into different segments according to their disposable income, time and attitude towards risk and motivation. The living style of the high net worth group is related to their belief and value, which will influence their decision-making when purchasing a product or acquiring a certain service. There are different luxury categories in the two main products offered by the hospitality industry: the hotel and the restaurant sector. Luxury hotels are usually classified by star rating, with the highest number of stars to be the most luxury one. Similarly, for restaurants, the Michelin Guide uses a star to classify restaurants awarding a maximum of three Michelin stars. The Slow Food movement is often associated with luxury food consumption if 'time' is taken into consideration as an important element.

The impact of Covid-19 in luxury food establishments was discussed using Singapore as an example. Although there was a decrease in in-store consumption during most of 2020, the increase in delivery for the luxury restaurant was notorious across all luxury restaurants, with many of them engaging in delivery services for the first time.

The future of luxury hospitality and tourism is linked with the number of HNWIs and their availability to spend freely. Behavioural changes in consumption of luxury products during the pandemic period will need to be monitored to understand whether these changes are temporary or permanent.

References

Ajitha, S., & Sivakumar, V. (2017). Understanding the effect of personal and social value on attitude and usage behavior of luxury cosmetic brands. *Journal of Retailing and Consumer Services*, *39*, 103–113.

Armoni, N. L. E., Ernawati, N. M., & Jendra, I. W. (2018). A service perspective of luxury tourism. *International Journal of Applied Sciences in Tourism and Events*, *2*(2), 118–124. doi:10.31940/ijaste.v2i2.1177

Batat, W. (2020). The role of luxury gastronomy in culinary tourism: An ethnographic study of Michelin-Starred restaurants in France. *International Journal of Tourism Research*, *23*(2), 150–163. doi:10.1002/jtr.2372

Bourdieu, P. (1990). *The logic of practice*. Cambridge, MA: Polity Press.

Bray, J. P. (2008). *Consumer behaviour theory: Approaches and models*. Discussion Paper. Unpublished.

Decrop, A. (2010). Destination choice sets: An inductive longitudinal approach. *Annals of Tourism Research*, *37*(1), 93–115. doi:10.1016/j.annals.2009.08.002

Gaytán, M. S. (2004). Globalizing resistance Slow Food and new local imaginaries. *Food, Culture and Society: An International Journal of Multidisciplinary Research*, *7*(2), 97–116. doi:10.2752/155280104786577842

GrabFood. (2020). GrabFood restaurants. Retrieved from https://food.grab.com/sg/en/restaurants

Hotelstars. (2020). European Hotelstar's Union classification. Retrieved from https://wwwhotelstars.eu/

Lalley, H. (2020). How will the COVID-19 crisis change consumer dining behavior? Retrieved from https://www.restaurantbusinessonline.com/consumer-trends/how-will-covid-19-crisis-change-consumer-dining-behavior. Accessed on June 29, 2020.

Lee, K. H., Scott, N., & Packer, J. (2015). The Fits-Like-A-Glove model and destination activities of Slow Food members. *Current Issues in Tourism*, *18*(3), 286–290. doi:10.1080/13683500.2014.909389

Liu, T., & Li, M. (2020). Leisure and travel as class signifier: Distinction practices of China's new rich. *Tourism Management Perspectives*, *33*, 100627. doi:10.1016/j.tmp.2019.100627

Loureiro, S. M. C., & de Araújo, C. M. B. (2014). Luxury values and experience as drivers for consumers to recommend and pay more. *Journal of Retailing and Consumer Services*, *21*(3), 394–400.

Michelin Guide. (2020). What the Michelin guide's symbols mean. Retrieved from https://guide.michelin.com/th/en/to-the-stars-and-beyond-th. Accessed on June 5, 2020.

Ni Made Rai, S., Ni Made, E., & Nyoman Mastiani, N. (2018). Luxury tourism: A perspective of facilities and amenities. *International Journal of Applied Sciences in Tourism and Events*, *2*(1), 32. doi:10.31940/ijaste.v2i1.900

Petrini, C. (2007). *Slow food nation: Why our food should be good, clean and fair*. New York, NY: Rizzoli Ex Libris.

Schneider, S. (2008). Good, clean, fair: The rhetoric of the slow food movement. *College English*, *70*(4), 384–402.

Shih, D. (1986). VALS as a tool of tourism market research: The Pennsylvania experience. *Journal of Travel Research*, *24*(4), 2–11.

Slowfood.com. (2020). Slow food international. Retrieved from http://www.slowfood.com/. Accessed on June 21, 2020.

Statista. (2020). Food & Beverage Singapore. Retrieved from https://www.statista.com/outlook/253/124/food-beverages/singapore

Statistics. (2019). Operating expenditure for food and beverage services, annual. Retrieved from https://data.gov.sg/dataset?q=food+and+beverage

Swarbrooke, J. (2018). *The meaning of luxury in tourism, hospitality and events*. Oxford: Goodfellow Publishers Ltd.

Wahlers, R. G., & Etzel, M. J. (1985). Vacation preference as a manifestation of optimal stimulation and lifestyle experience. *Journal of Leisure Research*, *17*(4), 283–295.

Chapter 24

The Pandemic of Tourism: How Tourism Has Become an Unsustainable Luxury

Clare Hindley,[1] Willy Legrand,[2] and Gabriel C.M. Laeis[3]

Abstract

This chapter aims to establish the relation of luxury tourism to sustainability and questions whether tourism in its current form is not itself a luxury. By analysing consumer travel motivation and demands of luxury tourism, we examine the impact of these perceptions and ask whether Anthropocene tourism does not by definition have a negative impact on the environment. A new concept of luxury has developed clearly illustrated by a move from Maslow's (1943) 'esteem' to the top tier of 'self-actualisation' as reflected in Pearce and Lee (2005) Travel Career Ladder and top tier of personal fulfilment. This move has led to a decline in physical trophy collection, but rather the desire for luxury is taking on a new definition more about a perception of environmental connection, personal fulfilment and finding a brand or experience that shares similar values to the consumer. The commodification of nature has led to new forms of tourism concentrating on connecting to places, people and causes. An analysis of tourism growth impact in the Global North and South, and neo-colonisation in tourism highlights the contradictions within sustainable goals and tourism. It is increasingly difficult to categorise tourism as sustainable or unsustainable, luxury or non-luxury, but rather this chapter questions whether tourism itself has become an unsustainable pandemic and an indefensible luxury.

Keywords: Luxury tourism; extravagance; Anthropocene tourism; Global South; sustainability; value perception

[1]https://orcid.org/0000-0002-4887-8803
[2]https://orcid.org/0000-0003-0637-3265
[3]https://orcid.org/0000-0003-3413-6419

The Emerald Handbook of Luxury Management for Hospitality and Tourism, 479–496

doi:10.1108/978-1-83982-900-020211024

Introduction

Luxury and extravagance have historically been seen as both a worthy aim and a crime. In 1559, Duke Frederick III was deposed and imprisoned for extravagance and oppression (Baring-Gould, 2020). Similarly, Ludwig II of Bavaria was deposed and declared insane partly due to extravagant luxury (Steinberg & Falkai, 2020). Today, one cannot deny a certain admiration of that extravagance. Ludwig's fairytale castle, *Schloss Neuschwanstein*, attracts 1.5 million visitors yearly and is thus the most popular excursion destination in Bavaria (Schloss, 2019). The display of opulence, luxury and extravagance are a magnet with tourists flocking to similar sites around the globe. Defining luxury as 'something expensive that is pleasant to have but is not necessary' (Cambridge, 2020a, 2020b) and extravagance as 'lacking restraint in spending money or using resources' (Lexico, 2020) whilst leaving 'resources' open to interpretation fits not only the examples above but also the modern-day consumer and traveller. Park, Kim, Kwak and Wyer, discussing the implications of extravagance for consumer decisions about luxury products, argue that 'people are often extravagant; that is, they often spend a considerable amount of money for luxury products that are objectively unnecessary for their well-being' (2013, p. 42). Buying strawberries out of season may be considered extravagant but equally a luxurious endeavour. Both concepts, extravagance and luxury, share similar features and contradictions and are explored here within the context of tourism and the pursuit of greater sustainability. This chapter questions, through an analysis of the development of both luxury and tourism and their impacts, whether tourism itself has become unsustainable and an indefensible luxury.

Luxury and Extravagance Metanarrative and the Tourism Industry

A Historical Perspective

Travail is often forgotten as the origin of the word 'travel' and the fact that 'travelling meant engaging in an activity that was troublesome and laborious' and commonly used in French to define 'work' (Boorstin, 1987 in Cohen, 2002, p. 31). Boorstin (1987) bewails the demise of the historical traveller who 'was active; [and] travelling required much planning, time and money' (in Cohen, 2002, p. 31). Ancient Romans were possibly the first to travel for pleasure with travel on roads relatively safe (Cohen, 2002) and a variety of villas or estate lodgings available to the wealthy to stay the night. For the less wealthy, there were strategically placed inns available. Travel deteriorated during the Dark Ages due to the decline of the transportation infrastructure only to pick up during medieval pilgrimage (Cohen, 2002). The extravagance of travel took a particular turn during the seventeenth and eighteenth centuries as the Grand Tour was established as a method 'to educate young rich English men to become future leaders' (Cohen, 2002, p. 32). The Grand Tour soon became very popular with estimations of 'over 40,000 Englishmen on the Continent at one point in time' (Cohen, 2002, p. 32).

Historical snapshots of luxury tourism conjure up images of the Grand Tour, a horse and carriage leisurely journeying across Europe, experiencing a wealth of culture and languages with all the romance and nostalgia associated. Those who took part in the Grand Tour were an elite group funded by seemingly unlimited financial means and time, with an ample supply of useful connections and forming a small minority of the population. From a twenty-first-century perspective, the Grand Tour may seem a luxury, but in fact, by the late eighteenth century, the 'Grand Tourists were looked upon sarcastically' (Cohen, 2002, p. 32), similar to today's European tours with over-packed itineraries based on 'a city a day' concept. Grand Tourists 'were referred to as members of the Macaroni Club' whereby Italy was considered a *must-see* destination within the journey (Cohen, 2002). Nevertheless, the eighteenth-century Grand Tours of Europe with the time and financial means they implicitly entailed do provide an example of people travelling for leisure purposes and were certainly, at least initially, viewed as an enviable undertaking. The luxury and extravagance involved was primarily time and money. Having the time and financial means to travel was an enviable privilege. Potentially this can be seen as a change in public perspective. The once admired extravagance turns into an unnecessary luxury reflected in the 'Macaroni Club'.

Throughout history, most travellers have tried to maximise their comfort with travel declining or even ceasing in times and places when dangers are high, and increasing when risks are lower. This demand spurs the development of tourist infrastructure, which further increases convenience and comfort and further increases demand. 'It appears that the contemporary tourist is not so much a new type of animal, but rather a decision-maker with a new array of options' (Cohen, 2002).

Luxury and Extravagance: Two Sides of the Same Coin

The basic premise of the theory of Conspicuous Consumption by Veblen provides insights into the concept of luxury whereby wealthy individuals tend to consume luxury goods and services to enhance their social standing (1899). Veblen argues that 'the consumption of luxuries, in the true sense, is a consumption directed to the comfort of the consumer himself' (Veblen, 1899, p. 35). So luxury implies status and consumption, but not solely (Yeoman & McMahon-Beattie, 2006). Kapferer, Klippert, and Leproux describe luxury as 'rare, hedonic objects and experiences beyond the necessities of life, therefore affordable mostly to those who have surplus money' (2014, p. 2). However, the authors also discuss the lack of agreement in defining the term, its subjectivity in conceptualisation as well as the challenges linked to the understanding of luxury depending on the role one takes. As the creator of *luxury*, one may focus on defining criteria such as 'exceptional quality, craftsmanship, handmade, rarity, noble ingredients, maintaining tradition' (2014, p. 2).

Conversely, as a buyer, the focus turns to concepts such as exclusivity, experiences, authenticity and hedonism (Dixit, 2020; Kapferer et al., 2014; Yeoman &

McMahon-Beattie, 2006). Besides the luxury creators and luxury buyers, there is yet a third category called the non-buyers who then 'equate luxury with conspicuousness, excess, waste' (Kapferer et al., 2014, p. 3). This offers a closer link to the concept of extravagance discussed above, which is defined as a general lack of restraint in expenditure or resource consumption. Luxury goods 'often possess extravagance related features that are aesthetically appealing' (Park et al., 2013, p. 42). However, as Davidson pointed out over a century ago, 'extravagance is not necessarily luxury' (1898, p. 57). In an article published in the *International Journal of Ethics*, Davidson argues that 'the only justification for the possession of great wealth is its wise and rational use; and extravagance is not a rational use' (1898, p. 58). Davidson argues that

> ...luxury is but the worst and most aggravated and aggravating form of extravagance where wealth is not transferred, but used up in the form of costly dishes and expensive wines, in rare exotics and splendid firework.
>
> (1898, p. 58)

In this sense, both luxury and extravagance are two sides of the same coin.

Luxury and Tourism: A Blurred Genre

Veblen discusses the 'Gentleman of Leisure' who can afford a non-productive consumption of time. And thus, beyond goods and services, luxury is understood as 'time [that] is consumed non-productively (1) from a sense of the unworthiness of productive work, and (2) as an evidence of pecuniary ability to afford a life of idleness' (1899, pp. 21–22). Bernstein revisited the concept of luxury within the hotel context and stated that 'perhaps the most fundamental definition of luxury is that it's a form of waste' (Bernstein, 1999, p. 48). Bernstein (1999), along similar lines to Veblen's idea of the Gentleman of Leisure, argues that both *time* and *prestige* are defining components of luxury. And while conspicuous consumption may have been the realm of the wealthy in the past, Yeoman and McMahon-Beattie argue that with 'increased affluence, luxury is a blurred genre which is no longer the preserve of the elite' (2006, p. 320). This is not to say that luxury is not about material consumption, but the past two decades have seen a 'cultural shift for personal fulfilment and aspiration through experience' (Yeoman & McMahon-Beattie, 2006, p. 320). This is particularly the case in tourism. Luxury has evolved to the top tier of 'self-actualisation' (Maslow, 1970), as reflected in Pearce's Travel Career Ladder and the top tier of personal fulfilment (Pearce & Lee, 2005).

This cultural shift has led to a decline in physical trophy collection. Luxury is more about the perception of environmental connection, personal fulfilment and finding a brand or experience that shares similar values to the consumer. This is accompanied by a search for both unique and personalised experiences. Consumers attach more importance to experiences compared to goods and services

(Pine & Gilmore, 2014). Hotels play a particular role enabling this shift 'through attention to detail and assisting in the efficient execution of functional activities, [they] can seemingly provide the time for the guest to enjoy an intrinsic luxury experience' (Bernstein, 1999, p. 50). Luxury in hotels also implies quality, aesthetics, excellence in services and high prices (Slattery and Games, 2010; Swarbrooke, 2018; Yang & Mattila, 2016). However, here too, we are experiencing a general lack of agreement on a precise definition of the concept of 'luxury hospitality' or 'luxury tourism' (Davidson, Guilding, & Timo, 2006). Bernstein (1999) argues that luxury in hospitality is often based and built on exquisite service in outstanding establishments. Luxurious hotel properties constitute a cornerstone of luxury tourism (Lee, 2010). Thus, luxury includes dimensions of tangibility (e.g. goods of quality) and intangibility (e.g. availability of time, unique services) as well as impacting the social status of luxury buyers (Daswani & Jain, 2011; Veblen, 1899). Increasingly, luxury consumption is also linked to various emotional factors such as 'pleasure', 'self-respect', 'achievement' or 'fulfilment' (Kapferer, 1997; Kapferer & Bastien, 2012; Lipovetsky & Charles, 2005; Seo & Buchanan-Oliver, 2015). Rather than considering the notion of luxury tourism, the more interesting question to ask is whether tourism itself is a luxury?

Luxury Tourism: The Twenty-first-century Tourist

The concept of pleasure with the least possible discomfort is core to tourism, as discussed previously. The question arises as to how much tourism itself is an extravagance and how much the term can be applied exclusively or at least increasingly to the twenty-first-century traveller.

The expansion of the European railways changed travel drastically from the days of the Grand Tour. The railways meant less time was needed to travel, and shorter, more frequent trips became the precursors of today's city breaks. The extravagance of time was no longer a vital part of extravagant travel. In many cases, the type of railway carriage provided luxury rather than the time that is taken to travel. The money available allowed Grand Tourists to travel for up to four years. However, money, in contemporary travel, is no longer necessary for extravagance. A weekend in Rome, a one-night stay in Budapest or a booze trip to Mallorca are all available at prices a vast number can afford without having to make significant savings elsewhere. This naturally leads to a weekend in Rome becoming the norm rather than an extravagance: a small luxury available to the masses. Yeoman and McMahon-Beattie (2006) refer to the 'luxurification' (2006, p. 321) in a society where increased access to discretionary income allows consumer spending on leisure and travel. Consequently, looking at the elements of 'extravagance', we can see how the word itself has changed, perhaps not in the dictionary definition but certainly in perspective.

A look at what is on offer in the field of tourism confronts us with newspaper articles headlining 'Is space tourism an extravagant idea?' (Guessoum, 2016). Guessoum puts forward one justification of space tourism, a sort of twenty-first-century Grand Tour, arguing that 'Humans have always been drawn to explore and go "where no man has gone before". People want to experience everything that can be experienced' (2016, para 9). Space travel has become more competitive as the ultra-rich look for new ways to part with their money (Pratley, 2019). Scroll forward to 2020, and we see the urge to go where no man has gone before has become more difficult for two major reasons. Firstly, pre-COVID-19, the travel industry boomed with record numbers of 1.5 billion international arrivals in 2019 (UNWTO, 2020a), and bucket lists were more easily achieved. Secondly, the travel industry came to a near-total halt for a few months during the 2020 COVID-19 crisis. With grounded travellers across the globe, many destinations experienced a breather from years of over-tourism. Reaching those destinations is difficult when unprecedented measures are put in place by national governments to mitigate the spread of the coronavirus, such as travel restrictions, border closures and lockdowns. However, the travel industry has traditionally shown resilience when faced with catastrophes of any sort, and thus, tourism is set to bounce back at all costs, financial, social and environmental.

Redefining Luxury in Tourism: The Anthropocene Extravagance

Following months of lockdowns during the COVID-19 health crisis, many will want to leave their four walls and experience the outdoors, the local coffee house or escape to more exotic destinations. Commonly known as *rebound* or *revenge travel* (Enger, Saxon, Suo, & Yu, 2020; Hentschel, 2020), individuals deprived or denied of travels for a certain amount of time will add yet more destinations to their bucket lists, catching up on missed travel opportunities once lockdowns are eased. As time spent for leisure and pleasure activities increases, 'the luxury of a time oasis becomes more important' (Yeoman & McMahon-Beattie, 2006, p. 323). However, beyond the concept of time, 'luxury is amid a metamorphosis, today it's become a way of being' (Shankman, 2018, para 1). Travelling has become a common luxury where 'air travel has actually become a commodity where the pressure is for prices to fall rather than rise' (Yeoman & McMahon-Beattie, 2006, p. 319) and thus allows tourists to travel 'with a budget carrier but stay in a five-star hotel' (Yeoman & McMahon-Beattie, 2006, p. 321). A five-star hotel is also about status, 'but luxury is more than monetary value' (Yeoman & McMahon-Beattie, 2006, p. 320). The physical representation of wealth has shifted towards the state of mind and spiritual expressions (Shankman, 2018). This shift in perceptions of luxury can be seen in Maslow's Hierarchy of Needs, where the second-highest tier *esteem* can be categorised as the *old* luxury-seeking recognition, being accepted by others and gaining status to increase self-esteem. The highest tier *self-actualisation* can be seen as the *new* luxury consumers are seeking. In the past few years, there has been an increase in demand for luxury and five-star hotels that can integrate themselves into natural surroundings or inside their natural landscape (TEH, 2018). The term *eco-resort* has been seen more often, leading to an Eco Hotel Summit in 2020 (Eco Hotel Summit, 2020). Eco hotels and eco-resorts align behind luxury resorts, where people choose a destination due to concern about the wellness of the entire environment (Powell, 2020). Therefore, luxury is taking on a new definition closer to environmental connection, personal fulfilment and finding a brand that shares similar values as the consumer, transforming their identity. Indeed,

> from small Caribbean destinations to high mountain areas, the Amazon rainforest or Indian ocean islands, the growing significance of the hidden luxury holidays, the possibility to escape from everyday life (and health crises) and to find oneself amidst pristine environment, a sort of inconspicuous luxury consumption (Makkar & Yap, 2018) will continue to gain attraction. (Legrand, 2020, p. 69)

Accordingly, luxury tourism has been described in contrasting ways, both as the most unsustainable tourism sector (e.g. Bernard & Cook, 2015; Legrand, 2020; Moscardo, 2017) and also as the only sector which thus far attracts consumers who can afford to be sustainable (Legrand, 2020). The criticisms of excessive water consumption and segregation from local communities, amongst others, are counterbalanced by arguments that only the *uber-rich* can afford to stay in luxury eco-resorts (DFTM, 2019), experiencing what is left of pristine environments. Amongst these experiences, accompanied by claims of protecting the environment, is what Fletcher (2019, p. 527) refers to as 'Anthropocene tourism'. Within the realm of sustainable eco-tourism, Anthropocene tourism attempts to create new experiences and products and 'expand into new areas and produce new clients' (Moore, 2015, p. 8, in Fletcher, 2019, p. 524). This commodification of nature has led to new forms of tourism concentrating on connection to places, people and also causes, including voluntourism, extinction tourism and develop tourism. All of these share a concentration on experiencing environmental and social issues rather than intact conditions. A clear comparison is evident with the origins and spread of the COVID-19 virus and the ensuing pandemic, feared to result from human kind's encroachment into ecosystems (Boni et al., 2020). It is increasingly difficult to categorise tourism as sustainable or unsustainable, luxury or non-luxury, but rather this chapter questions whether tourism itself has become an unsustainable and an indefensible luxury.

Extravagance and Luxury in Tourism and Sustainability Conundrum

COVID-19 Impact: Drop and Bounce Back

The tourism industry contributes significantly to global gross domestic product, and its growth rate outpaces that of many other economic sectors. Since the global financial crisis 2008/2009, there has been no downturn in tourism development (WTTC, 2019) until the 2020 COVID-19 health crisis. Predictions are difficult to make at the time of writing this article, but the year 2020 could see a drop of anywhere between 40% and 80% in international tourism arrivals (UNWTO, 2020b; WTTC, 2020). However, domestic travel is experiencing a sharp increase following months of various levels of lockdown, fuelling local and regional economies, especially in nature-based destinations such as coastal and mountainous regions. While international tourism arrivals may decrease to one billion or less in 2020 (similar to 2010 levels), looking towards the long term, all international and national agencies predict the return of tourism both in scale and scope (UNWTO, 2020b; WTTC, 2020). The COVID-19 crisis has put a dent in international travel and other human activities. During the first wave of lockdown and forced confinement in April 2020, the daily emissions of greenhouse gases (GHGs) plunged 17% compared with 2019 levels (Le Quéré et al., 2020). Along the same lines, data show that during the same period, over 50% of the world's population decreased travel by over 50% with Google mobility trends indicating that '>80% of the population in the 114 countries in the dataset (4 billion people) reduced their travel by >50%' (Forster et al., 2020, para 4). Neither unanimous pledges by politicians to fight climate change during COP21 (Kinley, 2017), nor the aftermath of climate activist movements, such as Fridays for Future (Gurría, 2019), had changed the growth of international travel. Yet, it has become apparent that tourism in its entire value chain is a more significant contributor to climate change than previously expected, accounting for 8% of global emissions (Lenzen et al., 2018). And with the lockdown conditions eased and travel slowly picking up pace again, at least domestically, a rebound in GHG emissions is expected.

Tourism Growth: An Extravaganza in Times of Climate Change

As mentioned earlier, 2019 saw a record-breaking 1.5 billion international tourism arrivals following a decade of continuous growth (UNWTO, 2020a). Considering the associated carbon footprint of transporting visitors across international borders, however, tourism at this scale is a key driver of climate change (Gössling, 2002). Recent studies emphasise a positive long-term relationship between tourism development and carbon emissions (Balli, Sigeze, Manga, Birdir, & Birdir, 2019; Koçak, Ulucak, & Ulucak, 2020). An older study conducted by the UNWTO and UNEP estimated that emissions from international and domestic tourism (including transport, accommodation and activities) accounted for about 5% (range: 3.9%–6.0%) of global emissions for the year 2005 (2008, pp. 33–34). Interestingly, the study concludes that,

> While aviation-based trips account for 17% of all tourism trips, they cause about 40% of CO_2 emissions from tourism. Long-haul travel by air between the five UNWTO world tourism regions represents only 2.2% of all tourist trips, but contributes 16% to global tourism-related CO_2 emissions. In contrast, international tourist trips (i.e., overnight tourist trips) by coach and rail, which account for an estimated 16% of international tourist trips, stand only for 1% of CO_2 emissions generated by all international tourist trips (transport emissions only).
>
> (UNWTO and UNEP, 2008, p. 34)

This puts weekend city escapes using low-frill airlines as well as the annual long-haul flights to remote holiday destinations at the top of what needs to be considered unsustainable. The findings presented above, however, disregard a number of resources that are also typically part of tourism products and further aggravate the situation.

Lenzen et al. (2018, p. 523) undertook a more comprehensive study, including further value chain emissions, such as food and goods purchased. They found that tourism's carbon footprint increased from 3.9 in 2009 to 4.5 gigatonnes of equivalent carbon dioxide ($GtCO_2e$) in 2013, equal to about 8% of global emissions as mentioned previously. The study concludes that the increase in emissions cancelled out any efficiency gains of tourism products and services during this time. It needs to be noted, however, that people consume food and other goods, whether on holiday or not. Looking ahead, Gössling and Peeters (2015) predicted that resources needed to fuel the anticipated tourism growth, such as energy, freshwater, land and food, will double until about 2050. This puts the tourism industry 'increasingly at odds with objectives to reduce global resource use, and in particular emissions of greenhouse gasses' (Gössling & Peeters, 2015, p. 655). This outlook goes against the urgent need to lower global GHG emissions and, thus, blatantly accepts the repercussions of a >2°C rise in global mean temperature (IPCC, 2014). It also exacerbates the loss of biodiversity, unequal development in the Global South and excessive pressure on particular destinations in the Global North.

One of the key threats of climate change are not only changing weather patterns and rising sea levels but also its repercussions on terrestrial and marine ecosystems. Sánchez-Bayo and Wyckhuys (2019) predict that within the oncoming decades, up to 40% of insect species may go extinct because of habitat loss, intensive agriculture, biological factors and climate change (see also Horváth, Ptacnik, Vad, & Chase, 2019). Directly and indirectly, tourism is a driver of this development through its need for land, freshwater and food (Gössling & Peeters, 2015) as well as GHG emissions (Lenzen et al., 2018). Arguably, this leads to a self-destructive mechanism whereby a growing tourism industry consumes ever more resources, directly or indirectly emitting increasing amounts of GHG and eventually contributing to the destruction of its own basis and that of its stakeholders. Tourism destroys what it seeks to find: intact physical and cultural environments (Holden, 2008). This development is already apparent in discourses on sustainable tourism development and overtourism. The often-assumed economic benefits of tourism, paired with a neoliberal attitude to governing economic growth, have produced visitor numbers that far exceeded societal and natural carrying capacities. Destinations such as Amsterdam, Barcelona, Budapest, Byron Bay, the Galápagos Islands, Greenland, Guanacaste, Hong Kong, Iceland, Milan, Palma de Mallorca, Paris, the Portuguese cities of Lisbon, Porto, and Aveiro, Reykjavik, the favelas of Rio de Janeiro and Venice are struggling to come to terms with the implications of increasing and badly managed visitor numbers (Milano, Cheer, & Novelli, 2019). It seems as if, peu à peu, the tourism industry is modelling the world to meet the needs of tourists, always fuelled by the economistic argument that tourism provides jobs, income and development. However, the true, long-term value of these benefits is doubtful, to say the least.

Case Study 24.1: Tourism Growth Impact: Global North versus Global South

A meta-study on tourism's contribution to economic growth (i.e. the so-called tourism-led growth hypothesis, TLGH) found that, on average, economies are likely to benefit from promoting inbound tourism (Brida, Cortes-Jimenez, & Pulina, 2016). This is particularly pertinent to 'those developing countries that possess natural beauty and the relevant infrastructure but lack other significant exploitable natural resources' (Asiedu & Gbedema, 2011, p. 30). However, some scholars assert that the TLGH effects are stronger in industrialised (read: Global North) than in developing countries (read: Global South) (Cárdenas-García, Sánchez-Rivero, & Pulido-Fernández, 2015). Again, others attest that with respect to developing countries, there is no significant relationship at all (Ekanayake & Long, 2011). In 2018, travellers from Europe, northern America and China dominated outbound tourism. Furthermore, economies of the Global North realised significantly more tourism revenue from international tourism (64.6 %) than economies in the Global South (35.4 %) and also saw higher arrival growth rates in 2017 (7.9% vs 6.3%) and 2018 (6.8% vs 4.3%) (UNWTO, 2019, p. 17). In other words, strong economies with high disposable income can afford to travel, spending the most money in the Global North, even though destinations in the Global South are gaining momentum.

There is agreement across academia, governments and international institutions that tourism can increase foreign exchange earnings, create jobs for various qualification levels, offer business opportunities and attract foreign investments to the Global South (Sharpley, 2015). Medina-Muñoz, Medina-Muñoz, and Gutiérrez-Pérez (2016), however, oppose this notion. Through a critical review of 172 studies on the effects of tourism on poverty alleviation, the authors found contradictory evidence about tourism's value for the poor. Oviedo-García, González-Rodríguez, and Vega-Vázquez (2018) are also critical about tourism benefits. In the Dominican Republic, mass tourism has neither alleviated poverty nor reduced inequality.

Two key problems emerge in this respect: the leakages of tourism-induced revenues and the control over tourism development. Leakages occur when income from tourism is passed on to non-local businesses for the provision of labour, services or goods (Sandbrook, 2010). Lacher and Nepal (2010, p. 82) outline four underlying reasons why leakages occur in a tourism industry context: non-availability of funds to set up local supply structures, lack of local ownership, lack of local employment and, lastly, the failure to establish linkages between primary tourism enterprises (e.g. a hotel) and supporting local businesses (e.g. farmers). Additionally, Dwyer and Thomas (2012) argue that the economic expansion of the tourism sector in developing countries leads to increasing domestic prices and, thus, expedites the rate of importation. The Worldwatch Institute (2003) estimated that economic leakages in Global South destinations amount to about 50% on average. In the case of small island development states, leakage factors reach up to 70% (Milne, 1990). The more economically diversified countries in the Global North, however, experience only 10–20% (Diaz Benevides, 2001, cited in D. Meyer, 2007, p. 561) because they are able to localise much of the tourism value chain. This is but a snapshot of the unequal relationship and development of tourism in the Global South. In this context, critical voices speak of a significant power imbalance in the global tourism industry, or even a 'neo-colonisation' through tourism.

Neo-colonisation: Domination through Tourism

The colonial character of tourism is not one of direct military and political interventions, but of indirect domination through 'the expansion of capitalism and economic and cultural globalisation so that the core powers exercise influence over the postcolonial periphery' (Hall & Tucker, 2004, p. 2). On this ground, some scholars argued that there is a need to carefully consider the 'ambiguous gift of capitalist modernity' (p. 6) and its cultural imperialisms as part and parcel of globalisation (Britton, 1991; various chapters of Hall and Tucker, 2004; Mowforth & Munt, 2016). To this end, Jaakson (2004) argued that,

> [c]apitalism in its quest for competitiveness seeks out and exploits differences in the cost of labour, favourable trade and permissive social and environmental regulations. The result is a global 'race to the bottom' between developing countries competing to be the recipients of outsourcing from developed countries of manufacturing and service provision. Corporate globalisation creates externalities, such as environmental degradation, pollution and resource depletion, which poor countries, already stretched fiscally, are unable to cope with.
>
> (pp. 171–172)

Mowforth and Munt (2016) argued that even supposedly more sustainable forms of alternative tourism (e.g. eco- or community tourism) speak of the power that Western tourists have over determining what forms of tourism should take place in developing countries. In this respect, Hall (1994) exemplified how the planning of eco-tourism ventures has neglected social dimensions. The imposition of 'Western notions of conservation' (p. 152) through tourism, in some cases, induced dependency on foreign capital, depletion of local resources and unequal wealth distribution. Moreover, multiple contributions to a special issue on violence in tourism in the *Journal of Sustainable Tourism* found 'recurring themes of enclosure and extraction, erasure and commodification, "destructive creation", and (neo) colonialism' (Devine & Ojeda, 2017, p. 605). Looking at the tourism encounters in Fiji, Fisher (2004) concluded that 'the entwinement of worldviews' (p. 137) went only one way: 'The local people have to accommodate the world-view of outsiders' (p. 137). Such findings reiterate the long-standing criticism of the Western assumption of holding the 'correct' worldview (Kapoor, 2004; Peet & Hartwick, 2015; Said, 1978).

Tourism is not only a significant contributor to global climate change and environmental degradation but also a driving force behind structuring destinations in the Global South to suit the needs of a particular set of visitors. Through its economic power, tourism may dictate what parts of culture are appreciated and how they are represented – in sum, what worldview governs tourism. This speaks critically to the goals of sustainable tourism, based on concepts by the UNWTO and UNEP (2005). For example, certain kinds of 'visitor fulfilments' and 'viable economic tourism ventures' might contradict the goals of 'local control', 'community wellbeing' and 'cultural richness'. Moreover, this kind of 'domination through tourism' may be at odds with concepts of a 'good life', such as esteem and freedom of choice (Goulet, 1968) within host communities. It appears that tourism as a form of self-actualisation of wealthy Global North tourists, as argued above, is extravagant with respect to the needs of host-destination communities in the Global South.

The Dangerous Extravagance: Unsustainable and Irresponsible Tourism

Tourism, as long as powered by fossil fuel – from food produced in large quantities for far away markets to cooling large hotel buildings in tropical destinations and cheap jet travels – is then a dangerous extravagance, a luxury at the cost of those facing the brunt of climate change today and future generations left with no other option than to adapt to more extreme conditions. And thus, as Davidson wrote more than a century ago, 'the question of luxury is ultimately the ethical question of the worth of the want satisfied' (1898, p. 66). And today, while luxury in tourism may equate to more intrinsic, lower impact experiences as discussed above, luxurious extravagance in this industry still resonates with outrageous offers as outlined in the article 'Luxury travel: 50 wealthy tourists, eight countries ... and one giant carbon footprint' (Neate, 2020). Here the author clearly puts the extravagance into question, stating that 'despite the climate crisis, "no emission spared" round-the-world holidays in private jets are selling out' (Neate, 2020, para 1). A group of 50 wealthy elite 'will board a privately chartered Boeing 757 to begin a 24-day guided tour of the globe' (Neate, 2020, para 2). This extravaganza includes '10 flights, 23 nights in five-star hotels or lodges, meals in some of the world's most famous restaurants, champagne and "every single tip"' (Neate, 2020, para 3). The environmental impact in terms of carbon emissions linked to transport adds up to 13 tonnes per passenger, but the company has 'paid to offset all flights'(Neate, 2020, para 5). Other emissions linked to travel, such as the sourcing of exquisite food and beverage at remote hotel destinations, is not mentioned. The above story naturally represents what form tourism can take for a small number of the world's super-rich. All dictionaries imply through the emphasis on wasting resources, financial and other and being implicitly unnecessary that 'extravagance' and 'luxury' are something not available to everyone but dependent on having the means to waste. The old extravagance of time and money has been replaced by the definition of lacking restraint in using money and resources. This seems to be inherently unsustainable. In an interview conducted in 2017, Taleb Rifai, at the time UNWTO Secretary-General, stated that '1.8 billion travellers by 2030 could be 1.8 billion opportunities or 1.8 billion disasters and it is all up to us [to choose]' (UN News, 2017). Extravagant or luxurious, sustainability must become a *sine qua non* of all forms and types of tourism.

References

Asiedu, A. B., & Gbedema, T. K. (2011). The nexus between agriculture and tourism in Ghana: A case of unexploited development potential. In R. M. Torres & J. H. Momsen (Eds.), *Tourism and agriculture: New geographies of consumption, production and rural restructuring* (pp. 28–46). London: Routledge.

Balli, E., Sigeze, C., Manga, M., Birdir, S., & Birdir, K. (2019). The relationship between tourism, CO2 emissions and economic growth: A case of Mediterranean countries. *Asia Pacific Journal of Tourism Research*, *24*(3), 219–232. doi:10.1080/10941665.2018.1557717

Baring-Gould, S. (2020). *Historic odities and strange events*. Frankfurt am Main: Outlook Verlag.

Bernard, K., & Cook, S. (2015). Luxury tourism investment and flood risk: Case study on unsustainable development in Denarau island resort in Fiji. *International Journal of Disaster Risk Reduction*, *14*(3), 302–311. doi:10.1016/j.ijdrr.2014.09.002

Bernstein, L. (1999). Luxury and the hotel brand: Art, science, or fiction. *Cornell Hotel and Restaurant Administration Quarterly*, *40*(1), 47–53. doi:10.1177/001088049904000122

Boni, M. F., Lemey, P., Jiang, X., Tsan-Yuk Lam, T., Perry, B. W., Castoe, T. A., … Robertson, D. L. (2020). Evolutionary origins of the SARS-CoV-2 sarbecovirus lineage responsible for the COVID-19 pandemic. *Nature Microbiology*, *5*, 1408–1417. doi:10.1038/s41564-020-0771-4

Brida, J. G., Cortes-Jimenez, I., & Pulina, M. (2016). Has the tourism-led growth hypothesis been validated? A literature review. *Current Issues in Tourism*, *19*(5), 394–430. doi:10.1080/13683500.2013.868414

Britton, S. G. (1991). Tourism, capital, and place: Towards a critical geography of tourism. *Environment and Planning D: Society and Space*, *9*(4), 451–478. doi:10.1068/d090451

Cambridge. (2020a). *Luxury*. Cambridge Dictionary. Retrieved from https://dictionary.cambridge.org/dictionary/english/luxury

Cambridge. (2020b). *Extravagance*. Cambridge Dictionary. Retrieved from https://dictionary.cambridge.org/de/worterbuch/englisch/extravagant

Cárdenas-García, P. J., Sánchez-Rivero, M., & Pulido-Fernández, J. I. (2015). Does tourism growth influence economic development? *Journal of Travel Research*, *54*(2), 206–221. doi:10.1177/0047287513514297

Cohen, J. (2002). The contemporary tourist: Is everything old new again? In S. M. Broniarczyk & K. Nakamoto (Eds.). *Advances in consumer research* (Vol. 29, pp. 31–35). Valdosta, GA: Association for Consumer Research.

Daswani, A., & Jain, V. (2011). Understanding luxury branding A conceptual framework. *Media Asia*, *38*(3), 130–139.
Davidson, J. (1898). Luxury and extravagance. *International Journal of Ethics*, *9*(1), 54–73.
Davidson, M., Guilding, C., & Timo, N. (2006). Employment, flexibility and labour market practices of domestic and MNC chain luxury hotels in Australia: Where has accountability gone? *International Journal of Hospitality Management*, *25*(2), 193–210. doi:10.1016/j.ijhm.2005.08.001
Devine, J., & Ojeda, D. (2017). Violence and dispossession in tourism development: A critical geographical approach. *Journal of Sustainable Tourism*, *25*(5), 605–617. doi:10.1080/09669582.2017.1293401
DFTM. (2019). Why ALL tourism should be sustainable tourism: Protecting the future of travel. Retrieved from https://www.dontforgettomove.com/sustainable-tourism-protecting-future-travel/
Dixit, S. K. (2020). *The routledge handbook of tourism experience management and marketing*. London: Routledge.
Dwyer, L., & Thomas, F. (2012). Tourism yield measures for Cambodia. *Current Issues in Tourism*, *15*(4), 303–328. doi:10.1080/13683500.2011.604405
Eco Hotel Summit. (2020). Growth and sustainability: Pot-COVID-19 for ecological hotels and resorts. Retrieved from https://www.ecohotelsummit.com/
Ekanayake, E. M., & Long, A. E. (2011). Tourism development and economic growth in developing countries. *Global Conference on Business and Finance Proceedings*, *6*(2), 520–526.
Enger, W., Saxon, S., Suo, P., & Yu, J. (2020). The way back: What the world can learn from China's travel restart after COVID-19. McKinsey and Company. Retrieved from https://www.mckinsey.com/industries/travel-logistics-and-transport-infrastructure/our-insights/the-way-back-what-the-world-can-learn-from-chinas-travel-restart-after-covid-19
Fisher, D. (2004). A colonial town for neocolonial tourists. In C. M. Hall & H. Tucker (Eds.), *Tourism and postcolonialism* (pp. 126–139). Oxon: Routledge.
Fletcher, R. (2019). Eco-tourism after nature: Anthropocene tourism as a new capitalist "fix". *Journal of Sustainable Tourism*, *27*(4), 522–535. doi:10.1080/09669582.2018.1471084
Forster, P. M., Forster, H. I., Evans, M. J., Gidden, M. J., Jones, C. D., Keller, C. A., … Turnock, S. T. (2020). Current and future global climate impacts resulting from COVID-19. *Nature Climate Change*. doi:10.1038/s41558-020-0883-0
Gössling, S. (2002). Global environmental consequences of tourism. *Global Environmental Change*, *12*, 283–302. doi:10.1016/S0959-3780(02)00044-4
Gössling, S., & Peeters, P. (2015). Assessing tourism's global environmental impact 1900–2050. *Journal of Sustainable Tourism*, *23*(5), 639–659. doi:10.1080/09669582.2015.1008500
Goulet, D. A. (1968). On the goals of development. *Cross Currents*, *18*, 387–405.
Guessoum, N. (2016, August 10). Is space tourism an extravagant idea? Gulf News. Retrieved from https://gulfnews.com/opinion/op-eds/is-space-tourism-an-extravagant-idea-1.1876919
Gurría, A. (2019). Climate: Reclaiming our common future. Fourth biennial lecture on climate change. OECD. Retrieved from http://www.oecd.org/environment/cc/climate-lecture-reclaiming-our-common-future.pdf

Hall, C. M. (1994). Eco-tourism in Australia, New Zealand and the South Pacific: Appropriate tourism or a new form of ecological imperialism? In E. Cater & G. Lowman (Eds.), *Eco-tourism: A sustainable option?* (pp. 137–158). Chichester: John Wiley & Sons.

Hall, C. M., & Tucker, H. (Eds.). (2004). *Tourism and postcolonialism*. Oxon: Routledge.

Hentschel, T. (2020). Recovery, relapse, rebounds and revenge travel. INDVSTRVS. Retrieved from https://indvstrvs.com/recovery-relapse-rebounds-and-revenge-travel/

Holden, A. (2008). *Environment and tourism* (2nd ed.). London: Routledge.

Horváth, Z., Ptacnik, R., Vad, C. V., & Chase, J. M. (2019). Habitat loss over six decades accelerates regional and local biodiversity loss via changing landscape connectance. *Ecology Letters*, *22*(6), 1019–1027. doi:10.1111/ele.13260

IPCC. (2014). *Climate change 2014: Synthesis report. Contribution of working groups I, II and III to the fifth assessment report of the intergovernmental panel on climate change* [Core Writing Team, R. K. Pachauri, & L. A. Meyer (Eds.)]. Geneva: IPCC.

Jaakson, R. (2004). Globalisation and neocolonialist tourism. In C. M. Hall & H. Tucker (Eds.), *Tourism and postcolonialism* (pp. 169–183). Oxon: Routledge.

Kapferer, J. N. (1997). Managing luxury brands. *Journal of Brand Management*, *4*(4), 251–259.

Kapferer, J. N., & Bastien, V. (2012). *The luxury strategy: Break the rules of marketing to build luxury brands*. London: Kogan Page Publishers.

Kapferer, J. N., Klippert, C., & Leproux, L. (2014). Does luxury have a minimum price? An exploratory study into consumers' psychology of luxury prices. *Journal of Revenue and Pricing Management*, *13*, 2–11.

Kapoor, I. (2004). Hyper-self-reflexive development? Spivak on representing the third world 'other'. *Third World Quarterly*, *25*(4), 627–647. doi:10.1080/0143659041000167889 8

Kinley, R. (2017). Climate change after Paris: From turning point to transformation. *Climate Policy*, *17*(1), 9–15. doi:10.1080/14693062.2016.1191009

Koçak, E., Ulucak, R., & Ulucak, Z. Ş. (2020). The impact of tourism developments on CO2 emissions: An advanced panel data estimation. *Tourism Management Perspectives*, *33*, 100611. doi:10.1016/j.tmp.2019.100611

Lacher, R. G., & Nepal, S. K. (2010). From leakages to linkages: Local-level strategies for capturing tourism revenue in northern Thailand. *Tourism Geographies*, *12*(1), 77–99. doi:10.1080/14616680903493654

Le Quéré, C., Jackson, R. B., Jones, M. W., Smith, A. J. P., Abernethy, S., Andrew, R. M., … Peters, G. P. (2020). Temporary reduction in daily global CO2 emissions during the COVID-19 forced confinement. *Nature Climate Change*, *10*, 647–653. doi:10.1038/s41558-020-0797-x

Lee, T. J. (2010). Role of hotel design in enhancing destination branding. *Annals of Tourism Research*, *38*(2), 708–711.

Legrand, W. (2020). Luxury tourism in remote destinations: An inquiry into sustainability. In J. Chen (Ed.), *Advances in hospitality and leisure 16* (pp. 65–81). Bingley: Emerald Publishing Limited.

Lenzen, M., Sun, Y., Faturay, F., Ting, Y.-P., Geschke, A., & Malik, A. (2018). The carbon footprint of global tourism. *Nature Climate Change*, *8*, 522–528. doi:10.1038/s41558-018-0141-x

Lexico. (2020). *Extravagant*. Lexico Dictionaries. Retrieved from https://www.lexico.com/en/definition/extravagant

Lipovetsky, G., & Sébastien, C. (2005). *The hypermodern times*. Cambridge: Polity.

Makkar, M., & Yap, S.-F. (2018). Emotional experiences behind the pursuit of inconspicuous luxury. *Journal of Retailing and Consumer Services*, *44*(September), 222–234.

Maslow, A. H. (1943). A theory of human motivation. *Psychological Review*, *50*(4), 370–396. doi:10.1037/h0054346 [Titel anhand dieser DOI in Citavi-Projekt übernehmen].

Maslow, A. H. (1970). *Motivation and personality* (2nd ed.). New York, NY: Harper & Row.

Medina-Muñoz, D. R., Medina-Muñoz, R. D., & Gutiérrez-Pérez, F. J. (2016). The impacts of tourism on poverty alleviation: An integrated research framework. *Journal of Sustainable Tourism*, *24*(2), 270–298. doi:10.1080/09669582.2015.1049611

Meyer, D. (2007). Pro-poor tourism: From leakages to linkages. A conceptual framework for creating linkages between the accommodation sector and 'poor' neighbouring communities. *Current Issues in Tourism*, *10*(6), 558–583. doi:10.2167/cit313.0

Milano, C., Cheer, J. M., & Novelli, M. (Eds.). (2019). *Overtourism: Excesses, discontents and measures in travel and tourism*. Oxfordshire: CABI.

Milne, S. (1990). The impact of tourism development in small Pacific island states: An overview. *New Zealand Journal of Geography*, *89*, 16–21.

Moscardo, G. (2017). Sustainable luxury in hotels and resorts: Is it possible? In M. Gardetti (Ed.), *Sustainable management of luxury. Environmental footprints and eco-design of products and processes* (pp. 163–189). Singapore: Springer.

Mowforth, M., & Munt, I. (2016). *Tourism and sustainability: Development, globalisation and new tourism in the Third World* (4th ed.). London: Routledge.

Neate, R. (2020, January 18). Luxury travel: 50 wealthy tourists, eight countries... and one giant carbon footprint. *The Guardian*. Retrieved from https://www.theguardian.com/environment/2020/jan/18/world-tour-luxury-jet

Oviedo-García, M. Á., González-Rodríguez, M. R., & Vega-Vázquez, M. (2018). Does sun-and-sea all-inclusive tourism contribute to poverty alleviation and/or income inequality reduction? The case of the Dominican Republic. *Journal of Travel Research*, *58*(6), 995–1013. doi:10.1177/0047287518789272

Park, J., Kim, K., Kwak, J., & Wyer, R. S., Jr (2013). Priming thoughts about extravagance: Implications for consumer decisions about luxury products. *Journal of Experimental Psychology Applied*, *20*(1), 40–54.

Pearce, P. L., & Lee, U.-I. (2005). Developing the travel career approach to tourist motivation. *Journal of Travel Research*, *43*(3), 226–237. doi:10.1177/0047287504272020

Peet, R., & Hartwick, E. (2015). *Theories of development contentions, arguments, alternatives* (3rd ed.). New York, NY: Guilford Press.

Pine, J., & Gilmore, J. (2014). A leader's guide to innovation in the experience economy. *Strategy & Leadership*, *42*(1), 24–29.

Powell, L. (2020). *Europe is shaping the luxury eco-resort movement*. Skift. Retrieved from https://skift.com/2020/02/13/europe-is-shaping-the-luxury-eco-resortmovement/

Pratley, N. (2019, October 28). Virgin Galactic wins space tourism race to float on stock market. *The Guardian*. Retrieved from https://www.theguardian.com/business/nils-pratley-on-finance/2019/oct/28/virgin-galactic-wins-space-tourism-race-to-float-on-stock-market

Said, E. W. (1978). *Orientalism: Western conceptions of the orient*. New York, NY: Pantheon Books.

Sánchez-Bayo, F., & Wyckhuys, K. (2019). Worldwide decline of the entomofauna: A review of its drivers. *Biological Conservation*, *232*, 8–27. doi:10.1016/j.biocon.2019.01.020

Sandbrook, C. G. (2010). Local economic impact of different forms of nature-based tourism. *Conservation Letters*, *3*(1), 21–28. doi:10.1111/j.1755-263X.2009.00085.x

Schloss. (2019). Schloss Neuschwanstein wieder auf Platz 1. Deutsche Welle. Retrieved from https://www.dw.com/de/schloss-neuschwanstein-wieder-auf-platz-1/a-42642968

Seo, Y., & Buchanan-Oliver, M. (2015). Luxury branding: The industry, trends, and future conceptualisations. *Asia Pacific Journal of Marketing and Logistics*, *27*(1), 82–98.

Shankman, S. (2018). *The definition of luxury is being reshaped by changing consumer behavior*. Skift. Retrieved from https://skift.com/2018/01/30/the-definition-of-luxury-is-being-reshaped-by-changing-consumer-behavior/

Sharpley, R. (2015). Tourism: A vehicle for development? In R. Sharpley & D. J. Telfer (Eds.), *Tourism and development: Concepts and issues* (2nd ed., pp. 11–34). Clevedon: Channel View Publications.

Slattery, P., & Games, I. (2010). The confused world of hotel categorisation. *Hotel Analyst*, *5*(3), 18–20.

Steinberg, R., & Falkai, P. (2020). Was king Ludwig II of Bavaria misdiagnosed by Gudden and his colleagues? *European Archives of Psychiatry and Clinical Neuroscience*, *271*, 799–807. doi:10.1007/s00406-020-01161-8

Swarbrooke, J. (2018). *The meaning of luxury in in tourism, hospitality and events*. Oxford: Goodfellow Publishers.

TEH, W.-L. (2018). Luxury hotels around the world that blend into its surrounding nature. Retrieved from https://www.lifestyleasia.com/https://www.lifestyleasia.com/kl/travel/hotels-resorts/luxury-hotels-around-theworld-that-blend-into-its-surrounding-nature/

UN News. (2017, December 27). World could see 1.8 billion tourists by 2030 – UN agency. *UN News*. Retrieved from https://news.un.org/en/story/2017/12/640512-world-could-see-18-billion-tourists-2030-un-agency

UNWTO. (2020b). *International tourism numbers could fall 60–80% in 2020, UNWTO reports*. Retrieved from https://www.unwto.org/news/covid-19-international-tourist-numbers-could-fall-60-80-in-2020

UNWTO and UNEP. (2005). *Making tourism more sustainable: A guide for policy makers*. Madrid and Paris: UNWTO and UNEP.

UNWTO and UNEP. (2008). *Climate change and tourism: Responding to global challenges*. Madrid and Milan: UNWTO and UNEP.

UNWTO. (2019). *International tourism highlights*. Madrid: UNWTO. doi:10.18111/9789284421152

UNWTO. (2020a). International tourism growth to outpace the global economy. Retrieved from https://www.unwto.org/international-tourism-growth-continues-to-outpace-the-economy

Veblen, T. (1899). *The theory of the leisure class*. Retrieved from http://moglen.law.-columbia.edu/LCS/theoryleisureclass.pdf

Worldwatch Institute. (2003). *Vital Signs 2003: The trends that are shaping our future*. New York, NY: W. W. Norton.

WTTC. (2019). Economic impact 2019: World. World Travel and Tourism Council. Retrieved from https://www.wttc.org/-/media/files/reports/economic-impact-research/regions-2019/world2019.pdf

WTTC. (2020). Travel and tourism recovery scenarios 2020 and economic impact from COVID-19. World Travel and Tourism Council. Retrieved from https://wttc.org/Research/Economic-Impact/Recovery-Scenarios-2020-Economic-Impact-from-COVID-19

Yang, W., & Mattila, A. S. (2016). Why do we buy luxury experiences? *International Journey of Contemporary Hospitality Management*, *28*(9), 1848–1867. doi:10.1108/ijchm-11-2014-0579

Yeoman, I., & McMahon-Beattie, U. (2006). Luxury markets and premium pricing. *Journal of Revenue and Pricing Management*, *4*(4), 319–328.

Chapter 25

Delivering Luxury Experiences in the Post-COVID-19 Reality

Barbara Czyzewska[1]

Abstract

As the world slowly emerges from the pandemic of the newly discovered COVID-19 virus, luxury hospitality providers are re-opening their businesses in a completely new reality. While they ensure that their unique brand identity and image are appropriately communicated to the customers, they must also strictly adhere to the hygiene and cleanliness standards imposed on businesses by local authorities. As a result, luxury hotels, restaurants, spas and resorts find themselves in a situation where they must adjust, if not re-design, experiences offered to their customers.

This chapter looks at the concept of experiences and examines key elements of creating authentic experiences. It discusses the role of embedding the culture of adaptability to change to ensure the business's longevity and resilience to crises. Finally, it deconstructs some of the industry's best practices to showcase the most creative and resourceful approaches to embracing the new reality and offering authentic experiences to discerning guests.

The chapter is concluded by an attempt to look into future customer expectations and businesses' opportunities to re-open and provide updated services to customers by actually embedding strategic foresight in both short- and long-term planning to build genuine resilience to future crises.

Keywords: Luxury hospitality; experiences; post-COVID-19 service; business resilience; strategic foresight; adaptability

[1]https://orcid.org/0000-0001-9700-5139

The Emerald Handbook of Luxury Management for Hospitality and Tourism, 497–511

doi:10.1108/978-1-83982-900-020211025

Setting the Stage

In early 2020, the world had been hit by the global pandemic of a new virus, soon named COVID19. Many industries suffered tremendous losses, but tourism and hospitality had been hit the hardest. This impact was felt on a global scale, as not only individual businesses were subject to closures but also whole countries entered lockdowns by closing their borders and cancelling international transportation. This unprecedented situation's economic and social impact will be, most likely, felt for years to come. Economists and strategists worldwide are trying to create recovery plans whose long-term aim will be to bring the economy back to the pre-COVID-19 state. However, the key element of this global revival will be re-building industries and individual companies so that they are more immune to these kinds of unpredictable situations in the future. Simon Sinek, a motivational speaker who became world famous for his ideas on leadership and business strategy and who describes himself as an 'unshakable optimist who believes in a bright future and our ability to build it together' (Sinek, 2020), describes how companies can ensure endurance. In his book entitled "The Infinite Game", he says: 'To succeed in the Infinite Game of business, we have to stop thinking about who wins or who's the best and start thinking about how to build organisations that are strong enough and healthy enough to stay in the game for many generations to come' (Sinek, 2019, p. 10). The COVID-19 crisis has very clearly exposed those companies that were concentrated on short-term goals and quick results instead of on building resilience for the unknown. If there is anything positive about this situation, it is the lesson that companies should learn. It is not stability that we should be striving for but a resilience, which is based on flexibility and adaptability to change. Sinek (2019, p. 16) further explains: 'A company built for resilience is a company that is structured to last forever. This is different from a company built for stability. Stability, by its very definition, is about remaining the same. A stable organisation can theoretically weather a storm, then come out of it the same as it was before'. A business playing infinite games embraces surprises and adapts to them, thus becoming more resilient, as opposed to stable. Consequently, building a resilient company automatically means accepting the possibility that the company's culture, strategy and existence might change. This kind of attitude might not be easy to adopt, but it is, ultimately, the only one that can ensure longevity. The few months of confinement caused by the outbreak of the new virus and millions of jobs lost as a result of global slowdown will certainly force many businesses to change. It is up to them, as well as their corporate culture, whether they will be able to embrace the changes that year 2020 brought about or not.

This chapter focuses specifically on luxury hospitality and on the ways in which luxury hospitality providers can change their operations and strategy in order to not only recover but also ensure more resilience for the future. While many scholars have attempted defining luxury, the only thing that they seem to agree on is that luxury is 'a subjective and multidimensional construct' (Wiedmann, Hennigs, & Siebel, 2007). Most researchers concentrate on the demand side when analysing the phenomenon of luxury by looking in-depth into concepts such as conspicuous consumption, hedonism and social status. As understanding of luxury, in general, has developed, we began to comprehend that its allure is not so much about possession, but rather about the experience which that possession, or access to a service as the case may be, creates. Kapferer and Bastien (2012, p. 47) argue that real luxury must be linked to heritage, be purposefully restricted, be priced higher than its functional value would suggest, be offered with personalised service, represent a social marker and create a qualitative hedonistic experience. Indeed, a great majority of scholars analysing not only the current state of luxury but also the drivers and trends in the sector suggest that experience is the key element (Dixit, 2020; Yeoman & McMahon-Beattie, 2018). As companies are spending millions for the perfect design of their properties and training of staff, Harkinson, Hemmington and Hyde (2018) point out a very interesting argument. They say: 'Although an organisation cannot guarantee that the customer will have a positive experience, organisations can create the circumstances and the environment in which such an experience is possible' (Harkinson et al., 2018, p. 1727). What they seem to suggest is that hotels can do everything for their guests, but it is, ultimately, down to each individual customer to co-create their own experience. In other words, the multidimensional impression of customers' experience is based on their willingness to be influenced by these interactive dimensions (Walls, Okumus, Wang, & Kwun, 2011; Harkinson et al., 2018). In times of global crisis, the unpredictability of the future and fear for personal safety, this willingness will be more crucial than ever.

Therefore, this chapter aims to provide not only a discussion on the delivery of luxury experiences but also, more importantly, practical solutions as to how luxury hotel managers can create the circumstances and the environments that will encourage guests to fully interact and immerse themselves in such experiences. A wide range of real-life examples and case studies will be brought into this discussion to illustrate the concepts as well as provide a platform for learning from some of the most successful players in the sector. The chapter will suggest how luxury brands can communicate with their existing and prospective guests to ensure their cooperation and active engagement. This will prove to be the key element, as only companies who have strong and lasting relationships with their clients will build resilience and, thus, cope with any kind of crisis that will face us in this infinite game of business.

Experiencing Confidence

Before we can even start discussing the creation and delivery of exceptional experiences, we need to ponder the concept itself. Pine and Gilmore (1998), who have coined the term 'experience economy', explain: 'An experience occurs when a company intentionally uses services as the stage, and goods as props, to engage individual customers in a way that creates a memorable event'. They further add: 'While prior economic offerings - commodities, goods, and services - are external to the buyer, experiences are inherently personal, existing only in the mind of an individual who has been engaged on an emotional, physical, intellectual, or even spiritual level. Thus, no two people can have the same experience because each experience derives from the interaction between the staged event (like a theatrical play) and the individual's state of mind' (Pine & Gilmore, 1998). The response of most people to a question of what is important in a luxury experience is 'personalisation'. Everybody knows that there is no luxury without personalisation. My feeling, however, is that the more service providers claim that they take a personalised approach, the less personal it actually feels. Standard operating procedures, scripts and worn-out courtesy slogans are used all too often, even by hotels that claim to provide genuine hospitality. Unfortunately, equally common is the request to fill in a satisfaction survey or post feedback on social media, which in my mind, fundamentally breaks one of the rules of luxury, that of not giving a client more work to do than it is absolutely necessary. The best way to imagine a truly individualised service is to translate it to human interaction. Nobody feels more loved by hearing a matter-of-factly dropped 'love you' comment. What can make us feel truly loved, however, is the time that the other person takes to read into our mood, feelings and needs. Hospitality companies who hire their team based on personality rather than skills seem to be demonstrating a good understanding of this principle. Certainly, procedures and standards are crucial in the delivery of consistent services, but what is more important is the authenticity of such interactions. Gilmore and Pine, in their second book, this time on delivering authentic experiences, highlight that such service cannot be created when staff are artificially friendly. They say: 'Candid, plainspoken words (...) as well as straightforward and unassuming actions – whether upfront or behind the scenes – generally render exceptional authenticity more effectively than going through the all-too-typical motions' (Gilmore & Pine, 2007, p. 64). Usually, this kind of authentic relationship can be built only when employees feel that they have enough time to invest in getting to understand the guest. It is, however, this abundance of time that is a major obstacle in the post–COVID-19 reality. In this context, time is indeed a luxury.

Experts in staging customer experiences claim that such experiences must be refreshed at least every five years (Pine & Gilmore, 2011, p. 145). The reason behind this requirement is that there is no real experience without surprising the customer, at least to some extent (Smith & Hanover, 2016). For many years hotels have been competing in how they can create the 'wow' effect for their guests and how they can exceed guests' expectations. This will be no different now, with the exception that customers' expectations as to experiences have somewhat evolved and, most likely, will continue evolving further in the coming months as the pandemic situation unravels. Today travel plans are often underlined with fear and concern for one's own health and safety. While travellers will, undoubtedly, still seek excitement and pampering, what can probably surprise them the most is the conviction that being away from home does not have to mean taking as much risk as they expected it to. They will want to be positively surprised by how confident they can feel that they are being taken care of. Indeed, we have come to the point when hotels and other service providers must refresh their offering. However, this time, this is the pursuit of stability and confidence, instead of an unexpected 'wow' moment, that guests want to experience. How their perceptions and expectations will be evolving further is difficult to predict and very much depends on the global state of affairs. What companies should, therefore, concentrate on is not addressing the current expectations but rather building resilience and preparing for flexible adjustment of the service offering in the future.

Most companies, from small players to huge multination corporations, managed to survive the crisis by laying off or furloughing a huge proportion of their employees. Some of them are now returning to work, but with dramatically reduced capacity and international travel picking up very slowly, many of these properties will still be working with teams limited to the bare minimum. The skills which recruiters are demanding from prospective employees right now are the ability to multitask, to seamlessly switch from one job to the next and be able to flexibly adjust to the needs of the business. Some companies go even further by saying that demarcation between roles is disappearing completely for the time being. This, naturally, comes at a cost. In light of such developments, we can hardly expect our staff to commit extra time and attention to build not simply rapport, but an honest relationship with customers. This is where, as pointed out by Jannes Soerensen, the General Manager of London's Beaumont Hotel, boutique hotels will be in a much better position than large corporations (Reynolds, 2020). Many of their employees have been already around for many years, as well as a large proportion of guests are those who already know the hotel very well. In such a small environment, as Soerensen says, every member of staff is in Guest Relations. This resonates very closely with the principles of Total Quality Management (Ciampa, 1992), as the quality of the experience should be everyone's responsibility, starting from the front of house staff to all supporting services. There is no doubt, however, that this kind of model is rather complex to implement and then sustain. All properties, but large chain hotels in particular, will need to figure out how to balance the requirement for multitasking with committing time for building a more meaningful relationship with the client.

Targeted planning and training will be of utmost importance as hotels re-open and try to deliver timeless service while ensuring social distancing and operating with limited staffing. Going back to the comparison between experiences and theatrical play, as well as the fact that we use the verb 'staging' to refer to experiences, it becomes clear that there is a connection between actors on stage and personnel delivering a hospitality experience. Just like actors feeding energy to one another during the performance, staff at a hotel work hand in hand to create a seamless experience for customers. This energy should flow throughout all levels of the organisation, starting from the customer-facing team, through behind-the-scenes, all the way to management and all support services. Now, more than ever, luxury companies should consider utilising models such as service blueprints or customer journeys to design the flow of experiences in their properties. The benefit of such models is that they put the customer in the centre of attention and allow for designing the process around the customer and their needs.

They, furthermore, make it possible to visualise the role of each and every single member of the team and thus predict how they can contribute to the overall experience. Once the company has clarity on the flow of activities, as well as on the role of each member of the team, it can begin training its people to make them prepared for dealing with all kinds of situations. Once again, here innovative approaches will most likely prevail, and so companies should consider a wide portfolio of methods. In their recent spotlight article dedicated to 'learning from the future', *Harvard Business Review* advocates for adopting strategic foresight to 'harness the power of imagination to build a dynamic link between planning and operations' (Scoblic, 2020, p. 40). The article lists a number of tools, including scenario planning, crisis simulations or tabletop exercises to not only predict the future but also, more importantly, to prepare the team for dealing with whatever it may bring. What is crucial, according to Scoblic, is the implementation of strategic foresight and ingraining it in the very culture of the organisation. Luxury hotels can utilise many training tools, including role-plays or war games, to bring all employees on board, regardless of their original role within the company, and make them more responsive to the fast-evolving customer expectations. Scoblic (2020) finishes his article with a very important reminder that organisations do not just prepare for the future but shape it. Such a proactive attitude will be best suited to engage staff whose roles might have changed or become broader.

To finish off this section on delivering experiences, let us go back to Gilmore and Pine (2007), who highlight that personalised offerings resonate as authentic when they match customer's self-image. While we can hardly influence what customers think about themselves, we certainly have an impact on how satisfied they are with the delivered experience. Pine and Gilmore (2011) define customer satisfaction as a difference between customers' expectation and what they perceive to get. I would advocate for reversing this equation, as we strive to narrow this gap by increasing the perceived experience or lowering the expectation (in luxury services, most likely the former). Whichever way we look at it, the key is that we influence not only what we actually deliver to the customer but also, equally importantly, how the customer perceives our efforts. Gilmore and Pine (2007) further discuss different types of customisation (collaborative, adaptive, transparent and cosmetic). These different service customisations refer to the actual level of service or product adaptation versus its representation, i.e. its packaging, marketing etc. While we accept that, in principle, luxury services are tailored to individual customers, when speaking about conveying the sense of cleanliness and hygiene, luxury hospitality providers should pay particular attention to cosmetic customisation. The idea behind this approach is that cleaning procedures must adhere to strict and clearly stated standards (so cannot be customised and must be heavily standardised), but what is now the challenge for hoteliers is to convince the guests that all these measures are performed personally for them, in order to create authentically safe and luxurious experiences. Therefore, the subsequent section of this chapter will discuss some of the practical solutions that luxury hospitality providers can adopt to deliver genuinely personalised experiences in light of new standards resulting from the global pandemic of COVID-19.

The Smile from behind the Mask

As the world is slowly re-opening after a few months of global lockdown, what we find ourselves in is not returning back to normal but rather entering a completely new reality. As previously discussed, those companies who are committed to stability might have survived the crisis but will inevitably find adjusting to the new world extremely challenging. On the other hand, those that had managed to develop resilience and had embraced change as part of their organisational culture will find it strategically easier to implement the necessary adjustments. This section will concentrate on some key ideas and best practice operational examples which are already being implemented across the luxury hospitality industry. What needs to be remembered, however, is that luxury is usually grounded in heritage and long tradition (Kapferer & Bastien, 2012). What service providers must strive for is not a revolution but rather a gradual evolution that will take into consideration new circumstances and customers' requirements as they emerge. This discussion will focus on the physical environment, as it sets the stage for any kind of experiences, as well as on behaviour and social interactions which must be adjusted in the face of the new virus being spread mainly through human contact. This, in no way, attempts to provide an exhaustive list of solutions, but rather serves as guidance as to how to approach this change to be able to navigate it fluently and, consequently, deliver experiences that are authentic and which remain true to the unique luxury brand identity.

One of the most visible changes to be noticed shortly after the re-opening of various customer-facing institutions was that of the physical environment in which services are being delivered. Restaurants and cafes, in pursuit of maintaining social distance, have moved tables further apart, and over-the-counter services have installed plexiglass screens to separate people from one another as much as possible. This has a dramatic impact on the overall experience of service because, as argued by Booms and Bitner (1981), physical environment, or servicescape, as it is referred to in service industries, plays a pivotal role in building the ambience of the place and in evoking specific emotions in people. It has the power to make people be more likely to interact with each other (by, for example, arranging seats to face one another) or to extend their visit (by providing soft, comfortable armchairs) or, as the case may be, to shorten their stay (by replacing comfortable armchairs with hard-surface bar stools). It can help people feel more energised or more relaxed, depending on the colours and background music used, or even to make them feel more connected to the natural environment by using earth-tone colours and natural material for décor. Even the choice of lighting plays a critical role in how people, customers and employees will feel and behave in a given environment. Consequently, the physical environment is the background to an experience, but it plays an active role in its creation and must be managed to perfection.

The power behind servicescape is that a well-themed room can draw guests into a fanciful, imaginary world and have the potential to touch the eye, mind and head of visitors (Pine & Gilmore, 2011, p. 72). Luxury hotels are spending millions of dollars on perfectly designed and decorated interiors to ensure appropriate communication of their brand identity, but they are now facing the challenge of adjusting them to the social distancing requirement. Changing space that has been meticulously designed is far from easy, but, as in most testing situations, it is up to us whether the change will be perceived as a threat or as an opportunity. Social distancing requirements can, indeed, be treated as an inconvenience and a lost revenue opportunity if one focuses on the fact that we work with less capacity and an increased number of unfamiliar procedures. On the other hand, it can be perceived as an opportunity to re-invent the space and turn it into a truly imaginary world. The sense of spaciousness and abundance of the room is one of the most vocal communicators of luxury (Godfrey, 2020). Whether modern or classic, European, Middle Eastern or Asian, luxury hotels tend to utilise large unoccupied spaces as communicators of exclusivity and opulence. While in the past, hotels had to trade between creating that generous feeling and turning space into a revenue generator by maximising sales opportunities, now they can concentrate on how that area can positively contribute to the overall theme of the hotel experience. Obviously, in the case of luxury hospitality services, we are not talking about a Disney-like kind of theme. However, it means that the entire space and service delivery must, together, contribute to a coherent and distinctive experience. Pine and Gilmore claim that, in fact, every experience has a theme. They say: 'Whether or not themed intentionally, whether or not designed well, and whether or not executed thoroughly and rigorously, a theme always emerges. Discovering a suitable theme is central to experience design. No matter what list or category prompts discovery, the key lies in determining what theme will actually prove to be compelling and captivating' (Pine & Gilmore, 2011, p. 73). It is indeed this subtle theme that is at the heart of establishing a sense of place and which is able to evoke emotions consistent with brand identity and create meaningful and memorable experiences. Most effective themes have the ability to complement the storytelling of the brand by becoming an integral part of its narrative through integrating space, matter and time into a cohesive, realistic whole.

The biggest challenge of the currently imposed health and safety standards is that they, more than anything, make businesses appear similar. If everybody has to wear the same masks and use the same hand sanitisers, the question which should be posed is that of how to make my service feel different from everybody else's. Yet, some hotels are already showing incredible creativity and are doing a great job differentiating their post–COVID-19 service from other providers. Some are offering uniquely designed and branded face masks made of premium materials. Dorchester Collection, for example, differentiate the experience by offering masks in each of the properties' signature colours. Others, like Ocean House in Rhode Island, launched a 'mobile bar', a cart which can be moved across the floors and which serves bar's specialities in the comfort of the guest's bedroom. The same hotel also offers to serve the meals in the form of a picnic, with a selection of not only the snacks but also the location, from the lawn to a beach cabana. Quite an innovative solution for dealing with social distancing was offered by The Inn at Little Washington, where the room is 'crowded' with mannequins dressed in designer costumes. This shows that, while we must be serious about health precautions, we can certainly take the implementation of them with a pinch of humour and let our guests enjoy a joyful surprise, all leading to a positive overall experience.

The new requirements can indeed create more opportunities than initially meet the eye. The obligation to wear personal protection, for example, creates an excellent opportunity to enhance the brand presence around the property. Forbes Travel Guide (Kester, 2020) recommends that hotels could enhance their service by offering pre-packaged health kits, including sanitiser, branded face mask and a multivitamin supplement. Companies could be very creative in how to package and brand these so that they come across as a thoughtful souvenir rather than just an irritating necessity. What is important, according to Forbes Travel Guide, is to think of all the customers, so these kits should have items specially designed for kids or youngsters if the hotel caters to such clientele. Comparable pre-packaged kits could be made available in the bar or pool-side area where sharing of snacks will no longer be available. These could include a signature nut mixture and a refreshing wipe or something else that will show the guest that we care about not only their comfort but also their safety.

Adjusting to the emerging situation and introducing a range of protective measures does not always have to mean sacrificing causes to which luxury hotels were committed to before. One of the great examples of continuing with a great initiative that had been undertaken before the outbreak of the pandemic is The Beaumont hotel in London which, aiming to reduce its plastic waste, introduced solid bar soaps and body wash in all its rooms. This was aimed as a practice adopted to combat global climate change and address the over-usage of plastic, but it turns out to be an excellent practice in the times when hotel guests would not feel comfortable using the same toiletry containers as previous residents of the room. The additional benefit of this kind of solid bar toiletries is that they can be packaged, taken away by guests and serve as a meaningful memoire after the stay.

All the above are excellent examples of embracing change. Pine and Gilmore (2011, p. 82) highlight, however, that in order to create a coherent experience, service providers must not only furnish the space and service process with positive cues but also, equally importantly, eliminate any negative cues. What this means is that a service provider, and particularly one in the luxury sector, must be careful not to accidentally reduce the overall experience by careless elements, which are usually rather small and easy to miss. Retail and hospitality businesses are, sadly, filled these days with badly smelling sanitising gels and plastic sanitiser dispensers with ugly medical stickers on them. Each time I see such a bottle, I cannot help but wonder how many people touched it before me. What goes without saying is that allowing such poor-quality tools in luxury boutiques and service providers is, simply, a grave mistake. While conducting research for this chapter, I came across something far worse than a cheap hand sanitiser. Towards the end of the COVID-19 lockdown, in one of the major European cities, I spotted a creased A4 note on the locked door of one of luxury fashion brands' boutiques. The note informed that the store was closed due to the pandemic and that the management apologised for the inconvenience. I could not help but be amazed that a brand of such profile could not afford a medium more appropriate than a piece of paper pulled out of a copier. What made the entire experience even more disappointing was the fact that, by that time, other, so-called mass-market brands have had their doors wide open and were gladly welcoming customers. The message which I got from that unfortunate note was that this particular company did not feel it profitable enough to open its boutiques when there were no international shopping travellers, known to be their main clientele. It took one piece of paper and approximately three seconds of reasoning to doubt in everything that this brand has been building for many decades. This creased note was exactly what Pine and Gilmore refer to when writing about negative cues. These are the kind of elements that are easily omitted when managers and staff concentrate too much on everyday operations. At the end of the day, there is nothing wrong with marking social distance with a tape on the floor, everybody is doing this, and we are getting used to this image, just like we are getting used to seeing people with surgical masks on their faces. It might even be the case that the customer will not see anything wrong with these kinds of solutions, but the fact is that they will inevitably have a negative impact on experiencing the brand's theme in a seamless way. Luxury hotels have a unique opportunity to think of imaginative ways of arranging furniture or décor to maximise style and comfort without compromising safety, so that even a waiting line does not feel like one, but instead, more like a relaxation area. It is these service providers who pay attention to the smallest of details who will come out of this operational and cultural transformation stronger.

The term 'servicescape' generally refers to the physical environment; however, it is crucial to take into consideration that people greatly contribute to the overall sense of place as well. One of the most disturbing elements in an experience which is supposed to be perfectly themed is when members of staff or service providers step out of their assigned roles. Famous is the strict rule adopted in Disney World, forbidding cast members from stepping out of character when at work. We can all imagine what impact a Mickey Mouse with his head off could have on a child. This mechanism will be no different in the case of luxury service being provided to grown-up customers. One dirty glove or one mask incorrectly placed under the nose will squander hours of work put into cleaning. We need to think of personal protective equipment as just yet another element of the uniform and treat them exactly the same way. In a luxury hotel, it is unforgivable to wear a dirty shirt or a creased jacket. It would be equally unforgivable to be seen in a face mask which is anything but perfect.

Now that a handshake is a gesture from the past, luxury hotels must also rethink how they will be greeting their guests to acknowledge every one of them with due attention. Let us establish right from the start that the anecdotal elbow-high-fives might be appropriate for greeting youngsters but will feel inappropriate in a great majority of other situations. Living in the globalised world, it is finally possible to make the most of international experiences and learn new skills from other cultures, in this case mainly those originating from Asia. A Japanese bow, Thai or Indian way of bowing will ensure personal attendance to every customer, maintain the social distance and remain elegant. As suggested by Sorensen, hotels have to ensure that the experience is not stripped down emotionally. He assures that 'whilst we cannot shake hands, we will remain emotionally close' (Reynolds, 2020).

To make delivering experiences even more complex, staff members are not the only people influencing the overall experience that our guest receives. Luxury hotels have long ago developed procedures for dealing with misbehaving or inappropriately dressed guests, though every hotel will have at least one dramatic story to tell about such clients. Today they find themselves in a situation where a sneeze or an outbreak of cough could make other guests feel rather uncomfortable and nervous. Once again, a creatively designed physical environment can be our ally in managing such moments which, inevitably, will occur. Skilful management of these situations will ensure not only safeguarding the experience of fellow guests but also offering positive experience to the client who found themselves in this distressing situation. Positive memories are often created in the moment of service recovery, and so we can really showcase how much we care, only when someone is actually in need of such care.

The COVID-19 pandemic has changed a lot in people's perception of various things and has greatly affected what people might consider normal or even appreciated. Just a few months ago, luxury service providers would have avoided situations where their guests can observe cleaning procedures. A guest would have expected to arrive in an immaculate space where they never have to see such down-to-earth activities as cleaning. Today, however, this has changed dramatically, at least for the first few months after the outbreak of the new virus, when travellers are greatly concerned about cleanliness. Today transparency about cleaning procedures is not only a bonus but a must in many service-providing properties. Some hotels have begun offering written information on what chemicals are used to clean the room. While this kind of practice would have seemed very bizarre just a few months ago, today it will be seen as building trust. Guests expect to see safety seals on their room doors, protective coverings on TV remote controllers and evidence of tabletops being regularly disinfected. People often do not believe in what they cannot see, so adding physical evidence to all the services that would normally be hidden from the eyes of the customers will now be vital.

We Can Do This Together

This chapter has begun with an assumption that in the post–COVID-19 world, customers will still be seeking experiences but of a different nature than just a few months before the outbreak of the pandemic. This is partially true. Indeed, customers might start paying attention to details which they previously ignored. Not many travellers had in the past requested sanitising bed mattresses and pillows or resisted the temptation of salted peanuts on the bar because of hygiene concerns. Expectations from these kinds of experiences might indeed change. What is not likely to change, however, is the very reason why people come to luxury hotels. Like in every experience-based business, customers visit these properties to be transformed. Luxury, by its very nature, is a means of upgrading a person's social status and helping them transport themselves into a different, better world, if only for one night. Understanding customers' motivations for a visit will never be easy, but what can certainly help is shifting our own focus to put the customer in the centre of attention. Once again, following Pine and Gilmore (2011), instead of thinking of our services, or even experiences, as the end products, we should be considering the customer to be the end product of all our efforts. At the end of the day, if customers come to us to be transformed, then whatever we do is merely a process leading to that transformation. In other words, we are creating the circumstances and the environment for our guests to engage in their own experiences (Harkinson et al., 2018; Sundbo & Dixit, 2020) or, as once proclaimed by Hilton, setting the stage so that guests can create their own stories. By focussing on the state in which our guests want to be in during and after their visits, we hand the control over to them and start acting as a team supporting customers' journeys instead of imposing our dream journeys on them. Thinking again of the concept of authenticity discussed earlier, such transformative experiences will bear the ultimate level of authenticity as they will be tailored perfectly to the customer's self-image.

One of the first things that each hotel manager should consider is that most travellers do not have insider's knowledge about the hospitality industry. Various practices and solutions might be obvious to the readers of this handbook, but they could be not so apparent for an average hotel guest. As customers are considering travelling again, they might feel anxious about new procedures or modifications in service. This is a slightly different type of anxiety, not strictly related to the feeling of safety, but more to the feeling of comfort of knowing how to behave in a given situation. To put guests at ease, it will be important to communicate with them prior to the arrival, if not even prior to booking. The underlying principle is that in times of uncertainty, trust is most important, and trust comes through communication. Potential guests must be informed about what is or is not available in the hotel (especially when spa services form a central part of the offering), what health-check procedures are put in place and what kinds of service modifications are expected, i.e. modified housekeeping procedures or changes to buffet-style dining. This will not only set the expectations on the right level, thus avoiding unnecessary disappointments, but also enhance the flow of guests in the hotel space. The important factor, however, is the quality of this communication. It has been stated previously that there is no luxury without personalisation. In this day and age, many brands assume that it is sufficient to communicate via social media when they want to reach out to their guests. This is not enough a channel for brands who want to provide luxury services. Individualised connection with each customer is necessary, especially in a situation of uncertainty and doubt, so direct mail or conversation will be a much more appropriate medium. As far as personalisation goes, let us not forget that most simple solutions are often best. In many cases, it is much better to ask a customer for their preferences instead of guessing and potentially getting them completely wrong. Soerensen describes that an element of the updated service in The Beaumont Hotel will be offering guests the choice of when and how often they want their rooms to be cleaned (Reynolds, 2020). Some customers might feel more secure when their rooms are cleaned more regularly, while others might prefer if no one enters their bedroom during their stay. This approach follows Gilmore and Pine's (2007: 67) recommendation to 'shape your offering around unique tastes or unusual preferences of customers, without becoming intrusive or inappropriate', if you want to render exceptional authenticity. In the volatile environment that we live in now, a simple question could, indeed, go a long way.

Come Fly with Me

We have spent a considerable time discussing how luxury businesses can respond to the ongoing and rather unpredictable crisis caused by the global pandemic of COVID-19; however, even in times of crisis, we must not lose sight of all other aspects that shape our day-to-day existence, as well as the future. As previously mentioned by Scoblic (2020), a truly resilient business will be able to plan for the future without losing sight of everyday operations; the key is the ability to embed strategic foresight into organisational culture. In the world where customers have free choice of the abundance of luxury brands, we can be certain that they will support only these ones which are meeting their high-set expectations. People have the capability of adjusting to new conditions very fast. As soon as they feel the sense of comfort and safety, they will be again paying attention to other matters. Let us not forget that the pre–COVID-19 problems have not disappeared; on the contrary, some of them have grown even further. The devastation of the natural environment, exploitation, gender and race inequalities are all very much topics that must be kept on our agendas alongside virus-free spaces. Customers see brands holistically and do not differentiate between various initiatives; they will either support the brand or not, sometimes without even realising the deep reasons for doing either. Summarising, as we concentrate on ensuring healthy stays and trying to come up with innovative experience offerings to excite our customers, we must not forget about all the issues that our guests care for. We have mentioned a couple of examples of how being ecologically responsible can go hand in hand with providing safe experiences to a guest, as long as hotels demonstrate creativity and resourcefulness. One of the key characteristics of luxury brands is that they communicate not only to their clientele but also to the wider audience (Kapferer & Bastien, 2012) and, thus, build the so-called dream and desire for purchase. Luxury companies that have been praised the most during the outbreak of the pandemic were the ones who were prepared to temporarily give up their luxury strategies and concentrate on what really matters in times of crisis. Moncler, Richemont Group, Ralph Laurent and Capri Holdings are among the many companies that provided financial contributions to health-care associations and hospitals. Other corporations, such as the Armani Group and LVMH, have converted their production sites into factories to create materials to help countries with medical supply shortages. Bulgari has turned its perfume factories into hand sanitiser manufacturers, whereas Armani Group was producing hospital gowns. Many fashion brands were sewing face masks which have been in high demand worldwide. These are all excellent initiatives, both from social responsibility and brand management points of view.

In a testing situation such as this one, it is most important to remain positive. The world around us is changing, but that does not mean that what people aspire to or dream of is changing as well. People will always seek the feeling of security and safety as this is one of the most basic human needs. They will also enjoy being taken care of and pampered, as this lies in our very human nature. Finally, they will be fascinated by the extraordinary and will be prepared to pay a premium to be granted access to such unique experiences. Robert Marks, Director of the Americas Les Clefs d'Or, is very optimistic as he says: 'I truly believe we are entering a new, golden age of personalised service. (…) There will be a million ways that concierges can amaze guests in the age of COVID19. And now, as always before, it will be through small gestures, authentic moments and the quiet shared understanding that we are all in this together that will allow guests to feel safe, valuable and wowed by the experiences their warm-hearted concierges can deliver' (Kester, 2020). I wish every reader that out of these million options, they find their one unique way to be truly themselves and deliver genuine and authentic experiences.

References

Booms, B. H., & Bitner, M. J. (1981). Marketing strategies and organisation structures for service firms. In J. Donnelly & W. R. George (Eds.), *Marketing of services*. Chicago, IL: American Marketing Association.

Ciampa, D. (1992). *Total quality: A user's guide for implementation*. Reading, MA: Addison-Wesley.

Dixit, S. K. (2020). *The Routledge handbook of tourism experience management and marketing*. London: Routledge.

Gilmore, J., & Pine, J. (2007). *Authenticity. What consumers really want*. Boston, MA: Harvard Business Review Press.

Godfrey, S. (2020, May 20). The meaning of space for luxury hotels & hospitality. *Hospitalitynet*. Retrieved from https://www.hospitalitynet.org/opinion/4098769.html. Accessed on July 17, 2020.

Harkinson, T., Hemmington, N., & Hyde, K. (2018). Creating the luxury accommodation experience: Case studies from New Zealand. *International Journal of Contemporary Hospitality Management, 30*(3), 1724–1740. doi:10.1108/IJCHM-05-2017-0247

Kapferer, J. N., & Bastien, V. (2012). *The luxury strategy: Break the rules of marketing to build luxury brands* (2nd ed.). London: Kogan Page.

Kester, J. (2020, June 26). What it's like to stay in a luxury hotel right now. *Forbes Travel Guide*. Retrieved from https://www.forbes.com/sites/forbestravelguide/2020/06/26/what-its-like-to-stay-in-a-luxury-hotel-right-now/. Accessed on July 15, 2020.

Pine, J., & Gilmore, J. (1998). Welcome to the experience economy. *Harvard Business Review*, July–August.

Pine, J., & Gilmore, J. (2011). *The experience economy* (Updated edition). Boston, MA: Harvard Business Review Press.

Reynolds, E. (2020, June 9). The Beaumont GM Jannes Soerensen on managing a luxury hotel. *Elite Traveler*. Retrieved from https://www.elitetraveler.com/leaders-in-luxury/the-beaumont-gm-jannes-soerensen-managing-luxury-hotel. Accessed on July 15, 2020.

Scoblic, J. P. (2020). Learning from the future. How to make robust strategy in time of deep uncertainty. *Harvard Business Review*, July–August, pp. 39–47.

Sinek, S. (2019). *The infinite game*. London: Penguin.

Sinek, S. (2020). About Simon Sinek. Retrieved from https://simonsinek.com/about/simon-sinek?ref=mainNav

Smith, K., & Hanover, D. (2016). *Experiential marketing*. Hoboken, NJ: Wiley and Sons.

Sundbo, J., & Dixit, S. K. (2020). Conceptualisations in tourism experience. In S. K. Dixit (Ed.), *The Routledge handbook of tourism experience management and marketing* (pp. 15–26). London: Routledge.

Walls, A., Okumus, F., Wang, Y., & Kwun, D. J. W. (2011). Understanding the consumer experience: An exploratory study of luxury hotels. *Journal of Hospitality Marketing and Management*, *20*(2), 166–197. doi:10.1080/19368623.2011.536074

Wiedmann, K., Hennings, N., & Siebels, A. (2007). Measuring consumers' luxury value perception: A cross-cultural framework. *Academy of Marketing Science Review*, *7*(7), 1–21. doi:10.1007/978-3-8349-4399-6_5

Yeoman, I., & McMahon-Beattie, U. (2018). The future of luxury: Mega drivers, new faces and scenarios. *Revenue Pricing Management*, *17*(4), 204–217. doi:10.1057/s41272-018-0140-6

Chapter 26

Luxury Tourism in the New Normal: In Search of the New Memorability

Diana Gavilan,[1] Adela Balderas-Cejudo,[2] and Gema Martinez-Navarro[3]

Abstract

This chapter explores in-depth the particularities of the luxury travel industry during the coronavirus pandemic. The chapter begins with a literature review on the concept of the experience referred to as luxury travel, followed by a brief overview of the measures adopted globally in the tourism sector for the prevention of COVID-19. Due to the impact of the crisis, high-net-worth individuals (HNWIs) have changed their motivations, preferences and behaviour. To meet the new demands prompted by the pandemic, the luxury travel industry has adopted a wide range of strategies that have been classified into three groups: (1) reducing new moments of friction; (2) experiencing protection and (3) building new memorable experiences. A selection of examples from the luxury travel industry illustrates each category. The chapter ends with a case study that describes the strategy adopted by Fiji to respond to the pandemic. The manner in which Fiji has tackled the pandemic may be inspiring for other tourism-dependent destinations as Fiji has been able to protect the health and safety of the inhabitants of its islands while preserving the local economy, which relies on tourism.

Keywords: Luxury tourism; COVID-19; pandemic; memorable experiences; HNWIs; innovation strategies

[1]https://orcid.org/0000-0002-5293-779X
[2]https://orcid.org/0000-0003-1703-068X
[3]https://orcid.org/0000-0002-9047-8010

The Emerald Handbook of Luxury Management for Hospitality and Tourism, 513–532

doi:10.1108/978-1-83982-900-020211026

Introduction

For the $1.54 trillion global luxury travel industry, the scale of the disruption caused by COVID-19 has no precedent (ILTM, 2020). The global luxury travel ecosystem consists of a collection of industries connected to travel and supported by global luxury travellers. As a result, restrictions in travel severely affect the millions of small businesses, communities and families that depend upon it. Unlike the 2008–2009 financial crisis, there is no new group of luxury consumers to compensate for the dip in spending by current consumers.

An increase in bookings is not likely until quarantines end, borders reopen and businesses have a suitable plan to ensure a successful tourism experience at their clients' disposal. However, experts predict that bookings will bounce back faster than other areas of the industry (Globetrender, 2020). Due to their relative wealth, high-net-worth individuals (HNWIs) are more likely to be in a financial position to make plans for their next trip. Some HNWIs desiring escape has already moved from crowded cities to secluded and glamorous locations, such as the Maldives, Seychelles or Fiji, to spend a luxurious lockdown. Luxury travel agents have noted an interest in bookings later in the year, but recently, there has been a shift in preferences, which now include flying domestically, luxury automobile travel and private villas (ILTM Survey, 2020). Privacy and safety are the number one priority at this time.

This chapter explores in-depth the particularities of the luxury travel industry during the coronavirus pandemic. This chapter begins with a literature review on the concept of the experience referred to as luxury travel, followed by a brief overview of the measures adopted globally in the tourism sector to prevent COVID-19. Due to the impact of the crisis, HNWIs have changed their motivations, preferences and behaviour. To meet the new demands prompted by the pandemic, the luxury travel industry has adopted a wide range of strategies that have been classified into three groups: (1) reducing new moments of friction; (2) experience protection and (3) building new memorable experiences (Globetrender, 2020). A selection of examples from the luxury travel industry is illustrated in each category. The chapter ends with a case study that describes the strategy adopted by Fiji to respond to the pandemic. The manner in which Fiji has tackled the pandemic may be inspiring for other tourism-dependent destinations as Fiji has been able to protect the health and safety of the inhabitants of its islands while preserving the local economy, which relies on tourism.

The Essence of the Luxury Tourism Experience

Over the last two decades, academic studies have analysed tourism in terms of creating memorable experiences, i.e. special events retained in our memory that take us away from everyday reality (Tung & Ritchie, 2011). Tourism experiences that are particularly memorable are significantly related to outcome factors, such as revisiting a destination and spreading positive word of mouth (Dixit, 2020; Woodside, Caldwell, & Albers-Miller, 2004). Experts agree that the delivery of positive, memorable experiences is fundamental to competitiveness and sustainability (Ritchie & Crouch, 2003). Certain features, such as the authenticity of experiences (MacCannell, 1973), the main outcome of satisfaction (Holbrook & Hirschman, 1982) and the deliberate management of an experience (Pine & Gilmore, 1998), even help explain the link between memory and experiences in the field of tourism.

While experience becomes synonymous with travel when experiences are referred to as 'luxury', its meaning gains more nuance and involves more than lasting memories and transformative events (Johnson, 2013). Luxury travel entails the exclusivity and uniqueness of enjoyable events tailored to the preferences of high-end clients. Given the specificity and the relevance of these positive, memorable and transformative experiences provided by luxury tourism, it is important to observe what previous research has yielded in order to gain a deeper understanding of the core of this concept (see Fig. 26.1).

The core concept of luxury travel experiences involves four features:

Experience as a source of competitive advantages: Based on the concept of experience in marketing (Schmitt, 1999), experts assume its relevance in creating competitive advantages. Experience refers to the new value-added to a commodity, good or service, and companies should focus on understanding it as it is the new engine of economic growth for organisations (Pine & Guilmore, 1998, 2011). Aman, an expert luxury travel agency, often uses the term 'experience' to describe the luxury villas in Thailand or the activities planned on the four-day tour camp at Amanwana in Indonesia (www.aman.com).

Experience as holistic: Consumption is meant to be a holistic experience. Therefore, luxury tourism experiences refer to all that tourist experiences related to a luxurious environment and the impressions they take away as a result of their interaction with the human or physical elements around them (Shaw, 2005). As an experience relies heavily on its context, luxury tourism experiences may

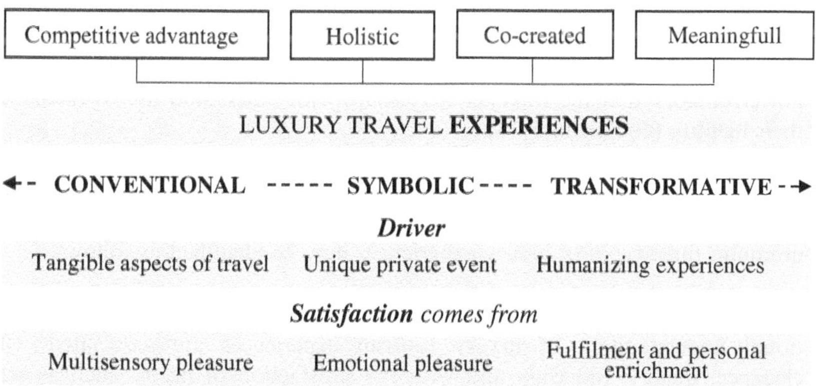

Fig. 26.1. The Core of Luxury Tourism Experience. *Source:* Gavilan (2020).

deliver sensory, emotional, cognitive-informational, behavioural and relational-social value to the traveller (Schmitt, 1999).

Experience as co-creation involving the consumer and their environment: Experience refers to a subjective perception as it is specific to an individual. For instance, an experience as popular as a safari through Uganda is, in fact, a subjective experience for each particular person on the safari. Experience depends on the service and facilities offered, but even when surrounded by the most luxurious and selective facilities and services, the final experience stems from the individual tourist's mood, state of mind, needs and wants. Thus, Mossberg (2007) states that experiences are always a co-creation that result from the blend of many elements such as emotional, intellectual and spiritual aspects of the customer. For certain customers, co-creation could even refer to tangible behaviour, such as participating in travel design with an expert planner.

Experiences have personal meanings and provide connections: In order to make an experience meaningful and memorable, luxury travel designers could harness stimuli provided to the tourist, such as sensory stimuli or memorabilia, since outstanding experiences have personal meaning. To design such experiences, a deep understanding of the client's motivations is needed. This is because the concept of luxury may vary for tourists depending on several factors, such as their specific needs (Bernstein, 1999). As such, the suitable design of high-quality memorabilia helps a tourist attach value and meaning to their experience. This can be seen in the case of the Como Uma hotel in Bhutan, which offers an eight-day film photography expedition. Participants get 10 rolls of Kodak Portra film and a 35mm SLR camera and learn how to take pictures with deliberation. The photos then showcase their learning progress and become relevant memorabilia. Additionally, this experience could provide a connection with others, with nature, and with the self, which in turn facilitates an escape from ordinary life. At the Zuri Zanzibar, visitors, paired with local fishermen from the Watumbatu tribe, learn traditional fishing techniques using nets and traps. High-value experiences deliver what people are unable to access in their daily lives: a 'once in a lifetime' event that is original and exclusive. When travel becomes a memorable experience, it creates an authentic and powerful story that the traveller may share with others and use to reconfirm their identity (Gieschen, 2018).

Thus, luxury tourism has always been experiential. However, experiential drivers in the luxury tourism industry have evolved. Along this experiential continuum, three types of luxury experiences may be identified as follows:

Conventional luxury experiences focusing on multisensory pleasure: The conventional understanding of luxury tourism experiences are associated with elegance, quality and class and with the satisfaction of needs, such as self-esteem (Dubois, Laurent, & Czellar, 2005; Truong & McColl, 2011). From this perspective, luxury tourism experiences emphasise the tangible aspects of travel, such as first-class facilities and services including five-star resorts, private jets and yachts, chauffeur-driven limousines, high-end consumer brands and Michelin-starred restaurants, which are always provided by trusted professionals (Cadotte & Turgeon, 1988; Sukmawati, Ernawati, & Nadra, 2018; Wu & Liang, 2009). According to Fulkerson (2014), our experience of the world involves a number of senses; sight, hearing, touch, taste and smell, which are not isolated from one another, providing a robust and coherent awareness of our environment. Thus, this kind of luxury experience mainly focuses on multisensory pleasure. Different senses create many distinct perceptual experiences.

Symbolic luxury experiences elicit emotional value: It cannot be denied that comfort factors are still in demand and that high standards of accommodation and catering will always be on the wish list of the luxury traveller. However, today, the concept of the luxury travel experience encompasses a new and wide variety of features aligned with symbolic dimensions of the luxury experience (Danziger, 2005; Wiedmann & Hennigs, 2013). These experiences can be described as tailor-made, exclusive, different and unique (Bakker, 2005; Petrick & Durko, 2015; Sukmawati et al., 2018; Sundbo & Dixit, 2020; Yeoman & McMahon-Beattie, 2006). Luxury forms an important part of the travel experience, enjoyed in safety and comfort (Chen & Peng, 2014), and this also has to do with emotional value (Fitzsimmons, 2017; Schensul, 2017). From this perspective, new elements of luxury travel emerge, such as access to a private art collection, extreme experiences in unexplored places or the organisation of a once-in-a-lifetime unique private event (Strong, 2006). The finer details, i.e. atmosphere, the attitudes and treatment of employees and the services and facilities, which may or may not be visible to the tourist, impact their experience by making it highly unique and relevant (Goeldner, Ritchie, & McIntosh, 2000).

Transformative luxury experiences satisfy needs: At the pinnacle of the symbolic luxury experience rests the so-called transformative travel experience, which is aimed at tourists fleeing a world homogenised by globalisation. The goal of such transformative experiences is to satisfy needs and offer tourists fulfilment by providing personal enrichment. Tourists in search of a transformative travel experience want to be inspired to be a better version of themselves (Warren, 2019). For instance, teaching English in the slums of New Delhi, India, could be favourable to a luxury tourist who craves authenticity and humanising experiences. By opting for a transformative travel experience, the tourist becomes a part of the process rather than merely enjoying the result. Moreover, this transformative experience upholds the tenets of exclusivity and authenticity, which are intrinsic to luxury travel (Chopra, 2012).

Measures Adopted Globally in the Tourism Sector for the Prevention of COVID-19

The COVID-19 outbreak has irrevocably impacted luxury tourism. As a result, stakeholders have to adapt their operational management to effectively address this pandemic. It is now crucial to adopt new safety protocols to ensure the health of staff and a safe customer experience in this new environment.

Despite the continuation of the pandemic, it may help to differentiate between the lockdown stage and the next stage. The latter has been called the 'next normal' or the 'new normal' based on the term coined by the tech investor Roger McNamee (Fast Company, 2003). In March, restrictions on people's mobility, border closures and social distancing measures were imposed by governments around the world to curb the contagion. This brought transport and tourism to a complete halt. During the lockdown, from 24 March to 20 April 2020, a total of 217 destinations – representing 100% of tourist destinations worldwide – introduced travel restrictions (WHO, 2020).

Once the lockdown was progressively lifted in many countries, the opening of borders commenced as well. In fact, from June onwards, the tourism sector has tried to resume its activity as much as possible while adopting the new health and hygiene standards in accordance with the guidelines of national and international organisations, such as the World Health Organisation (WHO), the World Travel and Tourism Council (WTTC) and the organisations responsible for disease control and prevention in each country. Since the pandemic is not controlled and its evolution differs among countries, the content of the guidelines and protocols is subject to modifications.

The WTTC, aiming at supporting the sector in achieving an effective recovery while keeping in mind that common rules are paramount, has established guidelines with common standards to ensure the safety of the workforce and travellers. The initial proposal covers the main stakeholders across travel and tourism: hospitality, outdoor retail, aviation, airports, cruises, tour operators, convention centres and meetings, incentives, conferences and exhibitions (MICE), car rentals, short-term rentals and insurance.

The WTTC identifies four fundamental areas of concern to ensure a safe travel experience in a changing scenario:

Operational and Staff Preparedness: This emphasises the development of prevention plans to guarantee staff safety and health by means of cleaning and disinfection (encouraging innovations for cleanliness and disinfection), ensuring physical distancing, integrating technologies to enable automation aimed at reducing contact procedures as well as staff training to prepare and execute operational plans regarding infection control according to the latest advice from public health authorities.

Ensuring a Safe Experience: This is focused on enhanced sanitation, disinfection and deep cleaning practices as well as ensuring physical distancing and limited interaction among clients.

Rebuilding Trust and Confidence: This is key to ensuring clear, consistent and enhanced communication with clients while ensuring adherence to health and hygiene safety protocols.

Implementing Enabling Policies: This suggests guidelines to enhance the stimulation of the tourism sector, collaboration with the industry and other governments, advisory measures that provide financial relief to the sector and destination promotion to boost demand, both domestically and internationally.

The New Mindset of Luxury Tourists after the COVID-19 Outbreak

Luxury experiences are targeted at elites. Elites are defined as upper-middle-class, high-net-worth and ultra-high net worth individuals who travel especially for non-work purposes to consolidate their social status. The luxury tourist seeks exclusivity, high quality, privacy and confidentiality (Yeoman & McMahon-Beattie, 2006) while being willing to pay for outstanding and unforgettable experiences (Jin, Moscardo, & Murphy, 2020). Elites express themselves through their taste for luxurious materials, success, 'the good life', personal lifestyle, power and high status (Birtchnell & Caletrio, 2014).

There are nearly 23 million individuals worldwide worth more than $1 million who can be defined as HNWIs and who are the most valuable group of travellers. HNWIs tend to book longer trips, more expensive rooms in luxury lodging and are significantly more likely to book premium seats on airlines. Their spending power is formidable, going beyond airlines, hotels and travel agencies and extending to leisure activities and experiences tailored to them. HNWIs contribute to just under half of the total expenditure of the global luxury travel sector (ILTM, 2020).

The increase in the number of HNWIs in emerging markets, the growth of disposable income and the rising standard of living among individuals has been driving a gradual growth of the luxury travel market. Before the COVID-19 outbreak, the global luxury travel market was growing at a compound annual growth rate of 7.9% (Allied Market Research, 2019).

On March 12, the WHO declared the coronavirus pandemic (2020). This represented a turning point for all business sectors, and luxury tourism was no exception. Since then, motivations and fears have gained new meanings as COVID-19 has multiple implications for the tourist experience (see Fig. 26.2).

The pandemic has prompted a shift in preferences, motivations and behaviour. As safety and privacy will now be offered at a premium (Sigala, 2020), priorities differing among those with wealth and travel goals have been redefined:

Seeking shelter: The urge to retreat to a safe space is followed by the wish to remain there, irrespective of whether the sense of safety or protection is experienced. Luxury tourists concerned about their health and safety will not spare any expense to minimise their interaction with the virus. Private planes and yachts, remote destinations in safe countries, trips to uninhabited islands, as well as customised itineraries, villas, houses and suites are thus in high demand.

Health is a priority: The COVID-19 pandemic has eroded trust in a larger sense. Every business now needs to take actions that effectively and quickly restore traveller confidence. Health and safety protocols are essential, and so as to enhance communication with customers on this matter, training staff to provide

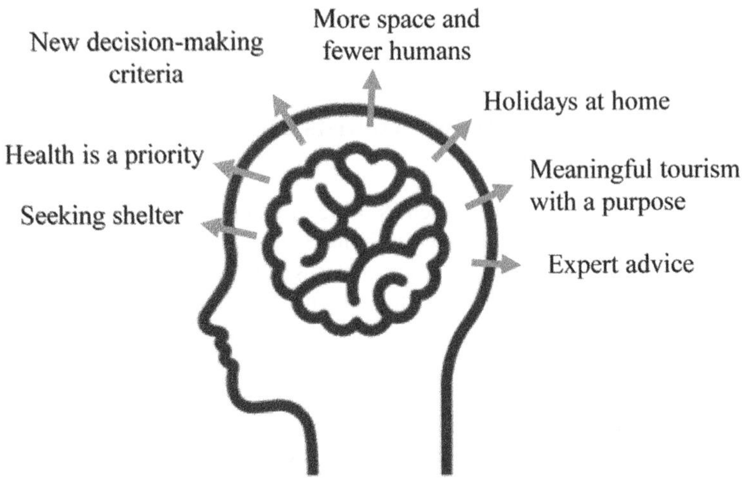

Fig. 26.2. The New Mindset of Luxury Tourists after COVID-19.
Source: Gavilan (2020).

recommendations is important. During and post the COVID-19 pandemic, health dominates tourist thinking, doing every business, in a way, a health business (Accenture, 2020).

New decision-making criteria: New demands will emerge, and companies should be prepared to meet the specific needs of a threatened, sometimes frightened and demanding customer. Destinations with low population density, suitable infrastructure, good health-care facilities and hygiene protocols, and which have passed the peak of the pandemic can offer the luxury tourist guarantees to a safe and enjoyable vacation. Hospital-rating systems are poised to be the new metric for destination choice, whereas traditional adventure destinations, such as the hinterlands of South America and East Africa, might be unappealing, especially for risk-averse clients.

More space and fewer humans: There have been shifts in consumer preferences from popular beaches, large resorts and busy cities to private villas, boats and boutique hotels, all with excellent cleaning standards and quiet coastal, lakeside, mountain and rural locations. Helicopter transfers, hotel takeovers and island buy-outs will score high among luxury traveller preferences, which focus on well-being (Globetrender, 2020)

Holidays at home: Staycations – holidays spent in one's home country rather than abroad – are key for those looking to get away and enjoy some rest and relaxation. There has been an increase in domestic travel to explore local areas, thus avoiding issues such as the fear of getting ill abroad, concerns about having to negotiate airports and planes, and border closures at arrival quarantines. The premium traveller wants to experience a feeling of being nurtured after this difficult time and its lockdowns (Simpson, 2020). Air Charter Service (ACS), a private jet charter that operates in China, took note of this. The company has experienced a significant increase in private jet charter demands within the country and the top three destinations of choice for post-lockdown are domestic (www.trip.com). Such clients also prefer luxury automobile travel for their first holiday after lockdown (ILTM Survey, 2020).

Meaningful tourism: Experts have warned about changes in travel decisions (Simpson, 2020). Fewer trips will be booked 'just because they can be', and more meaning will be put into the where, when and how of travel. In the near future, travellers will not take an intercontinental flight simply because of the lure of a beautiful hotel; the new travel trend will be to travel for meaning and, therefore, seek an answer to the following question: what are you searching for when you travel?

Purposeful travel: As a consequence of this approach to travel, involving a greater sense of purpose and meaning, it is expected that connecting with the local communities will become a relevant motivation for many tourists. The impact of the lack of tourism on areas that are wholly dependent on visitors has been devastating. Luxury tourists willing to travel to these destinations can participate in restoring a community or contributing to sustainability projects and local initiatives (Globetrender, 2020).

The pandemic has led to a new understanding of luxury travel as one that is life-enriching and value-based. Preferences will now focus on sustainability and conscious travel with health- and wellness-related travel itineraries (ILTM Survey, 2020).

Expert advice: Unlike budget travellers who tend to search and buy online, HNWIs are more likely to invest in the advice and service of consultants or agencies. Personal travel experts' help clients create tailor-made holidays, honeymoons and weddings and can provide advice on travel during the coronavirus pandemic. The value of the luxury travel agent has also been defined during the pandemic because booking through a reputable agent is much more reassuring than an online booking platform. Experts provide advice and help to arrange or postpone it if needed, and they do so quickly and easily. For example, Kuoni, an expert on tailor-made holidays, has launched a new video appointment service for customers looking to book holidays (www.kuoni.com).

Facing the Challenge

Luxury tourism felt the first rumblings of the storm when COVID-19 spread through China, a country whose citizens accounted for 90% of global luxury market growth in 2019. In addition, purchases of luxury goods and services by tourists will continue to be disrupted by travel restrictions and a lingering fear of possible contagion on planes and cruise ships.

Tourism is highly susceptible to the measures taken to counteract pandemics, and the early evidence on their impact on air travel, cruises and accommodations has been especially concerning (Gössling, 2020). However, today, there is a unanimous call to see and use COVID-19 as a transformative opportunity (Mair, 2020), to rethink and unthink certain pre-assumptions and mindsets, such as growth being the sole way to achieve success (Sigala, 2020). This crisis may be transforming the luxury tourism industry; however, it could be a transformation for the better. The luxury tourism industry needs to seize the opportunities that today's rupture points present and find innovative ways to surprise and delight their guests, especially as their wealthy clients are expecting worthwhile experiential luxury offers to escape from the threat of the coronavirus.

Based on both academic literature and sectorial innovation, three strategies adopted by the luxury tourism industry have been identified, which aim at providing memorable experiences within the framework of the New Normal. The first is 'reducing new moments of friction'. COVID-19 has turned irrelevant and taken-for-granted situations, such as cleaning rooms and touching doorknobs, into threats. This generates new and potentially conflicting situations. Thus, under the label 'reducing new moments of friction', this chapter brings together several practices adopted by the luxury industry to avoid the negative impact of the coronavirus or manage it successfully. Second, the section on 'experience protection' discusses several tactics aimed at preserving privacy and well-being when travelling during the New Normal and, finally, 'building new memorable experiences' deals with a range of measures that creative companies have taken to transform the crisis into an opportunity, thanks to innovation and technology. Therefore, by following the aforementioned steps, new memorable experiences can be created.

Reducing new moments of friction: Hygiene and cleanliness are the focus of attention everywhere. Improving sanitisation standards in flights and hotels has become a priority. Luxury brand hotels are already rolling out new initiatives to enhance cleanliness standards, including contactless service, guest service apps and new hygiene technology (Simpson, 2020).

However, luxury urban hotels and resorts will suffer enormously. Despite their keenness to understand the client's needs and having the creativity to address these needs, they are not designed for social distancing. Luxury resorts and hotels are thus opting for accredited certificates to ensure greater security and convey confidence to their guests. For instance, the Stay Safe with Melia program, certified by Bureau Veritas, guarantees the correct application of health and safety protocols in all its hotels (www.melia.com). The Global Biorisk Advisory Council (GBAC), a leader in the field of microbial and pathogenic threat analysis and recovery, is due to conduct an accreditation in all of Hyatt's hotels worldwide. Hyatt has even created a Global Care and Cleanliness Commitment, including hotel-level sanitisation specialists and a working group of trusted medical and industry advisors to further organise its operational guidance and resources around staff and guest safety. The process starts with an exhaustive audit critically examining the hotel experience from rooms to spas or dining. Hyatt's Global Care and Cleanliness Commitment implement rigorous protocols, integrating the latest research, technology, and innovation, to ensure that staff and guests feel confident. Every Hyatt hotel will have a Hygiene Manager who will guarantee the hygiene, safety and welfare of customers and hotel staff. The responsibility of this person is to supervise, certify and execute the new COVID-19 protocols with the aim of ensuring that guests can enjoy their experience without worry (www.hyatt.com).

Likewise, Four Seasons Hotels and Resorts has requested the collaboration with Johns Hopkins Medicine International to validate its new global health and safety program, Lead with Care. A dedicated advisory board has been established for COVID-19, bringing together the leaders of Four Seasons and leading experts from Johns Hopkins Medicine International. This board will create, enhance and review current procedures and provide online and in-person training to guide the implementation of Lead with Care across the Four Seasons global portfolio. Besides this, the program provides ongoing and real-time guidance on dealing with COVID-19 (www.fourseasons.com).

Airlines have adopted stringent measures to increase hygiene on board by using electrostatic sprays to kill germs and viruses in the cabin, removing in-flight magazines and boarding passengers from the back of the plane, i.e. from row one onwards. Emirates has also banned all cabin baggage beyond briefcases, handbags, laptops and baby items (www.emirates.com).

Experience Protection: Privacy and Well-being

Privacy and well-being are the overarching trends arising, and luxury hotels and resorts are perfectly positioned to satisfy these needs as travellers will actively seek out less crowded spaces, searching for privacy and a greater sense of calm and peacefulness. Demands for off-the-beaten-track and natural destinations have increased, and to respond to these demands, resorts and luxury properties have worked towards offering their guests exclusive stays with plans that guarantee health, convey safety and protection as well as maintain the experience that the luxury tourist is looking for. These properties, which are generally surrounded by extensive gardens, offer private swimming pools and are well situated, facilitate social distancing. Hôtel Particulier Villeroy in France, a member of the luxury group, The Collectionneur, has a private butler service. Further, since its restaurant is closed, bespoken menus are served to their guests in their suites (www.hotelvilleroy.com).

Along with privacy, there is a rise in wellness travel because people have realised, during this pandemic, how important and valuable good health is. As such, several travellers want to stay in places where they can improve their mental and physical well-being. Thus, luxury hotels have added specific services, such as resident doctors, helicopters or a telemedicine solution, to their packages. The luxurious Swiss hotel Le Bijou, in Zurich, offers its guests the 'COVID-19 Service', a package that includes food delivery ordered remotely via an iPad, with meals cooked by a personal chef, aside from 24/7 health monitoring and medical supervision, including in-room coronavirus testing (www.lebijou.com).

In the pursuit of privacy, clients sometimes take the lead in ensuring that they will be the only clients on the property. In the United States, wealthy groups have rented out entire hotels, such as the Relais and Châteaux's Blantyre country resort, a 24-room hotel located in Western Massachusetts, or the Caper Arundel Inn in Maine with 14 bedrooms with sea views. In this regard, the ultimate exercise in seclusion is to rent an entire island. Visitors of Sun Siyam Resorts can book one of two Maldives islands, the Iru Veli and the Vilu Reef (Globtrender, 2020).

Building New Memorable Experiences: Opportunities, Innovation and Technology

Pre-pandemic research had already highlighted the important role new technologies play in the tourist experience (Tussyadiah & Fesenmaier, 2009). In the COVID-19 scenario, this gains new meaning in terms of providing technological solutions to high-quality experiences. Experts agree that the key is finding a balance between the measures taken to face the pandemic and the experience the client is used to having. New technologies provide support in this regard, such as in cleaning processes.

In the fight against the contagion, robots may also become the next hotel room germ busters. With an extensive background in hospital disinfection, US-based tech company Xenex offers its LightStrike bots to hotels in order to sanitise rooms and common areas and help combat the threat of the coronavirus. LightStrike Robots have been proven to quickly destroy viruses, bacteria and fungi using intense pulsed xenon ultraviolet (UV) light. Marriott Hotels has adopted this new cleaning technology, which involves sanitising sprayers and UV light (www.marriott.com).

Given that surfaces are a major source of contamination, the demand for and the development of antibacterial and antimicrobial solutions has been promoted by the hospitality sector, including aviation, and has become critical since the COVID-19 outbreak. Extensive research has been conducted into the development of materials that meet both antibacterial and antimicrobial needs as well as sustainability requirements. It is a known fact that a virus' survival depends on the surface it is on. Unlike copper, upon which they survive for four hours, viruses on steel surfaces can survive up to two or even three days. To remedy this, Bio-Fence has patented an antimicrobial coating polymer that can be applied with a spray (www.bio-fence.com).

Sanitising, avoiding contagious diseases transmitted via surfaces and, eventually, low-touch living is the way forward. This means more automation, including walk-through scanners at airport checkpoints, contactless payments and hotel web check-ins or automatic check-in points installed in the lobby. QR codes can also be employed so that customers can consult the hotel's service directory or restaurant menus directly on their mobile. QR codes can even mean the use of a digital key, which is now widespread through mobile devices, to unlock hotel rooms. A system of intelligent toilets can also be used, controlled by sensors, which do not require human intervention for cleaning. All of this is possible with the aim of reducing the risk of contagion by contact.

As travel continues to be hit hard, luxury travel might have to embark on ventures surrounding mobile, digital-first experiences, thus making virtual reality the perfect ally of the luxury traveller. Such tourists will not need to explore faraway places in person and can instead have robots go for them and report back. Newme robots travel on behalf of humans and transmit 2K videos directly to the users, allowing them to interact with the bot's surroundings. In the near future, thanks to haptic technology, we will be able to convert sounds into textures and feel a tropical destination or appreciate the comfort of an airplane seat (www.avatarin.com).

The adoption of technology comprises a wide range of solutions that capture the most refined desires of clients. At the Ecclestone Square Hotel in London, UK, guests interact with tablet and telephone screens to control the intensity of light or music. In the shower, the water slides down the wall with the option of instant freezing. Flat TV screens are embedded in the bathroom mirror, and iPads allow guests to order whatever they want from the comfort of their beds. Innovation through technology guarantees the security of the clients while simultaneously investing in new experiences (www.ecclestonsquarehotel.com).

Automation can go even further: the Henn-na Hotel in Nagasaki, Japan, has already replaced staff with robots to provide reception service and even room service. Going the full distance in automation is the Pengheng Space Capsules in Shenzhen, China, where all the workers are robots.

Thus, while the traveller will always be at the core of the experience, the experience itself evolves. The coronavirus pandemic imposes strict health and safety standards, but the luxury travel industry is ready to implement them while providing unforgettable experiences.

Fiji, a Paradise

The Republic of Fiji is an island country in Melanesia in the South Pacific Ocean located about 1,100 nautical miles northeast of New Zealand's North Island. It consists of an archipelago of more than 330 islands, of which about 110 are permanently inhabited, and more than 500 are islets.

The two major islands, Viti Levu and Vanua Levu, account for 87% of the total population of 900,000. The capital, Suva, on Viti Levu, serves as the country's principal cruise ship port.

Fiji has one of the most developed economies in the Pacific due to its abundant forests and minerals; however, its economy heavily relies on tourism, which accounts for 40% of its gross national product (GDP).

The country's largest sources of international visitors are Australia (43%), New Zealand (21%), the United States (9%) and China (5%). Visitors are attracted by the white sandy beaches and aesthetically pleasing islands with year-round tropical weather, soft coral reefs and scuba diving activities. Thus, Fiji is a popular holiday and honeymoon destination offering a wide range of hotels and

Case Study 26.1: Managing the pandemic in Fiji: A retreat for billionaires

The luxury tourism sector is currently hardest hit by the coronavirus outbreak, with impacts on both travel supply and demand. To contain the virus and slow its spread, many destinations have implemented lockdowns and travel restrictions. Thus, tourism-reliant destinations have experienced severe economic impacts since tourists are unable to enter and visit attractions and places, such as hotels, restaurants and bars. This means that many jobs will be lost along with important economic resources.

However, there are significant differences in the ways some places have approached the coronavirus pandemic, aiming to protect their population while preserving the economy.

resorts. Most of these are accommodations in the luxury range, with some budget resorts in remote areas and a careful selection of world-class five-star hotels and resorts.

High-end Clients Seeking Refuge

In the months of February and March, after the WHO declared the coronavirus a pandemic, wealthy families from Europe, the United States, and China decided to leave their countries in search of a safe place to take shelter. Private jet companies were inundated with requests to take their clients to remote destinations.

This type of luxury tourist, called a pandemic tourist, was looking for a place where the coronavirus was under control, isolated and likely to avoid further outbreak as well as somewhere that would allow them to visit. However, the places where the coronavirus was under control and isolated did not allow even wealthy tourists access to their borders.

For a few countries whose economy depended on tourism, such as Fiji, the Bahamas or the Maldives, the risk of an outbreak with limited health-care infrastructure was high. Thus, European, American and Chinese visitors, despite their riches, were not allowed to stay.

Fiji's Lockdown

On 3 February, a few weeks before the first confirmed case of coronavirus, Fiji closed its borders to any foreigners who had visited mainland China in the past 14 days, in addition to screening the health of passengers on flights to Fiji. On 27 February, this ban was extended to travellers from Italy, Iran and parts of South Korea. On 17 March, cruise ships were stopped, and international events were cancelled.

Three days later, on 20 March, Fiji confirmed its first case of the coronavirus. A 27-year-old flight attendant who had flown in from San Francisco was infected. This case was followed by a chain of four community transmission cases in the city of Lautoka, and the prime minister urged people to avoid non-essential travel within Fiji.

By the end of the month, Nadi and Nausori, Fiji's international airports, were closed, and a curfew was decreed from 10 p.m. to 5 a.m.

Dealing with the Crisis

Fiji was the first Pacific island country to open a WHO-certified testing lab and the fourth country capable of testing for COVID-19. This enabled instant results rather than waiting for samples to be sent to and received from Australia.

Cases had been growing until they numbered 18. Suddenly, the transmission stopped for 78 days. Thus, in early July, Fiji had recorded a total of 21 infections, making it possible to trace every case. The low transmission rate combined with a recovery rate of 100% was proof that Fiji had successfully contained the outbreak. The early and economically painful decision to prevent an outbreak eventually put Fiji in a privileged position.

There have been no deaths from the virus in Fiji, and no new cases have been diagnosed since mid-April.

Reaping the Benefits

Despite Fiji being free of the coronavirus, the pandemic remains, and decisions regarding tourism are uncertain. Unlike other countries, such as Cuba, which allows visitors in certain areas (Rodríguez, 2020), or Hawaii, which allows them access if they provide a negative test result 72 hours before arrival, Fiji's approach has been different.

Since the risk of transmission is proportional to the number of visitors, but the number of visitors is not necessarily proportional to the benefits they bring, their goal has been to minimise risks while maximising income. Therefore, their strategy focuses on inviting only high-end visitors. This decision is based on the data of the study conducted by the Fiji Office of Market Development Facility in collaboration between AMSTEC Pty Limited, Port Denarau Marina and Vuda Marina (Economic Impact of International Yachting in Fiji, 2018) (Table 26.1):

This report stated that the 4,473 yachts that arrived that year injected $34.9m into the economy, including yacht equipment and fuel, yacht services aside from crew and guest living and recreation expenses. While yachts entail approximately the same risk of infection as cruises (2%), they contribute to 63% of the local economy. This analysis provides a clear course of action that Fiji has called the Blue Lanes Program.

Table 26.1. The Economic Impact of Tourists in Fiji.

Arrival by	**Average Spent**	**Contribution**	**Num. of Visitors**
Tourist arriving by plane	$116 per day plus $1,039 per trip		
Cruise ship passenger	$44 per port of call	$10.5 million	239,000 passengers
Visitor arriving by yacht	$5,052 during the average stay	$22.6 million	4,473 passengers

Blue Lanes Program

The Prime Minister of Fiji announced that yachts and leisure crafts would be allowed to bypass the country's coronavirus restrictions and enter one of its ports. An invitation to billionaires wishing to visit the country was even made from the Prime Minister's Twitter account on 24 June:

> So, say you're a billionaire looking to fly your own jet, rent your own island, and invest millions of dollars in Fiji in the process – if you've taken all the necessary health precautions and borne all associated costs, you may have a new home to escape the pandemic in paradise.

The Blue Lanes Program welcomes tourists back to Fiji's waters. The conditions are simple: pass a COVID-19 test before departure, spend 14 days at sea or make up the balance in quarantine on arrival on a remote private island (Sullivan, 2020).

However, Fiji is not alone in this move. Seychelles is also allowing visitors to arrive on private flights and then stay on an isolated island resort, and Thailand and the Maldives have similar plans to compete for the billionaire crowd. This move has not been in vain as a small number of 30 billionaires from a well-known company have announced their travel to Fiji and their plans to stay there for three months (Your Mileage May Vary, 2020).

The country's mission has thus been accomplished.

Conclusion

Luxury tourism has always been experiential, and it will continue to be so; however, the term experience will be referred to in different contexts, according to the moment in time and the motivations of the traveller. Luxury tourism offers a wide range of memorable opportunities, from the conventional experience elicited by the tangible luxury of services and amenities of high-quality hotels and resorts to the transformative travel experience where the tourist is in search of fulfilment and personal enrichment.

The coronavirus outbreak has deeply impacted tourism as a whole and the luxury tourism industry in particular. The main stakeholders across travel and tourism have to follow the guidelines of national and international organisations, such as the WHO, the WTTC and organisations responsible for disease control and prevention in each country. Taking into consideration the fact that the pandemic is not controlled and that its evolution differs among countries, the content of the guidelines and protocols is modified from day to day according to the rate of infection.

Despite the disruptive effect of the pandemic, the luxury tourism industry is expected to make an early recovery. HNWIs are looking for safe places to stay and are not likely to forgo holidays. Decisions will be made based on new criteria, such as the hospital-rating system of a destination, aiming at finding privacy on a lonely island instead of a popular beach and finding destinations in their own country rather than abroad.

The luxury travel industry has adopted several strategies to satisfy the new needs of these demanding circumstances. Enhancing cleanliness and safety to avoid the risk of contagion, protecting the pleasurable experience that the client is used to and building new memorable experiences based on innovation and technology will be the key to future tourism endeavours because life is indeed a quest for new experiences.

References

Accenture. (2020). The human experience: How organisations should respond to the experience implications of COVID-19. Retrieved from https://www.accenture.com/_acnmedia/Thought-Leadership-Assets/PDF-2/Accenture-COVID-19-New-Human-Truths-That-Experiences-Need-To-Address.pdf

Allied Market Research. (2019). Global luxury travel market. Retrieved from https://www.alliedmarketresearch.com/luxury-travel-market

AMSTEC and MDF. (2018). Economic impact of international yachting in Fiji. Retrieved from https://corporate.southpacificislands.travel/wp-content/uploads/2019/01/Economic-Impact-of-International-Yachting-in-Fiji-Low-Res.pdf

Bakker, M. (2005). *Luxury and tailor made holidays, travel and tourism analyst*. Chicago, IL: Mintel International Group Ltd.

Bernstein, L. (1999). Luxury and the hotel brand: Art, science, or fiction? *Cornell Hotel and Restaurant Administration Quarterly*, *40*(1), 47–53. doi:10.1177/001088049904000122

Birtchnell, T., & Caletrio, J. (2014). Introduction: The movement of the few. In T. Birtchnell & J. Caletrio (Eds.), *Elite mobilities* (pp. 1–20). London: Routledge.

Cadotte, E. R., & Turgeon, N. (1988). Key factors in guest satisfaction. *Cornell Hotel and Restaurant Administration Quarterly*, *28*, 44–51.

Chen, A., & Peng, N. (2014). Examining Chinese consumers' luxury hotel staying behavior. *International Journal of Hospitality Management*, *39*, 53–56. doi:10.1016/j.ijhm.2014.01.002

Chopra, N. (2012). What is experiential luxury travel and why I choose it as a Millennial. The Cultureur. Retrieved from https://thecultureur.com/what-is-experiential-luxury-travel/

Danziger, P. N. (2005). *Let them eat cake: Marketing luxury to the masses – as well as the classes*. New York, NY: Kaplan Publishing.

Dixit, S. K. (2020). *The Routledge handbook of tourism experience management and marketing*. London: Routledge.

Dubois, B., Laurent, G., & Czellar, S. (2005). Consumer segments based on attitudes toward luxury: Empirical evidence from twenty countries. *Marketing Letters*, *16*(2), 115–128.

Fast Company. (2003). The new normal. Retrieved from https://www.fastcompany.com/46387/new-normal

Fitzsimmons, A. (2017). Lux is in flux, says urban insider Annie Fitzsimmons: Here's what to know. *National Geographic*. Retrieved from http://www.nationalgeographic.com/travel/top-10/seven-luxury-travel-trends-hotel/

Fulkerson, M. (2014). Explaining multisensory experience. In R. Brown (Eds.), *Consciousness inside and out: Phenomenology, neuroscience, and the nature of experience* (pp. 365–373). Dordrecht: Springer.

Gieschen, J. (2017). *High value city travel report*. Madrid: IE Foundation Premium & Prestige Business Observatory. Retrieved from https://observatoriodelmercadopremium.ie.edu/wp-content/uploads/sites/59/2013/11/High-Value-City-Travel-Report-2017.pdf

Globetrender. (2020). Luxury travellers shun Covid-19 panic to book dream trips. Retrieved from https://globetrender.com/2020/04/10/luxury-travellers-coronavirus-dream-trips/

Goeldner, C. R., Ritchie, J. R. B., & McIntosh, R. W. (2000). *Tourism: Principals, practices, philosophies*. New York, NY: John Wiley & Sons.

Gössling, S. (2020). Integrating e-scooters in urban transportation: Problems, policies, and the prospect of system change. *Transportation Research Part D: Transport and Environment*, *79*, 102230. doi:10.1016/j.trd.2020.102230

Holbrook, M. B., & Hirschman, E. C. (1982). The experiential aspects of consumption: Consumer fantasies, feelings, and fun. *Journal of Consumer Research*, *9*(2), 132–140. doi:10.1086/208906

ILTM Survey. (2020). Luxury travel's true impact on business and employment. Retrieved from https://www.iltm.com/RXUK/RXUK_ILTM/2020/Portfolio/ILTM_Luxury_Travel_Universe.pdf

Jin, H., Moscardo, G., & Murphy, L. (2020). Exploring Chinese outbound tourist shopping: A social practice framework. *Journal of Travel Research*, *59*(1), 156–172. doi:10.1177/0047287519826303

Johnson, P. (2013). 10 definitions of 'luxury travel' from within the industry. Retrieved from http://www.aluxurytravelblog.com/2013/03/26/so-what-is-luxury-travel/

MacCannell, D. (1973). Staged authenticity: Arrangements of social space in tourist settings. *American Journal of Sociology*, *79*(3), 589–603.

Mair, S. (2020). What will the world be like after coronavirus? Four possible futures. *The Conversation*. Retrieved from https://theconversation.com/what-will-the-world-be-like-after-c…d=IwAR2wr9pzssSdBSxjaHaWba9-iHSF3flYgZ9BVI1jAx_Y4Y1XVAImcJcNdjM

Mossberg, L. (2007). A marketing approach to the tourist experience. *Scandinavian Journal of Hospitality and Tourism*, *7*(1), 59–74. doi:10.1080/15022250701231915

Petrick, J. F., & Durko, A. M. (2015). Segmenting luxury cruise tourists based on their motivations. *Tourism in Marine Environments*, *10*(3–4), 149–157. doi:10.3727/154427315X14181438892649

Pine, B. J., & Gilmore, J. H. (1998). Welcome to the experience economy. *Harvard Business Review*, *76*, 97–105.

Pine, B. J., & Gilmore, J. H. (2011). *The experience economy*. Boston, MA: Harvard Business Press.

Ritchie, J. B., & Crouch, G. I. (2003). *The competitive destination: A sustainable tourism perspective*. Cambridge: CABI Publishing.

Rodríguez, A. (2020, June 30). Cuba's new tourism system will separate visitors from locals. Retrieved from https://www.afar.com/magazine/cubas-tourism-system-to-isolate-visitors-from-local-population

Schensul, J. (2017). Vacations of the 1%: Trends from luxury travel agents. *USA Today*. Retrieved from https://www.usatoday.com/story/travel/destinations/2017/07/07/luxury-travel/455451001/

Schmitt, B. H. (1999). *Experiential marketing*. New York, NY: Free Press.

Shaw, C. (2005). The guiding principles. In C. Shaw (Ed.). *Revolutionise your customer experience*. London: Palgrave Macmillan.

Sigala, M. (2020). Tourism and COVID-19: Impacts and implications for advancing and resetting industry and research. *Journal of Business Research*, *17*, 312–321. doi:10.1016/j.jbusres.2020.06.015

Simpson, A. (2020). 6 ways luxury travel will change after COVID-19, as predicted by the experts. *Tatler*. Retrieved from https://hk.asiatatler.com/life/luxury-travel-covid-19-industry-experts

Strong, J. (2006). *Craving for travel: Luxury experiences for the sophisticated traveler*. Montreal: Luxury Travel Books.

Sukmawati, N. M. R., Ernawati, N. M., & Nadra, N. M. (2018). Luxury tourism: A perspective of facilities and amenities. *International Journal of Applied Sciences in Tourism and Events*, *2*(1), 32. doi:10.31940/ijaste.v2i1.900

Sullivan, R. (2020, June 28). Fiji markets itself as retreat for billionaires during pandemic. Retrieved from https://edition.cnn.com/travel/article/fiji-tourism-coronavirus-pandemic-scli-intl/index.html

Sundbo, J., & Dixit, S. K. (2020). Conceptualisations in tourism experience. In S. K. Dixit (Ed.), *The Routledge handbook of tourism experience management and marketing* (pp. 15–26). London: Routledge.

Truong, Y., & McColl, R. (2011). Intrinsic motivations, self-esteem, and luxury goods consumption. *Journal of Retailing and Consumer Services*, *18*, 555–561. doi:10.1016/j.jretconser.2011.08.004

Tung, V. W. S., & Ritchie, J. B. (2011). Exploring the essence of memorable tourism experiences. *Annals of Tourism Research*, *38*(4), 1367–1386. doi:10.1016/j.annals.2011.03.009

Tussyadiah, I. P., & Fesenmaier, D. R. (2009). Mediating the tourist experiences access to places via shared videos. *Annals of Tourism Research*, *36*(1), 24–40. doi:10.1016/j.annals.2008.10.001

Warren, K. (2019). Forget Champagne and infinity pools. *Business Insider*. Retrieved from https://www.businessinsider.com/luxury-travel-trends-authenticity-impact-inspiration-safari-2019-8?IR=T

WHO. (2020). Related travel restrictions – A global review for tourism. Retrieved from https://webunwto.s3.eu-west-1.amazonaws.com/s3fs-public/2020-05/TravelRestrictions-08-Mayo.pdf

Wiedmann, K., & Hennigs, N. (2013). *Luxury marketing: A challenge for theory and practice*. Dordrecht: Springer.

Woodside, A. G., Caldwell, M., & Albers-Miller, N. D. (2004). Broadening the study of tourism: Introduction to the special issue on the consumer psychology of travel/tourism behavior. *Journal of Travel & Tourism Marketing*, *17*(1), 1–6. doi:10.1300/J073v17n01_01

Wu, C. H.-J., & Liang, R.-D. (2009). Effect of experiential value on customer satisfaction with service encounters in luxury-hotels restaurants. *International Journal of Hospitality Management*, *28*, 586–593. doi:10.1016/j.ijhm.2009.03.008

Yeoman, A., & McMahon-Beattie, U. (2006). Premium pricing and luxury. *Journal of Revenue and Pricing Management*, *4*(4), 319–328. doi:10.1057/palgrave.rpm.5170155

Your Mileage May Vary. (2020, July 3). Fiji's brilliant way to financially survive through covid. Retrieved from https://yourmileagemayvary.net/2020/07/03/fijis-brilliant-way-to-financially-survive-through-covid/

Corporate Glossary

www.avatarin.com
www.bio-fence.com
www.ecclestonsquarehotel.com
www.emirates.com
www.fourseasons.com
www.hotelvilleroy.com
www.hyatt.com
www.kuoni.com
www.lebijou.com
www.marriott.com
www.melia.com
www.trip.com

Chapter 27

Conclusion: Way Forward for the Hospitality and Tourism Luxury

Saurabh Kumar Dixit[1] *and Anupama S. Kotur*[2]

Luxury is a buzzword used extensively in the modern hospitality and tourism business spheres. Luxury has always been associated with purchasing and consuming goods and services of outstanding quality, premium image, exclusivity and high price. The term luxury is used to express varied connotations, such as luxury consumption, luxury items, luxury services, luxury facilities and luxury goods. According to Knoll (2009), luxury is the inherent tendency of every human being to improve living conditions and, under certain circumstances, to concentrate consumption in a specific field. Ko, Costello, and Taylor (2019) define luxury goods or services as high quality, authentic, prestigious, premium-priced and stimulating association consumers. Besides the economic perspective, luxury has also been seen through the lens of social psychology theory, emphasising the conspicuous and social values inherent in the term 'luxury' (Vigneron & Johnson, 1999). Luxury is also delineated to have divergent perspectives, such as economic, social, psychological, marketing, supply and demand.

Managers and marketers frequently experiment with new luxury terms as true luxury, masstige, premium, ultra-premium, opuluxe, hyper luxe and so on. By reviewing the extant literature, one can conclude that the phenomenon of luxury is elastic and subjective and varies from one person to another. Therefore, the present Handbook serves as an essential resource to integrate the vivid domains of luxury management relating to hospitality and tourism. This final chapter presents a synthesis of the deliberations through various case studies on the emerging discourses relating to luxury management. The following sections offer an overall outlook of the core issues and important foci aroused in the 27 Handbook chapters (see Fig. 27.1). We hope that this Handbook has meaningfully advanced our comprehension of the conception and management of luxury in hospitality and tourism and will inspire further research.

[1]https://orcid.org/0000-0002-9478-084X
[2]https://orcid.org/0000-0003-4202-4716

The Emerald Handbook of Luxury Management for Hospitality and Tourism, 533–538

doi:10.1108/978-1-83982-900-020211027

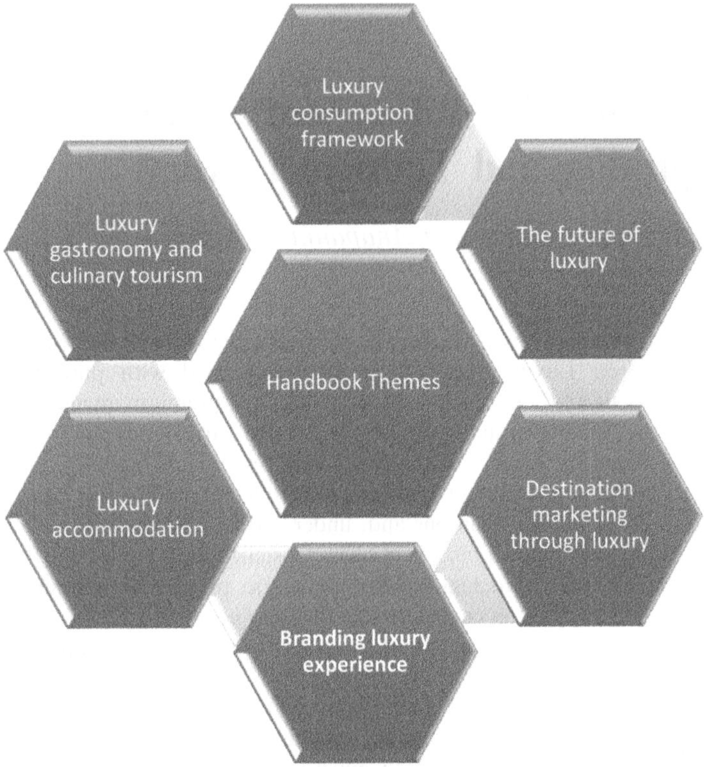

Fig. 27.1. Key Themes Pondered in the Handbook.

Luxury consumption framework: Due to rising disposable income with individuals, luxury consumption is getting acceptance in every sphere of the hospitality and tourism business. Luxury consumption builds pleasure and intense emotions that make the consumer feel special and distinguished. The two motives that characterise luxury consumption according to Vigneron and Johnson (1999) are 'interpersonal' and 'personal' motives, whereas, as per Truong (2010) are 'extrinsic' and 'intrinsic' motives. Various studies strived to respond to the question to explore consumers' luxury-seeking behaviour and purchase motivations. Pleasure, hedonism, self-enhancement, wellness and desire are the emerging motivations for luxury consumption (Carcano, Minichilli, & Corbetta, 2011). This Handbook strives to present varied dogma of luxury consumption. Chapter 1 offers historical evolutions and cross-cultural frameworks of global and Indian luxury tourism to construct the foundation of studying contemporary luxury business. Chapter 3 enlightens readers about the shades of grey within luxury lodging consumption in the Koyasan temple. Chapter 6 highlights the consumption patterns in the Czech market through three case studies on socio-demographic variables and spatial determinants. Chapter 20 informs the readers about the high yield luxury in tourism, demand/supply perspectives, experiential luxury and future research agenda for luxury tourism. The chapter, therefore, lays the necessary underpinning to have a basic acquaintance of luxury tourism.

Destination marketing through luxury: The luxury travellers plan their travel itineraries based on the destination attractiveness, accommodation and other facilities. Relaxation opportunity, gastronomic experiences, cultural visits, comfort, privacy, tailor-made excursions etc. motivate travellers to select a travel destination. Luxury marketers, therefore, need to make efforts to make the guests' travel experience extraordinary and unforgettable. Chapter 4 sheds light on the evolution of Vietnam as a luxury tourism destination during the Communist regime. The chapter also examines the relationship between Communism and luxury consumption. Chapter 5 focuses on the prospects and challenges associated with the promotion of Oman as a Luxury tourism destination. Moreover, the chapter also outlined the key strengths, attractions, public and private sector efforts for developing luxury tourism in Oman. Chapter 7 explains how marketing luxury travel may be conducted successfully in a destination that relies on unique experiences of nature. The chapter further delineated the potential of Norway as an upcoming luxury tourism destination due to the element of exclusivity in experiences. Chapter 19 discusses the challenges, frameworks and opportunities for creating sustainable indigenous community luxury experience. The chapter also discussed the case of Kimberley communities of Martuwarra (Western Australia) to develop a socio-cultural sustainable luxury tourism framework.

Branding luxury experience: Luxury is a growing field of research; most existing research examines luxury as goods, disregarding it as services and experiences (Batat, 2019). According to Atwal and Williams (2009), luxury is not merely about products and brands; it is primarily about a lived unusual experience that encompasses various elements. In the extant marketing literature, luxury is studied under three perspectives, i.e. product/brand management perspective, consumption perspective and purchasing motivations/consumer perspective (Fionda & Moore, 2009) (see also chapter 13). Chapter 11 details how a multi-brand hospitality organisation differentiates its luxury brand from other luxury brands in the portfolio using textual messages on its official websites. The chapter has also thrown light on the branding strategy of Marriott International Inc. Chapter 12 explores the complex cognitive process of assigning the relationship between travellers and luxury travel brands employing brand relationship scale, TraveLux. TraveLux was also found to establish a synch between the instrument constructs and manifested human cognition in a real-life situation. Chapter 25 suggests how luxury brands can communicate with their existing and prospective guests to ensure their cooperation and active engagement, especially during the post-pandemic environment. The chapter also offered practical solutions that luxury hospitality providers can offer to deliver personalised experiences in light of the new post-pandemic scenario.

Luxury accommodation: Luxury accommodation is a growing sector in the contemporary hospitality industry due to rising disposable income and movement. The luxury accommodations are differentiated based on the quality of services, prime locations, premium pricing, high-quality food and highly professional service staff. Chapter 8 provides the supply side of luxury accommodation by highlighting the role of employees in the luxury service quality. The overall satisfaction from luxury goods and services results from the pleasant service encounter with the staff. Chapter 10 throws light on the luxury hotels, hotel brands, memorable hospitality experience and co-creating hospitality experiences. The chapter also encompasses two case studies on a luxury hotel and lodge experience and how these experiences are created. Chapter 15 showcases the contribution of luxury room amenities in the hotel room booking and revisit intentions of the hotel guests. Based on its study of the select hotels of Delhi, the chapter discusses the top 10 most preferred and least preferred guest rooms amenities. Chapter 17 suggests a luxury brand framework to demonstrate how luxury hospitality organisations can create, communicate and sustainably manage their brands with the case studies of Four Seasons Hotels and Resorts.

Luxury gastronomy and culinary tourism: Appreciating gastronomic foods of a place is a growing trend in the modern tourism business, a leisure pursuit for thousands and one of the main motives for many tourists (Dixit, 2019). An emerging trend amongst the tourists to seek authenticity and novel narrative, local gastronomy and dining pattern has become a significant factor for tourists looking for co-creating an extraordinary experience at a destination (Dixit, 2020). Luxury gastronomy and culinary tourism is a thriving consumer market attracting business travellers and elites seeking to relish luxury cuisine. One expression of Luxury gastronomy is patronising Michelin restaurants. Chapter 16 deals with the interlinkages of gastronomy and luxury experiences using the interview of Chef Eneko Atxa from Michelin starred restaurant, Azurmendi. Chapter 23 explores the classifications of luxury hotels and Michelin restaurants besides discussing the influence of Covid-19 in the Luxury Food and Beverage Industry in Singapore through a case study. Chapter 9 discusses how luxury can be marketed in the context of gastronomic tourism experiences following the recent conceptualisations of experiential luxury as an affect. The chapter further presents the Hunter Valley Wine Tourism Association (HVWTA) case to understand the strategies used to market gastronomic experiences as luxury.

Paradoxes of contemporary luxury: The luxury consumption experience is not always smooth and free-flowing in the modern business arena. The changing market conditions and behavioural attributes compel organisations to seek innovative and authentic modes to connect with the consumers. The Handbook also presents a few chapters that peep into the paradoxes associated with luxury. Chapter 18 discusses the interlinkages between sustainability and luxury tourism. Luxury is criticised for lavish resource utilisation, whereas sustainability focuses on equitable distribution of resources. The chapter attempts to examine the existing knowledge about sustainable luxury tourism by implementing a systematic literature review. Luxury is often deemed to impart pleasure-seeking and positive values to the customers; contrary to that, chapter 21 explores the darker side of luxury hospitality, which deals with the illegal and unethical services provided to the guests by the hotels. It also examines the industry-wide problem of alcohol and drug abuse among employees and the possible reasons behind it. Chapter 24 conveys that luxury tourism has been described in contrasting ways, both as the most unsustainable tourism and the only sector that attracts consumers who can afford the sustainable product and services. Luxury consumption is considered closer to environmental connection, personal accomplishment and finding a brand that offers maximum values to modern buyers. The chapter, therefore, reconnoitres if tourism is an unsustainable luxury business. Chapter 26 presents the changing motivations, preferences and behaviour of luxury consumers due to the outbreak of COVID-19. To understand the post-pandemic luxury travel market needs, the chapter further sheds light on the strategies such as reducing new moments of friction, experience protection and memorable experience creation.

The future of luxury: The increasing use of technology, social media platforms and innovative ventures in hospitality and tourism has immensely impacted the luxury segment in hospitality and tourism. The rise in e-commerce, changing attributes of consumer purchase behaviour and the upsurge of novel luxury trends offer opportunities to new luxury business models and market entrants. This section will pinpoint some emerging themes deliberated in the present title, which would pave the upcoming avenues of research in luxury research in hospitality and tourism. Chapter 2 inspects the experiential nature of luxury, allied to a yoga retreat holiday in luxury hotel premises. It further apprehends how luxury experience is erected from the sensory and embodied perspective. Chapter 13 offers insights to the readers about the role of virtual reality (VR), augmented reality (AR), artificial intelligence (AI) and other technologies in offering luxury experiences in the tourism and hospitality business. The application of such business models is getting increasingly relevant during the COVID-19 pandemic era. Chapter 14 emphasised that *openness to change* and *status consumption orientation* play a critical role in increasing the perceived usefulness of voice assistants (VAs) in luxury hotels. Therefore, luxury hotel management should utilise these two factors to induce guests to make their stay experiences more luxurious. Hence in future, each hotel will equip rooms with VR services, games and artificial intelligence: the integration of Internet of things technologies, such as VAs, will be the first step towards such transformation. Chapter 22 elaborates the contributions of travel influencers in the transformations of luxury travel during the coronavirus pandemic. Based on an ethnographic investigation of travel influencers via three cases, the chapter argues that travel influencers showcase powerful representations of luxury as safe travel.

Besides the issues revealed in different *Handbook* chapters, we also wish to underscore some emerging avenues for luxury researchers in hospitality and tourism. In doing so, we have identified the following research gaps and new areas for undertaking luxury research.

(1) Creation of 'biosecurity bubbles' at destinations/countries and its impact on luxury travel
(2) Post-pandemic organisational/operational restructuring within luxury hospitality/tourism operators
(3) Role of robots, artificially intelligent chatbots in hospitality and tourism luxury
(4) Multisensory experiential marketing in luxury
(5) Emergence of 'tech-free zones' and 'digital detox' for luxury holidays

(6) Mercurial consumption behaviour of luxury tourists
(7) Crisis communication strategies adopted by luxury tourism and hospitality brands during the COVID-19 pandemic

We hope that this *Handbook* will offer significant research contributions for its readers and lay a foundation for future academic endeavours of luxury tourism research.

References

Atwal, G., & Williams, A. (2009). Luxury brand marketing: The experience is everything! *Journal of Brand Management*, *16*(5–6), 338–346. doi:10.1057/bm.2008.48

Batat, W. (2019). *Food and experiential marketing. Pleasure, wellbeing and consumption*. London: Routledge.

Carcano, L., Minichilli, A., & Corbetta, G. (2011). Why luxury firms are often family firms? Family identity, symbolic capital and value creation in luxury-related industries. *Universia Business Review*, *32*, 40–52.

Dixit, S. K. (2019). Gastronomic tourism: A theoretical construct. In S. K. Dixit (Ed.), *The Routledge handbook of gastronomic tourism* (pp. 13–23). Oxon: Routledge.

Dixit, S. K. (2020). Marketing gastronomic tourism experiences. In S. K. Dixit (Ed.), *The Routledge handbook of tourism experience management and marketing* (pp. 322–335). London: Routledge.

Fionda, A., & Moore, C. (2009). The anatomy of the luxury fashion brand. *Journal of Brand Management*, *16*, 347–363. doi:10.1057/bm.2008.45

Knoll, R. (2009). Luxury is when expectations are exceeded. *Goldsmith Newspaper*, p. 10.

Ko, E., Costello, J. P., & Taylor, C. R. (2019). What is a luxury brand? A new definition and review of the literature. *Journal of Business Research*, *99*, 405–413.

Truong, Y. (2010). Personal aspirations and the consumption of luxury. *International Journal of Market Research*, *52*(5), 653–671.

Vigneron, F., & Johnson, L. W. (1999). A review and a conceptual framework of prestige-seeking consumer behaviors. *Academy of Marketing Science Review*, *1*(1), 1–15.

Index

Printed and bound by CPI Group (UK) Ltd, Croydon, CR0 4YY

13/09/2026

14961029-0004